Advances in Soil Science

Structure and Organic Matter Storage in Agricultural Soils

Edited by

Martin R. Carter
Agriculture and Agri-Food Canada
Research Centre Charlottetown,
Prince Edward Island, Canada

B. A. Stewart
Dryland Agricultural Institute
West Texas A & M University
Canyon, Texas

Library of Congress Cataloging-in-Publication Data

Structure and organic matter storage in agricultural soils / edited by M. R. Carter, B. A. Stewart.
p. cm. — (Advances in soil science)
Includes bibliographical references and index.
ISBN 1-56670-033-7 (alk. paper)
1. Humus. 2. Soil structure. 3. Carbon sequestration.
I. Carter, Martin R. II. Stewart, B. A. (Bobby Alton), 1932– . III. Series: Advances in soil science (Boca Raton, Fla.).
S592.8.S78 1995
631.4'17—dc20 95-38348
CIP

Direct all inquiries to CRC Press, Inc., 2000 Corporate Blvd., N.W., Boca Raton, Florida 33431.

Lewis Publishers is an imprint of CRC Press

No claim to original U.S. Government works
International Standard Book Number 1-56670-033-7
Library of Congress Card Number 95-38348
Printed in the United States of America 1 2 3 4 5 6 7 8 9 0
Printed on acid-free paper

Preface

The interrelationship between structure and organic matter storage in soil has important implications for both agroecosystem and environmental studies, especially as the major pool of terrestrial carbon resides in soil. In recent years, the need to annotate the above interrelationship has been highlighted by increases in atmospheric carbon and the necessity to mitigate this phenomenon by conservation and retention of terrestrial carbon. This book explores the mechanisms and processes involved with the storage and sequestration of carbon in mainly agricultural soils.

Soil structure can be viewed as a major organizer for terrestrial ecosystems, and an important component of soil transport and permeability processes. In comparison, organic matter provides a major soil energy and nutrient reserve, functions as a soil structure forming agent, and plays an important role in system stability and sustainability. The interrelationship between soil aggregation and soil organic matter, where the latter binds soil particles while the former protects organic matter from decomposition, is a major theme of the book.

Many subject areas within soil science are well defined and major advances have been made, over the last few decades, in elucidating basic processes. However, information about soil structure and aggregation forming processes is limited. This is due in part to the multi-disciplinary nature of soil structure formation. For example, the underlying mechanisms involved with structure of soil are related to chemical, physical, biological, mineralogical, and ecological processes in soil. Further, the phenomenon of soil structure is operative at different scales and can be investigated at both the microstructure (μm) and macrostructure (mm) level. Generally, the latter is strongly related to the activity of soil microorganisms and fauna.

Many mechanisms, factors, and processes govern the degree of organic matter storage in soil. Soil aggregation is the main process whereby organic matter is retained in soil. Such retention can be characterized by both relatively short-term storage in macroaggregates or long-term sequestration in microaggregates. In addition, climate and soil type can strongly influence both the degree of soil aggregation and organic matter storage in soil. Differences and interactions between temperature and precipitation control plant productivity and the subsequent accumulation and decomposition of organic matter. Soil parent material and mineralogical differences can also influence primary production and dictate the level of potential soil aggregation and consequently the degree of organic matter retention. Land use and management, within a climatic or soil zone, can also impact on soil structure formation and storage of organic matter.

This book provides a synthesis of past and present research on soil structure and organic matter in agricultural soils, and points towards future research needs. The initial five chapters describe soil structure and elucidate the underlying mechanisms and processes involved with soil aggregation and organic matter retention. The next five chapters characterize the impact of varying climate, soil type, and soil management on soil aggregation and organic matter storage. These chapters cover a wide range of climate types in both temperate and tropical regions. The remaining three chapters address some current agronomic strategies to enhance organic matter storage in soil, and approaches to the modelling and measurement of organic matter storage.

M.R. Carter
Charlottetown, Prince Edward Island
Canada

B.A. Stewart
Amarillo, Texas
U.S.A.

About the Editors:

Dr. M.R. Carter holds degrees in Agriculture and Soil Science from the University of Alberta and obtained a Ph.D. degree in Soil Science from the University of Saskatchewan in 1983. Since 1977, he has held agricultural research positions with Agriculture and Agri-Food Canada and is currently a Research Scientist at the Research Centre, Charlottetown, Prince Edward Island. Dr. Carter recently edited *Soil Sampling and Methods of Analysis* (Lewis Publishers/CRC Press 1993) for the Canadian Society of Soil Science, and *Conservation Tillage in Temperate Agroecosystems* (Lewis Publishers/CRC Press 1994). He currently serves on the Editorial Board of *Soil & Tillage Research* and *Acta Agriculturæ Scandinavica*. Dr. Carter holds an appointment as Adjunct Professor in the Department of Chemistry and Soil Science, Nova Scotia Agricultural College.

Dr. B.A. Stewart is a Distinguished Professor of Soil Science, and Director of the Dryland Agriculture Institute at West Texas A&M University, Canyon, Texas. Prior to joining West Texas A&M University in 1993, he was Director of the USDA Conservation and Production Research Laboratory, Bushland, Texas. Dr. Stewart is past president of the Soil Science Society of America, and was a member of the 1990-93 Committee of Long Range Soil and Water Policy, National Research Council, National Academy of Sciences. He is a Fellow of the Soil Science Society of America, American Society of Agronomy, Soil and Water Conservation Society, a recipient of the USDA Superior Service Award, and a recipient of the Hugh Hammond Bennett Award by the Soil and Water Conservation Society.

Contributors

A. Albrecht, ORSTOM, (M.O.S.T.), 97259 Fort-de-France Cedex, France.

D.A. Angers, Agriculture and Agri-Food Canada, Soil and Crops Research and Development Centre, Sainte-Foy, Québec, G1V 2J3 Canada.

M.H. Beare, New Zealand Institute for Crop & Food Research, Canterbury Agricultural and Science Centre, Christchurch, New Zealand.

P. Becker-Heidmann, Institute of Soil Science, Hamburg University, 20146 Hamburg, Germany.

B.J. Bridge, CSIRO, Division of Soils, Toowoomba, Queensland 4350, Australia.

C.A. Cambardella, USDA-ARS, National Soil Tilth Laboratory, Ames, Iowa 50011-4420, U.S.A.

M.R. Carter, Agriculture and Agri-Food Canada, Research Centre, Charlottetown, Prince Edward Island, C1A 7M8 Canada.

B.T. Christensen, Danish Institute of Plant and Soil Science, Department of Soil Science, Research Centre Foulum, DK-8830 Tjele, Denmark.

R.C. Dalal, Queensland Wheat Research Institute, Toowoomba, Queensland 4350, Australia.

C. Feller, ORSTOM (L.C.S.C.), 34032 Montpellier Cedex, France.

E.G. Gregorich, Agriculture and Agri-Food Canada, Centre for Land and Biological Resources Research, Ottawa, Ontario, K1A 0C6 Canada.

R.J. Haynes, New Zealand Institute for Crop & Food Research, Canterbury Agricultural and Science Centre, Christchurch, New Zealand.

H.H. Janzen, Agriculture and Agri-Food Canada, Research Centre, Lethbridge, Alberta, T1J 4B1 Canada.

D.L. Karlen, USDA-ARS, National Soil Tilth Laboratory, Ames, Iowa 50011-4420, U.S.A.

M.J. Kooistra, DLO Winand Staring Centre for Integrated Land, Soil and Water Research, 6700 AC Wageningen, The Netherlands.

W.J. Parton, Natural Resource Ecology Laboratory, Colorado State University, Fort Collins, Colorado 80523, U.S.A.

E.-M. Pfeiffer, Institute of Soil Science, Hamburg University, 20146 Hamburg, Germany.

D.S. Ojima, Natural Resource Ecology Laboratory, Colorado State University, Fort Collins, Colorado 80523, U.S.A.

H.W. Scharpenseel, Institute of Soil Science, Hamburg University, 20146 Hamburg, Germany.

D.S. Schimel, Natural Resource Ecology Laboratory, Colorado State University, Fort Collins, Colorado 80523, U.S.A.

D. Tessier, INRA, Science du Sol, 78026 Versailles, France.

J.M. Tisdall, Institute of Sustainable Irrigated Agriculture, Ferguson Road, Tatura 3616, Australia.

M. van Noordwijk, International Centre for Research on AgroForestry, ICRAF S.E. Asia, Bogor 16001, Indonesia.

Contents

Introduction

Mechanisms and Processes

Impact of Climate, Soil Type, and Management

Assessment of Soil Organic Matter Storage

Introduction

Analysis of Soil Organic Matter Storage in Agroecosystems

M.R. Carter

I. Introduction

The purpose of this chapter is to provide a brief overview and characterization of the storage of organic matter in soil, as an introduction to the in-depth reviews provided in the following chapters.

Soil organic matter is a fundamental but transient component of the soil that controls many chemical, physical and biological properties affecting the ability of a soil to produce food, fibre or fuel. It is the primary source of energy for the soil ecosystem, and a major source of and a temporary sink for plant nutrients in agroecosystems. Soil organic matter is important in maintaining soil tilth, aiding the diffusion of air, the retention and infiltration of water, reducing soil erosion and controlling the efficacy and fate of applied pesticides (Gregorich et al., 1994).

Soil organic matter is the major terrestrial pool of carbon (C). It is a labile component of the soil but also a renewable resource. Changes in the soil environment leading to a decrease in plant productivity can cause a rapid decline in soil organic matter levels, while an increase in crop residue inputs or organic amendments can lead to a synthesis of new soil organic matter (Swift et al., 1991). Thus soil organic matter is a dynamic soil property, sensitive and responsive to ecosystem performance. Recent concerns with global changes in the atmospheric CO_2 balance have emphasized the possibility of increasing the storage of atmospheric CO_2-C in the soil by changes in land use and soil

1-56670-033-7/96/$0.00+$.50

management practices that increase the synthesis and retention of soil organic matter (Carter and Hall, 1995).

Climate and soil type can significantly influence the accumulation and storage of organic matter in the soil due to the interactions of temperature and moisture on plant productivity and the ability of the soil's mineral components to retain organic matter. Within any one soil type, however, the accumulation and storage of organic matter in soil are related to both the quality of the organic inputs and various interactions which separate C substrates from microbial enzymes. Thus quality of organic inputs and their interaction with soil components are the major controls on organic matter persistence and stability in soil as demonstrated by reduced rates of C mineralization, and longer turnover or mean residence times. The soil aggregation process provides the framework to characterize the storage, retention and sequestration of organic matter in soil.

Overall, soil aggregation and organic matter accumulation are interrelated: organic matter or fractions thereof are basic to the aggregation process, while organic matter sequestered within aggregates is protected against degradation. However, the degree of soil aggregation and extent of organic matter storage are influenced by land use, and soil and crop management practices.

II. Soil Capacity for Organic Matter Storage

The storage capacity of soil for organic matter is dependent upon climate, soil type and landscape, type of vegetation and soil management. The influence of climate on soil organic matter storage can be expressed by the relationship between mean annual temperature and annual precipitation. Wet, cool climates tend to slow organic matter turnover and subsequently favour organic matter accumulation in soil while moist, warm or hot climates favour rapid decomposition (Tate, 1992; Cole et al., 1993). Generally, soil organic matter decomposition processes are strongly dependent upon the interplay between temperature and precipitation.

The overall influence of climate, however, can be modified by soil type and landscape. Edaphic conditions, such as soil particle size, pH, quantity and type of clay minerals, and internal drainage can influence organic matter accumulation and storage. Such intrinsic properties can impact on organic matter storage and decomposition, either directly or indirectly, by modifying the soil chemical (Tate, 1992), physical and biological environment, and subsequently influencing the soil aggregation process (Oades and Waters, 1991; Robert and Chenu, 1992). Soil topography and drainage can also modify the macroclimate resulting in a range of microclimates across a landscape and subsequent differences in soil organic matter storage.

Within any one climatic zone, vegetation differences can have a major impact on soil organic matter accumulation. Differences in C fixing capacity and in C partitioning within plants resulting in concomitant differences in root biomass, shoot/root ratios, thickness of roots and amount of root exudates can influence organic matter mineralization and accumulation in soil (Juma, 1993). Shifts in use of vegetation can directly increase soil organic matter storage (Schlesinger, 1990). A combination of soil and vegetation factors can enhance organic C storage in some situations (e.g. Mollisols) (Scharpenseel et al., 1992). In many cases, however, vegetation effects are related to the high ratio of C to N in roots and other plant residues (compared to soil), and thus constitutes a temporary (i.e. non-sequestered) accumulation of organic matter subject to relatively rapid turnover and dependent on continual C inputs.

Soil management can also modify soil organic matter levels by affecting organic matter inputs and influencing to some extent the degree or potential for turnover. Evidence from long-term rotational experiments indicate that increasing C inputs (e.g. manure) can cause a gradual organic matter accumulation over time, especially in arable farming systems (Jenkinson, 1990). Soil management strategies used to enhance organic matter storage, where organic C inputs remain stable, involve practices that provide for cool wet conditions at the soil surface (i.e. change in soil microclimate) such as mulches and reduced tillage (Follett, 1993; Kern and Johnson, 1993). However, major short-term improvements in soil organic matter storage are dependent upon changes in vegetation or cropping practice.

A number of simulation models have been developed to describe the processes of soil organic matter accumulation and turnover (Parton et al., 1987; Jenkinson, 1990). In these models, soil organic matter cycles through pools having different turnover times. Rates of decomposition differ according to the chemical composition of the substrates and to the physical protection of substrates. The latter is the result of adsorption of substrates to mineral surfaces or due to the location of substrates within aggregates. Modelling of soil organic matter turnover and decomposition are essential to quantify changes in organic C storage over long time spans. In this regard, the classical long-term rotation experiments provide a unique time series of soil C data, and a means to assess rate of turnover for specific pools of organic matter (Jenkinson, 1990; Christensen, 1992; Paustian et al., 1992; Cole et al., 1993).

III. Benefits of Soil Structure for Organic Matter Storage

Soil structure can influence organic matter storage by providing physical barriers which separate C substrate from microbial enzymes (Kooistra, 1991). This phenomenon is related to both pore-size distribution and the formation of aggregates.

Soil structural form (Kay, 1990) or architecture (Van Veen and Kuikman, 1990) refers to the heterogeneous arrangement of particles and associated pore space (Oades, 1993). The size distribution of the latter is important for 'accessibility' of microbes and mesofauna to C substrates (McGill and Myers, 1987; Foster, 1988). Small pores (<1 μm effective diameter) can exclude bacteria, while slightly larger soil pores (1 to 2 μm) allow bacteria but exclude protozoa and nematodes (Juma, 1993). Such segregation not only separates substrate from microbes, but also separates microbes from predators. The separation of substrate and organism or separation of organism and predator tends to constrain soil biological activity and subsequently stabilizes organic matter (Ladd et al., 1993). The importance of soil porosity and pore-size distribution in regard to the physical protection of organic matter has been emphasized in recent studies (Hassink, 1992; Juma, 1993; Ladd et al., 1993).

The formation of soil aggregates with various degrees of stability is dependent upon both abiotic and biotic factors, the former being mainly related to soil clay content and the capacity for natural (e.g. alternating shrink-swell; freeze-thaw; wet-dry) structure-forming processes. The aggregation process is not totally random but occurs with some degree of order, in that for many soil types where biotic factors are the main aggregating agents small soil particles are apparently stabilized in large particles (Kay, 1990; Oades and Waters, 1991). Although soil specific, this ordering or arrangement of particles and aggregates can be viewed conceptually as a hierarchy consisting of three main orders: clay micro-structures (< 2 μm diameter), microaggregates (2 to 250 μm) and macroaggregates (> 250 μm). However, this order of aggregate categories is mainly derived from empirical studies using wet sieving or other fractionation procedures, thus at present it is not clear if the genesis of aggregate formation is mainly a sequential process (i.e. formation of microaggregates prior to macroaggregates) or a developmental process (i.e. formation of microaggregates within macroaggregates) (Beare et al., 1994). In soils (e.g. Oxisols) where inorganic compounds (i.e. Fe and Al oxides) are the main aggregating agents the above hierarchal order may not be evident (Oades, 1993). In clay micro-structures, clay-organic matter complexes are stabilized by humic acids and inorganic ions (e.g. Ca) (Tsutsuki et al., 1988; Theng et al., 1989). Microaggregates are stabilized directly by microbial materials such as polysaccharides, hyphal fragments, and bacterial cells or colonies (Tisdall, 1994). In comparison, the formation of macroaggregates and their temporary stabilization, is the result of a combination of mechanisms related to plant roots and the activity of soil fungi and fauna (Tisdall, 1991). For example, extracellular polysaccharides produced by the external hyphae of mycorrhizas, associated with plant root systems, are the means whereby microaggregates are bound together to form stable macroaggregates (Tisdall, 1994).

IV. Types of Organic Matter Storage in Soil

Organic matter itself can be characterized by various fractions (e.g. microbial biomass, carbohydrates, humic acids) with specific turnover times which can impact on soil C storage (Gregorich et al., 1994). Different constituents of soil organic matter will have relative stability, ranging from labile to stable forms (Theng et al., 1989). Rate of turnover of organic matter in soil will depend on both quality or kind of organic matter, and its location within the soil

Table 1. Characteristics of stored organic matter in agricultural soil

Type of organic matter	Proportion of total organic matter (%)	Turnover time (year)
Litter	-	1-3
Microbial biomass	2-5	0.1-0.4
Particulate	18-40	5-20
Light fraction	10-30	1-15
Inter-microaggregate[a]	20-35	5-50
Intra-microaggregate[b]		
Physically sequestered	20-40	50-1000
Chemically sequestered	20-40	1000-3000

[a]Organic matter stored within macroaggregates, but external to microagggregates: includes particulate, light fraction, and microbial C.
[b]Organic matter stored within microaggregates: includes sequestered light fraction and microbial derived C.

(Stevenson and Elliott, 1989). The factors influencing organic matter storage in soil, with emphasis on organic matter quality, type of organic matter input and interactions with soil material, ensures that the duration of storage will be variable. In many soils where structural stability is dominated mainly by organic matter, the dynamics of the hierarchical aggregate system will control organic matter storage (Oades and Waters, 1991). Table 1, based mainly on the hierarchical model, illustrates the relative size and stability of some of these 'pools' of stored organic matter.

Particulate or macroorganic matter, which includes a portion of the 'light fraction' organic matter (Gregorich et al., 1994), is derived from plant residues and organic amendments and can be found 'free' in soil or within macroaggregates. It can be composed of various kinds of material, that vary in both size and chemical composition (Cambardella and Elliot, 1992). The latter characteristics can control rate of turnover and consequently duration of storage in soil. Relatively large sized organic matter such as certain roots or litter may resist degradation even if labile in nature due to a restricted surface area for microbial attack. In addition, soil chemical (e.g. low pH) or aeration (e.g. waterlogging) conditions can further impede decomposition. Inherent chemical characteristics of the organic matter (e.g. lignin or polyphenol content, high C:N ratio) will also provide a form of chemical recalcitrance (Paustian et al., 1992; Jensen, 1994). Soil mixing either by cultivation or by the activities of macrofauna can greatly influence the storage duration of C and N derived from macroorganic matter. Jensen (1994) recently suggested that the rate of decomposition of large sized (e.g. 10 mm) residues is dependent on the C:N ratio, in that decomposition of high ratio residue may be N limited. However, small sized (e.g. 1 mm) residue of low C:N ratio may undergo less complete decomposition than coarser residues (of similar C:N ratio) due to a more thorough soil mixing of the former resulting in better protection of the residue derived microbial biomass and metabolites by clay minerals. Overall, size and type of organic matter will have a variable effect on soil organic matter storage.

Organic matter within macroaggregates can be conceptually separated into inter-microaggregate (i.e. within macroaggregates, but external to microaggregates) and intra-microaggregate (i.e. within microaggregates). The former consists mainly of macroorganic matter derived from living roots and fungal hyphae (including microbial structures and polysaccharides), and encrusted by inorganic components (Oades and Waters, 1991; Tisdall, 1991). It also includes a portion of the soil's 'light fraction' C (Gregorich et al., 1994). This organic matter is chemically labile but relatively physically protected within the macroaggregate structure. Its turnover time would be similar to that of the soil macroaggregates (1 to 10 years). Under grassland, up to 40% of total soil organic matter could exist as inter-microaggregate C. Cessation of root or fungal growth (due to change in cropping practice) would result in relatively rapid macroaggregate breakdown and initiate concomitant decomposition of inter-microaggregate organic matter. Thus, interaggregate organic matter in macroaggregates is of short term storage, especially in arable rotations (Waters and Oades, 1991).

Sequestered organic matter is protected, either physically or chemically, from rapid decomposition in microaggregates (i.e. intra-microaggregate). Organic matter can be retained or entrapped within and between clusters of microaggregates. Plant debris can be encrusted within 90 to 250 μm aggregates (Waters and Oades, 1991). Smaller microaggregates tend to be random associations of clay microstructures, biopolymers, and microorganisms. In clay microstructures, organic matter is mainly composed of microbial products (Oades and Waters, 1991) and humic substances (Theng et al., 1989). These can be considered chemically sequestered due to some degree of resistance against microbial decomposition, but mainly due to interactions with polyvalent cations, sesquioxides, and clay complexes (Tsutsuki et al., 1988; Theng et al., 1989). Generally, sequestered organic matter increases with increasing clay content. Studies with stable C isotopes have confirmed that organic matter sequestered in fine clay microstructures is a relatively long term pool of stored C (Balesdent et al., 1988).

V. Summary

Climate provides a distal control on soil organic matter storage, while soil type and vegetation are proximal controls. In addition, vegetation and cropping practice influence both the quantity and quality of C inputs, while soil type and vegetation control soil structural form and stability. The latter can impact on both the accumulation and duration of organic matter stored in soil. Generally, soil organic matter storage can be classified as short term, in the case of unprotected labile macroorganic matter and unprotected inter-microaggregate C, or long term for sequestered C in primary and secondary particles. The following chapters will elucidate the processes leading to soil structure and organic matter interactions, and characterize the impacts of climate and soil management on soil organic matter storage.

Acknowledgements

The author is grateful to Drs. D.A. Angers, B.T. Christensen, and E.G. Gregorich for helpful criticisms and comments in the preparation of this introductory chapter.

References

Balesdent, J., G.H. Wagner, and A. Mariotti. 1988. Soil organic matter turnover in long-term field experiments as revealed by carbon-13 natural abundance. *Soil Sci. Soc. Am. J.* 52:118-124.

Beare, M.H., P.F. Hendrix, and D.C. Coleman. 1994. Water-stable aggregates and organic matter fractions in conventional- and no-tillage soils. *Soil Sci. Soc. Am. J.* 58:777-786.

Cambardella, C.A. and E.T. Elliott. 1992. Particulate soil organic-matter changes across a grassland cultivation sequence. *Soil Sci. Soc. Am. J.* 56:777-783.

Carter, M.R. and D.O. Hall. 1995. Management of carbon sequestration in terrestrial ecosystems. In: M. Beran (ed.) *Prospects for carbon sequestration in the biosphere.* NATO ASI Series, Springer, Berlin (In press).

Christensen, B.T. 1992. Physical fractionation of soil and organic matter in primary particle size and density separates. *Adv. Soil Sci.* 20:1-90.

Cole, V.C., K. Paustian, E.T. Elliott, A.K. Metherell, D.S. Ojima, and W.J. Parton. 1993. Analysis of agroecosystem carbon pools. *Water, Air, Soil Pollut.* 70:357-371.

Follett, R.F. 1993. Global climate change, U.S. agriculture, and carbon dioxide. *J. Prod. Agric.* 6:181-190.

Foster, R.C. 1988. Microenvironments of soil microorganisms. *Biol. Fertil. Soils* 6:189-203.

Gregorich, E.G., M.R. Carter, D.A. Angers, C.M. Monreal, and B.H. Ellert. 1994. Towards a minimum data set to assess soil organic matter quality in agricultural soils. *Can. J. Soil Sci.* 74:367-385.

Hassink, J. 1992. Effects of soil texture and structure on carbon and nitrogen mineralization in grassland soils. *Biol. Fertil. Soils* 14:126-134.

Jenkinson, D.S. 1990. The turnover of organic carbon and nitrogen in soil. *Phil. Trans. R. Soc. Lond.* 1255:361-368.

Jensen, E.S. 1994. Mineralization-immobilization of nitrogen in soil amended with low C:N ratio plant residues with different particle sizes. *Soil Biol. Biochem.* 26:519-521.

Juma, N.J. 1993. Interrelationships between soil structure/texture, soil biota/soil organic matter and crop production. *Geoderma* 57:3-30.

Kay, B.D. 1990. Rates of change of soil structure under different cropping systems. *Adv. Soil Sci.* 12:1-52.

Kern, J.S. and M.G. Johnson. 1993. Conservation tillage impacts on national soil and atmospheric carbon levels. *Soil Sci. Soc. Am. J.* 57:200-210.

Kooistra, M.J. 1991. A micromorphological approach to the interaction between soil structure and soil biota. *Agric. Ecosyst. Environ.* 34:315-328.

Ladd, J.N., R.C. Foster, and J.O. Skjemstad. 1993. Soil structure: carbon and nitrogen metabolism. *Geoderma* 56:401-434.

McGill, W.B. and R.J.K. Myers. 1987. Controls on dynamics of soil and fertilizer nitrogen. p. 73-99. In: R.F. Follett, J.W.B. Stewart, and C.V. Cole (eds.), *Soil fertility and organic matter as critical components of production systems.* Soil Sci. Soc. America, Spec. Pub. 19, Madison, WI.

Oades, J.M. and A.G. Waters. 1991. Aggregate hierarchy in soils. *Aust. J. Soil Res.* 29:815-828.

Oades, J.M. 1993. The role of biology in the formation, stabilization and degradation of soil structure. *Geoderma* 56:377-400.

Parton, W.J., D.S. Schimel, C.V. Cole, and D.S. Ojima. 1987. Analysis of factors controlling soil organic matter levels in Great Plains grasslands. *Soil Sci. Soc. Am. J.* 51:1173-1179.

Paustian, K., W.J. Parton, and J. Persson. 1992. Modeling soil organic matter in organic-amended and nitrogen-fertilized long-term plots. *Soil Sci. Soc. Am. J.* 56:476-488.

Robert, M. and C. Chenu. 1992. Interactions between soil minerals and microorganisms. p. 307-404. In: G. Stotzki and J.M. Ballag (eds.), *Soil biochemistry, Vol. 7.* Marcel Dekker, N.Y.

Scharpenseel, H.W., H.U. Neue, and St. Singer. 1992. Biotransformations in different climatic belts: source-sink relationships. p. 91-105. In: J. Kubat (ed.), *Humus, its structure and role in agriculture and environment.* Elsevier Sci. Publishers, Amsterdam.

Schlesinger, W.H. 1990. Evidence from chronosequence studies for a low carbon-storage potential of soils. *Nature* 348:232-234.

Stevenson, F.J. and E.T. Elliott. 1989. Methodologies for assessing the quantity and quality of soil organic matter. p. 173-199. In: D.C. Coleman et al. (eds.), *Dynamics of soil organic matter in tropical ecosystems.* Univ. Hawaii Press, Honolulu.

Swift, M.J., B.T. Kang, K. Mulongoy, and P. Woomer. 1991. Organic-matter management for sustainable soil fertility in tropical cropping systems. p. 307-326. In: C.R. Elliot et al. (eds.), *Evaluation for sustainable land management in the developing world.* Vol. 2, Tech. Papers. IBSRAM Proc. No. 12, Bangkok, Thailand.

Tate, K.R. 1992. Assessment, based on a climosequence of soils in tussock grasslands, of soil carbon storage and release in response to global warming. *J. Soil Sci.* 43:697-707.

Theng, B.K.G., K.R. Tate, and P. Sollins. 1989. Constituents of organic matter in temperate and tropical soils. p. 5-31. In: D.C. Coleman et al. (eds.), *Dynamics of soil organic matter in tropical ecosystems.* Univ. Hawaii Press, Honolulu.

Tisdall, J.M. 1991. Fungal hyphae and structural stability of soil. *Aust. J. Soil Res.* 29:729-743.

Tisdall, J.M. 1994. Possible role of soil microorganisms in aggregation in soils. *Plant and Soil* 159:115-121.

Tsutsuki, K., C. Suzuki, S. Kuwatsuka, P. Becker-Heidmann, and H.W. Scharpenseel. 1988. Investigation on the stabilization of the humus in mollisols. *Z. Pflanzenernähr. Bodenk.* 151:87-90.

Van Veen, J.A. and P.J. Kuikman. 1990. Soil structural aspects of decomposition of organic matter by microorganisms. *Biogeochemistry* 11:213-223.

Waters, A.G. and J.M. Oades. 1991. Organic matter in water-stable aggregates. p. 163-174. In: W.S. Wilson (ed.), *Advances in soil organic matter research: the impact on agriculture and the environment.* Royal Soc. Chem., London.

Mechanisms and Processes

Soil Architecture and Distribution of Organic Matter

M.J. Kooistra and M. van Noordwijk

I. Introduction

Soil architecture deals with the spatial distribution and heterogeneity of the different components or properties of soils (Dexter, 1988). It determines and influences many soil processes: physical processes, such as water, air and temperature regimes, decomposition and transport processes, and the occurrence, survival and functioning of soil organisms and plant root systems. Soil architecture can be considered identical to soil structure. The term architecture, however, includes the aspect of a purposely made structure which fulfills special requirements. Several abiotic and biotic factors and processes can be distinguished to contribute to soil structure (Kooistra, 1991), but only the biotic ones are responsible for the architectural part of soil structure. The below-ground communities and plant roots are dependent and involved in the genesis and

1-56670-033-7/96$0.00+$.50

maintenance of soil structure for their functioning and growth. Yet, the various communities and plant roots have different requirements which can be mutually exclusive, mutually complementary or dependent, completely independent or mixtures of these. They also act at different hierarchical scales ranging from molecular level, via organisms level and pedon level to the fieldscale level (McGill and Myers, 1987). The impact of the biological component making up the soil architecture is therefore variable and depends on the land capability, i.e. the natural integration of soil properties with climate and plants, together with land use and management.

In cultivated land, human impact by soil tillage, traffic and machinery, application of pesticides, fertilizers and manures, regulates the overall soil structure throughout the year. In intensive agricultural systems, the natural physical aggregate formation and biological soil-structure formation processes generally play a secondary role. Farm management systems may try to increase the role of the natural processes, however, in as far as they contribute to the farmers purpose of growing and harvesting crops. In large parts of the tropics less intensive land use systems still predominate and the role of biologically mediated processes is larger. One of the major factors determining the abundance and the related impact of soil organisms is the presence of organic substrates as food, and thus the cropping pattern or agroforestry system used.

Organic carbon in soils normally originates from plants growing on site or in the direct neighbourhood, but considerable transport of organic material occurs as part of farming activities, by faunal action (e.g. termites) and in erosion/deposition cycles. Transformations and redistribution of organic carbon in soils from fresh organic inputs to release as CO_2 or incorporation in stable humus fractions can follow a number of pathways. A first distinction is between organic inputs which arrive via the soil surface, such as aboveground crop residues, tree litter or prunings, pollen and organic manures, and the below ground organic inputs to the soil from primary producers, such as exudates, root structural material and mycorrhizal hyphae. Structure-forming soil organisms normally play a role in physically fragmenting and redistributing organic inputs as well as modifying its chemical form and physical link with mineral particles in their excrements. Organisms which mainly follow existing soil structural voids or channels, such as mesofauna and microbes, modify the organic carbon pools, depending on the physical accessibility of the latter. Excrements of soil fauna are partly or completely composed of organic matter. Where mull, mull-like moder or moder humus forms occur in agricultural land, recognizable faunal excrements may contribute the bulk part of the organic carbon pools present in the soil.

The occurrence and distribution of organic matter in relation to the soil architecture thus depends on a number of processes. Generally, soil can be conceptualized in two ways:

a. As a perfectly mixed homogeneous medium responding uniformly to outside influences, and

b. As a highly structured, heterogeneous environment with large zones of little activity and scattered 'hot spots' full of activity.

The majority of existing models of soil physical and soil chemical processes and even of soil food webs and soil C transformations, are largely based on concept a, while reality may be more like b. The degree of heterogeneity and its consequences are still poorly understood, however. Three fields of research should be considered in this context:

1. The selection and development of methods to observe and quantify patterns of heterogeneity at a large range of scales.
2. Knowledge of the processes leading to the dynamics (rise and fall) of the pattern of 'hot spots' of activity. In order to understand these processes knowledge is also required on whether heterogeneous patterns for various phenomena coincide or not, both in time and space, i.e. what is their spatial correlation (synlocation) and their temporal correlation (synchrony).
3. Knowledge of the consequences of soil heterogeneity for processes at a larger scale, or higher level of complexity.

In this chapter some recent developments in these three fields will be reviewed as well as conclusions and recommendations for future research presented.

II. Soil Architecture and Distribution of Organic Carbon on a Range of Scales

A. General

Soil architecture can be studied on a range of scales. On a dm scale, pedality (presence of peds) or aggregation (Figure 1) can be observed in profile pits with the naked eye and can be characterized with a hand lens (Soil Survey Staff, 1975). Details on surface or interior of the aggregates and peds are beyond the resolution of the naked eye and hand lens, however. Most of the voids and other features occurring in the groundmass can only be analyzed and interpreted with the use of optical microscopy. The groundmass itself consists of primary particles which also can be arranged in different ways (packing). Separate 'grains' may occur and often finer material is present. To study the arrangement and organization of coarse and fine material other techniques such as electron microscopy and X-ray diffraction are often required (Kooistra and Tovey, 1994).

An overview of the occurrence of different kinds of organic carbon in an agricultural soil profile is given in Figure 2. Although the bulk part of the roots generally occur in the ploughed layer, roots are commonly found deeper in the soil. Crop residues and organic manure are normally restricted to the plough layer. After ploughing, the residues and manures which were present at the surface occur in inclined zones, due to the inversion and lateral transport of the soil material by the plough share. The occurrence of fungal hyphae is often

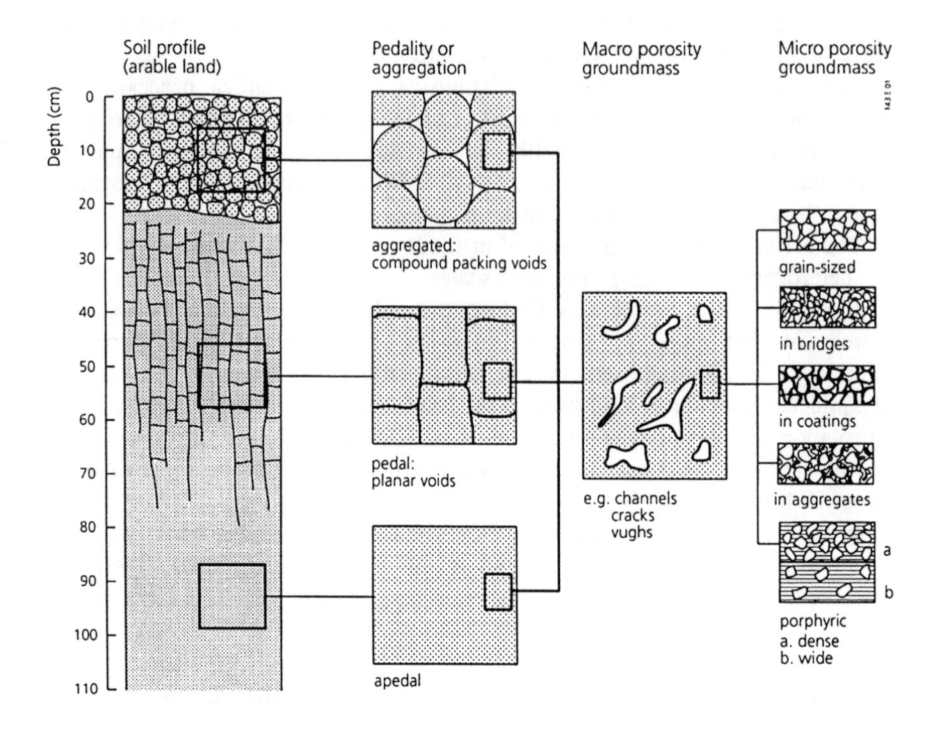

Figure 1. Overview of soil architecture scale range from field to microscopic level.

related to the locations of roots, but they may also be related to excrements, crop residues and organic manures, or be present in or adjoining voids. The bulk part of the meso- and macrofauna in agricultural land is present in the topsoil, but certain groups can also be found in the subsoil. Microorganisms can be found in the whole soil profile. Their greatest abundance is in the topsoil in and near accumulations of organic matter. The different kinds of organic carbon need to be observed and studied at different scales. An overview is given in the first columns of Table 1.

B. Methods to Observe Distribution Patterns of Single Phenomena

1. Profile Wall Observations

At a profile scale single phenomena can be marked with waterproof felt-pens on transparent sheets covering one of the cleaned and smoothened profile walls.

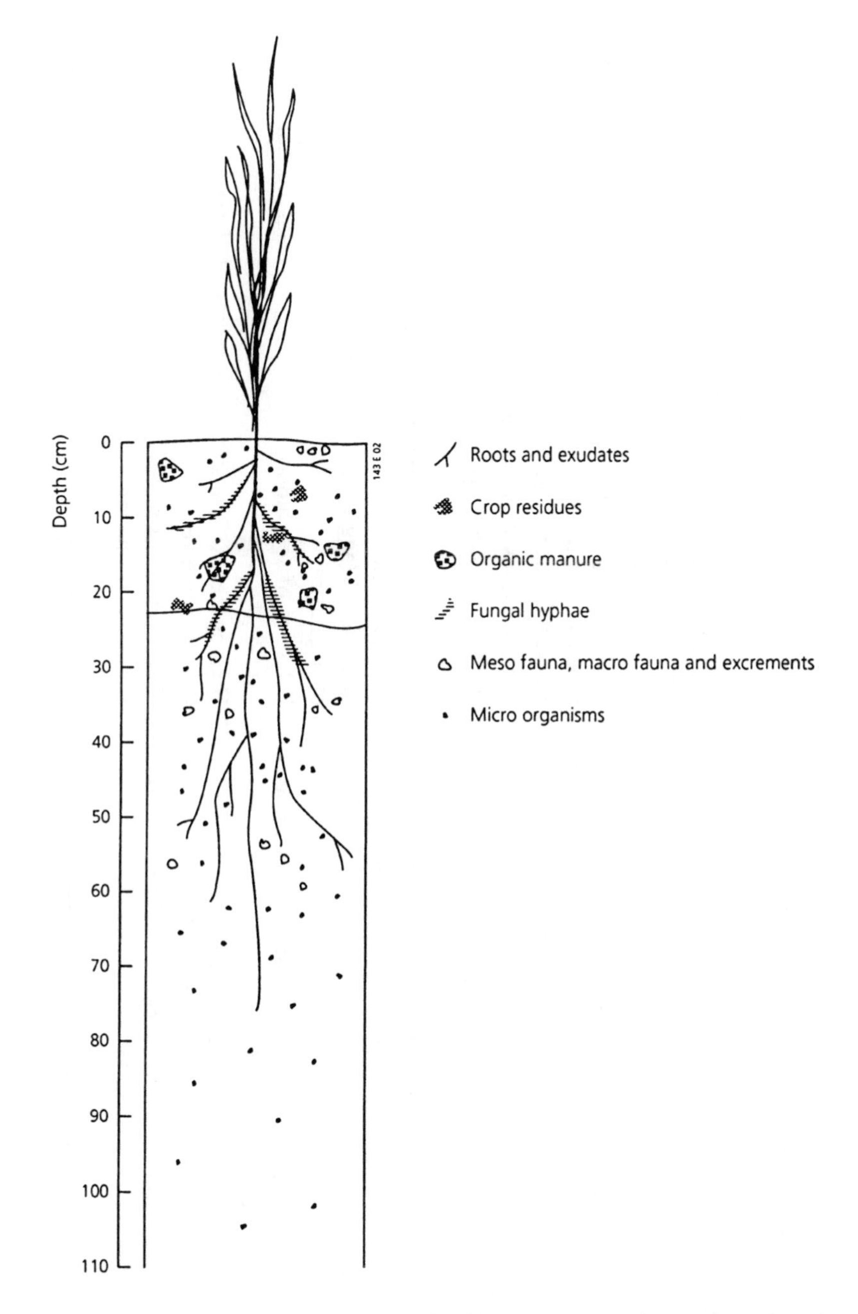

Figure 2. Overview of the occurrence of different sources of organic carbon in arable land.

Table 1. Patterns and processes influencing heterogeneity of soil organic matter sources at different scales

Organic matter source	Scale	Method to observe pattern	Process increasing heterogeneity	Process decreasing heterogeneity
Crop residues, green and brown manures	m, dm	Profile wall observations, stratified soil sampling, thin sections	Soil tillage and residue incorporation, (Previous) crop heterogeneity, traffic, slaking	Soil tillage, selective feeding, meso- and macrofauna
(Dead) roots and exudates	dm, cm, mm	Profile wall observations, stratified soil sampling, thin sections: light microscopy	Patchy root growth, due to heterogeneous soil structure, soil fertility, or inherent branching pattern	Decomposition by soil microflora and fauna
(Dead) fungal hyphae	cm, mm	Stratified soil sampling, thin sections: light and fluorescence microscopy, SEM	Heterogeneous organic matter sources, heterogeneous macrovoid spaces	Decomposition by soil flora and fauna
(Dead) meso- and macrofauna	mm, μm	Stratified soil sampling, thin sections: light microscopy	Heterogeneous food supply, temperature, aeration, etc., movements and distribution (aggregated or dispersed) of organisms	Decomposition by soil flora and fauna
Excrements	mm, μm	Thin sections: light microscopy, SEM	Behavior of soil biota	Deaggregation
Microorganisms	μm	Smears, thin sections: fluor. microscopy, SEM	Heterogeneous food supply, temperature, aeration, etc.	Preferential grazing by soil fauna, diffusion
Humified organic matter from above sources	μm	Thin sections: light microscopy, SEM	Heterogeneous sources, physical aggregate formation	Soil consumption by soil fauna

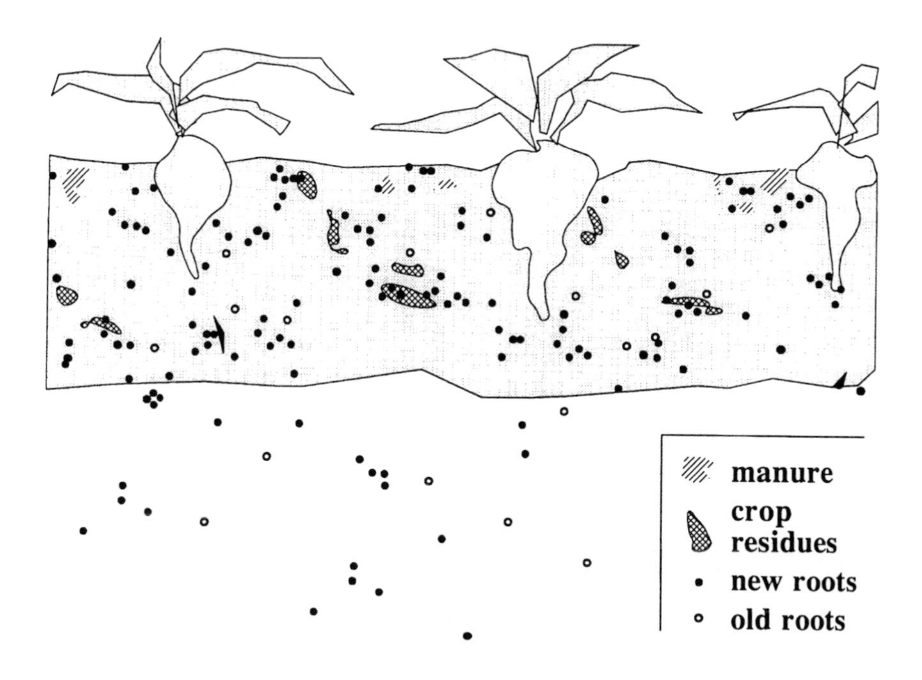

Figure 3. Schematic distribution of recent aboveground crop residues, recent organic manures, old and new roots in a sugar beet field in the plough layer.

The relevant organic features at this scale are roots, crop residues and organic manures. Figure 3 indicates schematically where aboveground crop residues, recent organic manures, old and new roots might be located in a sugar beet (*Beta vulgaris* L.) field under a certain tillage practice. Maps (1:1) of the real distribution of these features can be made in the field. The location of crop residues in the profile, as influenced by soil tillage operations, can be studied on profile walls made perpendicular to the direction of the main tillage operations. A length of several m may be needed, depending on the farm implements used, to cope with the spatial variability in depth and amounts of residues in the field.

Roots can be exposed by removing a few mm of soil from a fresh profile wall, by spraying with water, by blowing air over the surface or by carefully removing soil with a pin. The majority of roots will remain in place by such actions (Böhm, 1979). A considerable variation between laboratories exists in the way profile walls are prepared for this purpose and the absolute point density of observed roots should be calibrated with washed volumes of soil (Anderson and Ingram, 1993) to make quantitatively reliable conclusions.

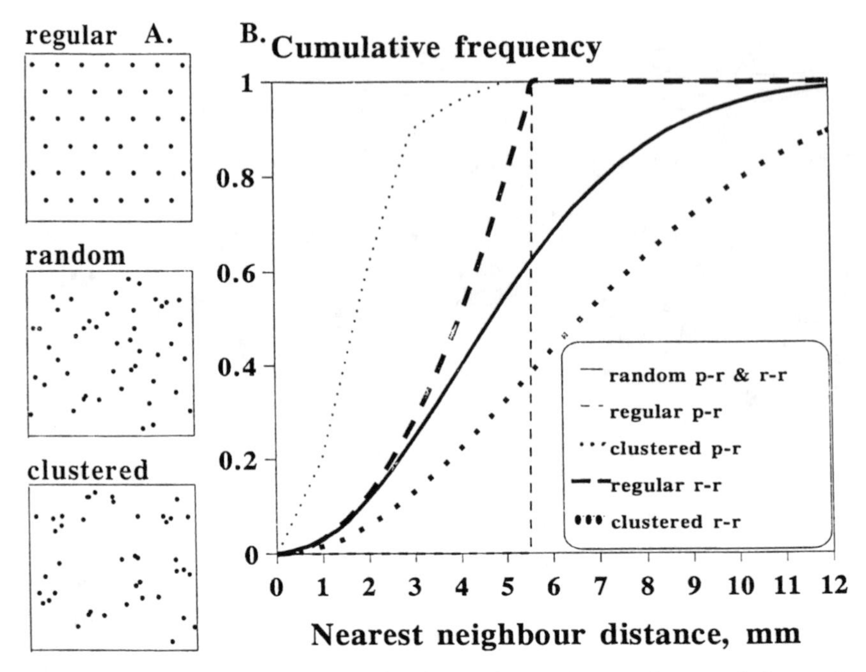

Figure 4. Basic types of point distributions - A: regular, random, and clustered, and B: cumulative frequency of nearest neighbour distances from root to root.

Figure 4 shows three basic types of point distribution patterns: regular, random, and clustered. Statistical tests of distribution patterns (Pielou, 1969) can conveniently be based on a comparison of the frequency distribution of nearest neighbour distances from roots to roots and from randomly chosen points to the nearest root. If roots are randomly distributed, the two frequency distributions are essentially the same (apart from sampling errors). In a regular pattern the root-root distances are larger, and the point-root distances on average smaller than in the random pattern. In a clustered pattern the root-root distances are smaller and the point-root distances larger. A simple test of the distribution pattern can be thus be based on the difference between the mean of the two types of nearest neighbour distances (Hopkins and Skellam test, quoted in Pielou, 1969). Analysis of root maps (or maps of any other point phenomenon) is normally based on digitization of the map, deriving the x,y coordinates of the points and a routine for selecting nearest neighbour distances (Diggle, 1983). In computerized image analysis, distance classifications of the whole map can routinely be based on a circular expansion around identified objects, as a circle of radius r contains all points within a distance r to the centre (Van Noordwijk et al., 1993a, b). Actual root distribution patterns may be random in local areas, while being clustered when larger areas are considered. The root distribution is on one hand determined by the branching pattern, with branch roots originating from main axes and thus clustered to some extent, and on the other hand modified by soil structure and heterogeneity of soil conditions, which restrict or stimulate local root development. A study of the pattern as such does not allow distinction between these two pattern forming processes, however (Diggle, 1983). A two-parameter description of a large range of theoretical distribution patterns can be based on a layered Poisson process, where a variable number of offspring (intensity) of the pattern is randomly located within a certain distance (grain size) of parent points (Diggle, 1983; Van Noordwijk et al., 1993a,b).

The spatial distribution of recognizable plant residues on the profile in arable soils can often be understood from the soil tillage implement used. The pattern can be mapped, or quantified by grid based soil sieving (Staricka et al., 1991). The presence of cracks and (tillage-induced) zones of loose and dense soil can be similarly mapped. Figure 5 gives an example of the distribution of anaerobic, blue zones in the plough layer of a clay loam soil in early spring, in relation to tillage cracks. Statistical tests of uniformity (or randomness) of the distribution first require identification of a sampling unit, which is not as straightforward as for the point patterns.

Using a hand lens or a simple field microscope soil material from selected depths can be studied either *in situ* in small boxes or disturbed on a hard surface. The material studied can be obtained from a profile wall or from augering. The organic features studied vary from humus forms, faunal channel systems to fungal hyphae and generally concern occurrence, type and quantity in relation with depth. This method is used in the Pleistocene sandy areas of The Netherlands to map the occurrence of moder humus and plaggen soils as part of the mapping procedure of the soil map scale 1:50 000, viz. sheet 34 W/O and explanation (Bodemkaart, 1979).

2. Microscopic Observations

Microscopic analyses are required when smaller scale features need to be studied *in situ* or more detailed information is to be obtained. Microscopic analyses are performed by means of visible light microscopy using polarized light, and by submicroscopic techniques.

Submicroscopy is used when morphological or chemical information is required that cannot be obtained using light microscopy. Submicroscopy involves the use of instruments that analyze emitted radiation of wavelengths shorter than that of visible light. The most common techniques are scanning electron microscopy (SEM), transmission electron microscopy (TEM) and energy dispersive X-ray analysis (EDXRA) (Kooistra and Tovey, 1994). The methods available for preparing samples for microscopic studies, including the commonly

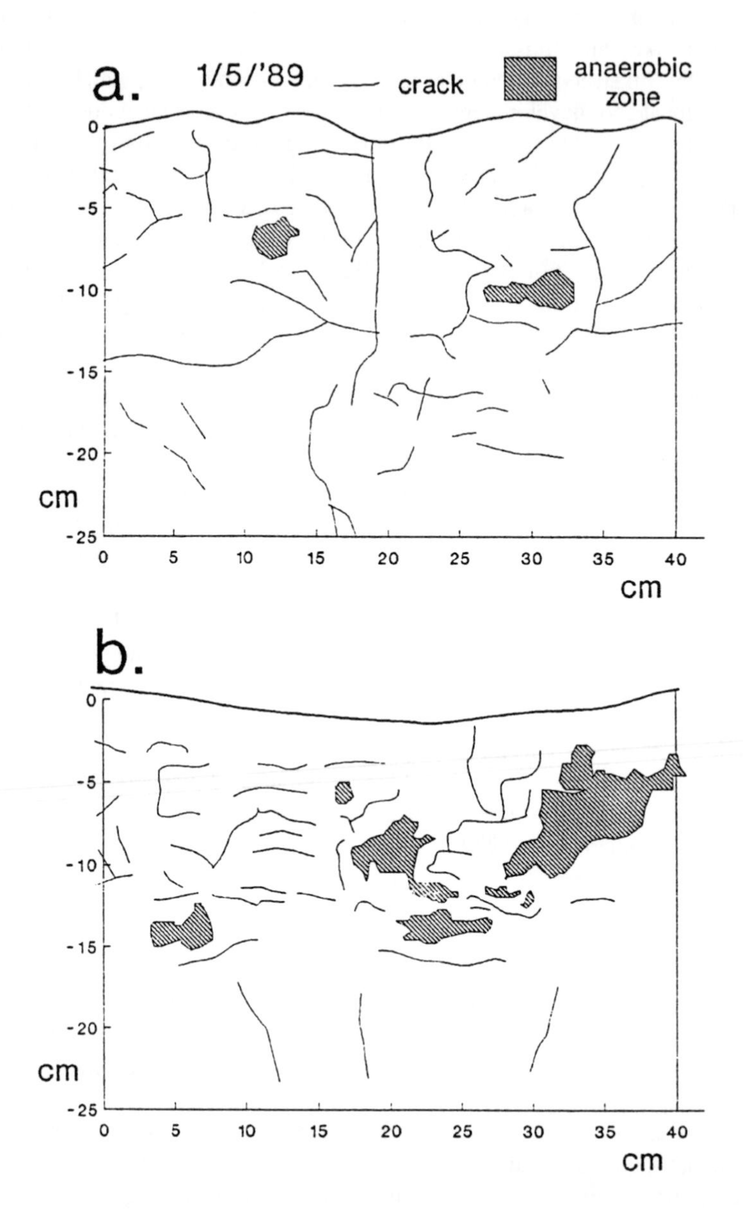

Figure 5. Maps of the plough layer in early spring of two plots (a. conventionally tilled, b. minimally tilled) of the Dutch programme on soil ecology of arable farming systems; hatched areas are anaerobic spots and lines the major cracks. Maps are made on polythene sheets.

used thin sections are summarized by Kooistra and Tovey (1994). The basic techniques for sample preparation in soil micromorphology are well documented in the literature (e.g., Jongerius and Heintzberger, 1975; Murphy, 1986; Whalley, 1979; Bennett et al. 1990). Specific sample preparations are required for research on organic materials and microstructure as they are very susceptible to shrinkage. Overviews of other techniques are given in Kooistra (1991) and Kooistra and Tovey (1994). The occurrence of organic materials can be studied qualitatively as well as quantitatively. Depending on the possibilities and research requirements, quantitative analyses can be performed using a microscope with a calibrated eye piece graticule or other types of grids, or digitized images can be analyzed using image analyzers. This permits the establishment of a number of characteristics, such as area/volume ratio, perimeter, number, length, width, orientation, position, shape factors and neighbour distances. An overview of the procedure used in light and submicroscopy is given in Kooistra (1991).

Two examples of the application of microscopic analyses are given here viz. the determination of water-conducting macrovoids and the study of the occurrence of bacteria in soil material.

A protocol was developed for studying the surface connectivity of macrovoids (Bouma et al., 1977, 1979). Large undisturbed columns of moist soil (16 cm x 15 cm x 8 cm) are carefully carved out in the field. They are slowly saturated with water in the laboratory from the bottom upwards to avoid entrapment of air. When the saturated water conductivity, K_{sat}, of these columns agrees well with the measurements of a larger group of columns, a solution of 0.5% methylene blue is ponded on top of the sample. Ponding is continued until the colour of the influent and the effluent has the same intensity. Afterwards the columns are cut into horizontal slices of 2 cm thickness each. Thin sections are prepared using the freeze-drying method to avoid contamination with methylene-blue stain on the walls of voids conducting water during the drying process. Quantitative analyses of the pore-size distributions are made on these thin sections using an image analyzer. Measurements were made on special photographs of the thin sections, obtained with crossed polarizing filters, causing voids to be black (Kooistra, 1991). Recently, however, measurements can be performed using a special CCD (Charge Coupled Device) camera whereby the thin section is placed between polarizing filters. A composite image of 5 separate images taken while crossing the polarizers from plane polarized light to crossed polarized light is computed on which all voids have the same gray level and are separated as such (pers. comm. D. Schoonderbeek). Voids with blue-stained walls are selected individually using an image-editing device or are computed from the previous images. The results for a layer from 35 to 45 cm depth in pasture land, arable land with a primary ploughpan and one with a recompacted ploughpan are given in Figure 6. Microscopic analyses of the

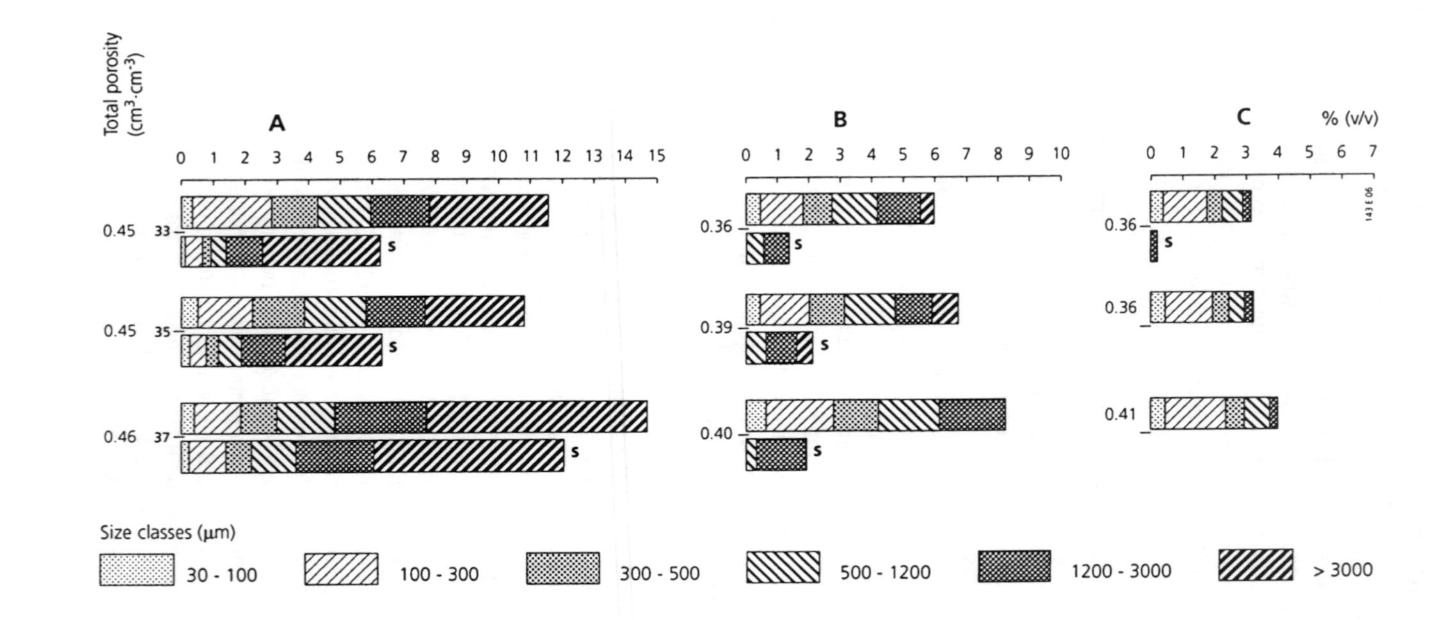

Figure 6. Total porosity and pore-size distribution by image analysis of horizontal thin sections of 33, 35 and 37 cm depth prepared from soil columns stained with methylene blue A. permanent pasture land; B. arable land with a primary ploughpan; C. arable land with a recompacted ploughpan, three years after deep tillage. s: proportion of voids with stained walls (adapted from Kooistra et al., 1985).

non-stained and stained voids explained the soil physical behavior of these layers and attributed to improved management of these soils (Kooistra et al., 1985).

During the last 10 years the staining of biological materials in soil thin sections has gained more attention to trace the presence of various groups of organisms. Non-fluorescent stains as well as fluorochromes are being developed (Altemüller and Haag, 1983; Tippkötter et al., 1986; Altemüller and Van Vliet-Lanoë, 1990; Postma and Altemüller, 1990). The latter aimed to examine the spatial distribution of certain bacteria strains and tested among other things the fluorescent brightener calcofluor white M2R. They developed a useful staining procedure, which is sufficiently documented for subsequent replication. The details are given in the referred publication. Bacterial cells, but also fungal hyphae and plant roots were clearly visible in relation to the groundmass. Most of the inoculated bacterial cells were detected, along surfaces of larger voids. Indigenous bacteria were less intensively stained and were found in smaller voids. The comparison to observations on stained soil smears suggested that some smaller cocoids, starving cells and bacterial spores remained unstained.

Single phenomena can also be studied when the soil material is disturbed. One of the most common phenomenon studied in disturbed soil material is the aggregate structure of soils. The size, quantity and stability of aggregates determined, reflect the local equilibrium of environmental forces which enhance aggregation or cause disruption. Kemper and Koch (1966) proposed a standard procedure for measuring aggregate stability, but modifications to the standard methodologies are increasingly being used. Nowadays the soil material can be disrupted with physical disruption as ultrasonic dispersion (North, 1976), with chemical dispersing agents (Tisdall and Oades, 1979), with gently misting and slaking (Elliott, 1986), dry-sieving (Gupta and Germida, 1988) and other methods. Beare and Bruce (1993) compared methods for measuring water stable aggregates and emphasized the value of comparing soil specific responses to different pretreatment procedures. They recommend that results of aggregate analyses be always accompanied by complete descriptions or references of the procedures applied, as these data can be critical to the interpretation of the data. Aggregate analyses are performed for different research questions varying from the structural stability in relation to faunal excrements to the physical separation of soil organic matter fractions.

A good example of the latter is the method for physical separation and characterization of soil organic matter fractions of different size classes developed by Cambardella and Elliott (1993). The method aimed to isolate and characterize soil organic matter fractions originally occluded within the aggregate structure to increase the current level of understanding about how the aggregate structure controls the turnover of soil organic matter. They improved the quantitative estimations of soil organic matter fractions in cultivated grassland systems. They concluded that adoption of reduced tillage management could be an important step towards the goal of sustainable production, which optimizes long-term profits to the farmer and minimizes damage to the environment.

Vittorello et al. (1989) combined a particle size fractionation (5 fractions sieved at different mesh sizes plus alkaline extractable C) with $^{12}C/^{13}C$ isotope ratios of the organic carbon for soil 12 and 50 years after conversion from forest to sugarcane. This isotope ratio allows distinction between organic material from plants with a C4 photosynthetic pathway (such as sugarcane) and material from plants with the more normal C3 photosynthesis. Their results show that 12 years after conversion, in the coarse sand fractions the majority of C was derived from sugar-cane and in the clay fractions 90% of the C still had a forest signature. Fifty years after conversion, about 70% of the clay fraction still had a forest signature. These data illustrate the importance of clay-organic matter linkages as C protection mechanism.

A recently developed method for fractionation of soil samples in suspended silica solutions of various physical densities, in combination with sieving (Meijboom et al., 1995; Hassink et al., 1993; Hassink, 1995) has a similar objective. Fresh plant material has a physical density of about 1 mg cm^{-3}; if organic material gets more and more associated with mineral soil particles (e.g. by faunal activity) it enters heavier fractions.

C. Spatial Correlation of Two or More Phenomena

All the methods dealt with in the previous section for describing the distribution of single phenomena *in situ* are also utilized to study the occurrence and spatial correlation of two or more phenomena. Depending on the research objectives more or less sophisticated qualitative methods are developed. The simplest way is a descriptive one for example noting that rare pollen grains as well as the few charcoal fragments, detected with microscopic analyses in thin sections of the subsoil, occurred in the partly infilled large root channels (Smeerdijk et al., 1994). A next step is the mapping of different phenomena on transparent sheets e.g. the occurrence of roots, cracks and crop residues in different soil horizons in a profile pit (see section B) or in thin sections. These results can be used for quantitative analyses, e.g. by means of an image analyzer.

Tests of spatial correlation can be based on the null-hypothesis of independent random distribution patterns. By quantifying the frequency of feature A in zones with increasing distance to feature B a simple test of synlocation is possible. Van Noordwijk et al. (1993c) gave examples of the distribution of roots and freshly introduced plant residues, quantified as a function of distance to the nearest macro voids in a soil profile. Two approaches are possible: enumerate all events x and determine their nearest neighbour of element y, or consider zones around all elements y with increasing distance and determine the density (number per unit area) of elements x. With image analysis computers the second approach is

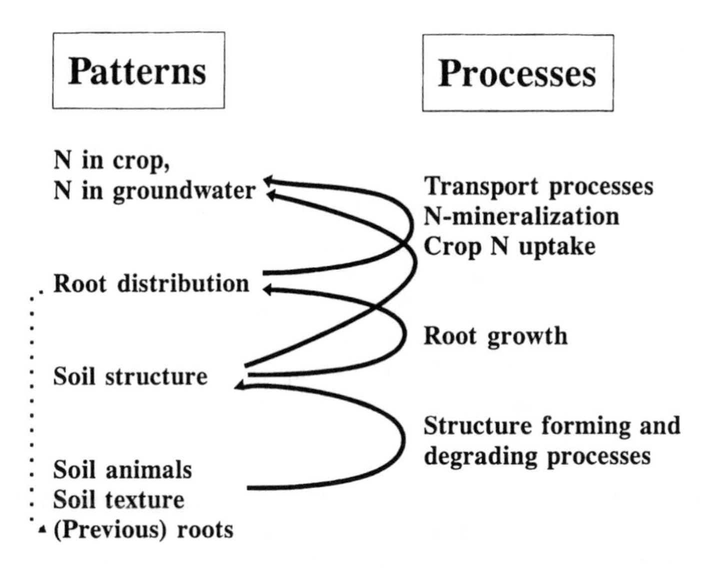

Figure 7. Chain linking patterns and processes from structure forming agents to N management.

preferrable and can run fully automatic, once elements x and y are properly identified and digitized.

III. Dynamics of Heterogeneity at a Range of Scales

A. General

The patterns of heterogeneity which can be observed with the methods of the previous section, are in a dynamic balance of rise and fall, due to counteracting processes. A causative chain of interrelations between "patterns" and "processes" can be developed (Van Noordwijk et al., 1993a), as patterns are derived from processes, but at the same time form the boundary conditions for such processes as well as those at a higher level of complexity. In general, patterns are more easily observed than processes, but processes can be better generalized and have a more universal value. In Figure 7 a possible chain linking the factors determining soil architecture with N management is summarized. This section is focused on the lower part of the chain, dealing with the processes (and underlying patterns) driving the dynamics of soil heterogeneity. These processes can lead to increasing or decreasing heterogeneity. The higher part of the chain is dealt with in the next section.

B. Processes Increasing Heterogeneity

Soil architecture is the dynamic result of many abiotic and biotic factors and processes. It may be difficult to imagine how heterogeneity, a non-uniform distribution, and patterns, a non-random distribution, can originate in a completely homogeneous environment. It is much easier to understand how existing heterogeneities are enlarged by processes taking place at different rates in the various micro-sites available. The basic, abiotic patterns with which biotic actors interact are: particle size distribution (texture) and gradients in soil water content, temperature and aeration between the soil surface and subsoil layers (with or without a water table). The latter group of patterns fluctuates in time with weather conditions. In non-cultivated land, the dynamics of heterogeneity in soil architecture are mainly governed by the weather regulating physical aggregate formation and biological processes. In cultivated land the human impact, viz. tillage, traffic and machinery, application of pesticides, fertilizers and manures, regulates the dynamics of heterogeneity in the topsoil throughout the year. The following subjects will be dealt with in more detail:
- physical aggregate formation,
- biological processes, especially plant roots and soil fauna,
- human impact (and related induced processes),
- preferential flow patterns.

1. Physical Aggregate Formation

The texture of the soil, especially the clay fraction (% $< 2\ \mu m$), and the water content determine the basic physical soil structure. Depending on clay mineralogy and clay content, the soil material swells and shrinks upon wetting and drying. Due to desiccation shrinkage cracks are formed, which once formed reappear at the same place. These cracks may form specific patterns resulting in a pedal soil. Pedality is defined as the occurrence of individual, natural, soil aggregates or peds (Soil Survey Staff, 1975). The individual peds can be classified according to their shape and arrangement into prisms, columns, plates, blocky peds, granules and crumbs, delineated by planar voids. With increasing clay content of the soil the swell/shrink potentials increase and different kinds of pedality occur. Figure 8 gives an example of the different pedalities occurring in the subsoil of marine and riverine deposits in The Netherlands with increasing clay content. The figure shows that the various pedalities differ in range, but there are overlaps where more than one type occurs. Besides a stronger desiccation, the biological impact also plays a role in these overlaps. Between 8 to 30% clay the biological impact reaches a maximum, but the structure is still related to the physical swell/shrink potential of the soil. Desiccation is related

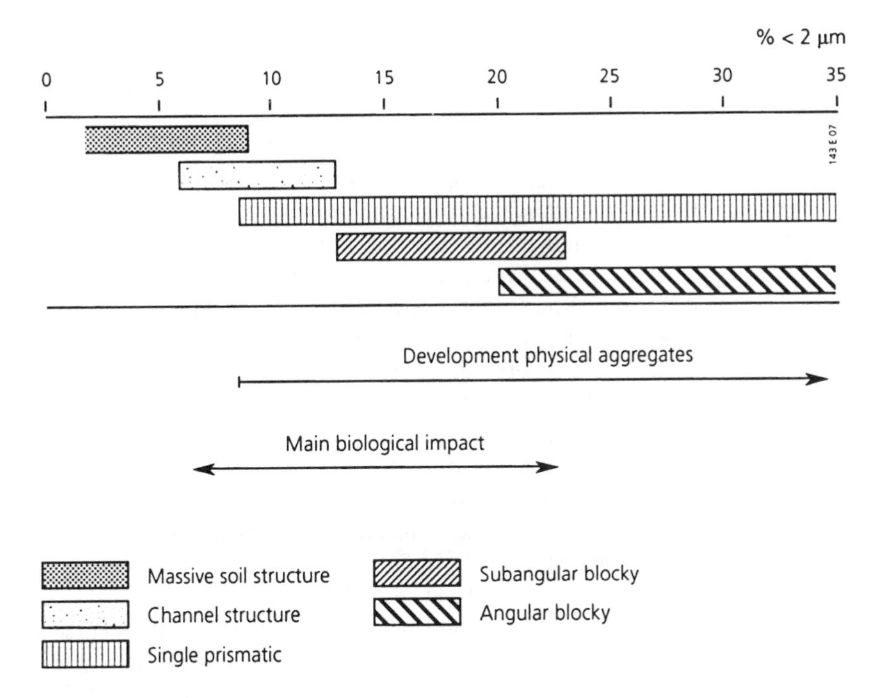

Figure 8. Pedalities occurring in the subsoil of marine and riverine deposits of The Netherlands with increasing clay content influenced by the physical aggregate formation and biological activity.

to the depth of water tables, the drainage and weather conditions. Soils with a high clay content, especially smectitic clays, strongly react on desiccation resulting in strong pedality. An example of this development, showing structure development on sedimentary deposits from sediment to vertisol is given by Blokhuis et al. (1990). The sequence in sedimentary deposits, subjected to a process of physical ripening (Pons and Zonneveld, 1965), starts with the development of widely spaced, wide and deep vertical desiccation cracks, followed by the formation of an angular blocky structure in the topsoil between the vertical cracks. Thereafter, large subhorizontal cracks develop in the ripening clays, which can be more horizontal, forming prisms that become subdivided into angular blocky structures (Inceptisols), or tend towards an oblique orientation, forming slickensides, that become subdivided into angular wedge-shaped structures (Vertisols).

2. Effects of Plant Roots

Roots not only give stability to the plant, they also regulate water, nutrient and oxygen uptake of the plant. Roots grow in such a way that they can fulfill these requirements. Roots, subsoil stem parts and filaments of algae produce channel systems by exertion of pressure on soil particles. Their diameter, more than their length, generally determines the effect on the soil. Roots and filamentous algae from about 40 μm in diameter can form lasting macropores.

They can produce systems of round channels which may be branched or not with different diameters. Roots often use available voids which they modify partly or completely by pressure during growth. In existing voids roots can adapt their own shape to a certain degree before modifying the void in which they grow (Figure 9a). Roots growing in larger voids can produce root hairs extending to the solid soil material to fulfill their nutrient requirements (Figure 9b). In the tilled zone of cultivated land, roots generally follow voids formed by tillage operations. Only a small percentage of the voids are primary root channels, which varies per crop and farm-management system. Sugar beet roots in a study of three cropping systems in The Netherlands in 1985 produced hardly any primary root channels in the tilled layer under conventional (i.e. moldboard ploughing, 20 to 25 cm) soil management, 0.4% in an integrated farming system (i.e. reduced N fertilization and pesticide use; and shallow tillage, 12 to 15 cm) and about 1% in a minimum tillage system. About 5% of the tillage voids in the conventional system were modified by the sugar beet roots and the percentage of modified voids in the integrated farming system and the minimum tillage system were not significantly higher (Kooistra et al., 1989). Below the tilled zone and in natural environments roots generally follow the cracks between the peds and faunal voids or former root channels which they all can modify to some extent.

If roots follow preexisting voids they will have no or only partial contact with the soil, except when they completely fill or expand a preexisting void. If, on the other hand, roots penetrate in the groundmass they will initially have a complete root-soil contact (100%). The degree of root-soil contact is an important parameter in studies on oxygen, water and nutrient uptake by plant roots (De Willigen and Van Noordwijk, 1987). Root-soil contact can be analyzed in thin sections which are freeze-dried before impregnation to avoid shrinkage of the roots. Analyses of maize (*Zea mays* L.) roots grown in pot experiments with soil material compacted to different bulk densities showed that roots do not shrink and that all root cross-sections are recognized (Van Noordwijk et al., 1992). In this pot experiment the degree of root-soil contact between the depths 9 to 14 cm was found to increase from 58 to 90% while soil porosity decreased from 60 to 44 % (Kooistra et al., 1992). The highest bulk density, corresponding with a total porosity of 44% reflect compacted layers in sandy loam soils. These soils are very susceptible to waterlogging after showers

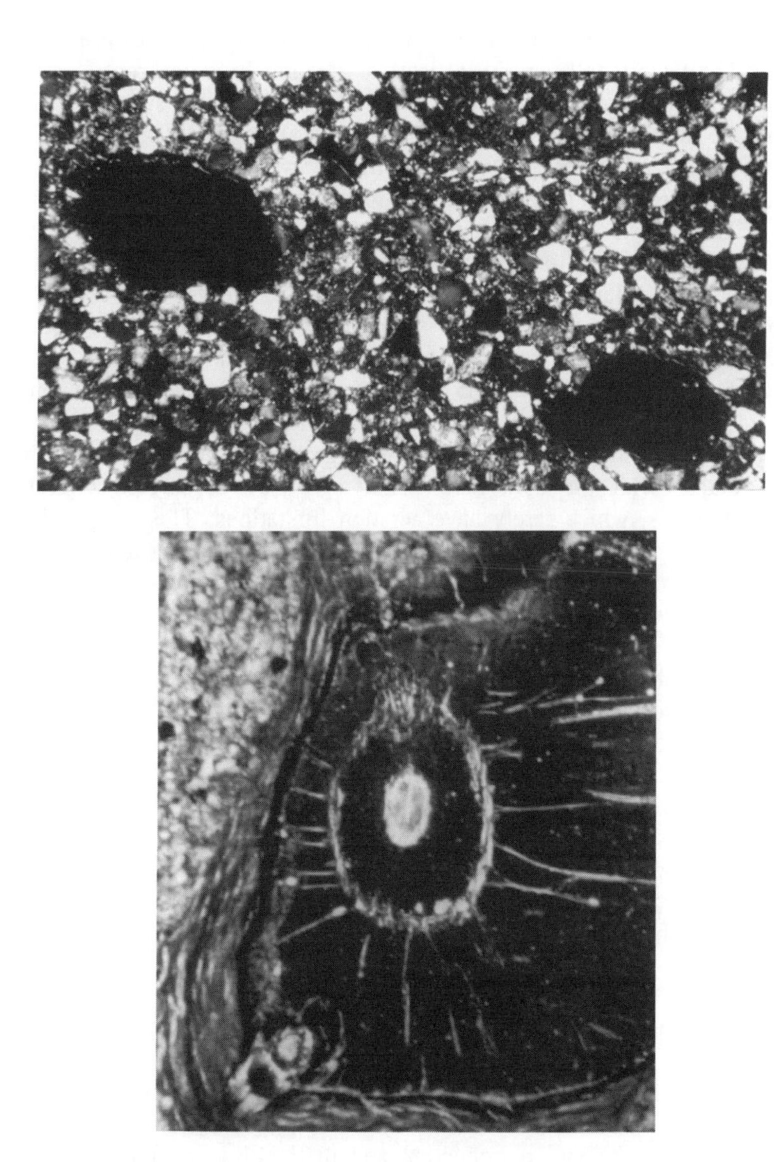

Figure 9. Roots in arable soils, a: modification of the root shape by growing into an existing void; b. production of roothairs when growing in large existing voids (magnification a: x 25; b: x 12).

and an average root-soil contact of 90% may be the limit for adequate oxygen supply. The lowest bulk density reflects a situation in which root growth is becoming limited in field situations. Here adequate nutrient supply seems to become the limiting factor, which is related to an average root-soil contact less than 58%. Root-soil contact of field grown winter wheat (*Triticum aestivum* L.) growing in 1990 in two of the above mentioned agroecosystems was quantified in horizontally oriented thin sections at 15, 25 and 45 cm depth (Van Noordwijk et al., 1993d). One day before sampling the soil surface was ponded with a methylene blue solution to stain surface-connected voids. The two fields, conventional and integrated farm management had different frequency distributions of root-soil contact in the plough layer. The percentage of roots with 100% root-soil contact was 65 and 37, and those with 0% root-soil contact 5 and 14, respectively. The average root-soil contact in the plough layer was 84% for the conventional and 66 % for the integrated system. The roots without direct contact with the soil were mainly growing in surface-connected, continuous voids, which rarely have aeration limitations. The roots with 50-100% root-soil contact were mainly the smaller roots occurring in non-stained voids. Below the plough layer, in the natural subsoil, the root-soil contact was less. In these two farm management systems, the soil structure in the plough layer was different. In the integrated system, the macroporosity of the soil was more than double (15% and 6%, respectively) than the conventional system and the biological impact by soil fauna was much larger resulting in a more open structure (Boersma and Kooistra, 1994). The existing porosity and its stability influences largely the root growth and root-soil contact of the crops in the plough layer. With increasing clay content, roots more and more follow existing pedal cracks, especially when 2:1 lattice clays are concerned.

Decaying root channels can be a major determinant in water infiltration patterns, e.g. where natural forest has been cut and the land is used for crop production in the humid tropics (Van Noordwijk et al., 1991).

3. Effects of Soil Fauna

The soil fauna is comprised of those animals which pass one or more active phases of their life cycle in the soil. They occur in the soil for several reasons e.g. protection, food and reproduction and their effects on the soil can be multifold. Their main impact on soil architecture is that they produce voids and excreta, whereby soil material, both organic and inorganic, can be dislocated, organic material fragmented, and mineral and organic material mixed. The basic products, voids and excreta, are considered in more detail below.

Many species produce voids, most of which are channels, but also other types of voids occur. Channels can be straight, curved or very convoluted, with or without branching or chambers. They can resemble root channels, but are rarely

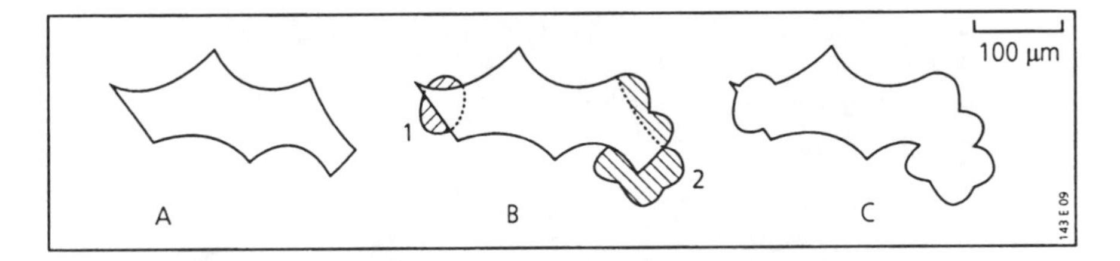

Figure 10. Modified void (c) developed from a void between earthworm excrements (a). In one end (b1) a root found its way; on the other end (b2) small mites modified the void.

as regular as these. The primary voids are formed in three ways: by pressure exerted on soil materials, by digging and removal of loose material and by soil consumption. Voids formed by pressure mainly occur in loosely packed, and in wet, plastic soil material and can be formed for example by worms and snails (Kooistra, 1978). Digging occurs in all kinds of soil materials. The result can be very regular channels as made by the dung beetle (Brussaard and Runia, 1984) or very irregular voids as made by termites. Also the digging purposes differ. Beetles dig to lay eggs at a specific depth and termites to collect fine-grained soil material for building and stabilization purposes. The same variety is found in voids due to soil consumption. Even within one group of soil organisms different void systems are produced. Earthworms, e.g. *Allolobophora longa*, can produce simple channel systems, while, e.g. *Hyperiodrilus* spp. and *Eudrilidae* can consume virtually all fine-grained soil material in specific zones in the soil in humid tropical forests (Kooistra et al., 1990). Soil animals also enlarge existing voids locally, which can be any type from cracks to root channels or other faunal voids. Necks in voids systems can be enlarged or cavities made in walls. Resulting voids can be very irregular (Figure 10). Many organisms can modify voids, especially mesofauna such as mites, collembola and enchytraeids. Void systems produced by soil fauna can start at the surface, but also deeper in the soil from existing voids. With increasing clay content, the effects of soil fauna become more restricted to modifications of existing cracks between the peds. Two features, produced by soil fauna, occur associated with faunal voids. These are wall plasters or coatings and cocoons. Plasters consist of fine-grained soil material or material from excrements. Termites generally use aggregates of fine-grained soil material, while earthworms use excreta, which are often darker coloured than the surrounding groundmass due to admixtures of organic material. The plasters occur in voids used for longer periods. Cocoons can locally be found in faunal void systems and consist of fine-grained soil material mainly derived from excrements. Earthworms can produce cocoons as retreat or hatching place for eggs (Figure 11a).

Recognizable excrements are only produced by soil fauna consuming solid materials, organic as well as inorganic. Excrements can be recognized by their shape, composition and/or organization. Many shapes can be distinguished, e.g. spheres, ellipsoids, cylinders, bacillocylinders, grooved plates, mammillated

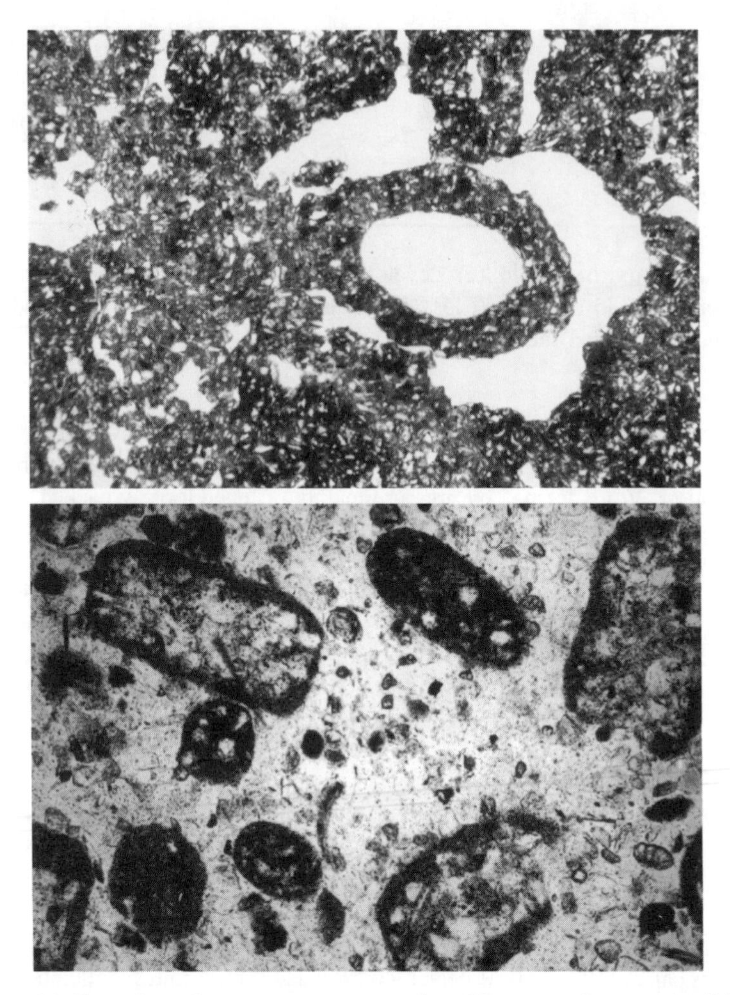

Figure 11. Faunal products, a. a cocoon produced by an earthworm; b. different shapes and composition of excrements in marine sediment, produced by pelecypods, worms and snails (magnification a: x 2.5; b: x 45).

excrements and threads (Bullock et al., 1985). Together with their size, some shapes can be ascribed to specific soil fauna. Excrements can be composed of only organic material or mixtures of organic and mineral material. The organic material is very varied and the mineral material is often fine-grained and has a distinct upper grain-size limit, which can be used for identification of the species (Figure 11b).

Shaped excrements can be organized as single units which either are loose or dense packed or form more or less coalescing units. Excrements of all shapes and organizations can partly or completely fill faunal voids. Mites, enchytraeids, diplopods and isopods generally produce accumulations of individual units, while earthworm excreta is more or less coalesced. Depending on the species, faunal voids can also be infilled with shapeless excreta (e.g. earthworm species), or be infilled with soil material by the soil fauna (e.g. dung beetle).

The impact of soil fauna is not only related to their population size but also to their mobility and activity. Sessile soil fauna, even if numerous, may have less impact on the soil architecture than mobile species, which always move around in search for food, protection or oviposition sites. On the other hand, a few distinct faunal channels produced by sessile soil fauna can have far more impact on soil physical parameters (Edwards et al., 1993) than the partly filled in tracks with soil material and/or excrements of the mobile soil fauna. In temperate climates and in forests the impact of the soil fauna is highest in the A horizon. In climates with extremes in temperature and dryness the highest impact is found in deeper horizons (Kooistra, 1982). Faunal voids often cross or follow for some distance cracks or root channels. In arable land soil fauna can produce more voids systems than plant roots, which mainly follow existing voids.

4. Human Impact

Human impact on soil architecture can be due to soil tillage, traffic (esp. at harvest time) and slaking. Soil tillage practices, especially the main or primary tillage, have the largest influence. During tillage the soil is loosened and partly pulverized as well as often mixed and inverted, whereby forming aggregates separated by rough-walled, irregular voids. Afterwards, by natural settlement, traffic and other cultivation activities, the aggregates merge. Thus, the continuity of the tillage-induced voids decrease, while isolated voids remain (see Figure 12). In the first month after ploughing and during the seedbed preparation most of the increased porosity disappears (by about 40% at both times) (Kuipers and Van Ouwerkerk, 1963). The soil is again loosened after the harvest. Tillage voids gradually become modified by soil organisms. Plant roots predominantly follow the irregular voids between the aggregates. With increasing bulk density and growth of crops their impact increases. Mesofauna enlarge necks between void systems or locally ingest fine-grained soil material. The macrofauna produce channel-like tracks on the surface of the aggregates and/or produce channel systems in aggregates, whether or not connected to the surface. The largest part of the voids occurring in arable topsoils are voids due to tillage (Figure 13). In topsoils of a conventional farming system in The Netherlands tillage voids during the years occupy about 90% of the total macroporosity

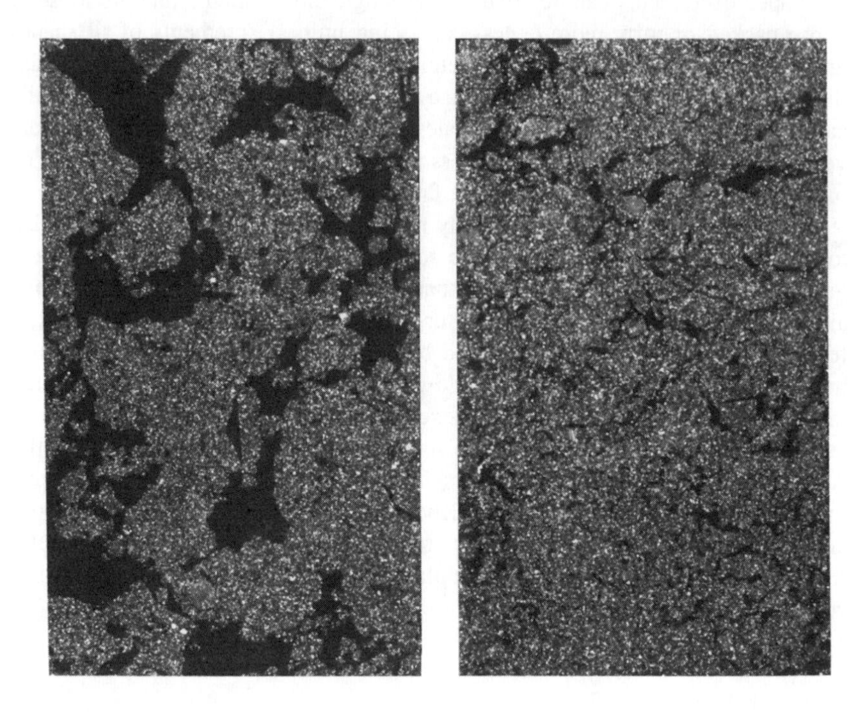

Figure 12. Settlement of soil aggregates after tillage. a: soil structure about a month after ploughing; b. soil structure after seedbed preparation (magnification a, b: x 2).

(> 30 μm diameter voids); in an integrated farming system they occupy 5-10% less as result of the increased impact of soil organisms (Boersma and Kooistra, 1994). Shrinkage cracks, starting from tillage voids, are included in this group, but they contribute only a minority of the volume occupied.

The equipment used in agricultural engineering has increased in size and weight over the last decades. This development has led to increased soil compaction and deformation of the groundmass and to rut formation. Existing voids are reduced in size and changed in shape during passage of wheels. They may become disrupted, closed or can disappear completely. A reduction of the volume occupied by voids with diameters > 100 μm by 3% (v/v) or more, and an increase on the volume of the voids < 100 μm is also common, resulting in an only slightly decreased total porosity (Kooistra, 1987). The stress exerted decreases with depth and continues below the tilled layer (Koolen et al., 1992). Maximum compaction, however, generally occurs between 4 to 8 cm depth

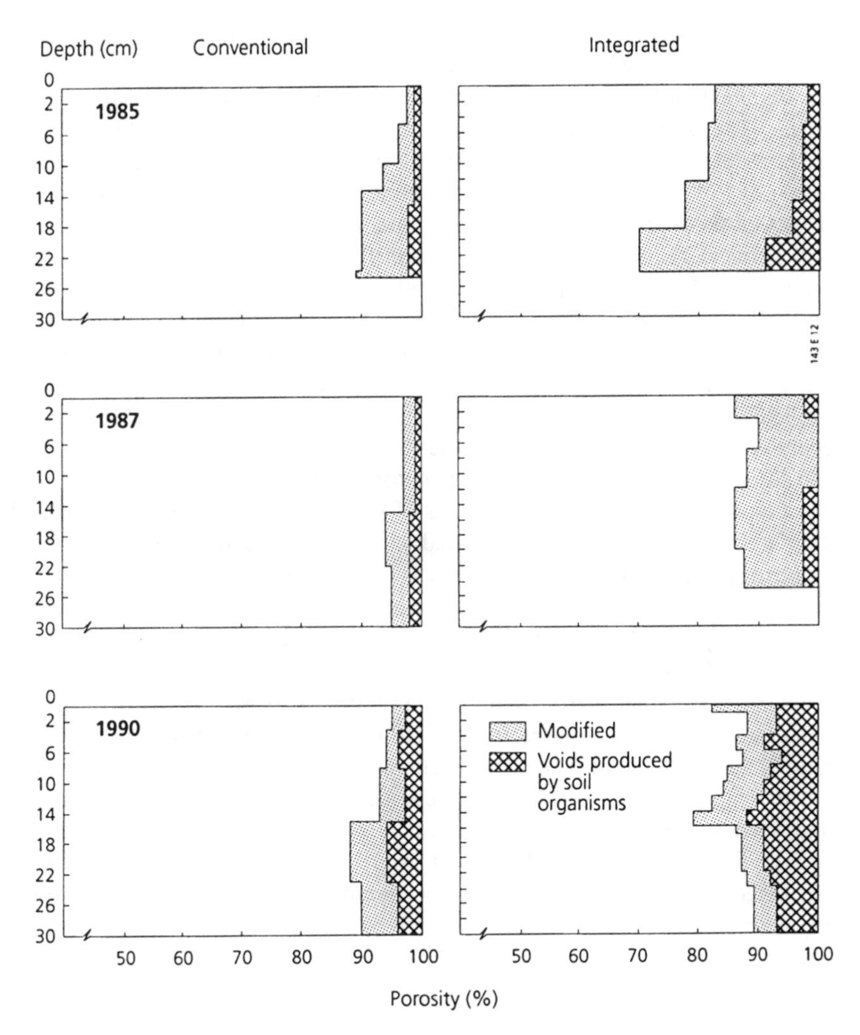

Figure 13. Proportion of biological induced voids as part of the total macroporosity (voids with diameters > 30 μm) in 1985, 1987 and 1990 in a conventional and integrated farming system.

(Tanaka et al., 1988), due to the deloading effect at the surface. This often results in the soil in the formation of sets of small horizontal cracks. These cracks can also be formed below the wheels driving over the furrow bottom during ploughing, which gives the platy appearance to ploughpans. Where these sets of cracks occur in the tilled zone most tillage voids have disappeared (Słowińska-Jurkiewicz and Domżał, 1991). When the soil is brought under cultivation, the surface soil is regularly tilled and the land is bare during several months of the year. When tilled and bare, the impact of wind and rain often lead to a disintegration of soil coherence. This starts with the aggregates at the surface. The disintegrated soil material can form crusts at the surface. When transport takes place, the soil material can move along the surface as soil erosion or flow downward into the soil as internal slaking. Microscopic studies reveal that there are many types of erosional and depositional crusts occurring in all climates (Courty, 1986; Kooistra and Siderius, 1986; Valentin and Ruiz Figueroa, 1987; Valentin and Bresson, 1992). Crusts cause restricted infiltration and limited plant germination due to their sorted, often laminated composition, lacking continuous voids. Transport of soil material along the surface due to wind and water erosion occurs on a large scale and is one of the main processes that takes place in areas with increased desertification. Also internal slaking occurs on a large scale. In the American soil classification "Soil Taxonomy" (Soil Survey Staff, 1975) a special diagnostic horizon, the agric horizon, has been defined to characterize the results of internal slaking. The illuviated soil material from the surface is deposited in voids where it forms coatings or complete infillings. The illuviated material can be deposited at any depth in the profile, depending on the present voids, quantities of water and soil material, and kind of soil material. Internal slaking results in a decrease of the total porosity and reduction of the continuous voids, which are essential for maintenance of water and air regimes and transport processes in the soil.

Repeated tillage also causes fragmentation, deformation and compaction of features due to the human impact. Although meant to loosen and equalize the soil, the aggregates formed by tillage can contain or consist of surface crusts, compacted layers, as well as soil material with infilled voids due to internal slaking. Moreover the distribution of crop residues etc. is also not evenly distributed. The resulting soil material is therefore often far more heterogeneous than natural soils.

5. Preferential Flow Patterns

In soils with large non-capillary sized voids, water and solutes flow via these voids. Large voids are cracks or planes formed by physical aggregate formation of soil tillage operations, or channels produced by roots or soil macrofauna. Coloured dyes and other traces are used by many investigators to determine and analyze large water-conducting voids (e.g. Ehlers, 1975; Bouma et al., 1977; Kooistra et al., 1985) The continuous water-conducting voids usually constitute only a few percent of the total void space and void necks have a substantial effect on the rate of flow. Dead-end voids and void necks are important locations where water with solutes can infiltrate into the soil matrix. Depending on the continuity of the macrovoids, water can bypass large volumes of soil (e.g. Bouma and Dekker, 1978; Bouma et al., 1982; Van Stiphout et al., 1987). Due to this bypass flow the effect of added fertilizers or pesticides can be restricted as their solutes can be leached deep into the soil (e.g. Dekker and Bouma, 1984; Steenhuis et al., 1990) and can cause contamination of groundwater.

Preferential flow of water in unsaturated soil need not to be restricted to large void systems. Also in zones without large void systems, the water and solutes often infiltrate in fingered patterns. Vegetation can cause uneven infiltration of rain water. Isolated trees with a 'funnel' shaped canopy, which have a high rate of stem flow, can cause deep water infiltration under their stem as found in African savanna vegetation (Knapp, 1973). These wetting patterns can, besides vegetation, also be caused by microtopography, soil layering and water repellency. When fingered flow occurs, the spatial variability in soil moisture content can be very high (e.g. Dekker and Ritsema, 1995). These patterns can be found in sandy soils as well as in clay and peat soils.

Also on a smaller scale preferential water flows have various consequences. Depending on the local soil architecture, increasingly smaller voids receive water that form films along walls or fill small voids. In these environments, microbial activities take place, depending on the presence of substrate, water tension and related diffusion of solutes and gasses, and the distribution of void diameters. McGill and Myers (1987) calculated that only about 40% of the volume of voids is small enough to retain water and is large enough to permit entry of microbes. Combined with the above mentioned decisive factors, microbial activity would take place in a smaller percentage of the occurring voids. Moreover, microorganisms are not equally sensitive to the water potential for their activities. Consequently, microenvironmental conditions are often so heterogeneous that in spots near to each other simultaneous apparently exclusive processes can take place, e.g. N mineralization or immobilization. In general, the presence and accessibility (which implies the distribution of continuous void-systems) of substrate, and the preferential wetting patterns are the main factors determining microbial processes.

C. Processes Decreasing Heterogeneity

Processes leading to heterogeneity can, when occurring at higher intensities and/or over a longer time-span, lead to homogeneity as well. This may be true, for example for the impact of soil fauna and for human impact by cultivation practices. A reduction of heterogeneity can be due to the fact that a process, such as channel formation, is gradually occurring in the whole soil matrix. A process which directly leads to a decrease of heterogeneity is diffusion of chemical compounds and organisms in the soil, which is driven by concentration gradients.

1. Impact of Soil Fauna

Activities of mobile as well as sessile soil fauna can reduce soil heterogeneity. Mobile soil fauna in search for food move through the soil following existing void systems, enlarging pore necks where the passage is hampered and/or creating new channels. They exploit local zones where food is available and produce excrements on their way. In arable land, food sources are related to the root systems of the crops, incorporated organic matter and organic matter deposited at the surface. As soil organisms are generally strongly related to specific zones where the same moisture and temperature conditions prevail, the same architecture can develop in whole zones. A characteristic example is a moder humus zone in a spodosol (Figure 14a), in The Netherlands, which often occurs below a plaggen epipedon. The organic matter together with the fine-grained mineral material becomes conversed into shaped organo-mineral excrements produced by small soil fauna, chiefly microarthropods such as enchytraeids and collembola. The excrements occur in open or close packed, individual or slightly welded clusters between the larger mineral grains. Nearly all fine-grained mineral material is present in the organo-mineral excrements.

Sessile soil fauna create their own channel system which is in use for prolonged periods. They, however, also move around in search for food, but retreat into their own channels. These animals deposit their excreta at the surface or, less often, into old side-branches of their channel system. With time and presence of enough food homogeneous zones are formed. A characteristic example of this type is a mull humus in a mollic epipedon (Figure 14b). In a mull humus the organic material is intimately mixed with the mineral material forming a clay-humus complex. This type of humus form is produced by earthworms, who consume organic as well as mineral material. The excreta form clusters or small aggregates of strongly welded or shapeless material.

2. Impact of Diffusion

Diffusion is driven by, but also reduces concentration gradients. Diffusion is thus one of the major processes leading to homogeneity of the soil, although complete uniformity is approached at an infinitely slow rate, and new disruptions will have taken place before the soil is completely uniform. To get a feeling for the speed at which such homogenization occurs, Figure 15 shows the time course of the average concentration of a sphere of soil of 1 cm radius, inside a second sphere of 20 cm radius (see appendix for equations). At time zero, the concentration in the inner sphere is 1.0 and in the surrounding sphere 0. When we take typical values for the diffusion coefficient representing nitrate in a soil at field capacity, diffusion is so rapid, that the soil will be uniform for all

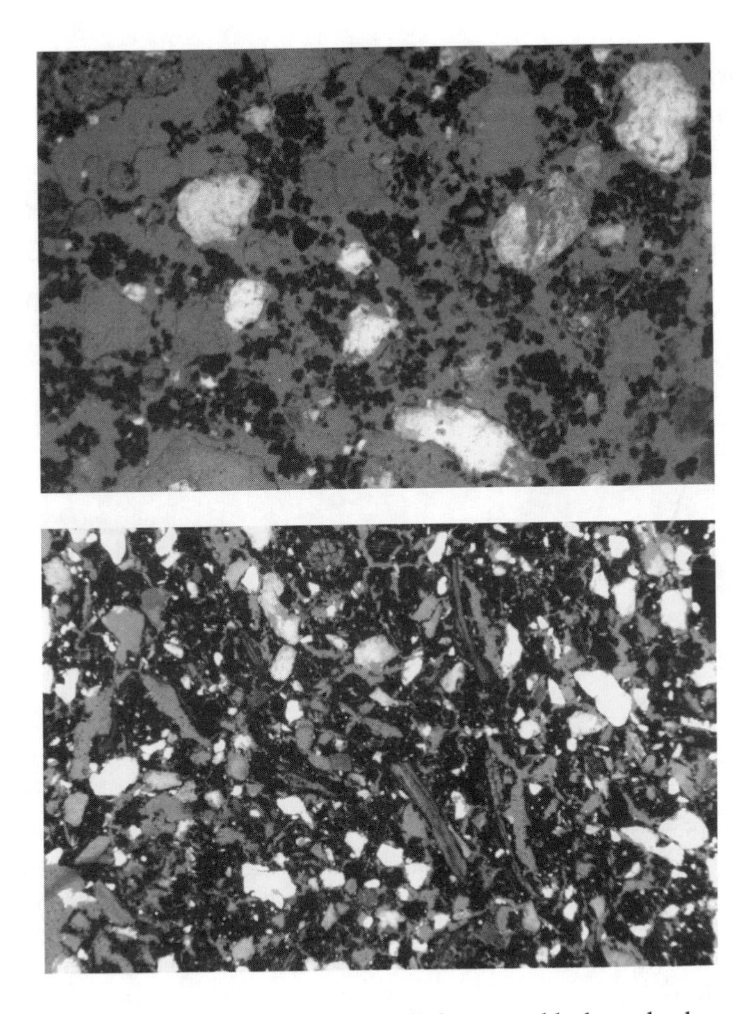

Figure 14. Humus forms produced by soil fauna. a. black moder humus; b. mull humus (magnification a: x 62; b: x 25).

practical purposes within a few days. For K it may be one or a few weeks, but for phosphate it will take weeks or months, depending on the P adsorption constant of the soil to reduce the concentration in the inner sphere to 50% of its original value. The time is scaled with the effective diffusion constant, and a tenfold reduction in mobility, leads to a tenfold increase in the time required for a set degree of homogenization. In drier soil conditions, the mobility of nutrients is reduced. So and Nye (1989) showed that for a tenfold decrease in D^* from its value at field capacity (pF = 2.0) a sandy loam has to dry out until pF 3.3

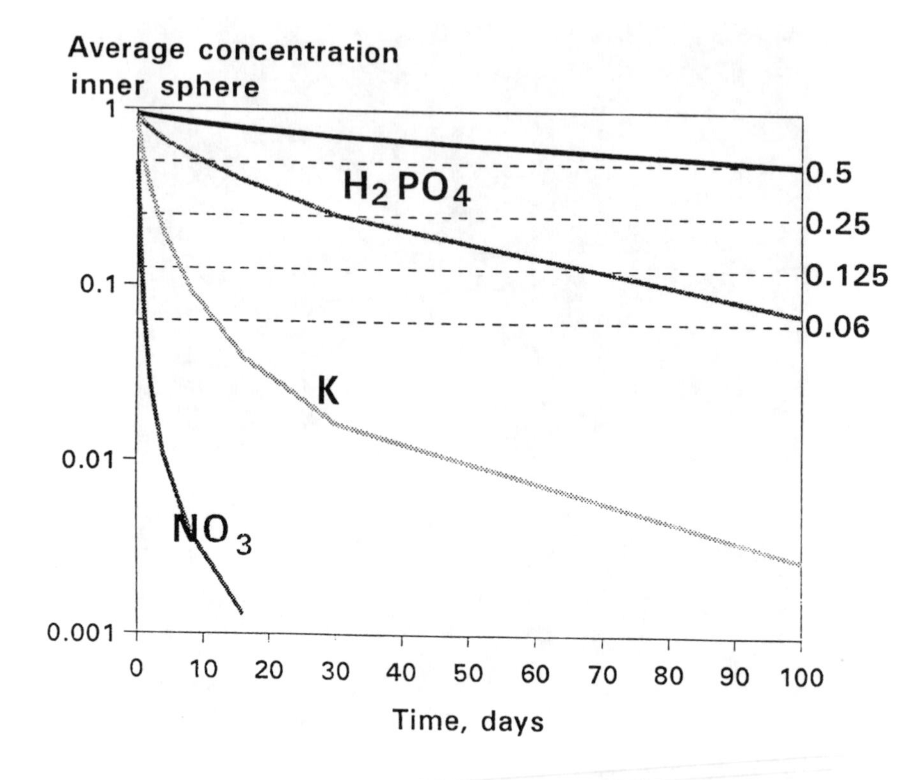

Figure 15. Concentration decrease by diffusion from a 'hot spot' (initial concentration 1), consisting of an inner sphere of radius 1 cm in a 'cold' sphere (initial concentration 0) of radius 20 cm, with values of the effective diffusion constant 1, 0.1 and 0.01-0.001 $cm^2\ day^{-1}$, taken to be representative for nitrate, potassium and phosphate, respectively. Calculations based on equation (3) in the appendix.

and a silty clay until pF 4.5. Such a decrease in soil water content renders the diffusion of NO_3 in a dry soil similar to that of K at field capacity.

Homogenization of concentration by diffusion also depends on the absolute size of the 'hot spot'; in the spherical model the time required for homogenization is (approximately) proportional to the second power of the radius. Hot spots of nitrate of 3.2 cm radius thus will follow the line indicated for K hot spots of 1 cm radius. Heterogeneities at short distances are thus of practical relevance for solutes with an effective mobility as low as, or lower than that of inorganic phosphate. The diffusivity of oxygen in water is 10^4 times slower than that in free air, but the diffusivity of oxygen in a water saturated soil is still similar to that of nitrate. Equilibration of oxygen concentration in a soil with aggregate sizes of 1 to 3 cm radius is thus a matter of hours to days (weeks), at least in the absence of oxygen consumption. With a continuing oxygen demand local differences in oxygen concentration and even local anaerobiosis can persist in (nearly) water saturated soil. The overriding importance of consumption on maintaining steep gradients was also shown for glucose diffusion into soil aggregates by Priesack and Kisser-Priesack (1993).

IV. Consequences of Heterogeneity for Processes and Resulting Patterns at Higher Levels of Complexity

This section is focused on the upper part of the chain as given in Figure 7. To what extent does soil structure and the distribution of soil carbon influence processes at a higher level of organization, such as water and nutrient use efficiency of agricultural systems? If a 'good' soil structure is to help in reducing negative environmental impact of agriculture, how does this work and how can it be used to the full extent?

Water and nutrient use efficiency of an agricultural system primarily depends on how well supply and demand are mutually adjusted. As long as demand exceeds supply, resource use efficiencies are generally high. It is difficult to achieve a high efficiency with a resource supply which does not limit productivity. Yet, highly productive but at the same time efficient agricultural production systems are needed for the next century. The 'efficiency gap' between well maintained experimental plots and current farmer's practice (Whitmore and Van Noordwijk, 1995) is at least partly due to difficulties in predicting rates of biologically mediated soil processes and to field scale spatial variability beyond the area exploited by an individual plant. Several of the components of the demand-supply balance are related to soil structure: root growth, mineralization, denitrification and leaching all are influenced by soil structure. To close the 'efficiency gap' a better quantitative understanding of the underlying patterns and processes is needed. In this context we may come back to the question of whether models can be based on the concept of a perfectly mixed homogeneous medium responding uniformly to outside influences or should be based on the recognition of soil as a highly structured, heterogeneous environment with large zones of little activity and scattered 'hot spots' full of activity. At first sight, the apparent success of many model descriptions of major soil processes, based on a homogeneous soil concept seems to indicate that the heterogeneity is more of academic than of practical interest.

The majority of transport models consider soil heterogeneity implicitly rather than explicitly. Relevant parameters are normally measured on the scale of an 'elementary soil volume' which can be thought to be representative. Thus, a soil may be described as if it were homogeneous, but with an average gas diffusivity assigned to it which is based on the macropores, a water diffusivity dominated by the mesopores, and a by-pass flow description, based on the continuous macro-voids. The way in which these overall parameters are influenced by soil structure, and by variations in soil structure due to management actions remains obscure in many instances.

Models of nutrient uptake are normally based on assumptions of a regular root distribution. The consequences for uptake processes of non-regular root distribution and incomplete root-soil contact imposed on the root architecture by a structured soil can now be quantified (Van Noordwijk et al., 1993a; Veen et al., 1992). To a considerable extent, such effects of soil structure can be expressed as a reduction factor on the measured total root length and can be overcome by the plant by making more roots. Such reduction factors can give an alternative explanation to the reduction factors employed to estimate the 'live' fraction of a root system, but the two aspects are as yet difficult to separate and quantify. In many cases, models are used to evaluate the consequences of root length densities actually found in the soil and do not attempt to predict how root development depends on soil conditions, as mediated by soil structure.

Mineralization patterns do depend on soil structure. The consequences for food web functioning of the restricted accessibility of bacteria in small sized pores for protozoan predators are a recent focus of attention (Hassink et al. 1993; Wright et al. 1993). Soil food web models (De Ruiter et al., 1994) so far are remarkably similar to those for aquatic systems, and do not explicitly consider spatial structure of a soil, and the 'hot spot' nature of organic substrates. With the existing food web models, the measured N mineralization patterns in field soils can be predicted within a factor-two error range and they thus form an alternative to the C pool based model approach. Again, their relative success does not implicate that soil heterogeneity is not important for the functioning of the soil food web, as the calculations are based on actually observed population densities and do not attempt to predict the population development as such. Some mineralization models based on C pool sizes do include the concepts of 'physical protection' of soil carbon against bacteria and of bacteria against grazers (Verberne et al., 1990) but the protection factors are not yet completely operationally defined. Present attempts to do so are largely confined to soil texture and do not yet attempt to include soil structure. In the next generation of models, the mobility of soil organisms and their relations with soil structure may be more fully described.

Leaching rates critically depend on 'by-pass flow' through macropores. Considerable progress in field measurement of these properties has been made (Edwards et al., 1993). Despite considerable progress at the single aggregate level (Leffelaar, 1993), our understanding of denitrification, and its relation with measurable aspects of soil architecture is still far from comprehensive. Development of anaerobic sites and denitrification depend on the presence of decomposable organic substrates under conditions of slow oxygen replenishment. Soil structure and soil water content determine the oxygen transport rates, but the spatial distribution of oxygen demanding bacteria is probably related to the hot spots of organic substrates.

Using larger amounts of organic inputs is not a simple recipe to obtain a better soil structure and a more efficient and yet productive type of agriculture. If they are integrated with other aspects of crop management, however, biological means of maintaining and improving soil structure can give a real contribution to overall resource use efficiency. Biological management of soil structure by earthworm and root channels remains an exciting option (Dexter, pers. comm.) for extensive land use systems; one should not expect miracles however and where the soil building has been severely mistreated, e.g. during deforestation in the tropics, the biological soil architects will take a very long time to reconstruct anything similar to the original structure.

V. Importance of Heterogeneity for Carbon Retention in Soil

The organic carbon present in the soil is derived from aboveground organic materials incorporated as part of tillage activities or below ground organic inputs from primary producers (see part I and Figure 2). The incorporated organic inputs have a heterogeneous distribution. The method of incorporation determines the spatial pattern: e.g. inverted soil surfaces deposited by plough shares (Figure 3), more finely mixed material with other tillage implements. The same is true for the below ground organic inputs from primary producers, viz. leaves pulled into wormholes or roots with a distinct distribution pattern in the soil, whether or not redistributed by tillage activities. This heterogeneous distribution by itself may not be very important for decomposition rates or its complement, carbon retention in soil. The same organic material can be decomposed via different pathways and in different time lapses. Thus it is not the location as such of organic carbon which matters, but its effects on accessibility of the organic input for decomposing bacteria and fungi, predators and water, gasses and solutes. If the accessibility is affected, e.g. because a crop residue is incorporated in groundmass, decomposition can be very slow, while residues a few cm away along a crack are quickly fragmented and partly consumed (Figure 16). The accessibility for incorporated organic matter differs strongly and cannot be predicted. Those of the original below ground organic materials, such as roots, exudates and mycorrhizal hyphae are in principle far more accessible, as their locations are determined by both the root distribution pattern itself and the root preference for existing void systems.

In more natural arable systems (e.g. in agroforestry), faunal activity is the major determinant of incorporation of above ground organic inputs. Considerable differences in litter layer exist between vegetation types, but also over small

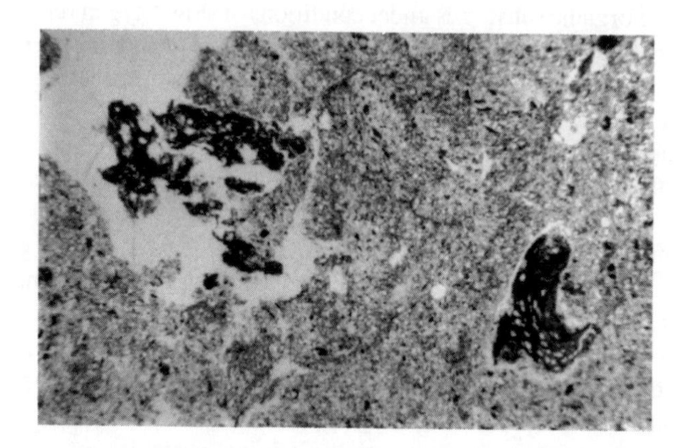

Figure 16. Difference in accessibility of crop residues in the plowlayer; the crop residue along the crack is partly consumed and fragmented by enchytraeids; the residue embedded in the groundmass remains intact (magnification x 2).

distances within a single forest type. Burghouts (1993) analyzed such patterns for a Bornean rainforest and concluded that differences in deposition rates rather than turnover rates were responsible for the apparent differences in carbon retention on the soil surface. Within a structured soil, however, possibilities for differences in turnover rates are substantially larger than on the soil surface.

Primary organic matter is decomposed faster than secondary organic matter present in excreta. The most decomposable excrements are those of the large macrofauna such as earthworm excreta. Mesofauna (enchytraeids, mites and collembola) consume coarser fragments of the incorporated organic material, sometimes leaving their excrements in the cavity they produced. Excrements can be conserved in the soils for centuries, in spite of tillage activities, without microscopic recognizable changes (e.g. Kooistra, 1978). Otherwise excrements disintegrate slowly forming shapeless organic coatings and bridges between mineral grains as in moder podsol soils.

VI. Summary and Conclusions

Soil structure has a biological component, the purposely made architectural part, which is highly variable in quality and quantity, occurs at different scales and varies throughout the year. The biological impact of belowground communities and plant roots depends on the different requirements and can be mutually exclusive, mutually complementary or dependent, completely independent or mixtures of these. The biological component of soil structure, therefore, is far less predictable than the physical part of soil structure dealing with swell/shrink and freeze/thaw cycles, and the human impact by tillage and management operations.

Nearly every observable/measurable biotically related soil phenomenon has a heterogeneous distribution in the soil. Increase and decrease of heterogeneity can be based on similar processes, but at different time scales, or on different processes. Recognition of and understanding of the dynamics of these heterogeneities appears to be essential for a 'real' understanding of what's going on in the soil. Yet, for many purposes models which treat soils as (a number of) black boxes, with average rates assigned to them have been highly successful. Such models are based on empirical 'fudge' factors, which are not always properly acknowledged in so-called 'process-based' models. Progress will be made when the empirically determined average parameter values for the black boxes can be explained on the basis of the underlying heterogeneity, the real actors and the variation in internal heterogeneity between sites. A number of new concepts and methods is now available for such an excercise.

The heterogeneous distribution by itself is not the main controlling factor in regard to decomposition rates or carbon retention in soils. Rather, it is the accessibility (not the location) of the organic input for decomposing bacteria and fungi, predators, water, gasses and solutes which is of primary importance. In studying these phenomena, new methods for analyzing and interpreting distribution patterns of single features across a range of scales are needed, especially methods to quantify the spatial correlations of patterns and their significance for turnover of organic inputs in the soil. As some of the products of turnover (e.g. anaerobically formed gasses such as CH_4 and N_2O, and solutes such as NO_3) are definitely undesirable when they reach other compartments of the environment (e.g. atmosphere, groundwater), research on soil structure may gain new relevance.

References

Altemüller, H.-J. and Th. Haag. 1983. Microscopische Untersuchungen an Maiswurzeln im umgestorten Bodenverband. *Kali-Briefe* 16: 349-363.

Altemüller, H.-J. and B. Van Vliet-Lanoë. 1990. Soil thin section fluorescence microscopy. p. 565-581. In: L.A. Douglas (ed.), *Soil Micromorphology*: a Basic and Applied Science. Developments in Soil Science, 19. Elsevier, Amsterdam.

Anderson, J.M. and J.S.I. Ingram (eds.), 1993. *Tropical soil biology and fertility, a Handbook of Methods*. 2nd edition. CAB International, Wallingford, 221 pp.

Beare, M.H. and R.R. Bruce. 1993. A comparison of methods for measuring water-stable aggregates: implications for determining environmental effects on soil structure. In: Brussaard, L. and M.J. Kooistra (eds.), Int. Workshop on Methods of Research on Soil Structure/Soil Biota Interrelationships. *Geoderma*, 56: 87-104.

Bennett, R.H., W.R. Bryant, and M.H. Hulbert (eds.), 1990. *Microstructure of Fine-Grained Sediments*. Springer, New York, U.S.A., 582 pp.

Blokhuis, W.A., M.J. Kooistra, and L.P. Wilding. 1990. Micromorphology of cracking clayey soils (Vertisols). p. 123-148. In: L.A. Douglas (ed.), *Soil Micromorphology: a Basic and Applied Science*. Developments in Soil Science, 19. Elsevier, Amsterdam.

Bodemkaart. 1979. *Bodemkaart van Nederland, schaal 1:50 000*. Blad 34 W/O met toelichting. Stiboka, Wageningen, The Netherlands, 166 pp.

Boersma, O.H. and M.J. Kooistra. 1994. Influences on soil structure of silt loam typic fluvaquents under various agricultural management practices. *Agric. Ecosyst. Environ*. 51:21-42.

Böhm, W. 1979. *Methods of studying root systems*. Springer, Berlin, 188 pp.

Bouma, J., A. Jongerius, O.H. Boersma, A. Jager and D. Schoonderbeek. 1977. The function of different types of macropores during saturated flow through 4 swelling soil horizons. *Soil Sci. Soc. Am. J.* 41 (5): 945-950.

Bouma, J. and L.W. Dekker. 1978. A case study on infiltration into dry clay soil. I. Morphological observations. *Geoderma* 20: 27-40.

Bouma, J. A. Jongerius and D. Schoonderbeek. 1979. Calculation of saturated hydraulic conductivity of some pedal clay soils using micromorphometric data. *Soil Sci. Soc. Am. J.* 43 (2): 261-264.

Bouma, J., C.F.M. Belmans and L.W. Dekker. 1982. Water infiltration and redistribution in a silt loam subsoil with vertical worm channels. *Soil. Sci. Soc. Am. J.* 46: 917-921.

Brussaard, L. and L.T. Runia. 1984. Recent and ancient traces of scarab beetle activity in sandy soils of The Netherlands. *Geoderma*. 34: 229-250.

Bullock, P., N. Federoff, A. Jongerius, G. Stoops, and T. Tursina. 1985. *Handbook for soil thin section description*. Waine Research Publ., Wolverhampton, U.K.

Burghouts, T. 1993. *Spatial heterogeneity of nutrient cycling in Bornean rainforest*. Free University Amsterdam, PhD Thesis, 156 pp.

Cambardella, C.A. and E.T. Elliott. 1993. Methods for physical separation and characterization of soil organic matter fractions. In: Brussaard, L. and M.J. Kooistra (eds.), Int. Workshop on Methods of Research on Soil Structure/Soil Biota Interrelationships. *Geoderma* 56: 449-457.

Carslaw, H.S. and J.C. Jaeger. 1959. *Conduction of heat in solids*. 2nd ed. Clarendon, Oxford.

Courty, M.A. 1986. Morphology and genesis of soil surface crusts in semi-arid conditions (Hissar region, Northwest India). p. 32-40. In: Callebaut, F., D. Gabriels, and M. De Boodt (eds.), *Assessment of soil surface sealing and crusting*. State Univ. Ghent, Belgium.

Dekker, L.W. and J. Bouma. 1984. Nitrogen leaching during sprinkler irrigation of a Dutch clay soil. *Agric. Water Manage.* 9: 37-45.

Dekker, L.W. and C.J. Ritsema. 1995. Fingerlike wetting patterns in two water repellent loam soils. *J. Env. Qual.* (in press).

De Ruiter, P.C., J.C. Moore, K.B. Zwart, L.A. Bouwman, J. Hassink, J. Bloem, J.A. de Vos, J.C.Y. Marinissen, W.A.M. Didden, G. Lebbink and L. Brussaard. 1994. Simulation of nitrogen mineralization in the belowground food webs of two winter wheat fields. *J. Appl. Ecol.* 30: 95-106.

De Willigen, P. and Van Noordwijk, M., 1987. *Roots, plant production and nutrient use efficiency*. Ph.D. Thesis, Agric. Univ. Wageningen, The Netherlands, 282 pp.

Dexter, A.R. 1988. Advances in characterization of soil structure. *Soil Tillage Res.* 11: 199-239.

Diggle, P.J. 1983. *Statistical Analysis of Spatial point Patterns*. Academic Press, London. 148 pp.

Edwards, W.M., M.J. Shipitalo, and L.B. Owens. 1993. Gas, water and solute transport in soils containing macropores: a review of methodology. In: Brussaard, L. and M.J. Kooistra (eds.), Int. Workshop on Methods of Research on Soil Structure/Soil Biota Interrelationships. *Geoderma*, 57: 31-49.

Ehlers, W. 1975. Observations on earthworm channels and infiltration on tilled and untilled loess soils. *Soil Sci.* 119: 242-249.

Elliott, E.T. 1986. Aggregate structure and carbon, nitrogen and phosphorus in native and cultivated soils. *Soil Sci. Soc. Am. J.* 50: 627-633.

Gupta, V.V.S.R. and J.J. Germina. 1988. Distribution of microbial biomass and its activity in different soil aggregate classes as affected by cultivation. *Soil Biol. Biochem.* 2 (6): 777- 786.

Hassink J., L.A. Bouwman, K.B. Zwart, J. Bloem, and L. Brussaard. 1993. Relationships between soil texture, physical protection of organic matter, soil biota, and C and N mineralization in grassland soils. In: Brussaard, L. and M.J. Kooistra (eds.), Int. Workshop on Methods of Research on Soil Structure/Soil Biota Interrelationships. *Geoderma* 57: 105-128.

Hassink, J. 1995. Density fractions of soil macroorganic matter and microbial biomass as predictors of C and N mineralization. *Soil Biol. Biochem* (in press).

Jongerius, A. and G. Heintzberger. 1975. *Methods in soil micromorphology; a technique for the preparation of large thin sections*. Soil Survey Paper no. 10, Netherlands Soil Survey Institute, Wageningen, 48 pp.

Kemper, W.D. and E.J. Koch. 1966. *Aggregate stability of soils from Western United States and Canada*. Techn. Bull., 1355, Agricultural Research Service, U.S. Dept. of Agriculture, Washington, D.C., 52 pp.

Knapp, R. 1973. *Die Vegetation von Africa*. Gustav Fischer Verlag, Stuttgart, 626 pp.

Kooistra, M.J. 1978. *Soil development in recent marine sediments of the intertidal zone in the Oosterschelde, The Netherlands. A soil micromorphological approach*. Soil Survey Paper no. 14, Netherlands Soil Survey Institute, Wageningen, 183 pp.

Kooistra, M.J. 1982. Micromorphology. p. 71-89. In: Murthy, R.S., L.R. Hirekerur, S.B. Deshpande, and B.V. Venkata Rao (eds.), *Benchmark Soils of India. Morphology, Characteristics and Classification for Resource Management*. National Bureau of Soil Survey and Land Use Planning (ICAR) in collaboration with The All India Soil and Land Use Survey (Government of India). Agricultural Universities and Soil Survey Organizations of the State Departments of Agriculture.

Kooistra, M.J. 1987. The effects of compaction and deep tillage on soil structure in a Dutch sandy loam soil. p. 445-450. In: Fedoroff, N., L.M. Bresson, and M.A. Courly (eds.), *Soil Micromorphology*, AFES, Plaisir, France.

Kooistra, M.J. 1991. A micromorphological approach to the interactions between soil structure and soil biota. *Agric. Ecosyst. Environ.* 34: 315-328.

Kooistra, M.J., J. Bouma, O.H. Boersma, and A. Jager. 1985. Soil-structure differences and associated physical properties of some loamy Typic Fluvaquents in The Netherlands. *Geoderma* 36: 215-228.

Kooistra, M.J. and W. Siderius. 1986. Micromorphological aspects of crust formation in Savanna climate under rainfed subsistence agriculture. p. 9-17. In: Callebaut, F., D. Gabriels, and M. De Boodt (eds.), *Assessment of soil surface sealing and crusting*. State Univ. Ghent, Belgium.

Kooistra, M.J., G. Lebbink, and L. Brussaard. 1989. The Dutch Programme on Soil Ecology of Arable Farming Systems II. Geogenesis, agricultural history, field site characteristics and present farming systems at the Lovinkhoeve experimental farm. *Agric. Ecosyst. Environ.* 27: 361-387.

Kooistra, M.J., A.S.R. Juo, and D. Schoonderbeek. 1990. Soil degradation in cultivated alfisols under different management systems in southwestern Nigeria. p. 61-70. In: L.A. Douglas (ed.), *Soil Micromorphology: a Basic and Applied Science*. Developments in Soil Science, 19. Elsevier, Amsterdam.

Kooistra, M.J., D. Schoonderbeek, F.R. Boone, and M. Van Noordwijk. 1992. Root-soil contact of maize, as measured by thin-section technique. II. Effects of soil compaction. *Plant Soil* 139: 119-129.

Kooistra, M.J. and N.K. Tovey. 1994. Effects of compaction on soil microstructure. p. 91-111 In: Soane, B.D. and Van Ouwerkerk, C. (eds.), *Soil compaction in crop production.* Elsevier, Amsterdam, The Netherlands.

Koolen, A.J., P. Lerink, D.A.G. Kurstjens, J.J.H. Van den Akker, and W.B.M. Arts. 1992. Prediction of aspects of soil-wheel systems. *Soil Till. Res.* 24:381-396.

Kuipers, H., and C. Van Ouwerkerk. 1963. Total pore-space estimationsin freshly ploughed soil. *Neth. J. Agric. Sci.* 11, 1: 45-53.

Leffelaar, P.A., 1993. Water movement, oxygen supply and biological processes on the aggregate scale. In: Brussaard, L. and M.J. Kooistra (eds.), Int. Workshop on Methods of Research on Soil Structure/Soil Biota Interrelationships. *Geoderma* 57: 143-165

McGill, W.B. and R.J.K. Myers. 1987. Controls on dynamics of soil and fertilizer nitrogen. p. 74-99. In: R.F. Follett, J.W.B. Stewart, and C.V. Cole (eds.), *Soil fertility and organic matter as critical components of production systems.* SSSA Spec. Pub. no. 19.

Murphy, C.P., 1986. *Thin section preparation of soils and sediments.* A.B. Academic Publishers, Berkhamsted, U.K., 149 pp.

Meijboom, F.W., J. Hassink and M. Van Noordwijk. 1995. Density fractionation of soil macroorganic matter using silica suspensions. *Soil Biol. Biochem* (in press).

North, P.F., 1976. Towards an absolute measurement of soil structural stability using ultrasound. *J. Soil Sci.* 27: 541-459.

Pielou, E.C., 1969. *An Introduction to Mathematical Ecology,* Wiley-Interscience, New York. 286 pp.

Postma, J., and H.-J. Altemüller. 1990. Bacteria in thin soil sections stained with the fluorescent brightener calcofluor white M2R. *Soil Biol. Biochem.* 22:89-96.

Pons, L.J., and I.S. Zonneveld. 1965. *Soil ripening and soil classification. Initial soil formation in alluvial deposits and a classification of the resulting soils.* Int. Inst. Land Reclamation and Improvement, Publ. 13, Wageningen, The Netherlands, 128 pp.

Priesack E., and G.M. Kisser-Priesack. 1993. Modelling diffusion and microbial uptake of 13C-glucose in soil aggregates. In: Brussaard, L. and M.J. Kooistra (eds.), Int. Workshop on Methods of Research on Soil Structure/Soil Biota Interrelationships. *Geoderma* 56:561-573.

Słowińska-Jurkiewicz, A. and H. Domżał. 1991. The structure of the cultivated horizon of soil compacted by the wheels of agricultural ractors. *Soil Till. Res.* 19:215-226.

Smeerdijk, D.G., T. Spek, and M.J. Kooistra. 1994. Anthropogenic soil formation and agricultural history of the open fields of Valthe (Drenthe, The Nethertlands) in medieval and early modern times. Integration of palaeooecology, historical geography and soilscience. *Mededelingen Rijks Geologische Dienst.* (in press).

So, H.B., and P.H. Nye. 1989. The effect of bulk density, water content and soil type on the diffusion of chloride in soil. *J. Soil Sci.* 40: 743-749.

Soil Survey Staff. 1975. Soil Taxonomy. *A basic system of soil classification for making and interpreting soil surveys.* US Government Printing Office, Washington, D.C., U.S.A., 754 pp.

Staricka, J.A., R.R. Allmaras, and W.W. Nelson. 1991. Spatial variation of crop residue incorporated by tillage. *Soil Sci. Soc. Am. J.* 55:1668-1674.

Steenhuis, T.S., W. Staubitz, M.S. Andreini, J. Surface,T.L. Richard, R. Paulsen, N.B. Pickering, J.R. Hagerman and L.D. Geohring. 1990. Preferential movement of pesticides and tracers in agricultural soils. *J. Irrig. Drain. Eng.* 116:50-66.

Tanaka, T., M. Yamazaki, and A.T. Abebe. 1988. Multiple pass effect of a wheel on soil compactability. In: Gee-Glough, D. and V.M. Salokke (eds.), *Proceedings of 2nd Asian- pacific Conference of the International Society for Terrain-vehicle Systems.* Asian Institute of Technology, Bangkok.

Tipkötter, R., K. Ritz and J.F. Darbyshire. 1986. The preparation of soil thin sections for biological studies. *J. Soil Sci.* 37:681-690.

Tisdall, J.M., and J.M. Oades. 1979. Stabilization of soil aggregates by the root systems of rye-grass. *Aust. J. Soil Res.* 17: 429-441.

Valentin, C. and J.F. Ruiz Figueroa. 1987. Effects of kinetic energy and water application rate on the development of crusts in a fine sandy loam soil using sprinkling irrigation and rainfall simulation. p. 401-409. In: Fedoroff, N., L.M. Bresson, and M.A. Courty (eds.), *Soil Micromorphology,* AFES, Plaisir, France.

Valentin, C. and L.M. Bresson. 1992. Morphology, genesis and classification of surface crusts in loamy and sandy soils. *Geoderma* 55: 225-246.

Van Noordwijk M., Widianto, M. Heinen, and K. Hairiah. 1991. Old tree root channels in acid soils in the humid tropics: important for crop root penetration, water infiltration and nitrogen management. *Plant Soil* 134: 37-44.

Van Noordwijk, M., M.J. Kooistra, F.R. Boone, B.W. Veen, and D. Schoonderbeek. 1992. Root-soil contact of maize, as measured by thin section technique. I. Validity of the method. *Plant Soil* 139: 109-118.

Van Noordwijk, M., G. Brouwer, K. Harmanny. 1993a. Concepts and methods for studying interactions of roots and soil structure. In: Brussaard, L. and M.J. Kooistra (eds), Int. Workshop on Methods of Research on Soil Structure/Soil Biota Interrelationships. *Geoderma* 56: 351-375.

Van Noordwijk, M., G. Brouwer, P. Zandt, F.W. Meijboom, and S. Burgers. 1993b. *Root patterns in space and time: procedures and programs for quantification.* IB-Nota 268, IB-DLO Haren.

Van Noordwijk, M., P.C. De Ruiter, K.B. Zwart, J. Bloem, J.C. Moore, H.G. Van Faassen, and S. Burgers. 1993c. Synlocation of biological activity, roots, cracks and recent organic inputs in a sugar beet field. In: Brussaard, L. and M.J. Kooistra (eds.), Int. Workshop on Methods of Research on Soil Structure/Soil Biota Interrelationships. *Geoderma* 56: 265-276.

Van Noordwijk, M., D. Schoonderbeek, and M.J. Kooistra. 1993d. Root-soil contact of field grown winter wheat. In: Brussaard, L. and M.J. Kooistra (eds.), Int. Workshop on Methods of Research on Soil Structure/Soil Biota Interrelationships. *Geoderma* 56:227-287.

Van Stiphout, T.P.J., H.A.J. van Lanen, O.H. Boersma and J. Bouma. 1987. The effect of bypass flow and internal catchment of rain on the water regime in a clay loam grassland soil. *J. Hydrol.* 95, 1/2: 1-11.

Veen B.W., M. Van Noordwijk, P. De Willigen, F.R. Boone and M.J. Kooistra. 1992. Root-soil contact of maize, as measured by thin-section technique. III. Effects on shoot growth, nitrate and water uptake efficiency. *Plant Soil* 139: 31-138.

Verberne, E.L.J., J. Hassink, P. De Willigen, J.J.R. Groot, and J.A. Van Veen. 1990. Modelling organic matter dynamics in different soils. *Neth. J. Agric. Sci.* 38: 221-238.

Vittorello, V.A., C.C. Cerri, F. Andreux, C. Feller, and R.L. Victoria. 1989. Organic matter and natural carbon-13 distribution in deforested and cultivated oxisols. *Soil Sci. Soc. Am. J.* 53: 773-778.

Wright D.A., K. Killham, L.A. Glover, and J.I. Prosser. 1993. The effect of location in soil on protozoal grazing of a genetically modified bacterial inoculum. In: Brussaard, L. and M.J. Kooistra (eds.), Int. Workshop on Methods of Research on Soil Structure/Soil Biota Interrelationships. *Geoderma* 56: 633-640.

Whalley, W.B. (ed.), 1979. Scanning Electron Microscopy in the Study of Sediments. *Geoabstracts*, Norwich, U.K., 414 pp.

Whitmore, A.P.M. and M. Van Noordwijk. 1995. Bridging the gap between environmentally acceptable and economically desirable nutrient supply. *Proceedings Long Ashton Symposium.* Sept. 1993. (in press).

Appendix

Diffusion out of a 'hot spot'

Consider a spherical 'hot spot' of radius R_0 and concentration C_0 inside a larger sphere of radius R_1 and uniform low concentration C_1. For the concentration $C_{r,t}$, a function of radial distance r and time t, the following equation can be derived from Carslaw and Jaeger (1959):

$$C_{r,t} = C_1 + (C_0 - C_1)\,[\frac{R_0^3}{R_1^3} + \frac{2}{rR_1} \sum_{n=1}^{\infty} \sin(r\,\alpha_n)\, e^{-D^* \alpha_n^2 t}\, F_{\alpha_n}] \qquad (1)$$

The function $F_{\alpha n}$ is equal to:

$$F_{\alpha n} = \frac{(1 + (\alpha_n R_1)^{-2})\ [\sin\,(\alpha_n R_0) - \alpha_n R_0 \cos(\alpha_n R_0)]}{\alpha_n^2} \qquad (2)$$

where the α_n are the roots of the equation cot $(\alpha\ R_1) = 1/(\alpha\ R_1)$.
For large t the term with the exponential function in equation (1) is approaching 0 and the concentration everywhere in the sphere becomes equal to the weighted average of the two initial concentrations. By integration the average concentration in the inner sphere can be derived as:

$$\overline{C_{i,t}} = C_1 + (C_0 - C_1)\,[\frac{R_0^3}{R_1^3} + \frac{6}{R_1 R_0^3} \sum_{n=1}^{\infty} e^{-D^* \alpha_n^2 t}\, F_{\alpha_n}] \qquad (3)$$

Time t and the effective diffusion constant D^* only occur as a product, and thus the solution for solutes with different mobility can be considered as shifts in time, directly proportional to D^*.

Formation of Soil Aggregates and Accumulation of Soil Organic Matter

J.M. Tisdall

I. Introduction

Soil structure is the arrangement of particles in soil and the particles of sand, silt, and clay, bound together into aggregates of various sizes by organic and inorganic materials. The structural stability is the ability of the aggregates and pores to remain intact when subjected to stress, e.g. when aggregates are wetted

1-56670-033-7/96/$0.00+$.50

quickly. In the field, the stability of these aggregates and the pores between them affect the movement and storage of water, aeration, erosion, biological activity, and the growth of crops.

This review discusses several models of aggregation in soil and the organic and inorganic bonds that stabilize microaggregates of various sizes (<250 μm diameter) and macroaggregates (>250 μm). The review emphasizes the importance of biota (especially bacteria, roots, fungal hyphae and earthworms) and the materials that they produce, on aggregation in soils that are mainly stabilized by organic matter. This discussion includes biological activity in the rhizosphere and the formation of stable bonds between particles of soil by polysaccharides and polyvalent cations. Finally the review discusses the importance of drying on aggregation, including localized drying around roots and organisms.

II. Models of Aggregation

Several authors have proposed models of aggregation which show that soils are not homogenous, but are made up of aggregates of different sizes held together by different organic and inorganic materials. Edwards and Bremner (1967) proposed that soils consist of microaggregates (<250 μm diameter) bound into macroaggregates (>250 μm) and that bonds within microaggregates are stronger than those between microaggregates. In this model, microaggregates less than 2 μm diameter consist of particles of clay attached to organic molecules (OM) by polyvalent cations (P). These microaggregates, represented as Clay-P-OM, are combined to form $(\text{Clay-P-OM})_x$ and $[(\text{Clay-P-OM})_x]_y$, probably through bridges of polyvalent cations. Particles Clay-P-Clay and OM-P-OM are also possible. The results of several other researchers support this hypothesis for alfisols and mollisols (Tisdall and Oades, 1982; Chaney and Swift, 1986b; Bartoli et al., 1988; Elliott and Coleman, 1988; Oades and Waters, 1991). For an alfisol, Tisdall and Oades (1982) proposed that microaggregates themselves are built up in stages with different types of bonds at each stage (Table 1). In this model, the stages of aggregation are: < 0.2 μm –> 0.2-2 μm –> 2-20 μm –> 20-250 μm –> > 2000 μm diameter.

Oades and Waters (1991) modified the model further for alfisols and mollisols, whose aggregates are mainly stabilized by organic materials. They suggested that it was not possible to distinguish steps of aggregation within aggregates less than 20 μm, but that aggregates 20 to 250 μm could be divided into aggregates 20 to 90 μm and 90 to 250 μm (Table 1). In this model, the stages of aggregation are: < 20 μm –> 20-90 μm –> 90-250 μm –> > 250 μm diameter.

In soils stabilized by organic matter, these stages of aggregation, or aggregate hierarchy, probably develop over many years, during the growth of roots, especially roots of grasses. The stages probably do not develop in young soils, such as polders (Oades, 1993).

Table 1. Models of aggregation and major stabilizing agents

Soil type	Stabilizing agent	Stage of aggregation (μm)	Reference
Alfisol	Inorganic materials, organic polymers, electrostatic bonds, coagulation	< 0.2	Tisdall and Oades, 1982
	Microbial and fungal debris	0.2 - 2 –> 2 - 20	
	Plant and fungal debris	2 - 20 –> 20 - 250	
	Roots and hyphae[a]	20 - 250 –> > 2000	
	Polysaccharides[b]	20 - 250 –> > 2000	
Alfisol, mollisol	Microbial debris, inorganic materials	< 20	Oades and Waters, 1991
	Plant debris	< 20 –> 20 - 90	
	Plant fragments	20 - 90 –> 90 - 250	
	Roots and hyphae	20 - 250 –> > 2000	
Oxisol	Oxides	< 20 –> > 250	Oades and Waters, 1991
Oxisol	Oxides	< 2 –> 100 - 500	Robert and Chenu, 1992
Vertisol	Organic matter	20 - 35 –> > 250	Collis-George and Lal, 1970
Andosols	Allophanes and amorphous aluminosilicates	0.001 - 0.01 –> 0.1 - 1	Robert and Chenu, 1992

[a]Soil with total organic carbon > 2%
[b]Soil with total organic carbon < 1%

Other soils, including those which are stabilized mainly by inorganic materials (e.g. oxisols, and andosols), do not always fit this model (Table 1).

III. Aggregation

Aggregation has been studied in at least two ways: a) the formation and stabilization of aggregates in suspensions of pure clay, and b) the separation of aggregates from soil disrupted by different amounts of energy. Larger aggregates contain larger pores than do smaller aggregates, and larger pores form planes of weakness (Dexter, 1988). Hence, the smaller the aggregate, a) the greater the contact between particles, b) the stronger the bonds between particles, and c) the higher the tensile strength of the aggregate (Braunack et al., 1979; Tisdall and Oades, 1980; Hadas, 1987; Dexter, 1988; Oades and Waters, 1991). Aggregates of different sizes can be separated from soil subjected to different amounts of energy. Hence, Oades and Waters (1991) separated aggregates of decreasing size from several soils by wet-sieving and sedimentation of soil after disruption by a) slow wetting, b) fast wetting, c) end-over-end shaking for 16 hours and d) ultrasound.

However, care must be taken in interpreting such results on types of organic matter associated with microaggregates of different sizes separated by ultrasound, since ultrasound may re-distribute organic materials among the various sizes. For example, more intense ultrasound may lead to more organic material in the finer fractions than does less intense ultrasound (Cambardella and Elliott, 1993). Results also depend on a) whether or not the free light fraction, which consists of slightly decomposed plant fragments which have not interacted with mineral particles, is removed before the sample is disrupted with ultrasound, and b) whether particles are separated by size, density or rate of sedimentation (Golchin et al., 1994).

Microaggregates of soil (<250 μm diameter) are highly stable, and in non-sodic soils are only dispersed by high energy such as ultrasound; in the field, heavy rainfall can disperse microaggregates. On the other hand, macroaggregates (>250 μm) are easily disrupted when wetted quickly or shaken end-over-end; in the field, their stability depends on management, so that macroaggregates are easily stabilized or destroyed by agricultural management (Tisdall, 1991).

Within stable aggregates of different sizes, the interaction of organic matter with particles of mineral soil protects the organic matter from microbial decomposition so that organic matter in soil can be over 1000 years old (Table 2; Oades, 1988; Van Veen and Kuikman, 1990). The age of organic matter within each fraction has been determined by radiocarbon dating (Table 2), and the degree of decomposition of the organic matter by the ratio of concentration of carbon to the concentration of nitrogen (C:N ratio) or the chemical composition of the organic matter (Anderson and Paul, 1984; Baldock et al., 1992). The C:N ratio of plant material ranges from 30 to 80, and of

Table 2. Radiocarbon age of several soils and their fractions

Particle size (μm)	Equivalent age (years)		
	Clay loam	Loam	Silty clay
Whole soil	Modern	480	795
5 - 50	Modern	645	800
2 - 5	Modern	495	965
0.2 - 2	Modern	200	1255
< 0.2	Modern	Modern	170

(From Anderson and Paul, 1984.)

microbial biomass in soil ranges from 8 to 12. Hence, the C:N ratio tends to decrease as the organic matter is decomposed, i.e. nitrogen tends to be recycled when carbon is mineralized and lost from the soil. High resolution solid state ^{13}C nuclear magnetic resonance (NMR) spectroscopy suggests that organic matter in soil which contains mainly O-alkyl C (hemicellulose, cellulose and proteins) is less decomposed than that which contains mainly aromatic-C (lignin) or the highly recalcitrant alkyl-C (waxes, fatty acids) (Baldock et al., 1992).

Christensen (1992) reviewed in detail the separation of aggregates of different sizes with the organic matter of different origins and different degrees of decomposition that they contained. Hence this will be discussed only briefly here.

IV. Microaggregates (Less than 250 μm Diameter)

Microaggregates, which can be further divided into different sizes, are held together by different organic and inorganic materials.

A. Microaggregates Less than 2 μm Diameter

Microaggregates less than 2 μm diameter have been studied in suspensions of pure clay, or have been isolated from soils subjected to ultrasound. For aggregates to remain intact when wetted or mechanically disturbed, the clay must remain flocculated in stable aggregates, i.e., 'aggregation is flocculation plus' (Quirk, 1978). The flocculated aggregates are domains, quasicrystals or assemblages, depending on the mineralogy of the clay, which remain intact in water (Rengasamy et al., 1984; Quirk and Murray, 1991).

In pure clay, calcium illite forms domains in which about 10 rigid plates are almost perfectly aligned, with 20% of the surfaces in contact (Emerson et al., 1986). Calcium montmorillonite forms quasicrystals, in which about 20 thin flexible plates or lamellae are aligned almost perfectly face-to-face, with about

80% of the surface area of individual lamellae in contact. Kaolinite forms assemblages in which large blocky particles are poorly aligned, with less than 10% of the surface areas of particles in contact; the assemblages are arranged in a card-house structure.

In pure clay or in soil, particles of clay interact due to a balance between a) the attractive London van der Waal's forces and ion correlation forces, and b) the repulsive forces due to ion hydration and hydrogen bonds between molecules of water near the interface between the clay and water (Quirk and Murray, 1991). Polyvalent cations such as Al^{+++} or Ca^{++} reduce the repulsive forces between negatively charged particles of clay and flocculate the clay. Flocculation is also increased by high concentrations of electrolyte in the soil solution, organic matter and minimum disturbance (Rimmer and Greenland, 1976; Rengasamy et al., 1984; Chenu, 1989). However, when aggregates are disrupted by rainfall, or tilled or smeared when wet, organic molecules adsorbed on the surface of the clay may prevent flocculation (Quirk and Murray, 1991).

Other aggregates less than 2 μm diameter in soil consist of particles of very fine clay (< 0.2 μm) held together by organic matter and polyvalent cations. Organic matter can fit between lamellae in suspensions of clay but rarely does in soils, and most enzymes and microorganisms are only adsorbed onto the external surfaces of domains, quasicrystals, or assemblages (Robert and Chenu, 1992).

Organic matter associated with fine clay less than 0.2 μm diameter, which was separated from the whole soil by ultrasound, was mainly fulvic acids, with aliphatic humic acids of high molecular weight that were rich in nutrients, such as nitrogen and sulphur (Anderson et al., 1981). These organic materials which held the particles together had probably originated from microorganisms, presumably from cytoplasm, or were metabolites of relatively young radiocarbon age, which were protected in complexes with the fine clay (Table 2; McGill and Paul, 1976). Organic matter in the fine clay (< 0.2 μm diameter) separated from the whole soil in several alfisols and mollisols had a low C:N ratio (e.g. Table 3) and contained a high proportion of alkyl C. Baldock et al. (1992) suggested that the alkyl-C was partly derived from the original plant material and microorganisms, and was partly synthesized by microorganisms from carbohydrates and/or aromatic fractions of the original material.

Humified cell walls (with a low C:N ratio and high proportion of O-alkyl C) from microorganisms were the main organic material in the particles 0.2 to 2 μm diameter separated from several soils, and probably held the particles together (McGill and Paul, 1976; Turchenek and Oades, 1979; Baldock et al., 1992; Table 3). Much of the ^{14}C and ^{15}N of the microbial biomass and organic residues in soil, which had accumulated from the decomposition of substrates labelled with isotopes, were present in the fraction 0.2 to 2 μm diameter (Ladd et al., 1993). Similarly, organic matter in particles less than 2 μm diameter, separated by ultrasound from several soils under old grassland, was amorphous with a low C:N ratio (mean 9.4) and did not stain with acridine orange (Hassink et al., 1993). This suggested that the organic matter which held the particles

Table 3. The C:N ratio of particles separated by size and density from the whole soil of Millicent mollisol disrupted by ultrasound, and of microbial biomass in the soil and of fresh plant material

Material	Particle size (μm)	Particle density (Mg m^{-3})	C:N
Soil	250 - 2000	< 2	39.6
		> 2	16.8
	50 - 250	< 2	20.4
		> 2	10.9
	20 - 50	< 2	17.1
		> 2	13.5
	2 - 20	< 2	11.5
		> 2	8.8
	0.2 - 2	-	8.1
	< 0.2	-	7.5
Microbial biomass in soil	-	-	8 - 12
Fresh plant	-	-	30 - 80

(From Baldock et al., 1992.)

together was highly degraded, had originated from microorganisms, and was protected by clay.

However, radiocarbon dating showed that organic matter in the particles 0.2 to 2 μm diameter isolated from a silty clay, but not from a clay loam or a loam, was relatively old (Table 2), suggesting that the organic matter in the silty clay was more decomposed than that in the other soils.

B. Microaggregates 2 to 20 μm Diameter

Microaggregates 2 to 20 μm diameter are held intact by inorganic and organic bonds. Floccules or aggregates of primary particles less than 2 μm diameter can sometimes form aggregates 2 to 20 μm diameter (Dexter, 1988). Other clusters (2 to 20 μm) consist of fine silt, coarse clay and fine clay held together by microbial debris, and bacterial cells and colonies encrusted with inorganic materials (Tisdall and Oades, 1982; Oades and Waters, 1991). The organic matter holding these aggregates together has a low C:N ratio, with the light fraction (density <2 Mg m^{-3}) less decomposed than the organomineral fraction (density >2 Mg m^{-3}) (Table 3), and electron micrographs showed that these aggregates contained little, if any, plant material (Oades and Waters, 1991; Baldock et al., 1992). Such clusters can be very stable, and are not disrupted by agricultural practice especially in mollisols and in soils from under old pasture; some aggregates 2 to 20 μm diameter from under old pasture withstood ultrasound for 5 minutes (Turchenek and Oades, 1978). These aggregates which were separated by ultrasound were rich in organic matter, and contained a large part of the microbial biomass, regardless of the texture of the soil (Hattori, 1973; Ladd et al., 1977; Jocteur Monrozier et al., 1991). Much of the ^{14}C and ^{15}N of the microbial biomass and organic residues in soil (which had accumulated from the decomposition of substrates labelled with isotopes), were present in the fraction 2 to 20 μm diameter (Ladd et al., 1993).

Oades and Waters (1991) suggested that it was not possible to distinguish in electron micrographs any steps of aggregation within aggregates less than 20 μm. Aggregates less than 20 μm diameter broke up when treated with hydrogen peroxide showing that they were partly held together by organic matter. Little extra clay may be released when the aggregates are treated with peroxide after ultrasound, but the surface area of clay is greatly increased suggesting that organic matter binds fine particles together (Burford et al., 1964; Turchenek and Oades, 1978; Oades, 1984). Aggregates less than 20 μm diameter contained little or no plant debris, but were stabilized mainly by microbial materials including polysaccharides, hyphal fragments and bacterial cells or colonies which were encrusted with particles of clay (Oades and Waters, 1991).

Dormaar and Foster (1991) described several early pathways for the formation of microaggregates 2 to 20 μm diameter. They grew ryegrass (*Lolium perenne* L.) and microorganisms in a suspension of attapulgite, a non-swelling clay. Their results suggested that the pressure exerted by roots compacted and reorientated the soil; the gel exuded by roots stabilized aggregates locally and reorientated particles which had accumulated, and were consolidated, around cells and colonies. Similarly, hyphae or bacterial cells produced microaggregates in pastes of kaolinite or montmorillonite in which they grew, without the whole sample having been dried (Dorioz and Robert, 1987). The microorganisms secreted large amounts of polysaccharide and compacted the clay platelets which were aligned parallel to the microbial surface. This was attributed to the pressure exerted by the microorganisms as they grew through the clay, and adsorbed water locally shrinking the clay. When the clay was dried and re-wetted, these microaggregates were separated from the mass of clay.

1. Bacterial Microaggregates

Electron micrographs of soils show that many bacterial cells and colonies are surrounded by a distinct capsule made of polysaccharide (Foster, 1983). The polysaccharide, which sometimes occupies more space than the cell does, is often surrounded by a layer of clay particles which are aligned parallel to the cell wall, and form a microaggregate; sometimes particles are radially orientated. It appears that as some colonies or cells grow or secrete polysaccharides, they compact and reorientate nearby clay minerals, which may be drawn closer together by surface tension as the clay dries. Sometimes the bacteria may have been trapped as the clay swelled or was compacted as roots or animals moved through the soil (Foster, 1988).

The clay particles which coat the bacterial cells or colonies to form microaggregates, appear to protect the bacteria physically from microbial degradation. Many bacteria occupy pores less than 3 μm diameter, so are also physically protected from protozoa and nematodes which occupy larger pores (Foster, 1988). Components of the cell wall are probably linked with phenolic polymers in the soil which would also protect the cells from microorganisms.

Since clay protects the bacteria, 40 to 60% of the microbial biomass may be associated with microaggregates 2 to 20 μm diameter, depending on the amount and type of clay mineral (Jocteur Monrozier et al., 1991). Electron micrographs showed that the coarse clay fraction and microaggregates separated by ultrasound from a tilled loess consisted mainly of bacterial microaggregates, with 90% of the counted bacteria coated with clay minerals (Robert and Chenu, 1992). Also the survival of bacteria in soil is related to the clay content of the soil (Marshall, 1976; Stotzky, 1986). For example, when plant residues, labelled with ^{14}C, were added to soil with neutral or alkaline pH and incubated for 66 weeks, the total ^{14}C which remained increased as the content of clay increased; this was mainly due to increased content of ^{14}C in the microbial biomass (Amato and Ladd, 1992).

Most microorganisms seen in micrographs of soil are probably starved or dormant, even though their osmotic barriers may remain intact (Foster and Martin, 1981). The biomass which is dormant or dead, but is protected from lysis, accumulates in soils over several years (Clark and Paul, 1970). The biomass appears to die as the soil dries out, and to re-grow when the soil is re-wetted (McGill et al., 1986).

Some empty clusters less than 20 μm diameter seen in micrographs of soil or earthworm casts were probably once bacteria encrusted with particles of clay (Foster, 1988; Lee and Foster, 1991). Once bacterial cells or colonies have autolysed, the remains of the colony with its capsule of polysaccharide cannot be identified as such, but the polysaccharide may continue to hold the microaggregates intact (Foster and Martin, 1981; Foster, 1988).

It has not yet been shown whether pili and flagellae of bacteria could help the bacteria move across a zone of repulsion and attach themselves to the surface of clay particles. The bacteria may then produce polysaccharides and strengthen the bonds between clay particles (Burns and Davies, 1986; Oades, 1993).

C. Microaggregates 20 to 250 μm Diameter

In surface soils stabilized mainly by organic matter, aggregates 20 to 250 μm diameter consist mainly of particles less than 20 μm held together by cements including plant and fungal debris encrusted with inorganic materials, crystalline oxides and highly disordered aluminosilicates (Tisdall and Oades, 1982). Microaggregates 20 to 250 μm diameter are stable to water in the field, but are disrupted by ultrasound (Oades and Waters, 1991), and are included in the stable microaggregates $((Clay\text{-}P\text{-}OM)_x)_y)$ of Edwards and Bremner (1967).

The organic matter in the particles 20 to 250 μm diameter separated from several whole soils was less decomposed, with a higher C:N ratio (e.g., Table 3) and a higher content of O-alkyl carbon, than those in the particles less than 20 μm diameter (Baldock et al., 1992). Much of the organic matter in the organomineral aggregates 20 to 250 μm diameter appears to be more decomposed than previously thought. Golchin et al. (1994) developed a method of separating from several soils two types of light fraction which differed chemically and in their location in soil. The two types were a) the free light fraction (separated by sedimentation from the whole soil not disrupted by ultrasound) which consisted mainly of undecomposed plant fragments 50 to 200 μm diameter, and some larger particles of plant origin encrusted with particles of clay, and b) the occluded light fraction (separated by sedimentation from soil after the free light fraction had been removed and the soil disrupted by ultrasound) which contained mainly round decomposed plant fragments 10 to 20 μm. The organic matter in the free light fraction contained more O-alkyl carbon, and less alkyl carbon and aromatic carbon than the occluded light fraction did.

Electron micrographs showed that the ratio of length: width of many microaggregates 100 to 200 μm was about two, and consisted mainly of plant debris often completely coated with inorganic material (Oades and Waters, 1991). Those microaggregates 100 to 200 μm diameter from the old arable soil, but not those from under old pasture, were completely disrupted and most of the clay was released when the soil was shaken for 16 hours; ultrasound disrupted microaggregates 100 to 200 μm from under the old pasture releasing plant debris. A few aggregates contained a few last remnants of plant debris or were empty cores that remained after the remnants had been completely used by microorganisms.

Microaggregates less than 100 μm diameter from two alfisols were stable to water in the field (Oades and Waters, 1991). These microaggregates were slightly elongated and some contained an organic core which probably consisted of highly degraded plant debris; others were empty, remaining after the organic debris had gone. Microaggregates less than 100 μm diameter often contain more organic matter, N, P and K than larger particles do. Up to 56% of aggregates in several tilled soils consisted of microaggregates less than 100 μm diameter and contained twice the amount of organic matter as aggregates greater than 100 μm (Dormaar, 1987). Also aggregates less than 100 μm diameter in untilled soil contained up to four times the amount of organic matter as similar aggregates in tilled soil. In some soils, particles of this size are easily eroded, when large amounts of organic matter and finer soil particles are removed, and the water holding capacity of the soil is reduced (Dormaar, 1987).

Faecal pellets of microarthropods (mites and collembola) found in soils are usually less than 100 μm diameter, but contain little or no mineral soil (Lee and Foster, 1991). In electron micrographs, these faeces are usually round, smooth and may be densely colonized by fungal hyphae and many densely packed bacteria. Where present, clay usually forms thin cutans associated with humified remnants of cell walls, which have probably been ingested accidentally with food such as plant residues.

V. Macroaggregates (Greater than 250 μm Diameter)

Unstable macroaggregates (>250 μm diameter) break down in water to microaggregates because the bonds between microaggregates are not strong enough to hold the macroaggregate together as the clay swells and entrapped air is released (Emerson, 1991). Macroaggregates which are held together by iron and aluminium oxides, such as in oxisols, are very stable and are only disrupted when subjected to high energy such as ultrasound (Oades and Waters, 1991). In many other soils, macroaggregates are mainly held together by fungal hyphae, fibrous roots and polysaccharides (Swaby, 1949; Tisdall and Oades, 1979; Miller and Jastrow, 1990).

Organic materials appear to stabilize the macroaggregates by stabilizing coarse pores. Quirk and Panabokke (1962) measured the rate of water uptake at different suctions by aggregates of an alfisol (20% clay, 28% silt) which had never been tilled (virgin; 2.7% organic carbon) and those which had been tilled in a rotation of wheat (*Triticum aestivum* L.)-fallow for 30 years (tilled; 1.3% organic carbon). At suctions of 0.2 kPa and 1 kPa, the tilled aggregates took up water faster than the virgin aggregates. The difference in uptake of water by the aggregates was attributed to the difference in the location of organic matter, and not to a difference in pore-size distribution. Tilled aggregates which had been wetted quickly to 0.2 kPa suction slaked in water, whereas aggregates wetted slowly to 0.2 kPa did not. This suggested that points of weakness in the tilled aggregates enabled the aggregates to collapse when wetted quickly, i.e. enabled *incipient failure*.

To test the hypothesis that the location of organic matter rather than the amount of organic matter determined the stability of aggregates, Quirk and Williams (1974) treated tilled aggregates at different suctions with polyvinyl alcohol (PVA), a soil stabilizer, to introduce the PVA into pores of different sizes, and then wetted the aggregates at 0.2 kPa suction. When the coarse pores, especially those 15 to 50 μm diameter, in the tilled aggregates were treated with PVA, the aggregates at 0.2 kPa suction wetted at the same rate as the virgin aggregates. Treated pores 15 to 50 μm diameter slowed the rate of wetting more than treated pores less than 7.5 μm. Pores 15 to 50 μm diameter are large enough to contain fungal hyphae, root hairs and fine roots, which supports the hypothesis that microaggregates 20 to 250 μm diameter are bound into macroaggregates by organic materials. If the organic matter is placed within coarse pores, small amounts may stabilize aggregates. This may partly explain why 0.02 to 0.2% of microbial polysaccharides in soil can stabilize aggregates (Foster, 1988).

Under growing plants, roots and hyphae of vesicular arbuscular mycorrhizal (VAM) fungi are probably the main stabilizers, but saprophytic fungi also stabilize aggregates especially when substrate such as straw is added to soil (Martin, 1945; Tisdall et al., 1978; Rasmussen et al., 1980; Lynch and Elliott, 1983; Lynch and Bragg, 1985). Emerson et al. (1986) suggested that ectomycorrhizal hyphae stabilized macroaggregates in several forest soils. Algal filaments, covered with slimy gels, bound particles of sandy soil into stable macroaggregates to form crusts on the surface, reducing the risk of erosion (Bond and Harris, 1964; Shields and Durrell, 1964). Aggregates also formed between clay and algae covered with mucilage in suspensions of bentonite (Avnimelech and Troeger, 1982).

Earthworm casts are also macroaggregates of soil and are probably also stabilized by organic matter, sometimes including fungal hyphae (Lee and Foster, 1991).

Extracellular polysaccharides produced by microorganisms when organic residues are added to soil, or polysaccharides from roots and microorganisms in the rhizosphere, also stabilize macroaggregates. These polysaccharides are produced quickly and are decomposed quickly, so do not persist in soil; they may be the main stabilizers in soil away from the rhizosphere (Sparling and Cheshire, 1985). Many polysaccharides in soil are negatively charged and are probably bound to clay particles by polyvalent bridges.

A. Roots and Fungal Hyphae

Electron micrographs show that roots and fungal hyphae form an extensive network in soil and are covered with extracellular polysaccharides to which microaggregates are firmly held (Low and Stuart, 1974; Tisdall and Oades, 1979; Foster, 1988; Oades and Waters, 1991). The network of encrusted roots and hyphae hold the macroaggregates intact, so that they do not collapse in water. The clay on the surface protects the roots and hyphae from microorganisms, so that the stability in many soils is related to the concentration of organic carbon (Chesters et al., 1957; Kemper and Koch, 1966; Hamblin and Davies, 1977; Oades, 1984; Stengel et al., 1984; Douglas et al., 1986; Tisdall and Oades, 1979; Chaney and Swift, 1984). When the roots and hyphae die and are disrupted by tillage or fauna, the decomposed fragments probably become the organic core in microaggregates 20 to 250 μm (Oades and Waters, 1991).

It is difficult to separate the effect of the physical compaction and reorientation of clay particles by roots and hyphae from that of cementation by extracellular polysaccharides. Martin (1945) estimated that at least 50% of the stability by fungi was due to mechanical binding and 50% was due to microbial materials. Allison (1968) suggested that as fine roots grow through the soil they dry the

Table 4. Relationships between roots, external hyphae and % water-stable aggregates greater than 2 mm diameter (%wsa)

Silty clay loams (Thomas et al., 1986)	r^2
%wsa ∝ mass of roots	0.54
%wsa ∝ external hypha length	0.20
Fine sandy loams (Tisdall and Oades, 1980)	r^2
%wsa = 0.22 (root length) + 0.13	0.93
%wsa = 1.45 (hyphal length) - 0.02	0.77
Silt loams and silty clay loams (Miller and Jastrow, 1992)	r^2
%wsa = 49.70 (fibrous root length)$^{0.15}$	0.81
%wsa = 3.91 (hyphal length)$^{0.73}$	0.74

soil locally helping to form aggregates which are then stabilized by microbial polysaccharides. Similarly, Elliott and Coleman (1988) suggested that roots and fungal hyphae entangle microaggregates to form macroaggregates which are then stabilized by extracellular polysaccharides or other organic materials. Saini and McLean (1966) and Chenu (1989) suggested that the fungi reorganize clay particles into aggregates and then stabilize these aggregates with extracellular polysaccharides. VAM fungi and several species of saprophytic fungi formed and stabilized macroaggregates, and fungal polysaccharides appeared to stabilize existing macroaggregates in several soils (Aspiras et al., 1971; Tisdall, 1991).

Several authors have attempted to separate the effects of roots and of VAM fungal hyphae on aggregation (Thomas et al., 1986; Miller and Jastrow, 1990; Thomas et al., 1993). Thomas et al. (1986) grew mycorrhizal and non-mycorrhizal onion plants (*Allium cepa* L.) in a calcareous silty clay loam in pots for 230 days. The mycorrhizal plants, with a total mass of roots and shoots of 18 gram, increased the percentage of water-stable aggregates by 72% when compared with non-mycorrhizal plants with a total mass of 4 gram. As the correlation was stronger with root length than with hyphal length (Table 4), Thomas et al. (1986) suggested that the hyphae only stabilized aggregates indirectly by increasing the growth of roots. It may be possible to separate the

Table 5. Direct and indirect effects of roots and VA mycorrhizal hyphae on the geometric mean diameter of aggregates of several silt loams and silty clay loams, as shown by path analysis

Variable	Path coefficient[a]	
	Direct effect	Indirect effect
Lengths		
External hyphae	0.54**	-
Colonised fine roots	-	0.36**
Colonised very fine roots	-	0.13
Fine roots	0.37**	0.40*
Very fine roots	-	0.12

[a] The larger the path coefficient, the stronger the relationship between two variables.
** $P < 0.01$, * $P < 0.05$
(From Miller and Jastrow, 1990.)

effects of roots and hyphae in similar experiments with cultivars of plants which become mycorrhizal and support external hyphae but are not dependent on VAM fungi (Baon et al., 1993).

Miller and Jastrow (1990) showed by path analysis that the length of external hyphae of VAM fungi had a greater direct effect than fine roots (0.2 to 1.0 mm diameter) or very fine roots (< 0.2 mm) on the geometric mean diameter of aggregates (a measure of stability of macroaggregates) of several silt loams and silty clay loams (Table 5). The length of external hyphae depended directly on the length of fine roots (0.2 to 1 mm diameter) colonized by VA mycorrhizal fungi (path coefficient = 0.76**), which in turn depended directly on the total length of fine roots (0.88**). Hence, the length of fine roots mainly affected the stability indirectly through their direct effect on the length of external hyphae. Prairie grasses (mainly *Andropogon gerardii* and *Sorghastrum nutans*) and perennial composites (mainly *Solidago altissima*, *Coreopsis tripteris*, *Ratibida pinnata*, *Solidago rigida* and *Silphium* spp.) which had mainly fine fibrous roots, indirectly affected the stability. On the other hand, non-prairie grasses (*Agropyron repens*, *Bromus inermis* and *Poa* spp.) with mainly very fine roots, had little indirect effect on the stability. The different effects of fine and very fine roots on stability were probably because plants with mainly very fine roots form fewer mycorrhizas than do those with mainly fine roots (Barea, 1991). Although it has not been shown yet, prairie grasses may also support more external hyphae than non-prairie grasses do, since prairie grasses are more dependent on VA mycorrhizal fungi for growth and uptake of nutrients (Miller and Jastrow, 1990). This may explain some of the relative effects of these grasses on the stability of aggregates.

Thomas et al. (1993) attempted to separate the relative effects of roots alone, VAM hyphae alone, and roots plus VAM hyphae on aggregation. They set up pots each with four chambers separated by a solid barrier or a 44-μm screen. Each pot contained a soybean (*Glycine max* L. Merrill) plant with roots split between two of the chambers, one chamber containing non-VAM roots only, and the other containing VAM roots and hyphae. The third chamber contained VAM hyphae only, and the fourth chamber (the control) contained no roots or VAM hyphae. After eight months, the macroaggregates in each chamber were less stable than those in the original soil. Within the pots, the macroaggregates in the chamber with both roots and VAM hyphae were the most stable, followed by VAM hyphae alone, or non-VAM roots, with those in the control the least stable. These results show that an interaction between roots and VAM hyphae led to greater stability of macroaggregates than did either roots or VAM hyphae alone. It was still not possible to separate the effects of roots and hyphae as a) VAM fungi stimulated the growth of roots compared with non VAM fungi, b) the concentration of non-VAM hyphae was only greater than the control in the presence of VAM roots. Presumably the soil in all chambers was always watered at the same time, yet the soil in the chambers with roots (especially with the longer mycorrhizal roots) probably dried out more between irrigations than those without roots did; this may have increased the stability of macroaggregates in the soil with roots (Allison, 1968; Kemper et al., 1987). However, VAM hyphae alone directly affected aggregation at least to about the same extent as roots, supporting the results of Miller and Jastrow (1990) and the hypothesis that roots and VAM hyphae can each stabilize macroaggregates. As shown above, where VAM hyphae increase the length of roots, an interaction between roots and VAM hyphae can further increase aggregation.

Miller and Jastrow (1992) suggested that VAM fungi contributed to the formation and stabilization of macroaggregates in three simultaneous processes.

1) External VAM hyphae produce a network which entangles primary particles of soil.

2) External hyphae physically bring mineral particles and organic materials together so that microaggregates can form.

3) External hyphae and roots enmesh microaggregates and smaller macroaggregates into larger macroaggregates.

In soils with little or no clay or organic matter, such as sand dunes, aggregates may never develop beyond the first process of Miller and Jastrow (1992) or may develop slowly (Koske et al., 1975; Clough and Sutton, 1978).

Tisdall (1991) suggested a similar model to that of Miller and Jastrow (1992) for the formation of a stable macroaggregate by fungal hyphae. The model is supported by electron micrographs of saprophytic hyphae grown in pure kaolinite or montmorillonite (Dorioz et al., 1993). The micrographs showed that fungal hyphae compacted and reorientated particles of clay so that they were parallel to the surface of each hypha for up to 20 μm from the surface. This was attributed to greater tension in the water as the hyphae dried the clay and pulled

Table 6. The effect of plant species, grown in artificial aggregates of an alfisol, on root length, organic carbon and water-stable aggregates > 250 μm diameter (%wsa)

Plant species	Root length (cm cm^{-3})	Organic carbon (%)	%wsa
Control	-	1.25	31.4
Pea	3.9	1.51	49.5
Ryegrass	81.6	1.73	64.4
Wheat	27.8	1.54	55.6

(From Materechera et al., 1992.)

the clay to the surface of the hypha. Extracellular polysaccharide of the hyphae then impregnated and stabilized the clay.

In the field, extracellular polysaccharides of fungi probably do stabilize existing aggregates (possibly formed by roots), and fungi and their polysaccharides can probably also form and stabilize new macroaggregates (Tisdall, 1991).

When sown to pasture, old arable soils differ in the rate at which stable macroaggregates form, possibly depending on the presence of highly stable microaggregates and on the rate at which the network of roots and fungal hyphae develop in the field (Elliott, 1986; Tisdall, 1991; Miller and Jastrow, 1992). Plant species such as ryegrass with large systems of fibrous roots and VAM hyphae which secrete large amounts of gel and dry the soil are effective stabilizers (Tisdall and Oades, 1979; Dormaar and Foster, 1991; Table 6). Little is known about the branching patterns of different species of hyphae, or the amount of gel that each species produces, yet these may also determine the rate that aggregation is restored (Tisdall, 1991).

Low (1955) and Miller and Jastrow (1992) showed that it took 5 to 50 years in several soils to restore the stability of macroaggregates to that of old pasture. On the other hand, incorporated lucerne (*Medicago sativa* L.) hay (20 tonne per hectare) and the growth of irrigated pasture for 12 months restored the stability of one old arable alfisol (40% water-stable macroaggregates) which already contained 70% stable microaggregates (20 to 250 μm diameter) to that of old pasture (95% macroaggregates) (Tisdall, unpublished data). The high level of decomposing organic matter probably contributed to the stability by encouraging the growth of both saprophytic and mycorrhizal hyphae in the soil (Harris et al., 1966; St. John et al., 1983).

Tillage breaks up the network of roots and hyphae so that macroaggregates are readily destabilized when soils from old pasture are used for crops. For example, tillage of several alfisols from under old pasture in the first year of production of tomato (*Lycopersicon esculentum* Mill.) decreased the stability of macroaggregates from 90% to 58% (Adem and Tisdall, 1984).

B. Earthworm Casts

Earthworms ingest and mix large amounts of soil and organic matter in the gut and deposit the material as casts or macroaggregates on the surface of the soil or in burrows, depending on the species. Each year in temperate pastures, earthworms produce an average of 40 to 50 tonne per hectare of casts on the surface of the soil with more below the surface (Lee, 1985). The tropical earthworm *Pontoscolex corethrurus* was reported to ingest 400 tonne of soil per year (Barois and Lavelle, 1986). Earthworm casts in agricultural soils are usually 2 to 10 mm diameter, and are usually rounder than aggregates of the same size from surrounding soil (Lee and Foster, 1991). The casts often contain higher concentrations of silt and clay, organic matter and cations (such as iron and calcium) than the uningested soil, and contain higher concentrations of mineral soil than faecal pellets (<100 μm diameter) from microarthropods (McKenzie and Dexter, 1987; Lee and Foster, 1991; Barois et al., 1993). The earthworms, especially when small, appear to avoid ingesting sand so that casts often contain more clay and silt than the surrounding soil. For example, casts of *Lumbricus rubellus* (earthworm mass 0.57 gram) contained 15.9% sand and 14.7% clay, and *L. terrestris* (earthworm mass 4.9 gram) contained 17.4% sand and 13.9% clay, whereas the surrounding loam contained 18.2% sand and 13.2% clay (Shipitalo and Protz, 1988).

The soil is usually destabilized as it passes through the earthworm's gut, with the fine clay often dispersed, so that fresh casts are usually less stable than the surrounding soil, and are susceptible to erosion (Lee and Foster, 1991; Barois et al., 1993). This is probably because in the gut, the material is both saturated and remoulded. For example, the material in the anterior part of the gut of *P. corethrurus* contained 150% water and 16% mucus compared with 35% and 0.12% respectively in aggregates of the surrounding vertisol (Barois and Lavelle, 1986). Although McKenzie and Dexter (1987) showed that *Aporrectodea rosea* exerted low pressures (260 Pa) on the material in the gut, they suggested that because the very wet material was sheared in all directions it was easily remoulded and dispersed, rather as soil is puddled in rice paddies. Organic acids or chelates in the gut (especially if the casts are sodic) may disperse the fine clay further (Gupta et al., 1984; Muneer and Oades, 1989b). Electron micrographs showed that some earthworm casts also contained bacterial colonies coated with clay (Lee and Foster, 1991). The microaggregates appeared not to be physically modified as they passed through the earthworm's gut.

The earthworm usually re-absorbs water and water-soluble organic materials in the posterior part of the gut (Lee, 1985), but fresh casts are wetter than the surrounding soil. This probably limits the stabilizing bonds which can form as the water keeps particles apart (Kemper et al., 1987). In a vertisol, the water content in the material in the anterior part of the gut of *P. corethrurus* was 150 per cent and dropped to 122% in the posterior part of the gut, whereas the fresh casts contained 90% water and the surrounding soil contained 35% (Barois and Lavelle, 1986).

Table 7. The dispersion index (DI) of treated casts produced by three species of the earthworm *Aporrectodea* from surrounding Hallam loam at 9 kPa suction

Treatment	DI
Fresh casts	0.5
Incubated moist casts	0.3
Air-dried casts	0
Surrounding soil	0.1

(From Hindell et al., 1994.)

Although fresh casts are usually less stable than the surrounding soil, they usually become less dispersible as they age, especially when air-dried (Shipitalo and Protz, 1988; Marinissen and Dexter, 1990). Hindell et al. (1994) showed that air-dried casts of three species of *Aporrectodea* were far less dispersible than fresh casts, incubated moist casts, or aggregates of the surrounding soil (Table 7). As they were dried, fresh casts immediately became stable, showing that microbial activity in the deposited casts was not necessary for the casts to be stabilized. Presumably, the water molecules in the fresh casts kept the clay particles apart, so that bonds could not form between clay and organic matter. As the clay dried, the greater tension in the water probably pulled the dispersed clay and organic matter into contact forming stable bonds (Kemper et al., 1987). The dispersed clay and organic material are well mixed in the gut which probably increases contact between particles in the dry casts. As the casts dry, they are often denser and harder than the surrounding soil, probably because the casts contain less sand than the soil and thus shrink more evenly (McKenzie and Dexter, 1987). This probably also increases contact between particles, and hence contributes to the stability of the casts.

Moist aged casts are also more stable than the surrounding soil (Hindell et al., 1994). This is probably due to the greater contact between soil particles and organic matter in the casts than in the surrounding soil. However, some stability is probably due to thixotropic hardening which can increase the stability of aggregates in the absence of microbial activity (Utomo and Dexter, 1981).

Some stability is probably also due to organic matter and microbial activity, and hence depends on the diet and species of the earthworm and on the content of organic carbon in the casts (Shipitalo and Protz, 1988). Lee and Foster (1991) suggested from electron micrographs that earthworm casts appeared to be held together mechanically by hyphae or remnants of cell walls, which interlocked with the soil particles rather than by the mucus secreted in the gut or intracellular polysaccharides. They suggested that strands of microbial polysaccharides connect parts of the casts over less than 1 μm, so are probably unlikely to be important in stability of the entire cast, but appear to stabilize microaggregates within the casts.

Some incubated casts appeared to be temporarily held together by fungal hyphae which grew in the cast or on its surface as it aged (Lee, 1985;

Marinissen and Dexter, 1990). Once deposited, casts of *Aporrectodea longa* became 50% more stable over the first 15 days, then lost the extra stability after 25 days (Parle, 1963). In the same casts, the length of fungal hyphae increased by 200 per cent after 15 days and decreased to the initial length after 25 days, suggesting that fungal hyphae may have temporarily stabilized the casts; numbers of bacteria and actinomycetes did not change. The hyphae probably invaded the casts from surrounding soil as the few hyphae in fresh casts were mostly dead (Lee and Foster, 1991; Barois et al., 1993). Fungi are strict aerobes, so they would not grow readily in the saturated conditions in the gut, but nutrients released from organic matter in the gut may be available for fungi in deposited casts as they gradually become aerobic (Lee, 1985).

Some casts deposited by *Millsonia anomala* remained stable in the field for at least 12 months after the earthworms had left (Lee and Foster, 1991). It is not known what mechanism stabilized the casts for so long.

VI. Polyvalent Cation Bridges

Some uncharged organic molecules such as aliphatic groups in polysaccharides, form H-bonds with water molecules in the H-shell of an exchangeable cation. Individual bonds are weak, but they are additive so that a long flexible molecule with many hydroxyl groups along its length may form many H-bonds and may be bound strongly to the surface of a clay particle. However, although extracellular polysaccharides from plants and microorganisms can be uncharged or positively charged, Foster (1982) showed by electron microscopy that most extracellular polysaccharides in the rhizosphere were negatively charged. Negatively charged organic materials such as polysaccharides or highly aromatic humic material can be adsorbed onto the surface of clay by polyvalent cations such as Ca^{++}, Fe^{+++}, Al^{+++}, aluminosilicates and hydroxyaluminium. In soils, 52 to 98% of the total organic matter consists of highly degraded, highly aromatic humic material associated with amorphous iron, aluminium and aluminosilicates forming part of the large organomineral fraction of the soil (Greenland, 1965; Hamblin, 1977; Tate and Churchman, 1978; Turchenek and Oades, 1978). The cation neutralizes a charge on the surface of the clay and another charge on the anion forming a bridge between the clay and the organic anion. When freshly produced, Fe and Al polycations form stable bonds with clay; once aged, these cations strengthen organic matter bonds (Emerson and Greenland, 1991). Edwards and Bremner (1967) suggested that the organo-mineral complexes include those bonds which link complexes of clay-polyvalent cation-organic matter, Clay-P-OM and (Clay-P-OM)x, both of which are less than 2 μm, into stable microaggregates $\{(\text{Clay-P-OM})_x\}_y$, which are less than 250 μm. Although bonds of Clay-P-C and OM-P-OM and of aluminium or iron oxide or H-bonds may also form, Edwards and Bremner (1967) suggested that bridges of polyvalent cation between the surface of clay particles or hydroxy polymers and organic polymers form the main and stronger bonds, within microaggregates. Hence, the stability of aggregates of soil was more persistent when polyvalent cations, e.g. Al^{+++} or Ca^{++} were present, suggesting that the negatively charged clay polysaccharides were bound to negatively charged clay particles by bridges of polyvalent cations (Martin, 1971). The polyvalent cations that strengthen bonds between clay and organic materials probably chemically stabilize organic matter against microbial degradation (Martin, 1971).

Although the polyvalent cations which form bridges are often attributed to Fe^{+++} and Al^{+++} (Hamblin and Greenland, 1977; Giovannini and Sequi, 1976), Muneer and Oades (1989a; 1989b) confirmed that Ca^{++} interacted with organic matter to stabilize both macroaggregates (>250 μm diameter) and microaggregates (<250 μm) in a soil which was mildly leached. They showed that this was due to the Ca^{++} forming a bridge between the negatively charged clay particles and the negatively charged organic molecules (Muneer and Oades, 1989c). Robert and Chenu (1992) suggested that the increased stabilization of aggregates by Ca^{++} during microbial activity may be because the Ca^{++} ions cross-link many anionic polysaccharides, forming gels which bind particles together. Organic linkages with Na^{+} are weak, so for sodic soils to be stabilized, Na^{+} must be replaced with polyvalent cations before organic matter is added (Rengasamy and Olsson, 1991).

Polyvalent cation bridges are not broken when treated with periodate but are broken when treated with acids, pyrophosphate or acetylene which disrupt microaggregates (Stefanson, 1971; Hamblin and Greenland, 1977; Giovannini and Sequi, 1976; Tisdall and Oades, 1980). Bacterial capsules are more stable to periodate treatment than are root mucilages but the most stable polysaccharides appear to be those secreted by fungal hyphae. In an old grassland soil and in forest soil, periodate or pyrophosphate alone did not disperse clay but combined treatment with periodate and pyrophosphate completely or partially dispersed the clay (Emerson et al., 1986). This was attributed to the Na^{+} from the Na pyrophosphate replacing polyvalent cation bridges as well as occupying exchange sites, increasing the negative charge of the clay. The extra swelling between the clay particles then allowed periodate ions to reach and oxidize the polysaccharides.

Humic materials in soil are persistent organic polymers resulting from microbial decomposition of the biomass, and differ from biopolymers, which are synthesized from organisms (Sposito, 1989). Humic materials can form stable complexes with polyvalent cations, forming stable microaggregates in soil (Tisdall and Oades, 1982). However, such interactions cannot be precisely determined as the chemical composition of humic materials is not specifically known.

Humic materials can be divided into humic acid, fulvic acid and humin fractions, according to their solubility in acid and alkali (Oades, 1989). The effects of humic acid and fulvic acid have been studied in several soils. The stability of macroaggregates was correlated with the concentration of humic acid in 120 soils (Chaney and Swift, 1984). When humic acid was extracted from soil, and then adsorbed onto soil minerals and incubated, the macroaggregates became more stable, especially when glucose had also been added (Chaney and Swift, 1986b). Chaney and Swift (1986b) suggested from these results that the humic materials formed microaggregates, which were then bound into stable macroaggregates by microorganisms. Humic and fulvic acids were also shown by micromorphology to increase macroaggregation in a sandy loam and in a clay soil (Fortun et al., 1990). However, these results and interpretations may be unreliable as artefacts may occur when humic materials are extracted from soil and purified (Oades et al., 1987).

Reid and Goss (1981) showed that the growth of roots of maize (*Zea mays* L.) or tomato, but not of ryegrass, dispersed clay from two loams. The increased dispersion was attributed to chelates and organic acids, exuded by roots, which removed polyvalent cations from the bonds between clay and organic matter (Reid et al., 1982). The accumulation of polyvalent cations in the rhizosphere does depend on the species of plant, which may explain the effects of different species on dispersion (Barber and Ozanne, 1970). Root exudates of high molecular weight from maize grown in axenic liquid culture were shown to combine with several different cations (Mench et al., 1987). Similar reactions in the rhizosphere of maize could remove polyvalent cations from bonds between organic matter and clay, supporting the hypothesis of Reid et al. (1982). However, dispersion by growing roots has not been shown in other soils, suggesting that different soils may react differently to roots (Monroe and Kladivko, 1987; Lynch and Bragg, 1985; Pojasok and Kay, 1990). Pojasok and Kay (1990) showed that root exudates of maize and bromegrass (*Bromus inermis* Leyss) incubated with macroaggregates of a loam increased the stability of the macroaggregates, rather than dispersing clay. They concluded that the exudates released unchelated polyvalent cations into the soil solution which strengthened bonds between organic matter and clay and increased the stability. This conclusion is similar to the hypothesis of Tisdall (1991) for the formation of bonds between fungal hyphae and clay and subsequent stabilization of macroaggregates.

On the other hand, it has been shown that organic matter can increase dispersion of clay. Leaf extract added to a red earth and straw added to a fine sandy loam each increased dispersion of clay (Emerson and Smith, 1970; Muneer and Oades, 1989b). This was attributed to organic acids released from decomposing straw blocking positive sites on the surface of colloids and reducing the attractive forces between clay particles (Gillman, 1974; Yeoh and Oades, 1981). Similarly, humic materials increased dispersion in sodic soils or in soils at high pH (Gupta et al., 1984).

VII. Polysaccharides

A. Stabilization of Aggregates

Although carbohydrates make up only 10 to 20% of the organic matter in soil, the stability of aggregates is often correlated with the concentration of carbohydrate in soil (Rennie et al., 1954; Oades, 1972; Burns and Davies, 1986). The carbohydrates include the cellular materials of plants, animals and microorganisms, and the mucilage secreted by roots, root hairs, fungal and algal hyphae, and bacteria. These mucilages are mainly polysaccharides but may also contain polyuronic acids and many amino compounds, and are usually negatively charged (Swincer et al., 1968; Foster, 1981a; 1981b; 1982). The polysaccharides probably do not move far in soil because they are strongly adsorbed onto the clay particles (Oades, 1984). Hence, clusters and microaggregates probably form when clay and silt are deposited from the soil solution rather than when the polysaccharides diffuse through soil.

In the rhizosphere, plant mucilages are quickly decomposed by microorganisms being replaced by microbial polysaccharides. The complex mixture of plant and microbial mucilages and embedded particles of clay is called mucigel (Jenny and Grossenbacher, 1963). Oades (1984) suggested that the relative proportions of plant and microbial polysaccharides, and hence those important in the stabilization of aggregates of different sizes, could be determined by the ratio of the concentration of galactose plus mannose to that of arabinose plus xylose (G + M) : (A + X). This ratio was greater than 2 for microbial polysaccharides and less than 0.5 for plant polysaccharides, since polysaccharides derived from plants contained mainly arabinose and xylose, whereas microbial polysaccharides contained mainly galactose and mannose and little, if any, arabinose and xylose (Oades, 1984). From this ratio, particle-size analyses showed that recalcitrant microbial polysaccharides were concentrated in the stable microaggregates 2 to 20 μm diameter which they helped to form (Turchenek and Oades, 1978; 1979; Tiessen and Stewart, 1983). Microaggregates 20 to 250 μm diameter contained a higher proportion of plant polysaccharides than microaggregates 2 to 20 μm did. This conclusion is supported by the electron micrographs of Oades and Waters (1991) and the results of Anderson et al. (1981) which suggested that the different types of organic matter stabilize aggregates of different sizes.

Baldock et al. (1987) showed that maize and bromegrass contained little galactose, and that a more accurate indicator for relative proportions of plant and microbial polysaccharides in soil was the ratio (M) : (A + X) . Baldock et al. (1987) used this ratio for aggregates of a silt loam under three different cropping treatments : continual bromegrass for 15 years (B_{15}), continual corn for 15 years (C_{15}), and continual corn for 13 years followed by continual bromegrass for 2 years ($C_{13}B_2$). They concluded that soil from B_{15} contained a higher proportion of carbohydrates derived from plant material than soil from C_{15} or $C_{13}B_2$ did (Table 8). This suggested that a smaller proportion of the plant material added to the soil in B_{15} was readily metabolized by microorganisms than

Table 8. Ratio of concentration of galactose plus mannose to arabinose plus xylose (G+M):(A+X), and of mannose to arabinose plus xylose (M):(A+X) for three cropping treatments in a silt loam

Treatment	(G+M):(A+X)	(M):(A+X)
B_{15}[a]	1.47	0.66
C_{15}[b]	1.54	0.78
$C_{13}B_2$[c]	1.58	0.81
LSD p=0.05	NS	0.086

[a]B_{15} Continual bromegrass for 15 years
[b]C_{15} Continual corn for 15 years
[c]$C_{13}B_2$ Continual corn for 13 years plus continual bromegrass for 2 years
(From Baldock et al., 1987.)

was in C_{15} or $C_{13}B_2$. The proportion of plant polysaccharides as determined by (M) : (A + X) increased in B_{15}, but decreased in C_{15} and in $C_{13}B_2$. These results for C_{15} and for $C_{13}B_2$ (but unexpectedly not for B_{15}) agree with the hypothesis of Tisdall and Oades (1982) that roots rather than polysaccharides stabilize macroaggregates in alfisols which contain more than 2% organic carbon.

Similarly, Carter et al. (1994) showed that the increased stability of macroaggregates in spodosols after the growth of perennial grasses for four years was not related to the concentration of carbohydrates (extracted either in hot water or in acid) in the soil. This result probably applies generally to soils with stable microaggregates, high concentrations of organic carbon and a high volume of rhizosphere (Elliott and Coleman, 1988). In such soils mainly roots and fungal hyphae stabilize macroaggregates (Tisdall and Oades, 1982; Miller and Jastrow, 1990).

Microbial polysaccharides, which often contain cellulose fibrils, often form a mucoid layer around the organisms on agar plates. Ultra-thin sections of soil show that these polysaccharides are often coated with aligned particles of clay. The mass of the polysaccharides near microorganisms may be much more than 1% of the mass of the surrounding clay (Chenu, 1993). Scanning electron micrographs showed that a significant part of the fungal polysaccharide, scleroglucan, which was adsorbed onto clay was not flattened onto the surface of the clay, but extended as strands into the spaces between particles (Chenu, 1989).

The above fibrils are of fine diameter (7 to 10 nm) so that a small mass of polysaccharide is made up of long fibres, which may explain why such small amounts of microbial polysaccharides (0.02 to 0.2%) are so effective at stabilizing clay microaggregates (Foster, 1988). This may also explain why the microbial biomass, which makes up 1 to 3% of the organic carbon in soil and occupies 0.001% of the volume of soil, has such a large effect on the stability of aggregates (Dormaar and Foster, 1991). Isolated pockets of granular gels,

Table 9. The effect of the fungal polysaccharide, scleroglucan, on the tensile strength of kaolinite

Scleroglucan content (mg g^{-1} clay)	Tensile strength (10^5 Pa)
0	1
1	2
9	6
13	14
21	18
24	21
30	19

(From Chenu and Guérif, 1991.)

some less than 0.5 μm diameter, are sometimes completely enclosed by clay (Foster, 1981a; 1981b). They are probably part of the organic matter which is protected in soil since microorganisms and their enzymes cannot enter such small pores. It is not known whether enzymes themselves can move far in soil without being adsorbed onto clay particles (Foster, 1988).

B. Other Physical Properties of Clay

Polysaccharides not only increase the stability of aggregates, they can also increase the tensile strength of clay and the amount of water that it retains. Particles of kaolinite or montmorillonite were treated with increasing concentrations of the uncharged fungal polysaccharide, scleroglucan, and the tensile strength measured (Chenu and Guérif, 1991). With kaolinite, which has few points of contact between particles, as little as 0.1% polysaccharide nearly doubled the strength of kaolinite. The tensile strength increased linearly with increased concentration of scleroglucan, to up to 20 times that of the original kaolinite (Table 9). It was calculated that at maximum strength of 21 x 10^5 Pa, scleroglucan was adsorbed onto about 60% of the surface area of the rigid particles of kaolinite. Once scleroglucan covered 60% of the surface, polysaccharides were assumed to form every bond between particles of clay. Scleroglucan had less effect on the strength of montmorillonite than on kaolinite. In untreated montmorillonite, 80 to 90% of the surface area of the particles is in contact so that the strength of the clay is high, and the area available for adsorbed polysaccharide is low. Chaney and Swift (1986a) suggested that bacterial alginate probably formed gels between microaggregates and then established bridges on a larger scale to form bigger aggregates. Adsorbed microbial polysaccharides, xanthan or scleroglucan, increased the amount of water retained by kaolinite or montmorillonite (Table 10) (Chenu, 1993). This was attributed in part to the water holding capacity of the polysaccharide itself,

Table 10. Water content at - 0.0032 MPa water potential of associations of clay and scleroglucan, a fungal polysaccharide

Scleroglucan (%)	Water content (g g^{-1})	
	Kaolinite	Montmorillonite
0	1.3	4.5
1	1.3	5.7
4	1.5	-
5	-	5.8
10	1.8	-
14	-	7.2

(From Chenu, 1993.)

and in part to a more open structure when strands of the polysaccharides bridged between particles of clay. Dudman (1977) proposed that the extra water stored by clays due to adsorbed polysaccharide could be used by microorganisms. However, as the water is held in small pores, it could only be extracted at high water suctions (<1 MPa) and only if accessible to the microorganisms (Chenu, 1993).

VIII. Rhizosphere

Living roots release many types of organic materials into the rhizosphere within 50 μm of the surface of the root. The types of bacteria in the rhizosphere usually differ from those in the soil outside the rhizosphere. In the rhizosphere there is usually a shortage of nitrogen, with a C:N ratio of about 40:1. Those bacteria which flourish in the rhizosphere are those which compete successfully for N and can use simple organic materials released by roots (Gosz and Fisher, 1984). As shown on agar plates or in electron micrographs, the bacteria which dominate the rhizosphere are Gram negative bacteria (e.g. *Flavobacterium*, *Arthrobacter*, *Pseudomonas*, *Rhizobium* and various bacteria from the N cycle) (Foster, 1983). The ratio of the number of bacteria in the rhizosphere to the number in the non-rhizosphere soil (i.e. the R:S ratio) of Gram negative bacteria ranges from 5 to 2000 as determined on agar plates, and Bowen and Rovira (1973) showed that *Pseudomonas* species grew faster close to the roots of *Pinus radiata* than in non-rhizosphere soil. Other unclassified bacteria, such as lobed forms, appear in electron micrographs of the rhizosphere only, so probably survive only when carbon is available (Foster, 1983). Outside the rhizosphere, species of *Bacillus*, *Azotobacter* and *Micrococcus* dominate (R:S <1). *Bacillus* species produce spores, which probably helps them to survive when carbon is limiting; these species were shown to grow more slowly on roots than in the non-rhizosphere soil (Bowen and Rovira, 1973).

On agar plates or in electron micrographs, many bacteria and colonies (especially of Gram negative bacteria) are surrounded by a distinct capsule made of polysaccharides (Hepper, 1975; Foster, 1983). Many of these bacteria also produce extracellular polysaccharides (often containing cellulose fibrils) which form a mucoid layer around the organisms on agar plates or as seen in electron micrographs (Hepper, 1975). These polysaccharides may be neutral or negatively charged polymers, and are shown in electron micrographs to interact with the surface of clay minerals, stabilizing microaggregates less than 20 μm diameter (Oades and Waters, 1991).

IX. Wetting and Drying Cycles

The water content of soil affects the bonds which form between particles of clay, and between organic matter and clay, and hence the formation and stabilization of aggregates of different sizes. When soils are kept at constant water content, especially at the plastic limit, particles of clay are rearranged to positions of minimum energy and are chemically bound together at points of contact (Semmel et al., 1990). As the soil dries, particles of clay, organic colloids and salts are deposited at points of contact, strengthening bonds between larger particles. Hence, much of the effect of roots on aggregation of soil may be due to localized drying around the roots (Allison, 1968; Perfect et al., 1990; Dexter, 1991). In dry soil, it is difficult to remove soil from roots (Foster, 1983). It has not been shown whether drying of the soil around roots increases points of contact between clay and polysaccharides and/or denatures the polysaccharides (Dormaar and Foster, 1991).

The clay particles are rearranged at several levels of aggregation. An outer skin (about 1 to 2 mm thick) formed on the surface of each artificial macroaggregate of a desert loess subjected to several cycles of wetting and drying (Semmel et al., 1990). The outer skin always contained higher concentrations of clay and salts than the inner part of the macroaggregate did. A similar dense layer of clay about 25 μm thick was found on the surface of earthworm casts collected from the field (Blanchart et al., 1993). Such a skin would probably be chemically more reactive than the inner part of the aggregate and would more readily combine with organic colloids. Similarly the clay moves as roots and microorganisms locally dry out and compact the surrounding soil (Foster, 1988; Dormaar and Foster, 1991). Fine clay (<0.2 μm diameter) has a greater surface area, and is more reactive and mobile, than coarser clay (Dixit, 1978; Robert and Chenu, 1992). Hence, electron micrographs of soil show that particles of fine clay are aligned parallel and are firmly attached to the surfaces of roots, fungal hyphae and bacteria, forming stable macroaggregates and microaggregates (Foster, 1988).

X. Protected Organic Matter

Much of the organic matter in soil which stabilizes aggregates of various sizes is physically and/or chemically protected from decomposition. The mechanism by which organic matter is protected may depend on soil type. Clay soils contain more fine pores, which potentially can physically protect organic matter, than soils with coarse texture (Hassink et al., 1993). For example, 72%, 40% and 22% of pores in clay soils, silty soils and sandy soils respectively were 3 μm or less in diameter (Juma, 1993). Mineralization of nitrogen was increased when clay soils, but not sandy soils, were passed through a fine sieve 1 mm in diameter, suggesting that the organic matter in clay soils was protected from decomposition in small pores, whereas in sandy soils the organic matter was protected by adsorption to clay minerals or encrustation by clay minerals (Hassink et al., 1993). Organic matter which was probably physically protected in microaggregates was more recalcitrant than the organic matter associated with macroaggregates (Elliott, 1986).

A. Physical Protection

Much of the polysaccharide in soil is labile, yet it is in pores that are small enough to provide protection from 30% of the bacterial population in soil that produce polysaccharases (Cheshire et al., 1974). It was calculated that in a clay loam, 10^8 bacteria per gram occupy 0.1% of the volume of pores, and most of the surfaces of soil are found in pores of less than one μm diameter (Sills et al., 1974; Adu and Oades, 1978). More than 90% of the surfaces in such soils are physically protected from microorganisms and their enzymes (Adu and Oades, 1978).

When polysaccharides of known chemical composition, and which were stable in soil, were extracted from the soil and then returned to the soil and incubated, they were readily decomposed; they were probably no longer protected (Cheshire et al., 1974). When simple soluble substrates were added to the soil and incubated, carbon dioxide was produced in a flush of microbial activity, presumably because the substrates were being metabolized (Sørensen and Paul, 1971; Van Veen et al, 1985). Within a few weeks, the microbial activity was greatly reduced even though a high proportion of the added carbon could be accounted for in the microbial cells and debris, and in metabolites which could be chemically defined (Sørensen, 1967; Sørensen and Paul, 1971). Presumably these new materials were now protected from microbial attack.

Adu and Oades (1978) made artificial aggregates of a fine sandy loam in which they placed ^{14}C-glucose or ^{14}C-starch in micropores and macropores. All the glucose diffused through the soil and was used by microorganisms so that labelled organisms and their metabolites were found in large pores on the outside of aggregates. On the other hand, the labelled starch was protected inside the

Table 11. The percentage of water-stable aggregates greater than 1 mm diameter in surface soils with added water, glucose or peptone incubated aerobically at 25°C for 14 days

Soil	Water-stable aggregates (% total soil)			
		Incubated		
	Control	Water	Glucose	Peptone
1	39	55	50	37
2	23	53	76	52
3	6	27	61	45
4	45	68	94	84
5	44	71	75	53
6	33	54	81	40

(From Skinner, 1979.)

aggregates. When the aggregates were disrupted, some starch was exposed and metabolized by microorganisms.

Reduced diffusion of oxygen and high C:N ratios of the polysaccharide in the fine pores, probably also help to protect the polysaccharides from decomposition (Skinner, 1979; Tiedje et al., 1984). Skinner (1979) added glucose, peptone or water to aggregates 1 to 3 mm diameter of several soils and incubated them for 14 days. Glucose almost always increased the stability of the incubated aggregates whereas peptone sometimes decreased the stability (Table 11). This was attributed to the nitrogen which was added in the peptone reducing the C:N ratio sufficiently to enable microorganisms to decompose polysaccharides or other binding materials. On the other hand, microorganisms appeared to produce binding materials, presumably polysaccharides, from added glucose, which stabilized the aggregates.

When the soil is disturbed, such as by tillage or soil fauna or when the soil is quickly wetted, C and N are often mineralized in a flush of microbial activity, probably because organic matter, which was physically protected in pores is exposed to microorganisms (Rovira and Greacen, 1957; Gregorich et al., 1989).

B. Chemical Protection

As soils dry, the bonds between clay and polysaccharides appear to become stronger, probably as there are more points of contact, and aggregates become more stable (Edwards and Bremner, 1967; Stefanson, 1971; Reid et al., 1982). When dried, the polysaccharides are probably also denatured and polymerized, and are chemically protected from decomposition (Dormaar and Foster, 1991). It was suggested that during the first 3 years of growth, grass roots increased the stability of aggregates by drying the soil rather than by adding organic matter (Perfect et al., 1990). Similarly, Allison (1968) and Dexter (1991) suggested that much of the effect of roots on aggregation of soil may be due to wetting and drying around roots. It is yet to be shown whether drying of the soil around roots increases points of contact between clay and polysaccharides and/or denatures the polysaccharides (Dormaar and Foster, 1991).

Polysaccharides may also be chemically protected from decomposition by forming complexes with polyvalent cations (such as Cu^{++}, Fe^{+++}, Zn^{++} and Ca^{++}) and with polyphenols (Martin, 1971; Griffiths and Burns, 1972; Martin et al., 1978; Muneer and Oades, 1989a). Although it has not been shown yet, the complex chemical composition of some polysaccharides in soil may protect them from decomposition; groups of enzymes from different microbial species may be needed to decompose some polysaccharides which contain two or more different sugar residues, and are branched or are complex fibrils (Burns and Davies, 1986).

C. Concentration of Organic Carbon in Aggregates of Different Sizes

The concentration of organic carbon was measured in various fractions of soils separated by ultrasound, size and density (Turchenek and Oades, 1979; Oades et al., 1987; Golchin et al., 1994). The organic carbon in each fraction was analysed chemically, or by solid state ^{13}C NMR spectroscopy, to determine the type and the degree of decomposition of the organic carbon. In Urrbrae fine sandy loam, an alfisol, the concentration of organic carbon generally decreased as the particle size increased (Turchenek and Oades, 1978; Table 12). The organic carbon was concentrated in all fractions of low density (<2 Mg m^{-3}). The coarser fractions (>53 μm diameter) contained high concentrations of O-alkyl carbon, i.e. carbohydrate, which was probably lignin originating from plants (Oades et al., 1987). The materials in the silt-sized fraction (2 to 20 μm) were more aromatic than those from other fractions, and made up 40 per cent of the total organic carbon in soil. The finer particles (<2 μm) contained high concentrations of alkyl-C, which appeared to be intimately associated with humic materials and/or clay materials. The alkyl-C probably consisted of highly degraded organic materials, plus some materials synthesized by microorganisms (Baldock et al., 1990).

Working with several soils, Golchin et al. (1994) showed that the light fraction (<1.6 Mg m^{-3}) collected after ultrasound and sedimentation of the whole soil consisted of two different kinds of organic materials. These materials were a) the free light fraction, which was mainly undecomposed and partly decomposed plant residues, and b) the occluded light fraction which was mainly highly decomposed organic material released from inside aggregates by ultrasound. Golchin et al. (1994) subsequently separated the free light fraction from each sample before disruption by ultrasound, and then the occluded light fraction from the same sample after ultrasound. NMR analysis of these free and occluded light fractions from Urrbrae fine sandy loam under old pasture contained 24.7%

Table 12. Carbon concentration of fractions of different size and density of Urrbrae fine sandy loam

Size (μm diameter)	Density (Mg m^{-3})	Carbon (%)
250 - 2000	< 2.0	37.5
53 - 250	< 2.0	36.1
20 - 53	< 2.0	8.3
2 - 20	< 2.0	18.7
2 - 20	2.0 - 2.2	4.7
0.2 - 2.0	< 2.0	22.5
0.2 - 2.0	2.0 - 2.2	11.5
0.2 - 2.0	> 2.2	3.7
0.2 - 2.0	ND[a]	5.0
0.04 - 0.2	ND	4.0
< 0.2	ND	4.6
< 0.04	ND	7.7
Whole soil	ND	2.30

[a]Not determined
(From Oades et al., 1987.)

and 39.6% organic carbon respectively, which was 7.7% and 9.7% of the total organic carbon in the soil. The free light fraction contained mainly O-alkyl carbon (carbohydrate), and the occluded light fraction contained mainly O-alkyl carbon and alkyl carbon.

XI. Summary

Soils consist of aggregates of different sizes, held together by organic and inorganic materials, and the pores between aggregates. Several models of aggregation divide soil into macroaggregates (>250 μm diameter) and microaggregates (<250 μm), often with several steps of aggregation within the microaggregates. The smaller the aggregate, the more energy is needed to disrupt the aggregates.

Microaggregates in alfisols and mollisols (which are stabilized by organic matter) form when polysaccharides, highly decomposed organic matter, bacterial cells or colonies, or fragments of plants or microorganisms are physically or chemically protected from microorganisms. The protected organic matter then stabilizes the microaggregates, which can persist for many years. Stable microaggregates form slowly in soil, but are not readily destroyed by agriculture in non-sodic soils.

Stable macroaggregates consist of microaggregates held together by roots and fungal hyphae; the macroaggregates include casts or faeces of earthworms. Macroaggregates form readily in soil under pasture or where organic residues have been added to soil, especially where the soil already contains large amounts of stable microaggregates. Macroaggregates are readily disrupted by tillage or heavy rain, when exposed organic matter is oxidized.

Polysaccharides in soil are secreted by roots and microorganisms, are usually negatively charged, and probably do not move far in soil. Microaggregates probably form when clay and silt are deposited from the soil solution, rather than when the polysaccharides diffuse through the soil. Polysaccharides not only stabilize aggregates, they can also increase the tensile strength of clay and the amount of water that it retains.

Much of the polysaccharide in soil is protected from microorganisms either a) physically within small pores or by adsorption to clay, or b) chemically because the polysaccharides are denatured or form complexes with polyvalent cations and polyphenols. Polyvalent cation bridges can form between negatively charged clay particles and negatively charged polysaccharides. Exudates from roots appear to release unchelated polyvalent cations into the soil which can then become bridges between clay and polysaccharides.

The rhizosphere is the soil within 50 μm from the surface of the root. Growing roots release many types of organic materials into the rhizosphere, which supports a large active microbial population. These microorganisms are often Gram negative bacteria, which can use the simple organic materials released by roots and which release polysaccharides into the soil. The polysaccharides interact with clay particles, stabilizing microaggregates.

The water content of the soil affects the bonds that form between particles of clay and organic matter, and hence the stability of aggregates. As soils dry, particles of soil, organic colloids and salts are deposited at points of contact, strengthening bonds between larger particles.

Acknowledgements

I thank Dr. P. Rengasamy for helpful discussion, and Dr. H. Downing for helpful criticism of the manuscript.

References

Adem, H.H. and J.M. Tisdall. 1984. Management of tillage and crop residues for double-cropping in fragile soils of south-eastern Australia. *Soil Tillage Res.* 4:577-589.

Adu, J.K. and J.M. Oades. 1978. Physical factors influencing decomposition of organic materials in soil aggregates. *Soil Biol. Biochem.* 10:109-115.

Allison, F.E. 1968. Soil aggregation - some facts and fallacies as seen by a microbiologist. *Soil Sci.* 106:136-143.

Amato, M. and J.N. Ladd. 1992. Decomposition of ^{14}C-labelled glucose and legume materials in soils : properties influencing the accumulation of organic residues C and microbial biomass C. *Soil Biol. Biochem.* 24:455-464.

Anderson, D.W. and E.A. Paul. 1984. Organo-mineral complexes and their study by radiocarbon dating. *Soil Sci. Soc. Am. J.* 48:298-301.

Anderson, D.W., S. Saggar, J.R. Bettany, and J.W.B. Stewart. 1981. Particle-size fractions and their use in studies of soil organic matter. I. Nature and distribution of forms of carbon, nitrogen and sulphur. *Soil Sci. Soc. Am. J.* 45:767-772.

Aspiras, R.B., O.N. Allen, G. Chesters, and R.F. Harris. 1971. Chemical and physical stability of microbially stabilised aggregates. *Soil Sci. Soc. Am. Proc.* 35:283-286.

Avnimelech, Y. and B.W. Troeger. 1982. Mutual flocculation of algae and clay: evidence and implications. *Science* 216:63-65.

Baldock, J.A., B.D. Kay, and M. Schnitzer. 1987. Influence of cropping treatments on the monosaccharide content of the hydrolysates of a soil and its aggregate fractions. *Can. J. Soil Sci.* 67:489-499.

Baldock, J.A., J.M. Oades, A.M. Vassallo, and M.A. Wilson. 1990. *Aust. J. Soil Res.* 28:213-225.

Baldock, J.A., J.M. Oades, A.G. Waters, X. Peng, A.M. Vassallo, and M.A. Wilson. 1992. Aspects of the chemical structure of soil organic materials as revealed by solid-state ^{13}C NMR spectroscopy. *Biogeochemistry* 15:1-42.

Baon, J.B., S.E. Smith, and A.M. Alston. 1993. Mycorrhizal responses of barley cultivars differing in P efficiency. *Plant Soil* 157:97-105.

Barber, S.A. and P.G. Ozanne. 1970. Autoradiographic evidence for the differential effect of four plant species in altering the calcium content of the rhizosphere soil. *Soil Sci. Soc. Am. Proc.* 34:635-637.

Barea, J.M. 1991. Vesicular-arbuscular mycorrhizae as modifiers of soil fertility. *Adv. Soil Sci.* 15:1-40.

Barois, I. and P. Lavelle. 1986. Changes in respiration rate and some physico-chemical properties of a tropical soil during transition through *Pontoscolex corethrurus* (Glossoscolecidae, Oligochaeta). *Soil Biol. Biochem.* 18:539-541.

Barois, I., G. Villemin, P. Lavelle, and F. Toutain. 1993. Transformation of the soil structure through *Pontoscolex corethrurus* (Oligochaeta) intestinal tract. *Geoderma* 56:57-66.

Bartoli, F., R. Phillipy, and G. Burtin. 1988. Aggregation in soils with small amounts of swelling clays. 1. Aggregate stability. *J. Soil Sci.* 39:617-628.

Blanchart, E., A. Bruand, and P. Lavelle. 1993. The physical structure of casts of *Millsonia anomala* in shrub savannah soils (Côte d'Ivoire). *Geoderma* 56:119-132.

Bond, R.D. and J.R. Harris. 1964. The influence of the microflora on physical properties of soils. I. Effects associated with filamentous algae and fungi. *Aust. J. Soil Res.* 2:111-122.

Bowen, G.D. and A.D. Rovira. 1973. Are modelling approaches useful in rhizosphere biology? *Bull. Ecol. Res. Comm. (Stockholm)* 17:443-450.

Braunack, M.V., J.S. Hewitt, and A.R. Dexter. 1979. Brittle fracture of soil aggregates and the compaction of aggregate beds. *J. Soil Sci.* 30:653-667.

Burford, J.R., T.L. Deshpande, D.J. Greenland, and J.P.Quirk. 1964. Influence of organic materials on the determination of the specific areas of soils. *J. Soil Sci.* 15:192-201.

Burns, R.G. and J.A. Davies. 1986. The microbiology of soil structure. *Biol. Agric. Hort.* 3:95-113.

Cambardella, C.A. and E.T. Elliott. 1993. Methods for physical separation and characterisation of soil organic matter fractions. *Geoderma* 56:449-457.

Carter, M.R., D.A. Angers, and H.T. Kunelius. 1994. Soil structural form and stability, and organic matter under cool-season perennial grasses. *Soil Sci. Soc. Am. J.* 58:1194-1199.

Chaney, K. and R.S. Swift. 1984. The influence of organic matter on aggregate stability in some British soils. *J. Soil Sci.* 35:223-230.

Chaney, K. and R.S. Swift. 1986a. Studies on aggregate stability. I. Reformation of soil aggregates. *J. Soil Sci.* 37:329-335.

Chaney, K. and R.S. Swift. 1986b. Studies on aggregate stability. II. The effect of humic substances on the stability of re-formed soil aggregates. *J. Soil Sci.* 37:337-343.

Chenu, C. 1989. Influence of fungal polysaccharide, scleroglucan, on clay microstructures. *Soil Biol. Biochem.* 21:299-305.

Chenu, C. 1993. Clay- or sand-polysaccharide associations as models for the interface between micro-organisms and soil:water related properties and microstructure. *Geoderma* 56:143-156.

Chenu, C. and J. Guérif. 1991. Mechanical strength of clay minerals as influenced by adsorbed polysaccharide. *Soil Sci. Soc. Am. J.* 55:1076-1080.

Cheshire, M.V., M.P. Greaves, and C.M. Mundie. 1974. Decomposition of soil polysaccharide. *J. Soil Sci.* 25:483-498.

Chesters, G., O.J. Attoe, and O.N. Allen. 1957. Soil aggregation in relation to various soil constituents. *Soil Sci. Soc. Am. Proc.* 21:272-277.

Christensen, B.T. 1992. Physical fractionation of soil and organic matter in primary particle size and density separates. *Adv. Soil. Sci.* 20:1-90.

Clark, F.E. and E.A. Paul. 1970. The microflora of grassland. *Adv. Agron.* 22:375-435.

Clough, K.S. and J.C. Sutton. 1978. Direct observation of fungal aggregates in sand dune soil. *Can. J. Microbiol.* 24:333-335.

Collis-George, N. and R. Lal. 1970. Infiltration into columns of swelling soil as studied by high speed photography. *Aust. J. Soil Res.* 9:107-116.

Dexter, A.R. 1988. Advances in characterisation of soil structure. *Soil Tillage Res.* 11:223-238.

Dexter, A.R. 1991. Amelioration of soil by natural processes. *Soil Tillage Res.* 20:199-238.

Dixit, S.P. 1978. Measurement of the mobility of soil colloids. *J. Soil Sci.* 29:557-566.

Dorioz, J.M. and M. Robert. 1987. Aspects microscopiques des relations entre les microorganismes ou vegetaux et les argiles: consequence sur la microorganisation et la microstructuration des sols. p. 353-361. In: N. Fedoroff, L.M. Bresson, and M.A. Courty (eds.), *Soil Micromorphology.* Proc. 7th Int. Working Group on Soil Micromorphology, Wageningen, The Netherlands.

Dorioz, J.M., M. Robert, and C. Chenu. 1993. The role of roots, fungi and bacteria on clay particle organisation. An experimental approach. *Geoderma* 56:179-194.

Dormaar, J.F. 1987. Quality and value of wind-moveable aggregates in Chernozemic Ap horizons. *Can. J. Soil Sci.* 67:601-607.

Dormaar, J.F. and R.C. Foster. 1991. Nascent aggregates in the rhizosphere of perennial ryegrass (*Lolium perenne* L.) *Can. J. Soil Sci.* 71:465-474.

Douglas, J.T., M.G. Jarvis, K.R. Howse, and M.J. Goss. 1986. Structure of a silty soil in relation to management. *J. Soil Sci.* 37:137-151.

Dudman, W.F. 1977. The role of surface polysaccharides in natural environments. p. 357-414. In: I. Sutherland (ed.), *Surface Carbohydrates of the Procaryotic Cell.* Academic Press, New York.

Edwards, A.P. and J.M. Bremner. 1967. Microaggregates in soils. *J. Soil Sci.* 18:64-73.

Elliott, E.T. 1986. Aggregate structure and C, N and P in native and cultivated soils. *Soil Sci. Soc. Am. J.* 50:627- 633.

Elliott, E.T. and D.C. Coleman. 1988. Let the soil work for us. *Ecol. Bull.* 39:23-32.

Emerson, W.W. 1991. Structural decline of soils, assessment and prevention. *Aust. J. Soil Res.* 29:905-921.

Emerson, W.W., R.C. Foster, and J.M. Oades. 1986. Organo-mineral complexes in relation to soil aggregation and structure. p. 521-548. In: *Interactions of Soil Minerals with Natural Organics and Microbes.* SSSA Spec. Publ. No. 17.

Emerson, W.W. and D.J. Greenland. 1991. Soil aggregates - formation and stability. p. 485-511. In: M.F.De Boodts, M.H.B. Hayes, and A. Herbillon (eds.), *Soil Colloids and their Associations in Aggregates.* Plenum Publishing Corporation, New York.

Emerson, W.W. and B.H. Smith. 1970. Magnesium, organic matter and soil structure. *Nature* 228:453-454.

Fortun, A., J. Benayas, and C. Fortun. 1990. The effects of fulvic and humic acids on soil aggregation: a micromorphological study. *J. Soil Sci.* 41:563-572.

Foster, R.C. 1981a. Polysaccharides in soil fabrics. *Science* 214:665-667.

Foster, R.C. 1981b. The ultrastructure and histochemistry of the rhizosphere. *New Phytol.* 89:263-273.

Foster, R.C. 1982. The fine structure of epidermal cell mucilages of roots. *New Phytol.* 91:727-740.

Foster, R.C. 1983. The plant root environment. p. 673-684. In: *Soils: An Australian Viewpoint.* CSIRO, Melbourne/Academic Press, London.

Foster, R.C. 1988. Microenvironments of soil microorganisms. *Biol. Fert. Soils* 6:189-203.

Foster, R.C. and J.K. Martin. 1981. In situ analysis of soil components of biological origin. *Soil Biochem.* 5:75-110.

Gillman, G.P. 1974. The influence of net charge on water-dispersible clay and sorbed sulfate. *Aust. J. Soil Res.* 12:173-176.

Giovannini, G. and P. Sequi. 1976. Iron and aluminium as cementing substances of soil aggregates. II. Changes in stability of soil aggregates following extraction of iron and aluminium by acetyl-acetone in a non-polar solvent. *J. Soil Sci.* 27:148-153.

Golchin, A., J.M. Oades, J.O. Skjemstad, and P. Clarke. 1994. Study of free and occluded particulate organic matter in soils by solid-state ^{13}CP/MAS NMR Spectroscopy and scanning electron microscopy. *Aust. J. Soil Res.* 32:285-309.

Gosz, J.R. and F.M. Fisher. 1984. Influence of clear-cutting on selected microbial processes in forest soils. p. 523-530. In: M.J. Klug and C.A. Reddy (eds.), *Current Perspectives in Microbial Ecology.* Amer. Soc. Microbiol., Washington.

Greenland, D.J. 1965. The adsorption of sugars by montmorillonite. I. X-ray studies. *J. Soil Sci.* 7:319-328.

Gregorich, E.G., R.G. Kachanoski, and R.P. Voroney. 1989. Carbon mineralisation in soil size fractions after various amounts of aggregate disruption. *J. Soil Sci.* 28:417-423.

Griffiths, E. and R.G. Burns. 1972. Interaction between phenolic substances and microbial polysaccharides in soil aggregation. *Plant Soil* 36: 599-612.

Gupta, R.K., D.K. Bhumbla, and I.P. Abrol. 1984. Effect of sodicity, pH, organic matter and calcium carbonate on the dispersion behaviour of soils. *Soil Sci.* 137:245-251.

Hadas, A. 1987. Long-term tillage practice effects on soil aggregation modes and strength. *Soil Sci. Soc. Am. J.* 51:191-197.

Hamblin, A.P. 1977. Structural features of aggregates in some East Anglian silt soils. *J. Soil Sci.* 28:23-28.

Hamblin, A.P. and D.B. Davies. 1977. Influence of organic matter on the physical properties of some East Anglian soils of high silt content. *J. Soil Sci.* 28:11-22.

Hamblin, A.P. and D.J. Greenland. 1977. Effect of organic constituents and complexed metal ions in aggregate stability of some East Anglian soils. *J. Soil Sci.* 28:410-416.

Harris, R.F., G. Chesters, and O.N. Allen. 1966. Dynamics of soil aggregation. *Adv. Agron.* 18:107-169.

Hassink, J., L.A. Bouman, K.B. Zwart, J. Bloem, and L. Brussaard. 1993. Relationships between soil texture, physical protection of organic matter, soil biota, and C and N mineralisation in grassland soils. *Geoderma* 57:105-128.

Hattori, T. 1973. *Microbial Life in the Soil*. Marcel Dekker, New York.

Hepper, C.C. 1975. Extracellular polysaccharides of soil bacteria. p. 93-110. In: N. Walker (ed.), *Soil Microbiology*. Butterworths, London.

Hindell, R.P., B.M. McKenzie, J.M. Tisdall, and M.J. Silvapulle. 1994. Relationships between casts of geophagous earthworms (Lumbricidae, Oligochaeta) and matric potential. II. Clay dispersion from casts. *Biol. Fert. Soils* 18:127-131.

Jenny, H. and K.A. Grossenbacher. 1963. Root-soil boundary zones as seen in the electron microscope. *Soil Sci. Soc. Am. J.* 27:273-277.

Jocteur Monrozier, L., J.N. Ladd, A.W. Fitzpatrick, R.C. Foster, and M. Raupach. 1991. Components and microbial biomass content of size fractions in soils of contrasting aggregation. *Geoderma* 49:37-62.

Juma, N.G. 1993. Interrelationships between soil structure/texture, soil biota/soil organic matter and crop production. *Geoderma* 57:3-30.

Kemper, W.D. and E.J. Koch. 1966. Aggregate stability of soils from Western United States and Canada. United States Department of Agriculture Technical Bulletin 1355.

Kemper, W.D., R.C. Rosenau, and A.R. Dexter. 1987. Cohesion development in disrupted soils as affected by clay and organic matter content and temperature. *Soil Sci. Soc. Am. J.* 51:860-867.

Koske, R.E., J.C. Sutton, and B.R. Sheppard. 1975. Ecology of *Endogone* in Lake Huron sand dunes. *Can. J. Bot.* 53:87-93.

Ladd, J.N., M. Amato, and J.W. Parsons. 1977. Studies of nitrogen immobilization and mineralisation in calcareous soil. III. Concentration and distribution of nitrogen derived from the soil biomass. Vol. 1, p. 301-312. *Soil Organic Matter Studies*. Vienna, IAEA.

Ladd, J.N., R.C. Foster, and J.O. Skjemstad. 1993. Soil structure: carbon and nitrogen metabolism. *Geoderma* 56:401-434.

Lee, K.E. 1985. *Earthworms - Their Ecology and Relationships with Soils and Land Use*. Academic Press, Sydney.

Lee, K.E. and R.C. Foster. 1991. Soil fauna and soil structure. *Aust. J. Soil Res.* 29:745-775.

Low, A.J. 1955. Improvements in the structural state of soils under leys. *J. Soil Sci.* 6:179-199.

Low, A.J. and P.R. Stuart. 1974. Micro-structural differences between arable and old grassland soils as shown in the scanning electron microscope. *J. Soil Sci.* 25:135-137.

Lynch, J.M. and E. Bragg. 1985. Microorganisms and soil aggregate stability. *Adv. Soil Sci.* 2:133-171.

Lynch, J.M. and L.F. Elliott. 1983. Aggregate stabilisation of volcanic ash and soil during microbial degradation of straw. *Appl. Environ. Microbiol.* 45:1398-1401.

McGill, W.B., K.R. Cannon, J.A. Robertson, and F.D. Cook. 1986. Dynamics of soil microbial biomass and water-soluble organic C in Breton L after 50 years of cropping to two rotations. *Can. J. Soil Sci.* 66:1-19.

McGill, W.B. and E.A. Paul. 1976. Fractionation of soil and ^{15}N nitrogen turnover to separate the organic and clay interactions of immobilized N. *Can. J. Soil Sci.* 56:203-212.

McKenzie, B.M. and A.R. Dexter. 1987. Physical properties of casts of the earthworm *Aporrectodea rosea*. *Biol. Fert. Soils* 5:152-157.

Marinissen, J.C.Y. and A.R. Dexter. 1990. Mechanisms of stabilisation of earthworm casts and artificial casts. *Biol. Fert. Soils* 9:163-167.

Marshall, K.C. 1976. *Interfaces in Microbial Ecology*. Harvard University Press, Cambridge.

Martin, J.P. 1945. Microorganisms and soil aggregation: I. Origin and nature of some of the aggregating substances. *Soil Sci.* 59:163-174.

Martin, J.P. 1971. Decomposition and binding action of polysaccharides in soil. *Soil Biol. Biochem.* 3:33-41.

Martin, J.P., A.A. Parsa, and K. Haider. 1978. The influence of intimate association with humic polymers on biodegradation of [14C]-labelled organic substrates. *Soil Biol. Biochem.*

Materechera, S.A., A.R. Dexter, and A.M. Alston. 1992. Formation of aggregates by plant roots in homogenised soils. *Plant Soil* 142:69-79.

Mench, M., J.L. Morel, and A. Guckert. 1987. Metal binding properties of high molecular weight soluble exudates from maize (*Zea mays* L.) roots. *Biol. Fert. Soils* 3:165-169.

Miller, R.M. and J.D. Jastrow. 1990. Hierarchy of root and mycorrhizal fungal interactions with soil aggregation. *Soil Biol. Biochem.* 22:579-584.

Miller, R.M. and J.D. Jastrow. 1992. The role of mycorrhizal fungi in soil conservation. p. 29-44. In: G.J. Bethlenfalvay and R.G. Linderman (eds.) *Mycorrhizae in Sustainable Agriculture*. ASA Special Publication No. 54.

Monroe, C.D. and E.K. Kladivko. 1987. Aggregate stability of a silt loam as affected by roots of corn, soybeans and wheat. *Commun. Soil Sci. Pl. Anal.* 18:1077-1087.

Muneer, M. and J.M. Oades. 1989a. The role of Ca-organic interactions in soil aggregate stability. I. Laboratory studies with ^{14}C-glucose, $CaCO_3$ and $CaS0_4.2H_20$. *Aust. J. Soil Res.* 27:389-399.

Muneer, M. and J.M. Oades. 1989b. The role of Ca-organic interactions in soil aggregate stability. II. Field studies with ^{14}C-glucose, $CaCO_3$ and $CaS0_4.2H_20$. *Aust. J. Soil Res.* 27:401-409.

Muneer, M. and J.M. Oades. 1989c. The role of Ca-organic interactions in soil aggregate stability. III. Mechanisms and models. *Aust. J. Soil Res.* 27:411-423.

Oades, J.M. 1972. Studies on soil polysaccharides. III. Composition of polysaccharides in some Australian soils. *Aust. J. Soil Res.* 10:113-126.

Oades, J.M. 1984. Soil organic matter and structural stability: mechanisms and implications for management. *Plant Soil* 76:319-337.

Oades, J.M. 1988. The retention of organic matter in soils. *Biogeochemistry* 5:35-70.

Oades, J.M. 1989. An introduction to organic matter in mineral soils. p. 89-159. In: J.B. Dixon and S.D. Weed (eds.), *Minerals in Soil Environments*. SSSA Book Series, no. 1.

Oades, J.M. 1993. The role of biology in the formation, stabilisation and degradation of soil structure. *Geoderma* 56:377-400.

Oades, J.M., A.M. Vassallo, A.G. Waters, and M.A. Wilson. 1987. Characterization of organic matter in particle size and density fractions from a red-brown earth by solid-state ^{13}C N.M.R. *Aust. J. Soil Res.* 25:71-82.

Oades, J.M. and A.G. Waters. 1991. Aggregate hierarchy in soils. *Aust. J. Soil Res.* 29:815-828.

Parle, J.N. 1963. A microbiological study of earthworm casts. *J. Gen. Microbiol.* 31:13-22.

Perfect, E., B.D. Kay, W.K.P. van Loon, R.W. Sheard, and T. Pojasok. 1990. Rates of change in structural stability under forages and corn. *Soil Sci. Soc. Am. J.* 54:179-186.

Pojasok, T. and B.D. Kay. 1990. Effect of root exudates from corn and bromegrass on soil structural stability. *Can. J. Soil Sci.* 70:351-362.

Quirk, J.P. 1978. Some physico-chemical aspects of soil structural stability - A review. p. 3-16. In: W.W. Emerson, R.D. Bond and A.R. Dexter (eds.), *Modification of Soil Structure*. Wiley, New York.

Quirk, J.P. and R.S. Murray. 1991. Towards a model for soil structural behaviour. *Aust J. Soil Res.* 29:829-867.

Quirk, J.P. and C.R. Panabokke. 1962. Incipient failure of soil aggregates. *J. Soil Sci.* 13:60-70.

Quirk, J.P. and B.G. Williams. 1974. The disposition of organic materials in relation to stable aggregation. Proc. 10th Conf. Int Soc. Soil Sci. Moscow.

Rasmussen, P.E., R. Allmaras, C.R. Rohde, and N.C. Roager Jr. 1980. Crop residue influences on soil carbon and nitrogen in a wheat-fallow system. *Soil Sci. Soc. Am. J.* 44:596-600.

Reid, J.B. and M.J. Goss. 1981. Effect of living roots of different plant species on the aggregate stability of two arable soils. *J. Soil Sci.* 32:521-541.

Reid, J.B., M.J. Goss, and P.D. Robertson. 1982. Relationship between the decreases in soil stability as affected by the growth of maize roots and changes in organically bound iron and aluminium. *J. Soil Sci.* 33:397-410.

Rengasamy, P., R.S.B. Greene, and G.W. Ford. 1984. The role of the clay fraction in the particle arrangement and stability of soil aggregates - a review. *Clay Res.* 3:53-67.

Rengasamy, P. and K.A. Olsson. 1991. Sodicity and soil structure. *Aust. J. Soil Res.* 29:935-952.

Rennie, D.A., E. Truog, and O.N. Allen. 1954. Soil aggregation as influenced by microbial gums, level of fertility and kind of crop. *Soil Sci. Soc. Am. Proc.* 18:399-403.

Rimmer, D. and D.J. Greenland. 1976. Effects of calcium carbonate on the swelling behaviour of a soil clay. *J. Soil Sci.* 27:129-139.

Robert, M. and C. Chenu. 1992. Interactions between soil minerals and microorganisms. *Soil Biochem.* 7:307-404.

Rovira, A.D. and E.L. Greacen. 1957. The effect of aggregate disruption on the activity of microorganisms in soil. *Aust. J. Agric. Res.* 8:659-673.

Saini, G.R. and A.A. McLean. 1966. Adsorption-flocculation reactions of soil polysaccharides with kaolinite. *Soil Sci. Soc. Am. J.* 30: 697-699.

Semmel, H., R. Horn, U., Hell, A.R. Dexter, and E.D. Schultze. 1990. The dynamics of aggregate formation and the effects on soil physical properties. *Soil Tech.* 3:113-129.

Shields, L.M. and L.W. Durrell. 1964. Algae in relation to soil fertility. *Bot. Rev.* 30:92-128.

Shipitalo, M.J. and R. Protz. 1988. Factors influencing the dispersibility of clay in worm casts. *Soil Sci. Soc. Am. J.* 52:764-769.

Sills, I.D., L.A.G. Aylmore, and J.P. Quirk. 1974. Relationships between pore size distributions and physical properties of clay soils. *Aust. J. Soil Res.* 12:107-118.

Skinner, F.A. 1979. Rothamsted studies of soil structure VII. The effects of incubation on soil aggregate stability. *J. Soil Sci.* 30:473-481.

Sørensen, L.H. 1967. Duration of amino acid metabolites formed in soils during decomposition of carbohydrates. *Soil Sci.* 104:234-241.

Sørensen, L.H. and E.A. Paul. 1971. Transformations of acetate carbon into carbohydrate and amino acid metabolites during decomposition in soil. *Soil Biol. Biochem.* 3:173-180.

Sparling, G.P. and M.V. Cheshire. 1985. Effect of periodate oxidation on the polysaccharide content and microaggregate stability of rhizosphere and non rhizosphere soils. *Plant Soil* 88:113-122.

Sposito, G. 1989. *The chemistry of soils.* Oxford University Press, NY.

Stefanson, R.C. 1971. Effect of periodate and pyrophosphate on the seasonal changes in aggregates stabilisation. *Aust. J. Soil Res.* 9:33-41.

Stengel, P., J.T. Douglas, J. Guérif, M.J. Goss, G. Monnier, and R.Q. Cannell. 1984. Factors influencing the variation of some properties of soils in relation to their suitability for direct-drilling. *Soil Tillage Res.* 4:35-53.

St. John, T.V., D.C. Coleman, and C.P.P. Reid. 1983. Association of vesicular-arbuscular mycorrhizal hyphae with soil organic particles. *Ecology* 64:957-959.

Stotzky, G. 1986. Influence of soil mineral colloids on metabolic processes, growth, adhesion, and ecology of microbes and viruses. p. 305-428. In: *Interactions of Soil Minerals with Natural Organics and Microbes.* SSSA Spec. Publ. No. 17.

Swaby, R.J. 1949. The relationship between micro-organisms and soil aggregation. *J. Gen. Microbiol.* 3:236-254.

Swincer, G.D., J.M. Oades, and D.J. Greenland. 1968. Studies on soil polysaccharides. II. The composition and properties of polysaccharides under pasture and under a fallow-wheat rotation. *Aust. J. Soil Res.* 6: 225-235.

Tate, K.R. and G.J. Churchman. 1978. Organo-mineral fractions of a climosequence of soils in New Zealand tussock grasslands. *J. Soil Sci.* 29:331-339.

Thomas, R.S., S. Dakessian, R.N. Ames, M.S. Brown, and G.J. Bethlenfalvay. 1986. Aggregation of a silty clay loam soil by mycorrhizal onion roots. *Soil Sci. Soc. Am. J.* 50:1494-1499.

Thomas, R.S., R.L. Franson, and G.J. Bethlenfalvay. 1993. Separation of vesicular-arbuscular mycorrhizal fungus and root effects on soil aggregation. *Soil Sci. Soc. Am. J.* 57:77-81.

Tiedje, J.M., A.J. Sextone, T.B. Parkin, N.P. Revsbech, and D.R. Shelton. 1984. Anaerobic processes in soil. *Plant Soil* 76:197-212.

Tiessen, H.J. and J.W.B. Stewart. 1983. Particle size fractions and their use in studies of soil organic matter. II. Cultivation effects on organic matter composition in size fractions. *Soil Sci. Soc. Am. J.* 47:509-514.

Tisdall, J.M. 1991. Fungal hyphae and structural stability of soil. *Aust. J. Soil Res.* 29:729-743.

Tisdall, J.M., B. Cockroft, and N.C. Uren. 1978. The stability of soil aggregates as affected by organic materials, microbial activity and physical disruption. *Aust. J. Soil Res.* 16:9-17.

Tisdall, J.M. and J.M. Oades. 1979. Stabilisation of soil aggregates by the root systems of ryegrass. *J. Soil Res.* 17:429-441.

Tisdall, J.M. and J.M. Oades. 1980. The effect of crop rotation on aggregation in a red-brown earth. *Aust. J. Soil Res.* 18:423-433.

Tisdall, J.M. and J.M. Oades. 1982. Organic matter and water-stable aggregates in soil. *J. Soil Sci.* 33:141-163.

Turchenek, L.W. and J.M. Oades. 1978. Organo-mineral particles in soils. p. 137-144. In: W.W. Emerson, R.D. Bond and A.R. Dexter (eds.), *Modification of Soil Structure*. Wiley, New York.

Turchenek, L.W. and J.M. Oades. 1979. Fractionation of organo-mineral complexes by sedimentation and density techniques. *Geoderma* 21:311-343.

Utomo, W.H. and A.R. Dexter. 1981. Effect of ageing on compression resistance and water stability of soil aggregates disturbed by tillage. *Soil Tillage Res.* 1:127-137.

Van Veen, J.A. and P.J. Kuikman. 1990. Soil structural aspects of decomposition of organic matter by microorganisms. *Biogeochemistry* 11:213-223.

Van Veen, J.A., J.N. Ladd, and M. Amato. 1985. The turnover of carbon and nitrogen through the microbial biomass in a sandy-loam and a clay soil incubated with [^{14}C (U)] glucose and [^{15}N] $(NH_4)_2SO_4$ under different moisture regimes. *Soil Biol. Biochem.* 17:747-756.

Yeoh, N.S. and J.M. Oades. 1981. Properties of clays and soils after acid treatment. *Aust. J. Soil Res.* 19:159-166.

Carbon in Primary and Secondary Organomineral Complexes

B.T. Christensen

I. Introduction

A. Soil Organic Matter Turnover

Advancing our knowledge on soil organic matter (SOM) structure and function is a prerequisite for a sustainable exploitation of inherent soil fertility and for assessing the impact of different land use strategies on soil carbon storage. The fate of mineral fertilizers, animal manure, crop residues and other organic wastes applied to agricultural soils are also intimately linked with SOM turnover. Further, SOM composition and dynamics are of significance to potentials for soil erosion, nutrient leaching, fluxes of trace gases and to the CO_2 balance of the atmosphere.

1-56670-033-7/96/$0.00+$.50

The increase in CO_2 and other radiatively active gases in the atmosphere is forecasted to induce a complex suite of changes in global climate, including a changed precipitation pattern and global warming. CO_2 is responsible for 50-60% of the increase in radiative forcing. Climate changes affect vegetation zones and agricultural production potentials, but also the exchange of CO_2 between soil and the atmosphere. Jenkinson et al. (1991) calculated that a global warming of 0.03°C yr^{-1} may cause an additional CO_2 release from soil organic matter of 61 GtC over a 60 year period, corresponding to one fifth of the CO_2 that would be released during the same period if fossil fuels were consumed at present rates. Terrestrial vegetation and soil organic matter show however significant spatial and temporal variability making it very difficult to assess changes in carbon storage.

Climate changes influence rates of carbon loss and build-up through changes in soil temperature, moisture and rhizodeposition. Inputs of plant residues to the soil may change in quantity and quality and the decomposition rate of inputs and soil carbon may be affected. Climate induced changes in total soil carbon levels will be manifest only over extended time spans. The carbon involved in relatively rapid turnover accounts however for a smaller part of the total carbon pool and may respond more quickly to changes in land use, atmospheric CO_2 and climate. Consequently the active soil organic matter pool is in focus when assessing changes in the net carbon flux between soil and atmosphere. Using modelling to up-scale results obtained over shorter time spans, changes in the turnover of the active soil organic matter pool can be invoked as an indicator for the direction of long-term changes in total carbon sequestration in soils.

Soil organic matter encompasses plant, animal and microbial residues in all stages of decay and a diversity of heterogenous organic substances intimately associated with inorganic soil components. A number of biotic and abiotic factors exert control over SOM turnover. Their effect appears to be most manifest in the early decomposition phases and is probably decisive for the size and turnover rate of the active SOM pool. The textural composition of soil and soil structure affect the decomposer micro-environment, the protection of decomposer organisms from faunal grazing, and the stabilization of substrates and decomposition products.

The interactions between soil inorganics, microbial activity and metabolic products, and SOM components have been reviewed in detail (e.g. Huang and Schnitzer, 1986; Oades, 1989; Coleman et al., 1989). More recent reviews have focused on aspects of substrate decomposition and SOM turnover in relation to soil structure (van Veen and Kuikman, 1990; Tisdall, 1991; Hassink et al., 1993; Ladd et al., 1993; Oades, 1993). Mechanisms responsible for stabilization of SOM may be divided into:

(i) chemical recalcitrance of the organic matter, the mechanisms being enzymatic, and also free radical and other non-enzymatic chemical reactions, e.g. catalyzed by abiotic factors. Thereby aliphatic, cyclic and aromatic moieties of varying molecular weight and derived from residual substrates and decomposition products may be incorporated into heteropolyaromatic structures. Recalcitrance may also be due to inherent chemical characteristics of substrates (e.g. aromatic plant polymers, fungal melanins).

(ii) chemical stabilization of otherwise decomposable compounds by chemical interaction of substrates with the mineral part of the soil, e.g. surface reactions with clay minerals and amorphous sesquioxides and intercalation between clay mineral sheets.

(iii) physical protection of otherwise decomposable substrates within micro- or macroaggregates by physical barriers created between substrates and decomposers. Similarly, physical barriers may protect microorganisms against faunal grazing.

It is important to recognize that the various stabilizing mechanisms operate simultaneously and affect substrates and decomposition products in all stages of decay.

B. Organomineral Complexes

This chapter focuses on carbon storage in primary and secondary organomineral complexes, including the distribution, chemical composition and decomposability of organic carbon associated with differently sized separates.

Primary organomineral complexes refer to the primary structure of soils as defined by the soil texture. Thus primary organomineral complexes are isolated after complete dispersion of soils and the distribution of organomineral complexes into clay-, silt- and sand-sized separates matches the corresponding size distribution obtained by a conventional textural analysis. Clay sized (<2 μm) primary organomineral complexes will however be a mixture of "true" primary particles and sub-microaggregates of finer clay mineral particles held together by organic and inorganic "cements" and electrostatic forces (Emerson et al., 1986).

Secondary organomineral complexes refers to the secondary structure of soils, which is defined by the degree of aggregation of primary organomineral complexes. Consequently secondary complexes are isolated by methods involving limited soil dispersion. Compared to secondary organomineral complexes, the size distribution of primary complexes represents a static property of a given soil. The distribution of secondary complexes is significantly affected by biotic and abiotic factors and soil management. In addition, soil sampling and fractionation procedures are often decisive for the size distribution of secondary organomineral complexes (Elliott and Cambardella, 1991; Jastrow and Miller, 1991; Beare and Bruce, 1993; Haynes, 1993), and therefore also for the interpretation of results acquired in subsequent analyses of the various aggregate classes. Needless to say, it is essential to recognize the dynamic nature of secondary organomineral complexes when assessing their potential for storing soil carbon. For the sake of convenience, secondary organomineral complexes will also be referred to as aggregates in accordance with common terminology.

To complete the hierarchical terminology adopted here, the tertiary structure of soils is defined as the soil matrix *in situ*. The tertiary soil structure thus integrates the effect of primary and secondary organomineral complexes, and on this hierarchical level, soil clods, macropores (voids and cracks), biological hot spots, mesofaunal and plant root activity, soil tillage and compaction become important phenomena for soil carbon turnover.

Studies of organomineral complexes rely on physical fractionations thereby emphasizing the role of soil structure in SOM turnover and stabilization. Physical fractionation techniques are considered chemically less destructive, and the results acquired from analyses of the soil fractions are expected to relate more directly to SOM *in situ*.

II. Primary Organomineral Size Separates

A. Fractionation

Separation of soil into primary organomineral complexes of different size is based on the concept that SOM associated with mineral particles of different size differs in structure and function.

A complete dispersion is an essential prerequisite for isolation of primary organomineral complexes. Separates obtained after limited dispersion may not represent the soil fraction which the results are intended to refer to. The recovered fraction will contain only the most easily dispersed part of the entire size fraction, most likely causing a selective isolation of SOM. Further, the recovered size fraction may consist of an unknown mixture of primary organomineral complexes and aggregates of the same size but made up of particles belonging to smaller size classes.

Conventional dispersion procedures in which samples are exposed to chemical pretreatments in order to remove SOM, sesquioxides and carbonates and other cementing agents before being dispersed by a combination of chemical and physical means are not feasible for isolation of intact organomineral complexes. In conflict with the concept of intact organo-mineral complexes, chemically assisted dispersion may introduce unintended in-process changes of SOM structure and distribution. The introduction of chemicals in various soil dispersion procedures should be avoided unless the action of the chemical is specific and well documented. Accordingly most studies on primary organo-mineral complexes have relied on ultrasonic vibrations in water to accomplish soil dispersion. Alternative procedures (e.g. shaking and resins) have been employed; and a number of studies have combined shaking and ultrasonic treatments. Procedures used for isolation of primary organomineral complexes have been discussed in detail by Christensen (1992). Other recent papers on this subject are Balesdent et al. (1991), Cambardella and Elliott (1993a), Elliott and

Cambardella (1991), Feller et al. (1991), Morra et al. (1991) and Raine and So (1993).

Size fractionation may be applied to whole soils and aggregate fractions, and to heavy fractions following density separation. Samples are usually dispersed in water, the fraction $>63\ \mu m$ being recovered and further separated by sieving. Primary complexes $<63\ \mu m$ are generally isolated by gravity sedimentation in water. Extended sedimentation times and effects of Brownian movements make gravity sedimentation less feasible for differentiation of clay sized particles. Centrifugation greatly speeds up processing time and this may be advantageous in order to reduce the possible effects of microbial activities in soil suspensions. The size limits for primary particles refer to equivalent spherical diameter (ESD), signifying the diameter of a spherical particle with a similar density and settling velocity as the actual particle under analysis. Stokes Law is applied when isolating size separates by sedimentation. However, most assumptions underlying Stokes Law are violated to some extent when applied to sedimenting soil particles (Christensen, 1992), and the isolation of clay and silt sized organomineral complexes is by no means theoretically straightforward. ESD is a practical but crude approximation, and it appears difficult to justify attempts to isolate several fractions over a narrow range of particle sizes.

The size limits of primary organomineral complexes basically follow the USDA or the ISSS classification schemes (Gee and Bauder, 1986). Clay is particles $<2\ \mu m$ in ESD, while silt is 2 to 20 μm in the ISSS and 2 to 50 μm in the USDA classification. Sand falls in the range 20 to 2000 μm (ISSS) or 50 to 2000 μm (USDA). Although most studies specify the adopted size limits, the use of non-equivalent limits in different studies reduces the comparability of results. Discussions of organomineral size separates regularly fail to realize the interpretative restrictions attached to different size limits. Often reference is made to clay and silt in general, disregarding that silt in one study may include material, which in another study is isolated along with the sand size separates. While clay separates are dominated by mineralogical clay-type minerals, the silt may be considered a transitory fraction containing clay-type minerals, mica and various proportions of the less weathered minerals that normally dominate the sand separate (quartz and feldspars). The size limits chosen for silt is therefore critical, especially since a large proportion of whole soil SOM is associated with silt sized organomineral complexes.

Generally no precautions are taken against microbial activity during the extended processing time involved in gravity sedimentation of clay sized separates. Although there are studies indicating metabolic activity during fractionations (Cheshire and Mundie, 1981; Baldock et al., 1990) this point has not been examined in detail. It may be envisaged that the microbial population present initially is relatively small due to the kill-off of microorganisms during ultrasonic dispersion. Moreover, the microbial population is rapidly diluted by the successive sedimentation-decanting cycles which also removes clay particles and soluble carbon and nutrients. The lack of more detailed information may warrant the addition of microbial inhibitors to soil suspensions. Baldock et al. (1990)

proposed the use of sodium azide or mercuric chloride. Alternatively, isolation of clay by centrifugation should be favoured.

Solubilized SOM may be included in the finest size separate or may be isolated separately from the supernatant left after centrifugation of flocculated clay suspensions. The means of concentrating and drying the size separates should be adjusted to the type of analyses subsequently imposed on the sample. When exposed to biological tests, size separates should be inoculated because an unknown but probably large proportion of the soil microorganisms is killed during ultrasonic dispersion. Excessive concentrations of salts added to flocculate clay and residual contents of microbial inhibitors may also interfere with the outcome of biologically founded tests. Temperatures used to dry size separates should be kept below 40°C in order to minimize changes in biomacromolecules that might affect SOM decomposability.

To sum up, a complete soil dispersion is essential for isolation of primary organomineral complexes. Chemically assisted dispersions are generally not feasible, and most studies have relied on ultrasonic vibrations in water. Size limits for organomineral complexes most often follow the USDA or the ISSS classification schemes, size limits referring to equivalent spherical diameters (ESD). The ESD approximation does not warrant the isolation of several size classes over a narrow range of diameters. If size separates are exposed to biological test after fractionation, drying temperatures should be kept below 40°C, excessive concentrations of flocculants and microbial inhibitors should be removed and incubated samples inoculated.

B. Carbon Distribution

The degree of soil dispersion is decisive for the resulting distribution of carbon among primary organomineral size separates. Gregorich et al. (1988) examined the distribution of carbon across size separates after application of step-wise increased levels of ultrasonic dispersion. When soil was dispersed completely (1500 J ml^{-1} soil suspension), clay (<2 μm), silt (2 to 50 μm) and sand (50 to 2000 μm) contained 60, 30 and 5% of the whole soil carbon, respectively, whereas limited dispersion (e.g. 100 J ml^{-1}) allocated 28, 54 and 13% to these separates. Changes in the distribution of carbon between size separates resulted from changes in fraction weight and carbon content (Figure 1). Similar observations were reported by Balesdent et al. (1991). Using relatively low energy levels, Morra et al. (1991) found a corresponding trend for coarse clay (0.2 to 2 μm) whereas medium clay (0.08 to 0.2 μm) in their study decreased in carbon with increasing sonication energy.

To compare carbon distributions across clay and silt isolated from different soils, a carbon enrichment factor (E_C= mg C g^{-1} separate/mg C g^{-1} whole soil) was calculated for each separate (Table 1). The use of enrichment factors excludes effects of different carbon levels in whole soils. The highest concentrations of carbon were found in fine silt and in clay sized separates. Fine silt was

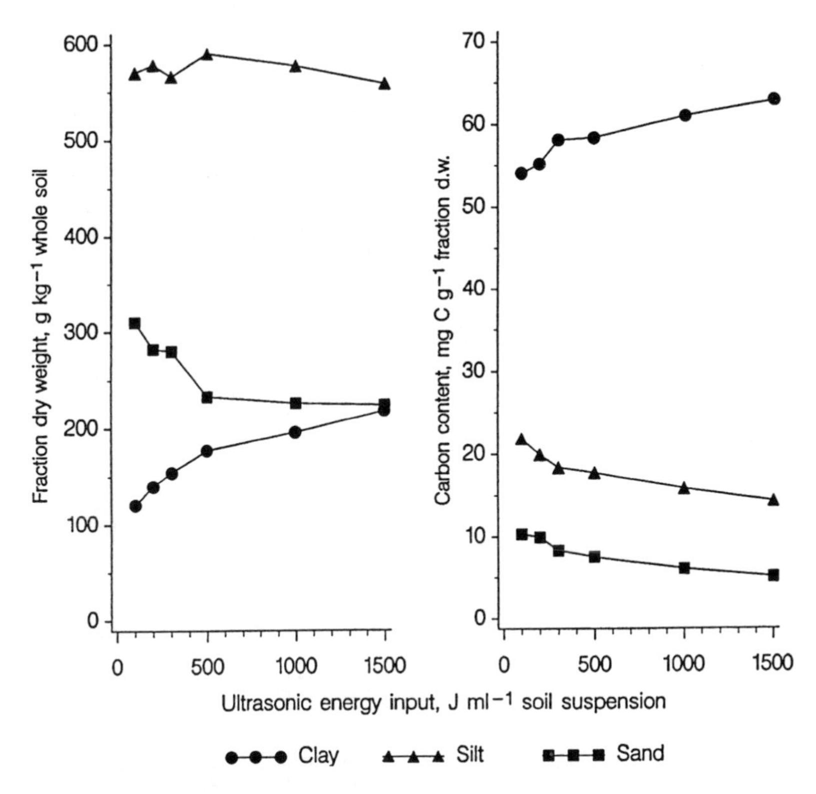

Figure 1. Distribution of clay (<2 μm), silt (2 to 50 μm) and sand (50 to 2000 μm) dry weight and carbon content after dispersion of 1 to 2 mm aggregates at different ultrasonic energy levels. Aggregates were from the 0 to 15 cm layer of a cultivated Aquic Eutrochrept (Gleyed Melanic Brunisol). (Data from Gregorich et al., 1988.)

1.3 to 4.0 times higher in carbon than whole soils, whereas carbon enrichments ranged much wider for clay. Other studies demonstrate even greater E_C values, especially for soils low in clay and silt (Leinweber and Reuter, 1988; Elustondo et al., 1990). Excluding soils high in macro-OM, sand sized separates usually show E_C <0.1. In Mollisols, carbon concentrations often peak in fine silt while in Alfisols the clay sized separates generally show the highest contents. Differences between soils do not appear to be related to whole soil pH or carbon content. Since incomplete disaggregation is indicated for some soils, differences observed between soils may reflect both inherent soil type characteristics and the ease with which different soils are dispersed when exposed to ultrasonic vibrations.

Table 1. Carbon enrichment (% C in separate/% C in whole soil) in different silt and clay size classes isolated by ultrasonic dispersion from predominantly cultivated soils. The separate showing the highest enrichment within a soil is underlined.

					Carbon enrichment								
					Silt (μm)						Clay (μm)		
Ref.[a]	USDA Soil Taxonomy	Location (sampling depth)	% C	pH	50-20	50-5	50-2	20-5	20-2	5-2	2-0.2	<2	<0.2
1	Typic Argiudoll	Iowa, USA (0-25 cm)	2.18	5.1	0.2				1.1		2.2		2.3
2	Typic Argiboroll	Sask., Canada, Ap-horizon	1.71	nd		0.6				3.8	2.0		1.6
2	Udic Haploboroll	Sask., Canada, Ap-horizon	3.32	nd		0.7				3.9	2.5		2.0
3	Typic Cryoboroll	Canada, A-horizon	2.41	nd[b]		0.6				2.5	2.2		
	1.3												
4	Typic Haploboroll	Sask., Canada, Ap-horizon	2.50	5.6	0.1			1.1		2.4	2.6		2.1
4	Boralfic Cryoboroll	Sask., Canada, Ap-horizon	4.60	5.0	0.2			1.0		1.6	1.4		1.4
4	Aridic Natriboroll	Sask., Canada, Ap-horizon	2.10	5.3	0.1			1.2		2.9	2.8		2.4
5	Haploboroll	Sask., Canada (0-30 cm)	6.29	6.2	0.3			1.0		1.5	1.1		1.0
5	Calciaquoll	Australia (1-10 cm)	5.35	8.0	0.2			1.2		1.3	0.9		0.5
1	Typic Hapludalf	Iowa, USA (0-20 cm)	1.51	5.4	0.2				1.3		2.8		3.8
6	Udollic Ochraqualf	Missouri, USA (0-10 cm)	1.13	6.1	0.1				0.6		2.9		2.6
5	Vertic Rhodoxeralf	Australia (0-10 cm)	1.04	6.0	0.2			0.8		1.9	2.6		3.0
7	Typic Hapludalf	Denmark (0-20 cm)	1.63	6.8	0.1				3.0			5.1	
3	Vertic Cryorthent	Canada, A-horizon	2.44	nd		0.9				1.5	1.3		0.6
7	Mollic Fluvaquent	Denmark (0-20 cm)	1.80	6.8	0.1				2.0			3.3	
8	Typic Eutrochrept	Bavaria, Germany, Ap-horizon	2.46	6.4					1.0			2.6	
9	Aquic Entrochrept	Canada (0-15 cm)	2.31	6.8			0.6					2.7	
10	Chromic Pelloxererts	South Australia (0-10 cm)	1.14	8.2			2.1				1.6		1.0
11	Typic Haplorthox	Sao Paulo, Brazil (0-10 cm)	1.41	nd	0.7				2.2			1.5	
7	Orthic Haplohumod	Denmark (0-20 cm)	1.73	5.6	0.3				7.0			11.1	

[a] Ref.: 1. Zhang et al. (1988); 2. Anderson et al. (1981); 3. Tiessen and Stewart (1983); 4. Curtin et al. (1987); 5. Turchenek and Oades (1979); 6. Balesdent et al. (1988); 7. Christensen (1985); 8. Guggenberger et al. (1994); 9. Gregorich et al. (1988); 10. Amato and Ladd (1980); 11. Bonde et al. (1992).

[b] nd: not determined.

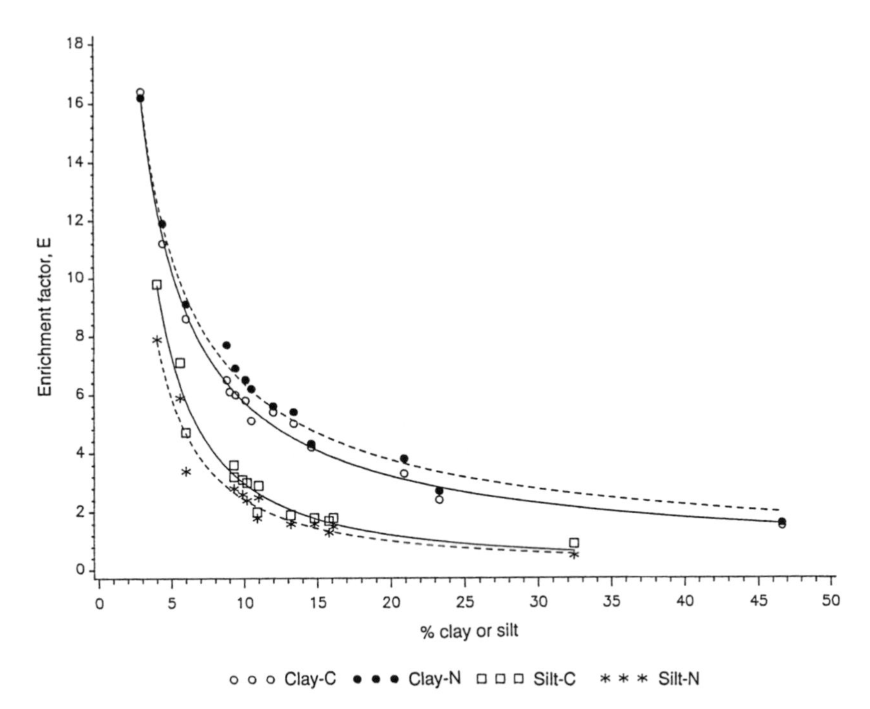

Figure 2. Relationship between fraction size and carbon (E_C) and nitrogen (E_N) enrichment factors (% in fraction to % in whole soil) for clay (<2 μm) and silt (2 to 20 μm) isolated from a range of Danish arable soils. Clay $E_C = 39.7\ x^{-0.84}$ ($r^2 = 0.99$); Clay $E_N = 36.6\ x^{-0.76}$ ($r^2 = 0.99$); Silt $E_C = 59.0\ x^{-1.30}$ ($r^2 = 0.97$); and Silt $E_N = 46.0\ x^{-1.28}$ ($r^2 = 0.95$). (Data from Christensen, 1985, 1986, 1987, 1988; Christensen and Sørensen, 1985, 1986; Christensen and Bech-Andersen, 1989; Christensen and Christensen, 1991; and Christensen et al., 1989.)

The carbon enrichment of clay and silt has been found to be inversely related to the proportion of these size separates in whole soil (Christensen, 1985; Christensen and Sørensen, 1985; Leinweber and Reuter, 1988; Gregorich, 1989), whereas carbon enrichment of sand does not relate to sand yields. Leinweber (1988) and Elustondo et al. (1990) observed that E_C in clay and silt sized organomineral complexes declined following a power function. Figure 2 is based on results from Danish arable soils. Clay shows higher E_C than silt for similar yields of dry weight, and clay tends to be more enriched in nitrogen than in carbon, the opposite being true for silt size separates. The differences between carbon and nitrogen enrichments of clay and silt are reflected in the C/N ratio, the clay associated SOM having a lower C/N ratio than SOM in silt.

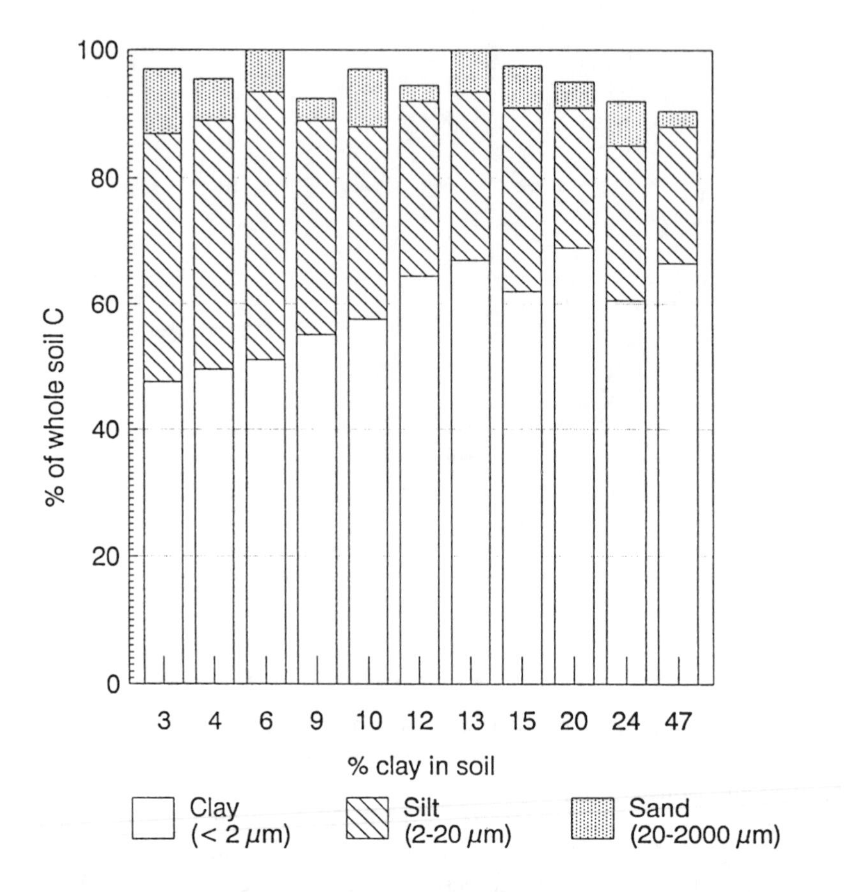

Figure 3. Distribution of whole soil carbon between clay (<2 μm), silt (2 to 20 μm), and sand (20 to 2000 μm) size separates from the Ap-horizon of Danish arable soils. (Based on data from references included in Figure 2.)

Most of the carbon in arable soils low in macro-OM is found in clay and silt, clay sized separates generally accounting for more than 50% of whole soil carbon (Tiessen and Stewart, 1983; Christensen, 1985; Curtin et al., 1987; Balesdent et al., 1988; Gregorich et al., 1988; Bonde et al., 1992). Figure 3 illustrates the relative distribution of carbon among size separates from long-term arable soil differing in clay content. Between 48 and 69% of the carbon was in the clay, while silt and sand accounted for 21 to 43% and 2 to 10%, respectively. For soils with <20% of clay, the proportion of whole soil carbon associated with clay tends to increase with increasing clay content while that in silt decreases. Balesdent et al. (1991) also found that carbon in fine (<0.2 μm) and coarse (0.2 to 2 μm) clay of arable soils increased linearly with increasing yields of these separates. The slope of the regression line for fine clay was found to be lower than that for coarse clay, but the distribution pattern did not relate to whole soil carbon content.

Microbially synthesized amino acids account for a significant proportion of the metabolites formed during decomposition and their retention in soil has been found to increase linearly with increasing soil clay+silt content. Sørensen (1975) found that the proportion of added ^{14}C labelled cellulose recovered in amino acids was 7 to 19% after 10 days and 6 to 10% after 1 year of incubation, corresponding to 9 to 30 and 27 to 56% of the labelled carbon remaining in soil, respectively. A similar pattern was found for a variety of ^{14}C labelled substrates (Sørensen, 1981, 1983). Thus with time, amino acids accounted for an increasing proportion of residual substrate derived carbon in soil, indicating microbial recycling and a higher stability of amino acid containing compounds.

Consequently SOM in size separates have been characterized by analyses of hydrolysable amino acids. Watson and Parsons (1974a) found similar proportions of 6 *N* HCl hydrolysable amino acid-N in clay and silt and the distribution of individual amino acids was generally similar for clay and silt. The suite of amino acids in SOM extracts was almost similar to that of intact clay SOM (Watson and Parsons, 1974b). These results were confirmed by Christensen and Bech-Andersen (1989). Amino acid-N made up 31 to 39% of the total nitrogen in clay and silt, while values for sand were lower (21 to 30%). Amino acid contents in clay and silt were 5 to 10 and 2 to 5 times higher than contents in whole soils, respectively, sand being 10 to 20 times lower. But the relative distribution of individual amino acids was almost the same in all separates. Annual straw incorporation increased the concentration of most amino acids and the hydrolysability of sand SOM but was not reflected in the distribution of individual amino acids. The contribution of microbial cell wall constituents to the SOM associated with different size separates was examined by analysing for diaminopimelic acid, which is confined to the cell wall peptidoglycans of procaryotes (Table 2). The clay-to-silt ratio for diaminopimelic acid was 2.6 to 3.0 in loamy sand and 4.7 to 4.8 in sandy loam soil. Although the stability of diaminopimelic acid in soil remains unknown, a higher content of diaminopimelic acid in clay may reflect a larger (past or present) bacterial population of clay relative to silt.

Carbohydrates are abundant in plants and are also produced in significant amounts during decomposition. The proportion of carbon present in carbohydrates generally increase with decreasing particle size (Table 3). Soils rich in macro-OM may show a bimodal distribution pattern, silt being lower than clay and sand size separates. Soil carbohydrates are a mixture of polysaccharides derived from plants and microorganisms. Because pentoses (e.g. arabinose and xylose) are not synthesized to any great extent by microorganisms but are quantitatively important constituents of plant residues, hexose-to-pentose ratios

Table 2. Distribution of 6 N HCl hydrolyzable diaminopimelic acid (DAPA) in whole soil, clay, silt and sand sized separates from the Ap-horizon of two arable soils

	μg DAPA g^{-1} separate		% of total amino acid-N	
Separate, μm	Loamy sand	Sandy loam	Loamy sand	Sandy loam
Clay, <2	170	115	0.6	0.9
Silt, 2-20	61	25	0.4	0.5
Sand, 20-2000	1	1	0.5	0.6
Whole soil, <2000	15	20	0.5	0.7
Mean (s.d.)			0.5 (0.1)	0.7 (0.2)

(From Christensen and Bech-Andersen, 1989.)

may be used to indicate the origin of soil carbohydrates (Turchenek and Oades, 1979). Typically the ratio between (galactose+mannose) and (arabinose+xylose) is adopted, galactose and mannose being minor constituents of plant polysaccharides. The hexose-to-pentose ratio is generally higher for clay than for coarser separates (Table 3), indicating a greater microbial input of carbohydrates to clay than to silt and sand.

Lignin is considered as one of the most resistant components of plant residues (Zeikus, 1981). Guggenberger and Christensen (1993) and Guggenberger et al. (1994) examined lignin derived phenols present in primary organomineral complexes from A-horizons of Typic Eutrochrepts. The sum of vanillyl, syringyl and cinnamyl (VSC) units following alkaline CuO oxidation was taken to indicate the lignin content, the ratio of acid-to-aldehyde of vanillyl $(ac/al)_V$ and of syringyl-to-vanillyl (S/V) illustrating the degree of microbial lignin breakdown. The sum of VSC in size separates corresponded well with the amount of VSC determined for bulk soils. VSC was most abundant in SOM associated with sand size separates, while reduced yields were obtained for silt and clay (Figure 4). The proportion of VSC in SOM and the enrichment factor for VSC (E_{vsc} = mg VSC g^{-1} C in separate/mg VSC g^{-1} C in bulk soil) declined with decreasing particle size. The $(ac/al)_V$ ratio increased with decreasing particle size indicating a progressing lignin degradation from sand over silt to clay. Increasing $(ac/al)_V$ ratios reflect increased side-chain oxidation of the phenylpropane unit. The S/V ratio invoked to describe changes in soil lignin material showed a steep decline with decreasing particle size, confirming the $(ac/al)_V$ data. Generally syringyl units are degraded preferentially to vanillyl units.

Decreasing yields of CuO oxidation products, increasing $(ac/al)_V$ ratios and decreasing S/V ratios from the sand over silt to clay are directly related to the decomposition processes of lignin. Clearly lignin degradation in clay-bound SOM is much more advanced than in silt SOM. The key reactions are oxidative

Table 3. Hexose to pentose ratio of SOM and the proportion of total carbon present as hydrolyzable carbohydrate-C in sand (2000-20 μm), silt (20-2 μm) and clay (<2 μm) isolated from Ap-horizons of different soils

Reference	Site	% Clay	Carbohydrate-C (mg C g^{-1} separate-C)			Galactose + Mannose / Arabinose + Xylose		
			Sand	Silt	Clay	Sand	Silt	Clay
Cheshire et al., 1990	Askov sand, Denmark	5.7	28	77	91	0.71	0.92	1.24
	Ll. Valby, Denmark	9.0	40	97	86	1.29	1.39	1.50
	Højer, Denmark	24.3	42	69	90	0.95	0.95	1.47
	Ribe, Denmark	45.9	-	29	101	-	1.34	1.43
Cheshire and Mundie, 1981	Insch, Scotland	13.5				0.58	1.31	1.52
	Countesswells, Scotland	11.9				0.61	1.25	1.64
[a] Angers and Mehuys, 1990	Kamouraska, Canada	47	69	51	73	0.61	1.40	1.61
[b] Angers and N'Dayegamiye, 1991	Neubois, Canada	12	28	49	56	0.47	1.67	1.96
Guggenberger and Christensen, 1993	Bavaria, Germany	25	76	44	80	0.40	0.70	1.15

[a] Sand was 6000-50 μm and silt 50-2 μm; [b] sand was 2000-50 μm and silt 50-2 μm.

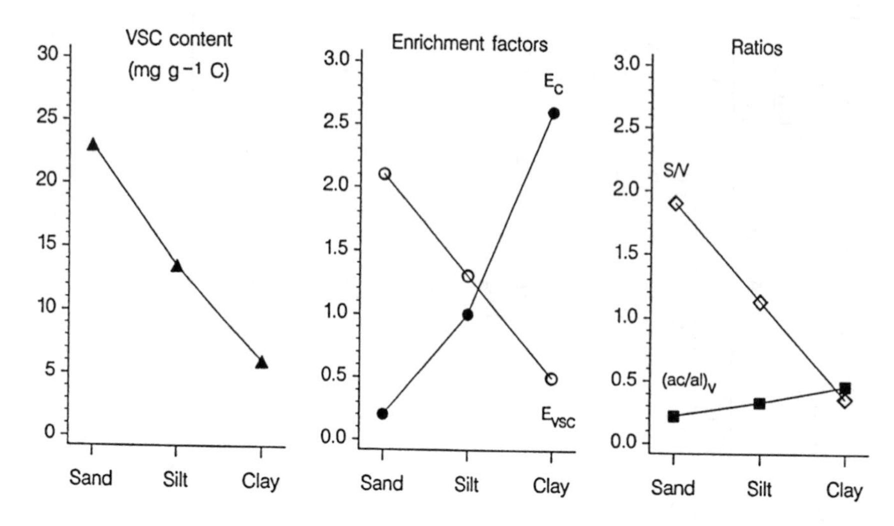

Figure 4. Content of VSC (sum of vanillyl (V), syringyl (S) and cinnamyl (C) derived from lignin by alkaline CuO oxidation), enrichment factors (content in separate divided by content in whole soil) for total C (E_C) and VSC (E_{VSC}), and S/V and $(ac/al)_V$ ratios for SOM in clay (<2 μm), silt (2-20 μm) and sand (20-2000 μm) isolated from the A-horizon of an arable Typic Eutrochrept in Bavaria, Germany. (Data from Guggenberger and Christensen, 1993 and Guggenberger et al., 1994.)

cleavage of phenylpropanoic side-chains, demethylation of methoxyl groups, hydroxylation of aromatic rings, and cleavage of aromatic rings while they arestill in the polymers. Zech and Kögel-Knabner (1994) provide more detailed information on the decomposition of lignin.

^{13}C-NMR performed on alkaline extracts of a Vertisol showed that alkyl material was concentrated in clay sized separates (Skjemstad et al., 1986). Alkaline SOM extracts of clay from an Aquoll were also more aliphatic in nature than SOM extracted from any coarser separate (Catroux and Schnitzer, 1987). Theng et al. (1986) using solid-state ^{13}C-NMR found little evidence for the occurrence of aromatic structures in clay, the ^{13}C-NMR spectra being indicative of substituted polymethylene chains. Almost two-thirds of the SOM in clay examined by Oades et al. (1987) was found to be alkyl in nature, and an accumulation of long-chain material was indicated. This material was considered to include natural waxes strongly associated with the clay minerals. Natural waxes extracted from various size separates with chloroform and hexane have been found to account for 1 to 7.5% of the SOM in size separates (Schnitzer and Schuppli, 1989b), but no consistent trend was observed in the extractability of various size separates. Similar extractions of fine clay (<0.2 μm) from an Aquoll, a Udic Boroll and a Typic Boroll showed that 3.4, 2.5 and 8.9% of the SOM was in wax-like material, respectively (Schulten and Schnitzer, 1990), but the chemical composition of the extracts appeared very similar. Schnitzer et al. (1988) concluded that at least 50% of the SOM in intact fine clay from a Typic Haplorthod was made up of long-chain aliphatics.

Catroux and Schnitzer (1987) found the aromaticity of SOM extracted from silt (2 to 45 μm) to be higher than that extracted from clay (<2 μm). Solid-state ^{13}C-NMR on size separates from a Rhodoxeralf (Oades et al., 1987, 1988) confirmed that the proportion of SOM present in aromatic structures generally was greatest in silt (2 to 20 μm). Schnitzer and Schuppli (1989a) examined alkaline extracts of an Aquoll and a Udic Boroll and found aromaticity of medium silts (5 to 20 μm) to be almost twice that of coarse clays (0.2 to 2 μm). Aromaticities of medium silt from Boroll and an Aquoll were 58 and 30%, respectively.

Employing solid state ^{13}C-NMR and soil incubated with ^{13}C labelled glucose for up to 34 days (whereby 65% of the glucose was mineralized), Baldock et al. (1989, 1990) demonstrated that little aromatic material was produced from glucose, while native SOM contained appreciable amounts of aromatic structures. Alkyl, O-alkyl, acetal and carboxyl structures formed during glucose decomposition accumulated predominantly in clay sized (<2 μm) organomineral complexes.

Using pyrolysis-field ionization mass spectrometry, Schulten and Leinweber (1991) showed that contents of lignin mono- and dimers decreased with decreasing particle size while contents of phenolic, carbohydrate and heterocyclic structures increased. Leinweber and Schulten (1992) confirmed the results except for lignin dimers which decreased with decreasing particle size. Fatty acids also declined with decreasing particle size and based on pyrolysis - chemical ionization tandem mass spectrometry, Baldock et al. (1991) reported a similar distribution pattern for fatty acids. For saturated fatty acids, larger sized fractions were rich in odd number carbons (particularly C_{15}) while smaller sized separates were dominated by fatty acids in the C_{10}-C_{13} range. The results suggest that fatty acids in larger size separates originate from plant residues whereas fatty acids in smaller sized separates are derived primarily from decomposers and partly degraded substrates. Phenolics/lignins tend to increase with decreasing particle size, while proteinaceous materials peaked in silt (2-20 μm). Saccharides generally increased with increasing particle size. The proportion of chitin (derived from fungi) in organomineral complexes did not appear to have any correlation with particle size.

Baldock et al. (1991, 1992) summarized their data on the chemical composition of SOM as revealed by solid state ^{13}C-NMR (CP/MAS-^{13}C-NMR). Results obtained from a Millicent Mollisol and an Urrbrae Alfisol (Baldock et al., 1991) showed that coarser size fractions were dominated by O-alkyl carbon presumably due to carbohydrates in plant residues. O-alkyl carbon decreased and alkyl carbon increased with decreasing particle size. The high content of alkyl in finer separates was ascribed to gradual accumulation of plant derived waxes associated with clay minerals and to alkyl materials produced by microbes engaged in carbohydrate decomposition. Aromatic carbon peaks in intermediate size fractions. Changes in the chemical structure of SOM across size separates were less prominent in Andosol, whereas for Mollisols and Oxisols, O-alkyl, aromatic and alkyl structures were greatest in the coarse, intermediate and fine organomineral complexes, respectively (Table 4). The largest differences in SOM composition were observed between soil orders while soils within orders differed less.

To summarize, the highest carbon concentrations in soils low in macro-OM are observed in separates <5 μm. The decline in carbon enrichment of clay and silt has been found to follow a power function as the proportion of these separates increases in whole soil. Clay generally accounts for $>50\%$ of the carbon in these soils, clay and silt (<20 μm) together accounting for $>90\%$. The proportion of carbon in carbohydrates generally increases with decreasing particle size and a greater microbial input to clay than to silt is indicated by hexose to pentose ratios. Analyses of CuO oxidation products reveal that lignin degradation advances from sand over silt to clay. Clay associated SOM is more aliphatic in nature than SOM in coarser separates while aromatic structures peak in silt. O-alkyl carbon decrease and alkyl carbon increase with decreasing particle size. Baldock et al. (1991, 1992) suggest three successive stages in the decomposition of plant residues: 1) an initial loss of O-alkyl carbon (hemicellulose, cellulose and protein), 2) a subsequent exposure and decomposition of aromatic carbon (lignin), and 3) finally a loss of recalcitrant alkyl carbon.

C. Surface Area and Activity

The surface properties of primary organomineral particles interact with substances in the soil solution, influence the environment of microbes in the vicinity of or adhering to the surface, and affect the chemical reactivity of SOM attached to the mineral surface.

The cation exchange capacity (CEC) of primary organomineral complexes isolated from predominantly Ap-horizons by ultrasonic dispersion in water has been found to decrease with increasing particle size (Table 5). Much of the whole soil CEC can be assigned to clays, but silt sized organomineral complexes also account for a substantial CEC. In contrast, the CEC of sand is negligible. For separates >2 μm, CEC was highly correlated with carbon content, correlation coefficients (r) ranging from 0.91 to 0.99 (Bremner and Genrich, 1990; Leinweber et al., 1993). Although the CEC of clay was correlated with carbon concentration, r values were markedly lower (0.52-0.79). The differences in

Table 4. Distribution of carboxyl, aromatic, O-alkyl and alkyl spectral regions of the CP/MAS ^{13}C-NMR spectra for particle size and density fractions

Fraction	% whole soil d.w.	% whole soil Carbon	% total signal intensity Carboxyl	Aromatic	O-alkyl	Alkyl
Millicent Mollisol						
250-2000 μm (LF)[a]	0.2	2.0	6.3	16.0	56.7	21.0
52-250 μm (LF)	1.2	7.0	12.6	17.9	46.1	23.4
20-53 μm (LF)	0.1	0.6	13.7	21.4	42.1	22.8
2-20 μm (LF)	2.7	17.7	17.0	21.8	35.3	25.9
0.2-2 μm	18.9	28.1	17.4	20.1	36.6	26.0
<0.2 μm	26.1	18.5	14.3	12.7	36.7	36.3
Recovery[b]	49.2	73.9				
Henongjiang Mollisol						
53-2000 μm (LF)	0.5	4.5	11.6	23.6	49.1	15.6
20-53 μm (LF)	1.0	2.4	12.7	26.7	38.2	22.3
2-20 μm (LF)	4.1	23.4	16.3	33.8	32.3	17.5
0.2-2 μm	19.6	34.2	8.8	19.4	44.2	27.6
<0.2 μm	9.4	11.6	17.3	17.9	42.6	22.2
Recovery[b]	34.6	76.1				
Mount Schank Andosol						
250-2000 μm (LF)	0.2	0.7	11.7	25.8	45.0	17.5
53-250 μm (LF)	0.1	0.4	18.2	26.1	39.1	16.6
2-20 μm	33.3	53.4	18.6	25.8	39.4	16.2
<2 μm	12.6	25.5	16.1	16.1	47.0	20.8
Recovery[b]	46.2	80.0				
Malanda Oxisol						
53-2000 μm (LF)	1.3	10.4	10.5	27.7	38.8	23.0
20-53 μm (LF)	0.2	1.3	8.2	34.5	31.8	25.6
2-20 μm (LF)	2.1	12.4	14.7	27.5	30.0	27.8
0.2-2 μm	14.5	17.0	9.7	12.9	37.8	39.6
<0.2 μm	41.1	23.9	8.9	11.9	33.1	46.2
Recovery[b]	59.2	65.0				
Guangdong Oxisol						
53-2000 μm (LF)	0.3	4.1	6.8	24.6	42.8	26.2
20-53 μm (LF)	0.2	1.3	12.8	30.0	31.0	26.1
2-20 μm (LF)	0.5	5.4	11.3	30.9	26.6	29.9
0.2-2 μm	20.7	17.6	12.4	21.4	38.9	27.3
<0.2 μm	49.8	53.4	9.2	14.9	39.3	36.6
Recovery[b]	71.5	81.8				

[a](LF) = light fraction <2.0 g ml^{-1} (<1.0 g ml^{-1} for Mount Schank Andosol).
[b]Recovery = dry weight and carbon recovered in analyzed size and density separates. (From Baldock et al., 1992.)

Table 5. Cation exchange capacity (CEC) of primary organomineral complexes isolated from Ap-horizons of Mollisols (USDA Soil Taxonomy, ref. 1 and 2) and Luvisols, Greyzem, Phaeozem, Chernozem, and Cambisol (FAO Soil Units, reference 3)

			CEC, mmol kg^{-1}					
	Number		---------Clay--------		---------Silt--------		---------Sand------	
Reference[a]	of soils		Fine	Coarse	Fine	Coarse	Fine	Coarse
1	4	mean	621	457	330	135	34	
		range	(574-677)	(336-561)	(204-414)	(66-181)	(25-37)	
2[b]	10	mean	938		163		69	
		range	(851-1100)		(86-345)		(13-231)	
3	18	mean	668	525	332	138	40	38
		range	(498-813)	(367-749)	(202-587)	(63-345)	(12-128)	(10-156)

[a] Ref.: 1, Curtin et al., 1987, fine and coarse clay and silt, and fine sand size limits are <0.2, 0.2-2, 2-5, 5-20 and 20-50 μm, respectively; 2, Bremner and Genrich, 1990, clay, silt, and sand size limits are <2, 2-50, and 50-2000 μm, respectively; 3, Leinweber et al., 1993, fine and coarse clay, silt, and sand size limits are <0.63, 0.63-2, 2-6.3, 6.3-20, 20-63, and 63-2000 μm, respectively; [b] values given in meq kg^{-1}.

correlation coefficients suggest that the CEC of silt and sand rests primarily on their SOM content, while SOM and the mineral part both contribute substantially to the CEC of clay sized organomineral complexes. Applying linear regression (CEC = a×%C+b) Leinweber et al. (1993) found that the CEC of clay minerals (a=0) was for 391 mmol kg^{-1} whereas coarser minerals accounted for 9-86 mmol kg^{-1}. The CEC of SOM in different separates ranged from 700 to 1500 mmol kg^{-1} carbon. SOM associated with fine clay exhibited the lowest CEC while coarse clay, fine and coarse silt had the highest SOM derived CEC. Dudas and Pawluk (1970) found that the CEC of coarse (1 to 2 μm), medium (0.2 to 1 μm) and fine clay (<0.2 μm) was reduced by 50, 40 and 11 to 29% when SOM was partly removed by hydrogen peroxide, demonstrating a differential contribution of SOM to CEC in the various clay subfractions.

Curtin et al. (1987) observed that the pH values of the 0.2 to 20 μm separates were 0.3 to 0.4 pH-units lower than whole soil, fine clay (<0.2 μm) and the >20 μm separates. Titratable acidity peaked in coarse clay (0.2 to 2 μm), but 37% of whole soil titratable acidity was accounted for by silt (2 to 20 μm). Titratable acidity was significantly correlated with SOM and with citrate/dithionite/bicar-bonate extractable iron and aluminium. The results clearly indicate that separates coarser than clay are of great quantitative importance to the physicochemical reactions of soils.

Nkedi-Kizza et al. (1983) examined the adsorption of the herbicides diuron and 2,4,5-T on primary organomineral complexes isolated from a sandy clay loam (Typic Haplaquoll). The Freundlich adsorption coefficient (K) showed a marked increase with decreasing particle size. When normalized with respect to carbon content, K_C (= K/C) varied much less, suggesting that the adsorptivity of SOM associated with organomineral complexes was essentially the same. The pooled data for diuron showed K_C = 682, indicating that soil carbon was primarily responsible for diuron adsorption. For 2,4,5-T the K_C of sand was 85, adsorption to finer sized separates being described by K_C = 217. Huang et al. (1984) found that adsorption of atrazine peaked in fine silt (2-5 μm), clays (<2 μm) and coarse silt (5-20 μm) showing somewhat smaller adsorptivity. Separates >20 μm were 5 to 10 times lower in atrazine adsorption.

Using ^{75}Se-labelled sodium selenite, Christensen et al. (1989) examined the effect of SOM on selenite fixation to size separates from an Alfisol. Intact clay (<2 μm) as well as silt (2 to 20 μm) showed high fixation capacities, the fixation capacity of sand (20 to 2000 μm) being significantly lower. Hydrogen peroxide treatments reduced the selenite fixation capacity of silt and sand to 17 and 1% of that of intact samples. In contrast, clay maintained a fixation capacity corresponding to about 50%. The fixation capacity of peroxide treated 'natural' clay was in accordance with values obtained previously for mined 'pure' clay minerals (Hamdy and Gissel-Nielsen, 1977). The content of dithionite/citrate extractable iron and aluminium in size separates was not affected by the peroxide treatments. It was concluded that the fixation capacity of sand and silt was mainly related to the SOM content, while only half the capacity of clay was accounted for by SOM.

The specific surface area increases with decreasing particle size, values for sand, silt and clay ranging from 1 to 39, 7 to 110 and 24 to 203 m^2 g^{-1} fraction, respectively (Table 6). Feller et al. (1992) found that the weighed sum of fraction specific surface areas corresponded well with the surface area obtained for whole soil samples, indicating that soil dispersion per se did not affect specific surface areas. Partial removal of carbon by treatments with H_2O_2 increased the surface area of clay sized organomineral complexes, while surface areas of silt and sand were less effected. To account for increased surface areas following SOM removal, Oades (1984) suggested that clays were embedded in a matrix of SOM. The greatest response to H_2O_2 treatment was observed for the finer clay separates, and an increased specific surface area of clay sized organomineral complexes following SOM removal may be ascribed to dispersion of <2 μm sub-microaggregates (Burford et al., 1964; Turchenek and Oades, 1978; Oades, 1984; Feller et al., 1992).

Accumulation of carbon in clay sized separates induces a decrease in surface area, the SOM induced reduction in specific surface area of silt and sand being much less significant. Thus the dramatic effects of SOM removal observed in sorption studies do not correspond to changes in the specific surface area of these size separates. It should be added, however, that the concept of SOM having a very large specific surface area has been challenged by Chiou et al. (1990). Using the BET method, they measured specific surface areas of less than 1 m^2 g^{-1} for SOM preparations and soils very high in SOM.

Relating the amount of SOM in size separates to the corresponding SOM-free specific surface area, Broersma and Lavkulich (1980) observed that the weight of SOM per unit area increased with decreasing surface area and accordingly with increasing particle size. Coarse silt (20 to 50 μm) and fine clay (<0.2 μm) had SOM loads of 4.9 and 2.5 mg C m^{-2} , respectively. Calculations based on data from Jocteur-Monrozier et al. (1991), Oades and Waters (1991), and Feller et al. (1992) provide a similar pattern, the area specific carbon load increases with increasing particle size. Values for clay and silt are typically in the range 0.5 to 1.5 and 1 to 5 mg C m^{-2}, respectively. Significant proportions of particulate SOM not associated with organomineral complexes may however inflate calculations of area specific carbon loads for coarser separates.

SOM-free clay and silt may catalyze oxidative polymerization of various phenolic compounds (Wang et al., 1986). The catalytic power of silt was 23 to 67% of that of clay size separates, and removal of sesquioxides by citrate/dithionite/bicarbonate reduced the oxidative power of clay and silt by 1 to 45% and 7 to 25%, respectively (Wang et al., 1983). The catalytic polymerization of phenolics differed among various mined 'pure' clay minerals and quartz also exhibited some catalytic power (Wang et al., 1978). Since samples were treated repeatedly with hydrogen peroxide to remove SOM, the catalytic abilities of intact primary organomineral separates remain unknown. Mineral particles embedded in SOM may well possess considerably less catalytic power than SOM-free preparations.

For arable soils, the CEC, the adsorption capacity and the specific surface area of primary organomineral complexes decrease with increasing particle size. While the CEC of particles coarser than 2 μm depends on their carbon content, the mineral and the organic part both contribute substantially to clay CEC.

Table 6. Specific surface area (N_2-BET, EGME for Reference 2) of primary intact organomineral complexes (+) and complexes where carbon was (partly) (-) removed by H_2O_2 (or NaOCl, reference 2) treatments

			Specific surface area, m^2 g^{-1} fraction											
			2000-20 μm		50-20 μm		20-2 μm		<2 μm		2-0.2 μm		<0.2 μm	
Ref.[a]	Soil	%C	+	-	+	-	+	-	+	-	+	-	+	-
1	Urrbrae PP, Alfisol (0-10 cm)	2.7							16	77	24	54	36	113
	Melfort, Mollisol (A_h)	6.3									24	85	25	172
2	Sombric Brunisol/Podzol (6 sites, 10 A_h samples)					12		43				113		169
3	Gleyed Molanic Brunisol	2.3							203	225				
4	Alfisol Ap-horizon	1.6			3		14				28			
	Alfisol Bf-horizon	0.4			13		54				95		168	
	Vertisol A1-horizon	1.9			39		110				215			
5	Urrbrae PP, Alfisol (0-10 cm)	2.8					9	11	39	70				
	Urrbrae WF, Alfisol (0-10 cm)	1.2					11	10	54	87				
6	Ferric Lixisol	0.9	1				7				24		48	
	Ferralic Cambisol	2.2	8	7			32	35			42	53	73	95
	Rhodic Ferralsol	4.3	3				12				37		48	
	Rhodic Ferralsol	1.5	2	2			32	36			53	58	63	70
	Rhodic Ferralsol	4.1	20				44				59		72	

[a]Ref.: 1. Turchenek and Oades (1978); 2. Broersma and Lavkulich (1980); 3. Greforich et al. (1988); 4. Jocteur-Monrozier et al. (1991); 5. Oades and Waters (1991); 6. Feller et al. (1992). For Ref. 1, coarse and fine clay were 2-0.4 and <0.4 μm, respectively; for Ref. 4, coarse and fine clay were 2-0.1 and <0.1 μm, respectively.

Partial removal of SOM increased the specific surface area of clay sized particles while coarser fractions were slightly affected.

D. Carbon Dynamics

1. Mineralization During Incubation

Christensen (1992) reviewed studies in which mineralization of nitrogen during incubation was used to assess the decomposability of SOM associated with primary particle size separates. Regardless of aeration status, the proportion of nitrogen mineralized from various size separates was generally found to increase with decreasing particle size. Only information on carbon mineralization during incubation of primary size separates will be considered here.

Christensen (1987) measured the CO_2 evolution from clay (<2 μm), silt (2 to 20 μm) and sand size (20 to 2000 μm) separates incubated aerobically for 7 weeks. The decomposability of SOM decreased in the order sand, clay, whole soil and silt (Table 7). The CO_2 evolution rate of clay was twice that observed for silt, and while straw incorporation increased the decomposition rate of sand associated SOM, minor effects were noted for clay and silt. The relatively high decomposition rate of SOM in sand was due to the presence of macro-OM. The high lability of carbon does not conflict with the low nitrogen mineralization rate generally observed for sand. Elevated CO_2 release from macro-OM in sand indicates a high microbial activity which may well be coupled with immobilization of nitrogen considering the high C/N ratios observed for SOM in light fractions of sand (Christensen, 1992). Adams (1980) found that the addition of macro-OM to soil lead to an increased carbon and a decreased nitrogen mineralization.

For primary size separates isolated from 1 to 2 mm aggregates and incubated for 20 days, Gregorich et al. (1989) found that 5 and 1.5% of the carbon in sand (2000 to 50 μm) and silt (50 to 2 μm) respectively was evolved as CO_2, whereas less than 1% of the clay carbon was mineralized. Pools of potentially mineralizable carbon (C_0) and associated rate constants (k) were calculated for sand and clay sized separates by $C_t = C_0 (1-e^{-kt})$, C_t being carbon mineralized in time (t). The decomposition rate was found to be 2.4 times higher for clay than for sand. The higher CO_2 evolution from sand was ascribed to a larger pool of potentially mineralizable carbon in this size separate, the C_0 being 8 times higher for sand than for clay. Correlations between calculated pools and rates may however inflate the significance of regression analyses. The ranking of size separates with regard to CO_2 production during incubation apparently differed from that obtained by Christensen (1987). However, silt included the 20 to 50 μm particles which in the study of Christensen (1987) was allocated to sand, and the CO_2 evolved from size separates summed up 60% of that from whole soil, leaving more than 40% of the potential CO_2 production unaccounted for. Moreover, the incubation period was shorter (20 days) and included the initial CO_2 flush which was left out in the study of Christensen (1987).

Table 7. Carbon mineralization from samples of whole soil and size separates incubated aerobically for 7 weeks at 20° C; the decomposition rate constant (k) was obtained from the regression equation y = a + kx, where y is the cumulative CO_2 production for the 7-49 day period, x is the incubation time and a is a constant; correlation coefficients (r) ranged from 0.983 to 0.999

		Decomposition rate constant, k ($\mu g\ CO_2$-C g^{-1} sample-C day^{-1})		Respiration loss after 7 weeks at 20° C			
				% of sample-C		mg C kg^{-1} whole soil	
Separate	Straw[a]	CS-soil[b]	SL-soil[c]	CS-soil	SL-soil	CS-soil	SL-soil
Sand (2000-20 μm)	burned	190	490	0.9	2.4	16	10
	incorporated	400	1300	2.0	6.5	36	40
Silt (20-2 μm)	burned	80	230	0.4	1.2	40	38
	incorporated	90	280	0.5	1.4	43	49
Clay (<2 μm)	burned	180	350	0.9	1.8	106	130
	incorporated	170	390	0.9	2.0	109	148
Sum of separates	burned			0.7	1.6	162	178
	incorporated			0.8	2.0	188	237
Whole soil (<2000 μm)	burned	110	280	0.6	1.4	129	164
	incorporated	170	380	0.9	1.9	205	241

[a] Straw (4 t ha^{-1}) incorporated or burned annually for 18 years; [b] CS-soil; coarse sand soil (pH 7.0; 6% clay; 2.4% C); [c] SL-soil; sandy loam soil (pH 7.1; 15% clay; 1.2% C). (Data from Christensen, 1987.)

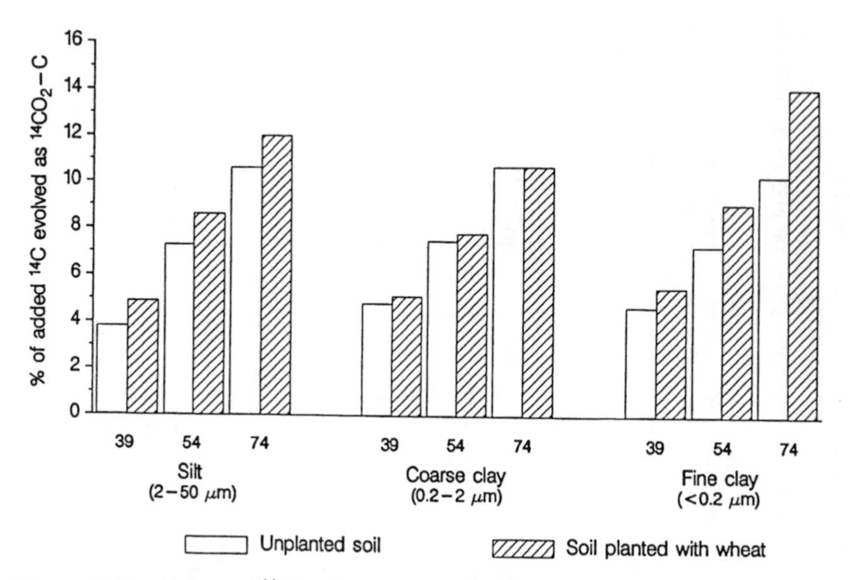

Figure 5. Evolution of $^{14}CO_2$ from soils (28% clay, 43% silt, 1.5% C and pH 7.6) amended with silt, coarse or fine clay enriched in ^{14}C-labelled bacterial residues. Size separates were isolated from a similar soil previously incubated with ^{14}C-labelled *Klebsiella coccytoca* for a 30 week period during which 60% of the ^{14}C had been evolved as $^{14}CO_2$ (Cortez, 1989). Soil samples amended with silt, coarse or fine clay were either unplanted or planted to wheat during the 74 day incubation period. (Data from Cortez and Cherqui, 1991.)

Treatment with H_2O_2 reduced the carbon content of sand, silt and clay with 55, 87 and 74%, respectively, but H_2O_2 resistant SOM was more labile than SOM in intact size separates (Gregorich et al., 1989). The CO_2 evolved during incubation equalled 7.7, 7.9 and 2.3% of the carbon remaining in sand, silt and clay, respectively.

Cortez and Cherqui (1991) incubated soil with ^{14}C labelled bacterial residues. After 30 weeks of incubation, size separates were isolated. Almost equivalent amounts of ^{14}C in size separates were then added individually to unlabelled whole soil samples, and the $^{14}CO_2$ evolved during a subsequent 74 days incubation period was determined (Figure 5). For unplanted treatments, the proportion of ^{14}C evolved as $^{14}CO_2$ was almost similar for different size separates. The presence of wheat plants during incubation enhanced the $^{14}CO_2$ release and introduced a differentiation in the turnover of ^{14}C added in different size separates. Evolution of $^{14}CO_2$ now decreased in the order fine clay > silt > coarse clay.

Christensen and Christensen (1991) examined the bioavailability of carbon by measuring the N_2O evolution from clay (<2 μm) and silt (2 to 20 μm) during

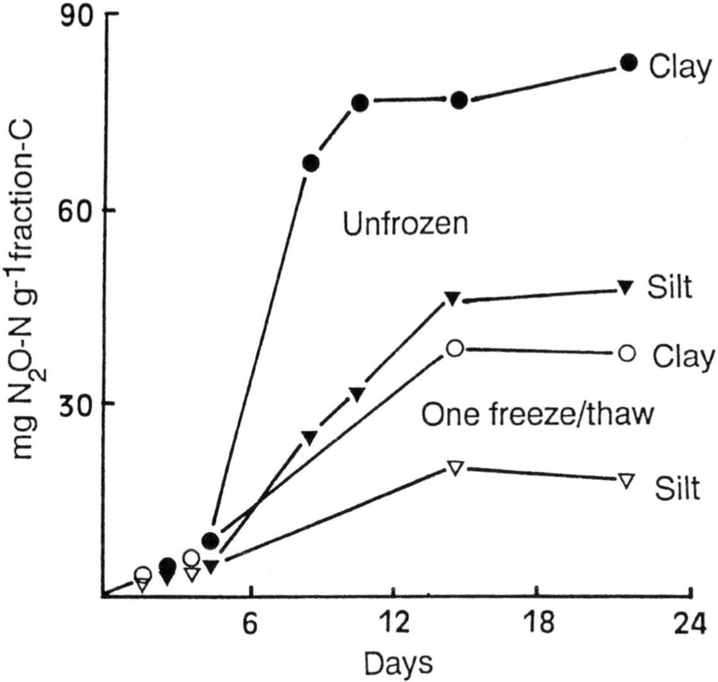

Figure 6. Bioavailability of SOM in clay (<2 μm) and silt (2-20 μm) measured by N_2O evolution from samples incubated anaerobically with excess nitrate at 21°C. Samples were incubated directly (unfrozen, closed symbols) or exposed to one freeze/thaw cycle (open symbols) before incubation. (Data from Christensen and Christensen, 1991.)

anaerobic incubation with excess nitrate. Size separates were used unfrozen or exposed to one freeze/thaw cycle before incubation. The availability of clay SOM for denitrification was twice that observed for silt associated SOM, and the N_2O production following a lag period increased more steeply for clay than for silt (Figure 6). One freeze/thaw cycle halved the N_2O evolution from clay and silt compared to unfrozen separates. The reduced availability of SOM was ascribed to formation of microaggregates in which SOM may be less accessible to the denitrifyers.

2. Distribution of Substrate Derived Carbon

The distribution and stabilization of decomposition products across primary organomineral size separates has been studied using soils incubated with isotope labelled materials.

Table 8. The relative distribution of dry weight (d.w.), native and labelled carbon in Northfield Vertisol incubated with ground (<1 mm) ^{14}C labelled *Medicago littoralis* leaves at 25°C for 34 and 130 days

	Relative distribution[a]			
			Labelled C	
Separate	d.w.	Native C	Day 34	Day 130
Sand (50-2000 μm)	26	3	0.5	0.2
Silt (2-50 μm)	22	39	33	33
Clay (<2 μm)	52	57	66	67
Coarse (0.2-2 μm)	30	38	25	33
Fine (0.04-0.2 μm)	15	13	31	21
Very fine (<0.04 μm)	7	6	11	13

[a] % of material recovered in all separates. (Data from Amato and Ladd, 1980.)

Extending previous studies (Ladd et al., 1977a, b, c) on soils incubated with ^{15}N-nitrate and glucose or wheat (*Triticum aestivum* L.) straw, Amato and Ladd (1980) incubated clay soil with ^{15}N and ^{14}C labelled leaves of *Medicago littoralis* (L). After 34 and 130 days, samples were fractionated according to particle size.

In accordance with their earlier work, the proportion of labelled carbon found in clay after 34 days of incubation exceeded that of native carbon, the opposite being true for silt (Table 8). During day 34 to day 130, the proportion of labelled carbon in fine clay decreased while that in coarse clay increased. In contrast, the proportion associated with silt remained unchanged. It was concluded that SOM formed during the decomposition of added substrates became distributed among most size separates, but that the newly formed SOM in clays was more readily transferred or decomposed.

In the longterm, the proportion of substrate derived SOM in clay will continue to decline while silt will tend to accumulate higher proportions of the more stable residual SOM. Christensen and Sørensen (1985) isolated size separates from four soils differing in clay content and incubated for 5 to 6 years with ^{14}C-glucose, -hemicellulose and -straw, and from one soil incubated for 18 years with ^{14}C-straw. After 5 to 6 years, carbon enrichments (% C in fraction/% C in whole soil) of clay were higher for labelled than for native SOM, silt enrichments being lower (Figure 7). The distribution pattern was not related to kind of organic amendment. The highest specific activity was in clay, and in accordance with studies based on short-term incubations, the proportion of carbon in clay was greater for labelled than for native carbon. The opposite trend was seen for silt. In contrast, similar specific activities were found for carbon in clay and silt from the 18 year incubation. Moreover, both clay and silt showed similar proportions of labelled and native carbon (Figure 8).

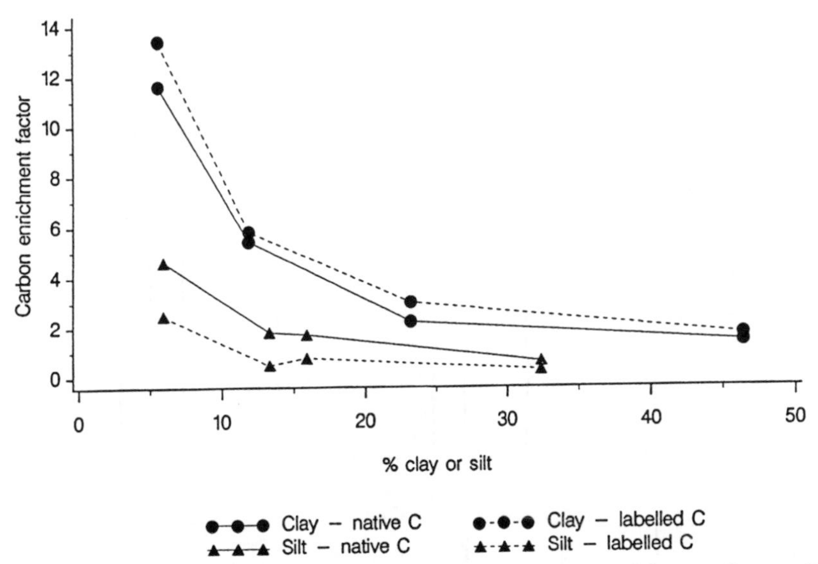

Figure 7. Relationship between fraction size and carbon enrichment factors (% C in fraction to % C in whole soil) for labelled and native SOM in clay (<2 μm) and silt (2 to 20 μm) isolated from soils incubated for 5 to 6 years with ^{14}C-labelled glucose, hemicellulose and barley straw. (Redrawn from Christensen and Sørensen, 1985.)

Cheshire and Mundie (1981) examined the distribution of carbohydrates in soils incubated with ^{14}C-labelled glucose (28 days), ryegrass (*Lolium perenne* L.) (1 year), and straw (*Secale cereale* L.) (4 years). Clay (<2 μm) was enriched and silt (2 to 20 μm) depleted in labelled carbohydrates produced during decomposition of glucose. Soils incubated with plant materials had substantial proportions of labelled glucose, xylose and arabinose (sugars dominating plant materials) in the sand separates (20 to 2000 μm), but the distribution of microbially derived sugars (mannose, rhamnose and fucose) resembled that found after glucose addition. Excluding plant related sugars, a relatively similar distribution pattern was therefore seen for labelled sugars regardless of substrate. For soils incubated with ^{14}C labelled straw for 6 to 18 years, Cheshire et al. (1990) found that the total ^{14}C activity of individual sugars was greater in clay than in silt. The specific activity of xylose remained higher than that of other sugars even after 18 years of incubation, indicating that straw polysaccharides were still present. However, differences between clay and silt in the specific activities of individual sugars revealed no consistent trend, and the long-term turnover of straw was not reflected in the sugar composition of SOM associated with differently sized primary organomineral complexes. This finding matches

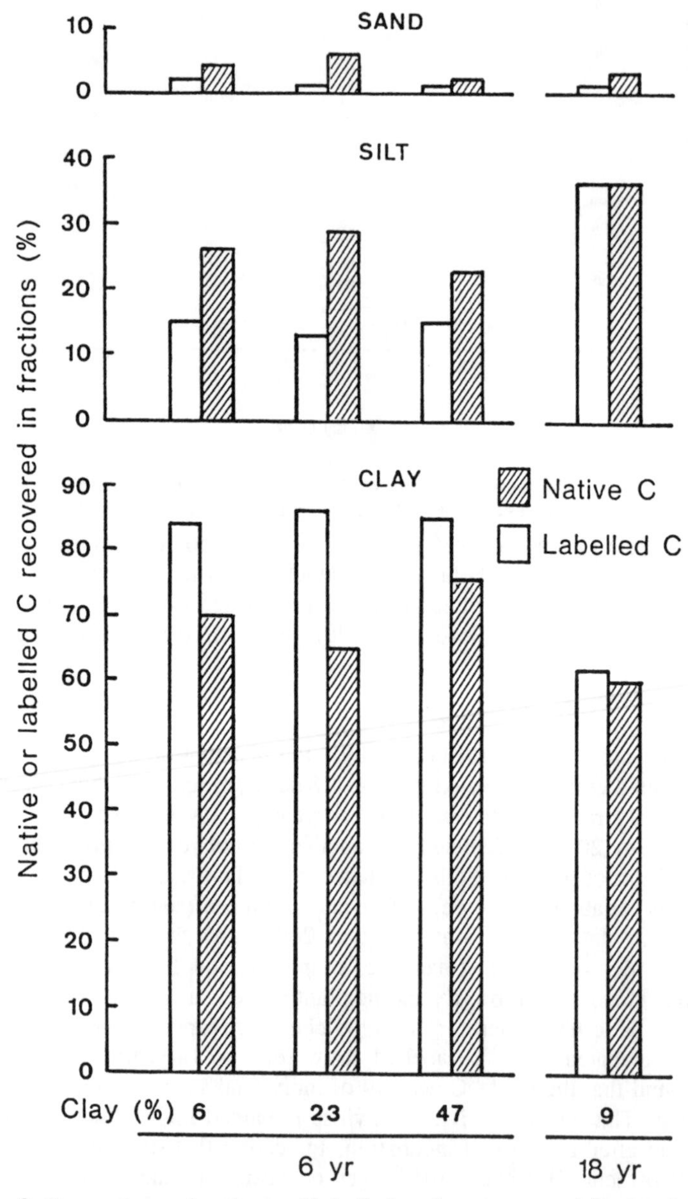

Figure 8. Proportions of native and labelled carbon recovered in clay ($<2\ \mu m$), silt (2 to 20 μm), and sand (20 to 2000 μm) size separates isolated from soils incubated for 6 or 18 years with ^{14}C-labelled barley straw. (Redrawn from Christensen and Sørensen, 1985.)

results on the distribution of individual amino acids following annual incorporation of straw in the field (Christensen and Bech-Andersen, 1989).

Using soil incubated for 30 weeks with ^{14}C labelled bacterial fractions, Cortez (1989) examined the distribution of labelled carbon among various soil fractions. During incubation, 60% of the added ^{14}C evolved as $^{14}CO_2$. Initially labelled material appeared in all size separates. During the first 2 weeks of incubation silt (2 to 50 μm) accumulated labelled material while the ^{14}C of other fractions declined rapidly. After 30 weeks, 7 and 4% of the ^{14}C retained in soil was in the light (<1.4 g ml^{-1}) and water soluble fractions, respectively. The distribution of labelled carbon among organomineral complexes was similar regardless of type of amendment, the sand and silt holding 4 and 26% of the substrate derived carbon. Clay retained 59% with 34% in coarse (2 to 0.2 μm) and 25% in fine clay (<0.2 μm).

A preferential accumulation of labelled SOM in clay was also reported for soil incubated with ^{13}C-glucose for 5 weeks (Baldock et al., 1990, 1991). Thus 72% of the recovered residual ^{13}C was in the clay, while 19 and 9% was found in light (<2.0 g/ml) and heavy fractions >2 μm. Alkyl and O-alkyl were the dominating carbon structures as seen by ^{13}C-NMR. The composition of residual glucose derived ^{13}C found in clay and light fractions was similar, the major differences being a lower alkyl and a higher acetal and aromatic content in the clay separate. Comparison with spectra collected for fungal and bacterial soil isolates suggested that the ^{13}C-glucose was utilized predominantly by soil fungi.

For 10 soils with the same suite of clay minerals and under the same management regime but ranging in clay content from 19 to 38%, Gregorich (1989) found that following 90 days of incubation with ^{14}C labelled glucose, clay enrichment factors for both native and labelled carbon showed a close negative relationship with clay yield. But the slope of the linear regression was significantly greater for labelled than for native SOM. Thus clay from more sandy soils became relatively more enriched in new SOM than clay isolated from heavier soils. Since the distribution pattern of native and substrate derived SOM becomes similar eventually, the results suggest that clay SOM formed during the initial and rapid decomposition phase is less resistant to further turnover in sandy soils than in more clayey soils. Enrichment factors for native SOM in silt (2 to 50 μm) was also linearly correlated with silt yields, whereas enrichment factors for labelled SOM were not. A similar trend appears for silt data in Figure 7.

Most of the microbial biomass has been found to be associated with clay and silt (Amato and Ladd, 1980) or with discrete organic particles (Kanazawa and Filip, 1986), and the levels of microbial biomass in soils has been found to increase with increased soil clay content (Ladd et al., 1985). Employing the same set of soils as Gregorich (1989), Gregorich et al. (1991) reported that the microbial biomass derived from ^{14}C labelled glucose was maintained at higher levels in more clayey soils during the first 45 days of incubation. The native microbial biomass was significantly correlated with the amount of clay in soil at all samplings. The ratio of living to non-living ^{14}C was consistently higher in soils with more clay. This was ascribed to stabilization of non-living labelled carbon by clay and to subsequent turnover of ^{14}C labelled microbial metabolites.

Sørensen (1983) also observed that native biomass carbon was higher in clay rich soils whereas the labelled biomass formed during the decomposition of simple and complex substrates for up to 2 years could not be related to soil texture. The specific death rate of the microbial population (μg microbial-C lost per μg microbial-C present initially per time unit) has been found to be negatively correlated with clay content in soils stored for 18 months at 15°C (Anderson, 1991), the specific death rate being 3 times higher in soil with 7% clay than in soil with 32% clay.

Generally the carbon mineralization rate decreases in the order: sand, clay, whole soil and silt. Following addition of simple substrates, new SOM is found in most separates although clay sized organomineral complexes most often show greater accumulations. In the longer term, the proportion of substrate derived SOM in clay declines more rapidly than that in silt indicating a higher stability. It is indicated that newly formed clay SOM is less stable in sandy than in clayey soils.

E. Effects of Land Use and Soil Management

1. Land Use and Cultivation

It is well documented that cultivation of native soils and permanent pastures causes substantial reductions in carbon levels. Cultivation effects are through complex interactions of the physical, chemical and biological soil processes, leading to increased decomposition rates and redistribution of carbon. Reduced inputs of plant residues and increased soil disturbance usually follow cultivation, but the exact nature of the changes induced by cultivation depends on the particular agronomic practice adopted and on properties of the virgin soil.

The effect of cultivation and agronomy on the amount and composition of light fraction SOM (non-complexed or macro-OM) has been addressed previously (Christensen, 1992). Generally, light fraction carbon is much reduced upon cultivation, and agronomic practices with little return of organic materials to the soil provide little light SOM. In soils cultivated for long periods, light fractions or macro-OM in sand usually account for <10% of whole soil carbon.

Tiessen and Stewart (1983) examined the effects of cultivation on the carbon distribution among size separates from Canadian prairie soils of different textures (Cryoborolls). Floatable macro-OM in sand was physically disintegrated upon cultivation and either released as CO_2 or transferred to finer separates, predominantly coarse silt. However cultivation reduced the carbon concentration in all size separates. Fine clay and sand showed the greatest decline with much smaller reductions in fine silt and coarse clay. After the initial 4 year cultivation period, the proportion of whole soil carbon present in coarse clay increased and that in coarse silt declined (Figure 9). The proportion residing in fine silt

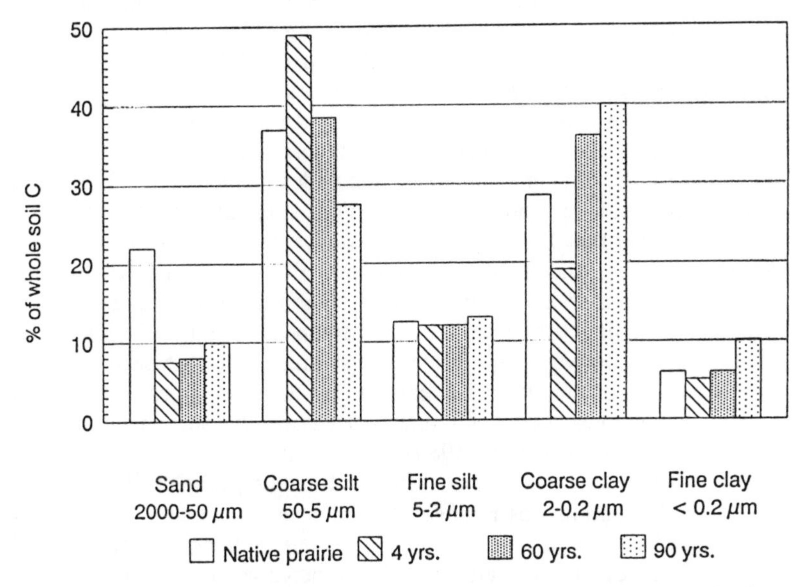

Figure 9. Carbon distribution among particle size separates from A-horizons of Typic Cryoborolls under native prairie or cultivated for 4, 60, or 90 years. (Based on data from Tiessen and Stewart, 1983.)

remained almost constant. Losses of carbon from a sandy loam followed a similar pattern, while a heavy clay soil (65% clay) showed greater stability of silt and clay carbon. In this soil, reductions in sand + macro-OM made up 40% of whole soil carbon loss due to cultivation. Cultivation caused only a slight decrease in carbon associated with coarse clay whereas that in coarse silt increased. Zhang et al. (1988) also found that cultivation induced a relative proportional shift of carbon from coarser (sand + macro-OM and coarse silt) to the finer size separates. The proportion of whole soil carbon in fine and coarse clay of Typic Hapludalfs increased and that in coarse (50 to 20 μm) and fine (20 to 2 μm) silt decreased upon cultivation. In Typic Argiudolls fine clay showed increasing proportions only, while coarse clay, fine and coarse silt remained constant in carbon.

The distribution of carbon in subtropical Vertisols under native vegetation or cereal cropping for up to 70 years was examined by Dalal and Mayer (1986). For virgin soils, no single size separate was consistently enriched in carbon as compared to whole soil concentration. The relative proportion of carbon in sand (2000 to 20 μm) declined rapidly from 26 to 12% while that in clay (<2 μm) increased from 48 to 61%. The proportion of carbon in silt (20 to 2 μm) showed no consistent trend with time of cultivation. The proportion in clay generally increased exponentially, the rate of increase being inversely related to soil clay content. In contrast to Tiessen and Stewart (1983), the rapid initial decline in sand SOM apparently occurred without a concomitant increase in silt and clay SOM. The sand size separate (2000 to 20 μm) of Dalal and Mayer (1986) included much material which in the study of Tiessen et al. (1983) was assigned to coarse silt (50 to 5 μm). Moreover, soil fractionation methods were very different, making direct comparison of results less feasible.

Cerri et al. (1985) examined a Haplorthox under native forest (C_3-photosynthetic pathway) and two adjacent clearfelled sites planted to sugar cane (*Saccharum officinarum* L., C_4-photosynthetic pathway) for 12 and 50 years. Size separates were isolated from 0 to 20 cm and analyzed for $^{12}C/^{13}C$ ratios. Carbon levels decreased substantially in all separates during the first 12 years of sugar cane cropping. About 60% of the carbon left in the <50 μm separate after 50 years was derived from the previous forest. The fine sand (50 to 200 μm) contained less of the whole soil carbon, but turnover times of SOM in the two >50 μm separates were both about 40 years. The sites were studied in greater detail by Vitorello et al. (1989). Samples from 10 to 20 cm were freed of recognizable plant material and extracted with sodiumpyrophosphate. Thereby 33 to 37% of the carbon was removed. Concentrations of carbon in fine clay (<0.1 μm) and coarse clay + silt (0.1 to 50 μm) remained constant, but the proportion of total carbon in coarse clay + silt increased from 38 to 55% during cultivation. Most changes took place during the initial 12 year after deforestation. The proportion of carbon derived from sugar cane decreased progressively with particle size, and after 12 and 50 years the 0.1 to 50 μm separate contained 89 and 69% forest derived carbon, respectively. The corresponding values for fine clay were similar, suggesting that the SOM in fine clay did not differ qualitatively from that of the coarse clay + silt. Alkali extractable carbon was similar to whole soil carbon in isotopic composition, and does therefore not appear to represent a qualitatively distinct SOM pool.

While Vitorello et al. (1989) relied on alkali extracted soil from 10 to 20 cm depth and chemical/mechanical dispersion, Bonde et al. (1992) applied ultrasonic dispersion in water to intact soil from 0 to 10 cm. Fine particles were in this study separated into clay (<2 μm), silt (2 to 20 μm) and fine sand 1 (20 to 63 μm). Silt had the highest concentration of carbon (31 to 93 mg C g^{-1} separate), while clay contained the highest proportion of whole soil carbon (Figure 10). The proportion held in silt, fine sand 2 (63 to 200 μm) and coarse sand (200 to 2000 μm) decreased after forestation while that in clay increased. Turnover times for carbon in clay, silt and sand sized separates were 59, 6 and 4 years.

Martin et al. (1990) studied the ^{13}C abundance of a ferrasol (FAO) in the humid tropics whose vegetation had shifted from grass savanna (C_4-species) to a dense woodland (C_3-species) due to protection against fires. After 25 years, 29% of the savanna derived carbon remained. Losses of carbon from the >50 μm separates were much greater (89 to 97%) than losses from clay (<2 μm), fine (2 to 20 μm) and coarse (20 to 50 μm) silt (43 to 48%). Savanna derived carbon accounted for 28, 56 and 70% of the total carbon in coarse silt, fine silt and clay, respectively, indicating a higher protection of residual savanna SOM

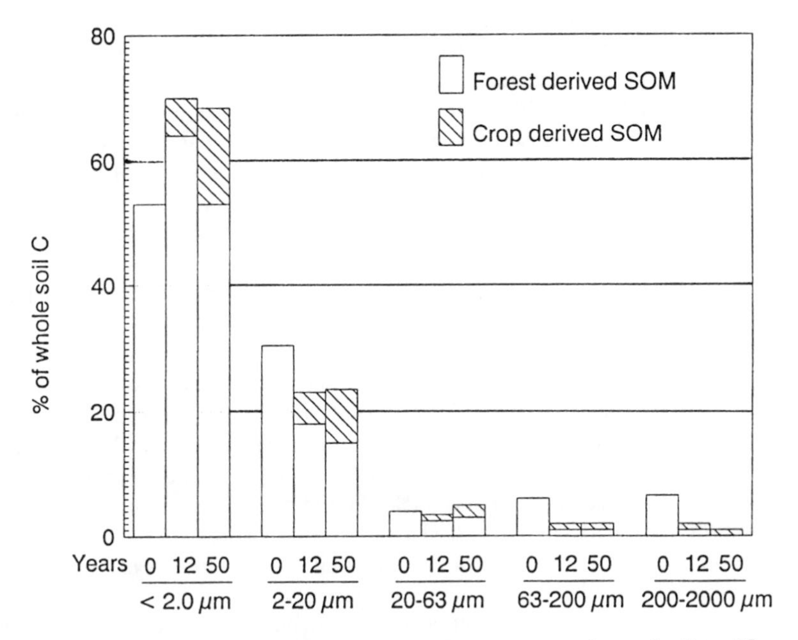

Figure 10. Distribution of SOM between size separates from the 0 to 10 cm of a Haplorthox under native forest (0) or cultivated with sugar cane for 12 and 50 years after deforestation. (Based on data from Bonde et al., 1992.)

in finer sized separates. The studies of Martin et al. (1990) and Bonde et al. (1992) both suggest that in tropical soils the most stable carbon is in clay sized organomineral complexes.

Studying the natural ^{13}C abundance in size separates from the 0-30 cm of a temperate Hapludalf planted to maize (*Zea mays* L.) (C_4) for 23 years after pine forest (C_3) clearing, Balesdent et al. (1987) found that silt (2 to 20 μm) contained the smallest proportion of maize derived carbon (12%) and thus had the slowest turnover. Coarse sand (200 to 2000 μm) had 61% of its carbon from maize, the turnover of fine sand separates (20 to 50 and 50 to 200 μm) being much slower. Coarse clay (0.2 to 2 μm) and fine clay ($<$0.2 μm) had 18 and 26% of the carbon from maize, indicating that carbon in clay turned over faster than carbon in silt. These results thus conform to the general ranking of size separates with regard to SOM stability in temperate soils.

Oades et al. (1988) sampled the 0 to 10 cm of a Rhodoxeralf carrying a virgin pasture, a permanent pasture on previously cropped area and a long-term wheat-fallow rotation. Compared to pasture soils, cultivation decreased carbon concentrations in sand ($>$250 μm) by almost 60% while carbon in silt (20 to 2 μm), coarse (2 to 0.2 μm) and fine clay ($<$0.2 μm) decreased 43, 24 and 40%, respectively. The chemical composition of SOM from the different sites as detected by ^{13}C-NMR was however rather similar, particularly for SOM associated with clay. Differences were associated with SOM in separates >20 μm which contained most of the aromatic carbon. A rapid loss of phenolic carbon originating in lignin was indicated. It was concluded that the chemical composition of SOM in a particular soil was not influenced significantly by changes in amount and nature of plant residue inputs, the composition of SOM being controlled by interactions between the microbial biomass, decomposition products and the soil matrix. Similarly, Dalal and Henry (1988) found that cultivation effects on soil carbohydrates generally matched those of total carbon. Also the distribution pattern for monosaccharides in size separates showed no consistent trend with period of cultivation.

To assess the effect of contrasting land use on the chemical composition of SOM, Guggenberger et al. (1994, 1995) examined size separates from the A-horizon of an Eutrochrept under long-term spruce forest, mixed deciduous forest, permanent pasture and arable rotation. All sites were within a small geographical area with uniform elevation, geological parent material and climate. Clay and silt yields averaged 25 and 35%. Arable use caused reduced contents of carbon, while spruce maintained high concentrations (Figure 11). The proportion of carbon present in carbohydrates was reduced under forests and enhanced under pasture. Silt sized organomineral complexes were lower in carbohydrates than clay and sand, especially under spruce and deciduous forests. The amount of carbon present in lignin-derived phenols decreased in the order sand, silt and clay. Except for spruce, land use regimes did not significantly affect clay associated phenols.

To eliminate effects of the different contents in whole soils, enrichment factors were calculated for the various size separates. Arable rotation caused a significant increase in the carbon enrichment of clay while that of sand decreased (Figure 12). The other three land use regimes produced similar enrichment patterns across size separates. Land use was not reflected in clay and silt enrichment factors for lignin-derived phenols, whereas spruce and pasture reduced the phenol enrichment in sand. Clay was enriched and sand depleted in microbially derived sugars under arable use. For plant derived sugars, differences were small although forests tended to cause a lower enrichment in silt. However enrichment factors for silt were generally least affected by land use regime.

The ratio of microbial to plant derived sugars declined in the order clay, silt and sand (Figure 13). The ratio maintained the same ranking with respect to particle size regardless of land use. However the ratio for a particular size separate decreased in the order spruce, deciduous, pasture and arable soil. Qualitative changes in the structure of lignin were illustrated by the $(ac/al)_V$ ratio (acid to aldehyde ratio of vanillyl units) and S/V ratio (syringyl to vanillyl ratio). The $(ac/al)_V$ ratio increased from sand over silt to clay reflecting an increased side-chain oxidation of phenylpropane. Again the same pattern was seen regardless of land use but ratios generally decreased in the order forests, pasture

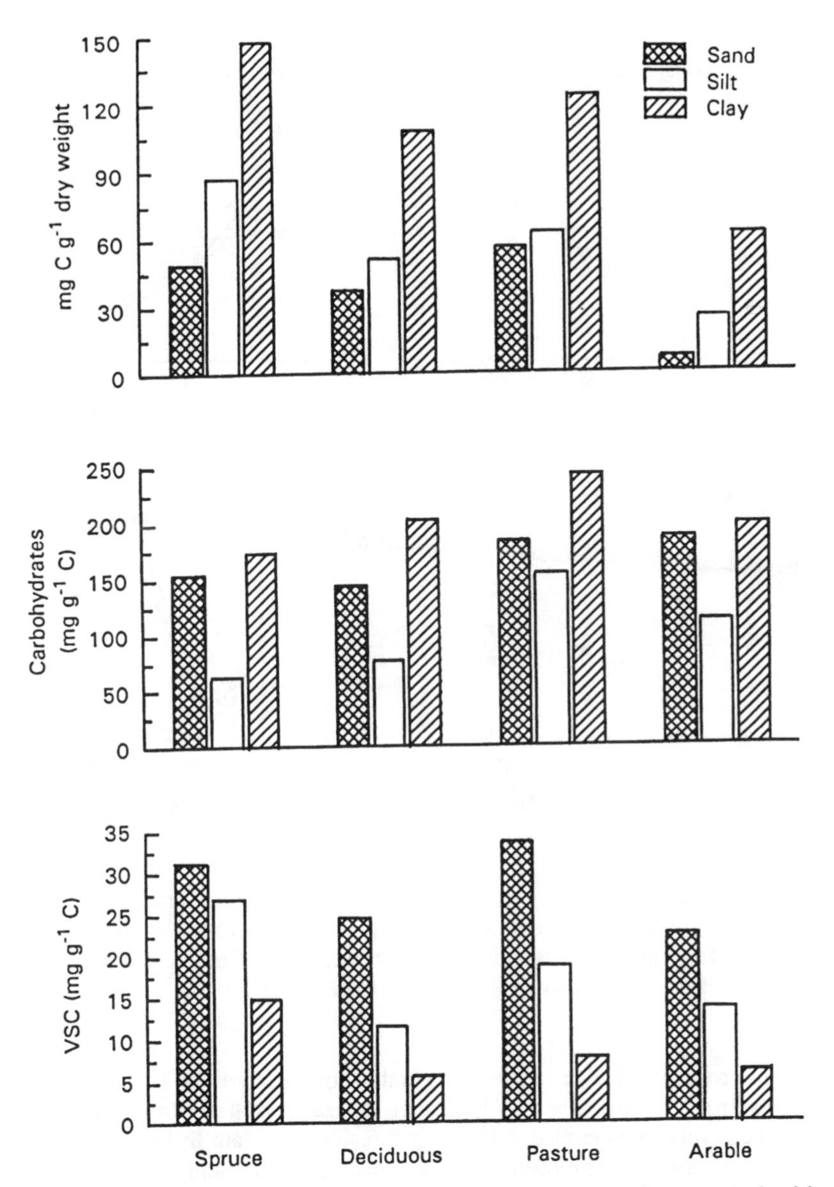

Figure 11. Content of total carbon, carbohydrates (sum of monosaccharides released by TFA hydrolysis) and lignin-derived phenols (sum of vanillyl, syringyl and cinnamyl units obtained by alkaline CuO oxidation) in clay (<2 μm), silt (2 to 20 μm) and sand (20 to 2000 μm) isolated from A-horizon samples (Eutrochrepts) under four different land use regimes. (Redrawn from Guggenberger et al., 1994.)

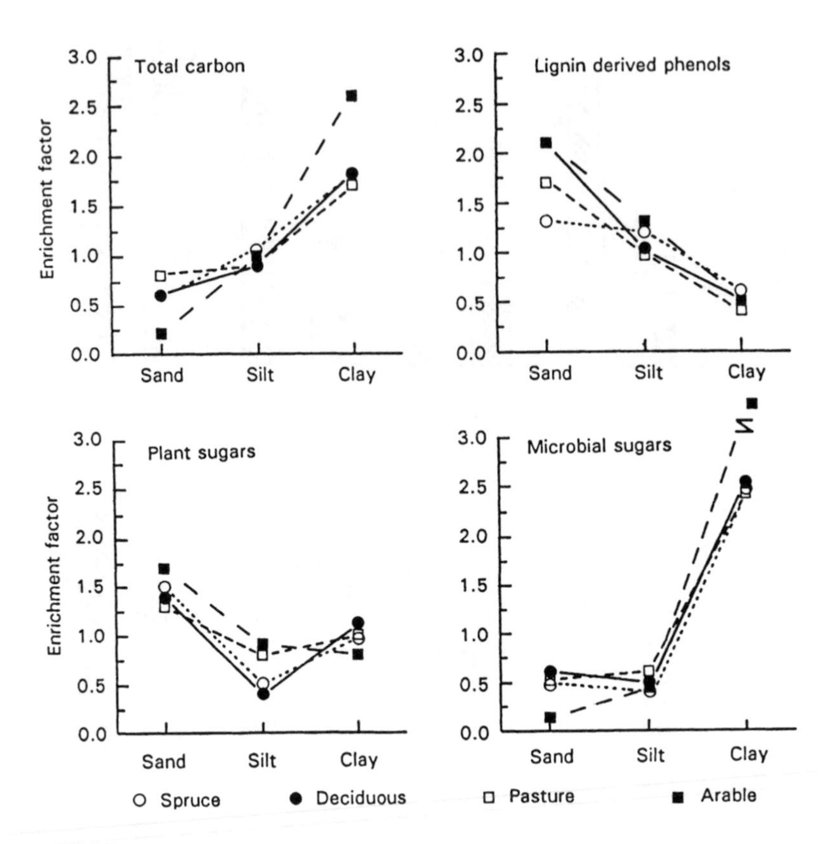

Figure 12. Enrichment factors (content in separate/content in whole soil) for total carbon, lignin-derived phenols (vanillyl + syringyl + cinnamyl), plant (arabinose + xylose) and microbial (mannose + galactose + rhamnose + fucose) derived sugars in organomineral size separates from A-horizon under different land use. (Redrawn from Guggenberger et al., 1994.)

and arable rotation. Size separates from arable soil differed less in $(ac/al)_V$ than separates from forests. The S/V ratios of separates isolated from soil under spruce were very low because gymnosperm residues contain little syringyl. For the other land uses, S/V ratios declined with decreasing particle size. Decreasing yields of lignin-derived phenols, decreasing S/V and increasing $(ac/al)_V$ ratios from sand over silt to clay indicate oxidative decomposition and transformation of plant lignin material, the microbial lignin degradation being most advanced in clay sized organomineral complexes regardless of land use. The results from wet degradative chemical analyses were generally confirmed by solid state ^{13}C NMR spectroscopy operated on size separates. Table 9 shows the enrichment

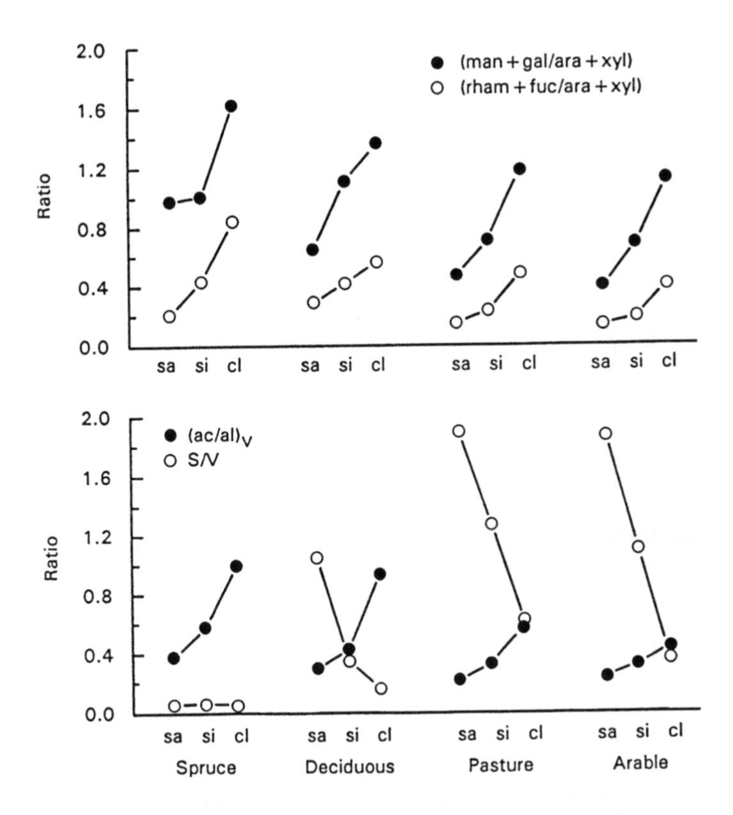

Figure 13. Ratios of microbial to plant derived sugars, vanillyl acid to aldehyde $(ac/al)_V$ and syringyl to vanillyl (S/V) of SOM associated with sand (sa = 20 to 2000 μm), silt (si = 2 to 20 μm) and clay (cl = <2 μm) from A-horizons under different land use. (Redrawn from Guggenberger et al., 1994.)

factors for carboxyl, aromatic, O-alkyl and alkyl carbon. Silt showed rather similar enrichment factors for the various carbon species, except for arable soil where silt was more enriched in aromatic carbon. Sand was generally depleted in alkyl-carbon while clay showed a greater enrichment in alkyl carbon than in carboxyl, aromatic and O-alkyl carbon. Differences between the various land uses were only manifest for arable soil.

Table 9. Enrichment factors for carboxyl-C, aromatic-C, O-alkyl-C and alkyl-C calculated from CP/MAS-^{13}C NMR spectra; size separates were isolated from A-horizons under different land use regimes; enrichment factors were calculated as content in separate divided by content in whole soil

		Enrichment factor			
Fraction	Land use	carboxyl-C	aromatic-C	O-alkyl-C	alkyl-C
Sand (20-2000 μm)	Spruce	0.6	0.9	0.8	0.3
	Mixed deciduous	0.5	0.8	0.8	0.4
	Pasture	0.9	0.9	1.0	0.5
	Arable	0.2	0.3	0.3	0.1
Silt (2-20 μm)	Spruce	0.9	0.9	1.0	1.0
	Mixed deciduous	1.0	1.0	0.8	0.8
	Pasture	0.8	1.0	0.8	0.9
	Arable	1.0	1.4	1.1	0.7
Clay (<2 μm)	Spruce	1.9	1.4	1.5	2.2
	Mixed deciduous	1.8	1.4	1.6	2.2
	Pasture	1.7	1.1	1.4	2.2
	Arable	2.6	1.8	2.4	3.1

(After Guggenberger et al., 1995.)

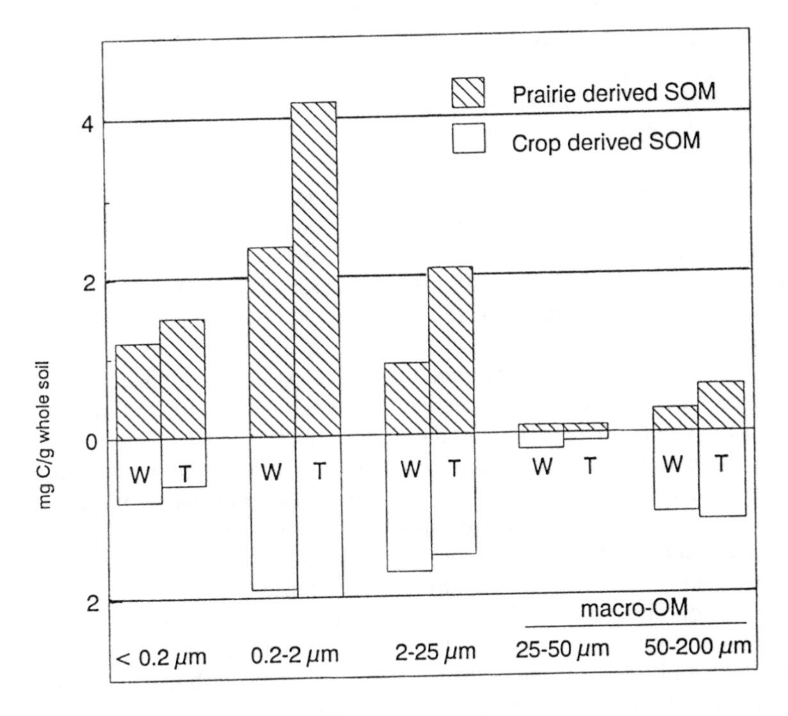

Figure 14. Distribution of prairie- and crop-derived SOM between size separates from Sanborn Field (Udollic Ochraqualf) plots under timothy (T) and wheat (W) for 97 years. (Data from Balesdent et al., 1988.)

2. Soil Management and Agronomy

Carbon associated with clay from a Udollic Ochragualf prairie (C_4) soil cultivated to timothy (*Phleum pratense* L.) or wheat (C_3) for 97 years was found to be most persistent (Balesdent et al., 1988). The major loss of carbon occurred during the initial 27 years and was from macro-OM in separates >25 μm and SOM in fine silt (2 to 25 μm). After 27 years, the quantity of carbon in fine silt declined more than that for other size separates. Timothy showed a greater stabilizing effect on prairie derived carbon than wheat (Figure 14), the effect being most marked in coarse clay (0.2 to 2 μm) and fine silt. In contrast, the stabilization of carbon derived from crop residues differed less between crops, suggesting that the higher level of carbon under timothy was due to a greater protection of old SOM rather than a greater retention of crop residue derived carbon. Wheat derived carbon made up 44% of the SOM in fine and coarse clay, while 67% of the carbon in fine silt was from wheat residues. The major replacement of prairie carbon by crop derived carbon took place during the initial 27 year period of cultivation. Since the prairie carbon left after this period was relatively stable subsequent carbon turnover mainly involved SOM of crop origin. In accordance with Vitorello et al. (1989), Skjemstad et al. (1990) and Cambardella and Elliott (1992), macro-OM fractions were found to include highly stable materials. Of the macro-OM in the 50 to 200 μm separate 36% was of prairie origin after 97 years of cultivation.

Cropping a Typic Haplaquept clay soil to barley (*Hordeum vulgare* L.) or lucerne (*Medicago sativa* L.) for two years produced significantly larger carbon and carbohydrate contents in sand (50 to 6000 μm) compared to fallow reference treatment (Angers and Mehuys, 1990). In contrast only small effects of cropping system were registered for silt (2-50 μm) and clay (<2 μm) size separates. For Quebec soils under continuous lucerne or maize for more than 5 years, Elustondo et al. (1990) also observed that the cropping system had no consistent influence on the carbon content in clay whereas sand (50 to 2000 μm) and silt (2 to 50 μm) were 61 and 15% higher, respectively, under lucerne than under maize. Thus the initial effects of introducing crops with a higher return to soil of plant residues are an increase in carbon associated with coarser particles. This carbon is in particulate plant remains undergoing decomposition. The sand associated SOM derived from cropping systems with elevated return of plant residues shows a high decomposability. Christensen (1987) found that straw incorporation more than doubled the specific decomposition rate (μg CO_2-C/mg fraction-C) of sand SOM whereas the decomposability of clay and silt SOM was less affected (Table 7). Sand SOM averaged 6% of the carbon content in whole soil, but contributed 18% to the CO_2 evolution.

The relative increase in clay and silt carbon after annual incorporation of straw and after repeated additions of equivalent amounts of nutrients in mineral fertilizer (NPK) and animal manure (FYM) is shown in Figure 15. Following straw incorporation or use of NPK, the relative increase in silt SOM (2 to 20 μm) was generally twice that observed for clay (<2 μm). Mineral fertilizers influence SOM levels by increasing crop productivity which causes a greater return of plant residues to soil. The effect of NPK on SOM may thus be similar to that of straw. The FYM treatment caused similar relative increases in clay and silt SOM, probably because animal manure provides an additional input of biologically processed material.

Recalling the greater stability of silt than of clay associated carbon in these soils the results suggest that in a particular soil where similar carbon contents have been reached either by the use of mineral fertilizer or animal manure, animal manure will cause a higher proportion of labile SOM than dressings of mineral fertilizers.

Annual dressings of organic manures cause increased carbon content in most size separates (Leinweber et al., 1990, 1991). The distribution of whole soil carbon across size separates shows a relative shift following organic manures (Leinweber and Reuter, 1990), a higher proportion of whole soil carbon being found in separates >63 μm and in medium silt (6.3 to 20 μm). While coarse (20 to 63 μm) and fine silt (2 to 6.3 μm) show no consistent trend, clay (<2

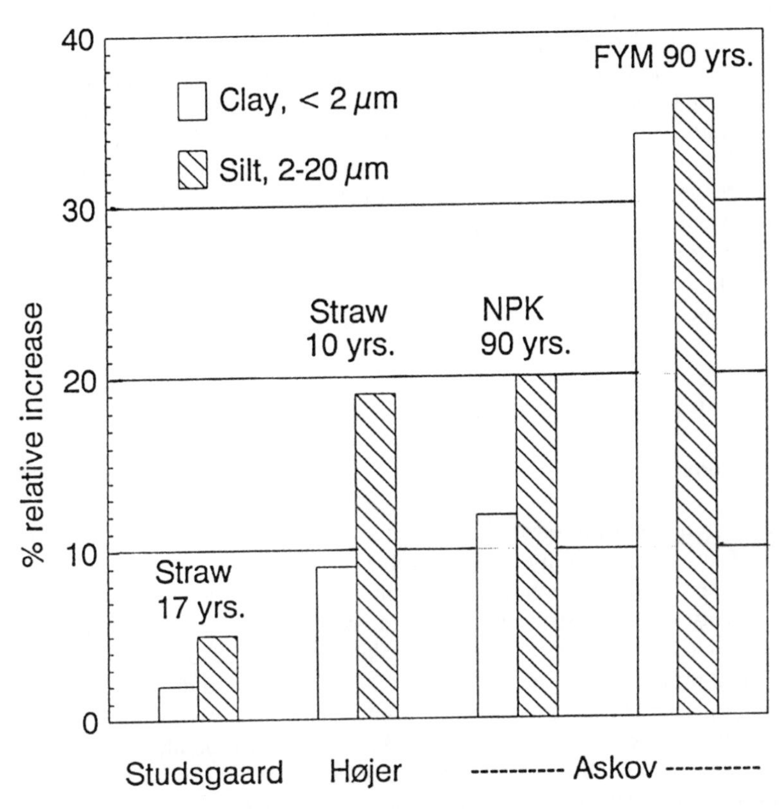

Figure 15. Relative increase in carbon content of clay and silt following straw incorporation (reference: straw removed) and use of mineral fertilizer (NPK) and animal manure (FYM) (reference: unmanured). Studsgaard, Højer, and Askov soils contained 6, 24 and 9% of clay, respectively. (Data from Christensen, 1986, 1988.)

μm) contains a significant smaller proportion of whole soil carbon in soil receiving organic manures. Angers and N'Dayegamiye (1991) also found that animal manure applications over a period of 10 years significantly increased soil carbon and carbohydrate contents. Differences in contents were apparent in all size separates. For carbohydrates differences were only quantitative, the relative sugar composition being almost similar in whole soil and size separates regardless of manure treatments. Using pyrolysis-field ionization mass spectrometry, Schulten and Leinweber (1991) observed that annual addition of farmyard manure caused an increase in lignin fragments and lipids such as fatty acids in the silts. Unfertilized soil were higher in carbohydrates and nitrogen containing heterocyclic compounds in separates <20 μm.

Light fraction carbon associated with coarse organomineral complexes is much reduced upon cultivation of virgin soils, forests and grasslands. The loss of carbon is due to mineralization and transfer of SOM to silt and clay sized separates. In tropical soils, clay SOM may contain the most stable carbon while in temperate soils SOM in silt appears more stable than clay SOM. Available data suggest that the chemical composition of SOM is not drastically influenced by changes in amounts and nature of plant residue inputs. Contrasting land uses do however reveal some differences between permanently vegetated soils and arable rotation in terms of SOM quality. The sand associated SOM responds more rapidly to changes in land use and soil management than separates smaller than 20 μm.

III. Secondary Organomineral Size Separates

A. Fractionation

Any determination of the size distribution of aggregates incorporates a determination of aggregate stability, because some disintegrating force is necessary to break the tertiary soil structure into secondary organomineral complexes.

Obviously the size distribution obtained on aggregates reflect both the forces that stabilize the aggregate against breakdown and the abrasiveness of the fractionation procedure (Beare and Bruce, 1993). Early studies on aggregation and aggregate stability attempted to apply disintegrating forces that would simulate the forces operative in the field. Although results obtained in laboratory studies may be valuable for the understanding of processes in the field, the relationship between aggregate behaviour in the field and laboratory results remains empirical.

Aggregates are generally isolated by dry or wet sieving techniques (Kemper and Rosenau, 1986). Dry sieving rests on the mechanical breakdown of structure when dry soil is exposed to relatively short periods of rotary sieving in air. In contrast, wet sieving procedures also include abrasion of soil structure due to wetting processes. The mechanical forces imposed on the soil during wet sieving are considered less abrasive than those related to dry sieving, because collisions of aggregates with each other and with the sieve are alleviated by the viscosity of water. Initial sample moisture content (field moist or predried), the wetting procedures employed and actual sieve set-up are all decisive for the outcome of a specific wet sieving procedure. Abrasive forces operating during wet sieving procedures cause slaking and dispersion. Slaking is the break up of larger aggregates into smaller ones while dispersion is the release of primary size particles from aggregates.

For studies targeting the relationship between SOM turnover and secondary soil structure, methods are often designed according to the specific purpose of the individual study (Elliott, 1986; Elliott et al., 1991; Beare and Bruce, 1993; Cambardella and Elliott, 1993a, 1993b).

Secondary organomineral complexes are usually separated into size classes within the range >0.02 mm to <6-8 mm, the terms micro- and macro-aggregates referring to aggregates smaller and larger than 0.25 mm, respectively. During fractionation, aggregate size classes will accumulate similarly sized non-aggregated primary particles. Corrections for "loose" primary particles (Kemper, 1965; Coughlan et al., 1978; Elliott, 1986; Elliott et al., 1991; Beare and Bruce, 1993; Cambardella and Elliott, 1993b) e.g. by calculating the content of true aggregates in a given size class (Christensen, 1986; Black and Chanasyk, 1989) are considered important when addressing questions related to carbon turnover.

B. Composition of Aggregates

Macroaggregates isolated by dry sieving from a sandy clay loam with almost no sand >250 μm were found to be enriched in clay (<2 μm) and silt (2 to 20 μm) sized particles whereas the fraction <0.25 mm was comparatively low in these particles (Christensen, 1986). For a loamy sand with 35% sand coarser than 250 μm, 0.25 to 2 mm aggregates appeared to be lower in clay and silt than whole soil, the 2 to 20 mm aggregates and the <0.25 mm fraction containing relatively more clay and silt. But after correction for 'loose' sand particles which had accumulated in its own size fraction during the dry sieving operation, the distribution pattern for clay and silt in true aggregates followed that of the sandy clay loam soil, macroaggregates being enriched in <20 μm particles.

Dormaar (1983) observed small and non-consistent differences between sand contents of wet sieved macroaggregate classes from capillary wetted samples of a cultivated Typic Haploboroll. The 0.10 to 0.25 mm fraction was however significantly higher in sand than macroaggregates. Similarly, Elliott (1986) found that wet stable macroaggregates, retrieved from a cultivated Pachic Haplustoll and slaked by submersion of airdry samples in water, were substantially lower in sand than fractions <0.21 mm. However, in this study, the content of sand in macroaggregates decreased gradually with decreasing size of aggregates. This trend was confirmed in a subsequent study (Cambardella and Elliott, 1993b) which compared wet stable macroaggregates isolated from slaked and from capillary wetted samples of the same soil. Following slaking, large (>2 mm) aggregates from cultivated soils appeared higher in sand (62%) than 0.25 to 2 mm aggregates (49%). In contrast to Elliott (1986), macroaggregates were richer in sand than the 0.05 to 0.25 mm fraction (37% sand). Compared to slaking, capillary wetting of soil caused macroaggregates to accumulate less sand, and large and small macroaggregates did not differ markedly in sand content. The 0.05 to 0.25 mm fraction of capillary wetted samples held more sand than the macroaggregates.

The depletion in sand observed both for dry and for wet stable macroaggregates may well be apparent only, because sand particles located on the exterior surface of macroaggregates may be detached during sieving operations (Coughlan et al., 1978). Since the ratio of external surface to aggregate volume increases as aggregate sizes decline, the relative loss of sand particles will increase with decreasing aggregate size, a trend apparent in the studies of Christensen (1986), Elliott (1986), and Cambardella and Elliott (1993b).

If the detached sand particles are smaller than the lower limit of the macroaggregate class, the 'loose' sand particles escape the aggregate fraction and accumulate in the <0.25 mm fractions (Coughlan et al., 1978; Dormaar, 1983; Christensen, 1986; Elliott, 1986). Consequently macroaggregates will appear enriched in clay and silt. Alternatively, if loose sand from larger aggregate fractions accumulate in similarly sized aggregate classes, aggregates will appear depleted in clay and silt. Unless the soil is very low in sand particles >250 μm, both phenomena probably occur simultaneously. Thus the application of procedures for normalizing the sand content becomes essential for interpreting results on aggregate composition correctly.

Partly in line with Coughlan et al. (1978) who stated that larger dry aggregates in heavy clay soils were simply random agglomerates of the particles that make up the soil, the studies referred to in this section suggest that the distribution of primary organomineral complexes is similar for differently sized wet and dry stable macroaggregates. Reviewing early literature on mechanical properties of aggregates, Braunack and Dexter (1989) concluded that many workers had found no significant differences in the clay, silt and sand contents of different aggregate size classes.

The similar content of primary particles does not necessarily conflict with Tisdall and Oades (1982) who stated that water stable macroaggregates are not simply a random arrangement of the various particles responsible for the texture of the soil. Assuming that aggregate hierarchy exists in soils where aggregate stability is controlled by organic materials (Oades and Waters, 1991), macroaggregates can be viewed to consist of microaggregates, sand grains and particulate organic matter not intimately associated with primary organomineral particles. Similar proportions of clay, silt and sand in macroaggregates may merely indicate that incorporation of sand particles into macroaggregates is directly related to the amount of sand present in a given soil, the sand playing a passive role in the formation of macroaggregates, and that silt is integrated into microaggregates in accordance with the proportion they make up in whole soil.

It is difficult to establish directly the particle composition of microaggregates. Obviously primary particles >250 μm are excluded by definition. Experimental evidence from studies based on limited soil dispersion is confounded by size classes containing mixtures of true microaggregates and primary organomineral complexes. Apparently studies have not attempted to distinguish experimentally between free primary particles and particles engaged in microaggregates. Therefore circumstantial evidence and scanning electron microscopy is often invoked. In accordance with Edwards and Bremner (1967), Tisdall and Oades (1982) and Oades and Waters (1991) conclude that water stable 0.02 to 0.25 mm aggregates consist largely of 2 to 20 μm particles which in turn is made up of <2 μm particles, suggesting that fine sand and silt particles do not form part of microaggregates and are only accidentally entrapped during the aggregate forming process. Whereas this may be true for sand size primary particles which generally have very little SOM directly associated with the minerals (most SOM isolated in primary sand size separates being light fraction organic matter), it seems untenable for silt sized primary organomineral complexes. Silt particles generally account for similar or higher proportions of the whole soil than clay, and primary silt particles show significant contents of reactive mineral associated SOM. Thus in a study of 189 British soils, Williams (1971) found that silt (2 to 20 μm) was nearly as important as clay in making aggregates stable. Small proportions of fine sand had a similar effect as silt and clay but when fine sand was a major fraction soils became less stable. The most stable groups of soils were found to have 10 to 15% of particles in the 0.2 to 6 mm range.

Dormaar (1983) found that macroaggregates isolated from capillary wetted samples of a Typic Haploboroll under wheat-fallow rotation were higher in carbon than whole soil while those of a continuous wheat contained about the same. Carbon content generally peaked in 1 to 2 mm macroaggregates. Carter (1992) also found that 1 to 2 mm aggregates from a Podzol were enriched in carbon compared with whole soil. In accordance with data of Tisdall (1980, ref. Oades, 1990), microaggregates (0.08 to 0.25 mm) were markedly lower in carbon than macroaggregates, the lower carbon content most probably being due to accumulation of 'loose' sand particles which were low in carbon and therefore caused a dilution in the carbon content of true microaggregates. Similarly, Water and Oades (1991) found that 0.02 to 0.09 mm microaggregates isolated by wet sieving from Alfisol and Mollisol samples very low in >250 μm sand were depleted in carbon. Carbon contents in macroaggregates did not differ greatly within a soil, but macroaggregates appeared richer in carbon than whole soils. For an Oxisol, contents of carbon in macro- and microaggregates >0.02 mm were almost similar but somewhat lower than whole soil.

Figure 16 shows the carbon content of dry stable macroaggregates from two Danish arable soils with a long record of annual straw incorporation (Christensen, 1986). In the loamy sand, carbon peaked in 2 to 20 mm aggregates while 0.25 to 1 mm aggregates were markedly lower in carbon than whole soil. For the sandy clay loam, carbon in macroaggregates increased gradually with decreasing aggregate size, the highest carbon content being recorded for 0.25 to 0.5 mm aggregates. This carbon distribution pattern reflected the differential accumulation of sand particles in the various aggregate classes, the carbon content being closely correlated with the clay + silt content of the aggregates ($r = 0.814$-0.982).

Clay and silt isolated from macroaggregates by ultrasonic dispersion in water were analyzed for carbon content (Figure 17). The carbon concentration was almost similar for clay from whole soils and from macroaggregates. Compared to silt in whole soil, silt from the <0.25 mm fraction of the loamy sand appeared to be depleted in carbon while silt from 0.25 to 1 mm was enriched.

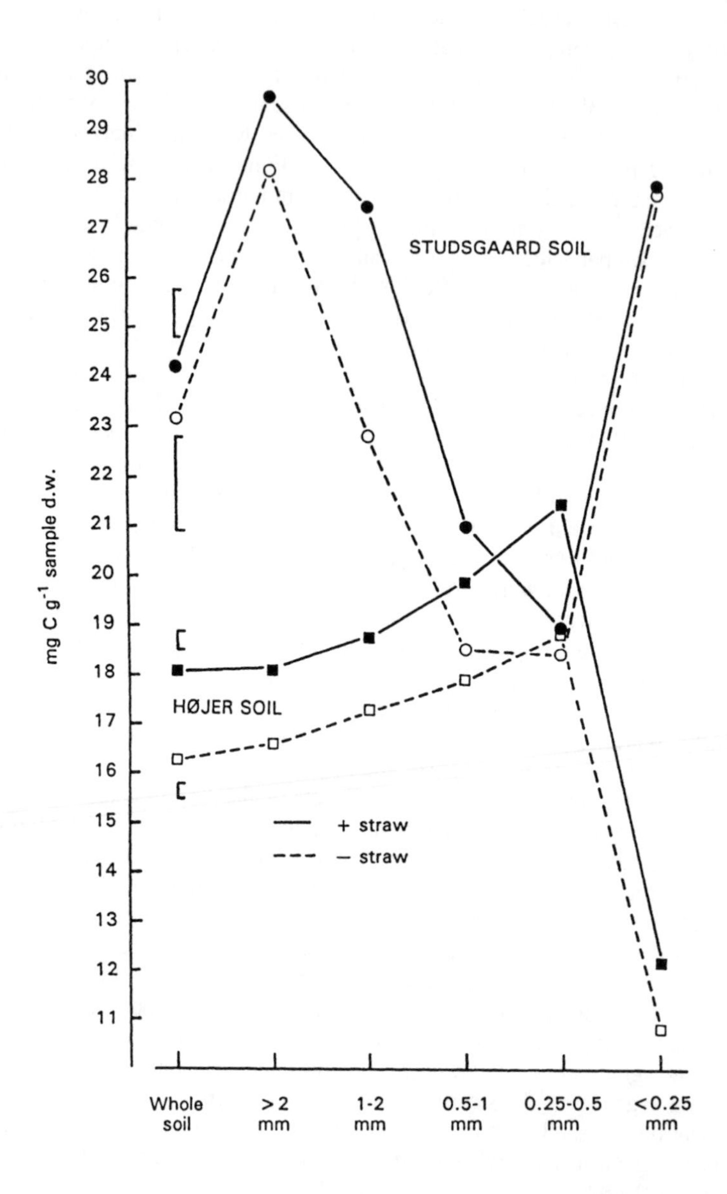

Figure 16. The content of carbon in whole soil and dry-sieved aggregates from Studsgaard loamy sand (circles) and Højer sandy clay loam (squares). Soil was sampled in long-term field experiments where straw had been incorporated (closed symbols) or removed (open symbols). Bars indicate ±1 standard error. (Data from Christensen, 1986.)

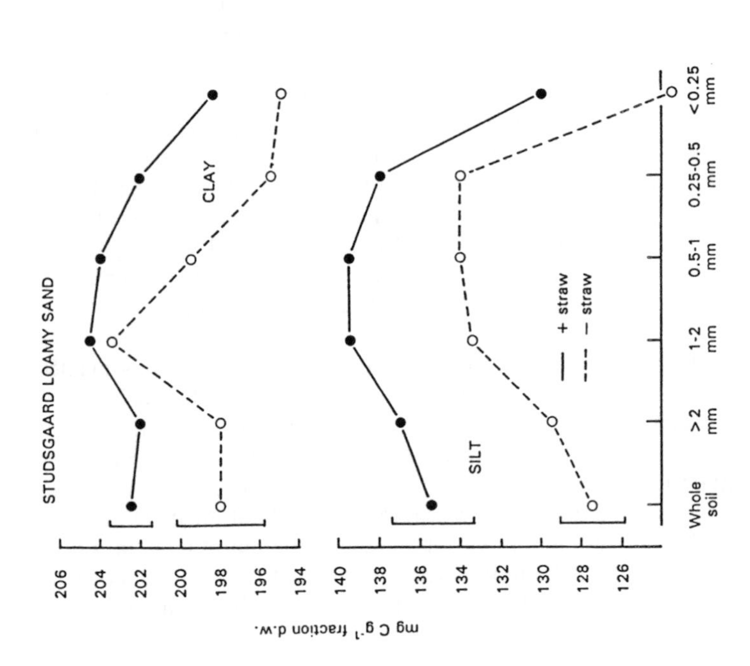

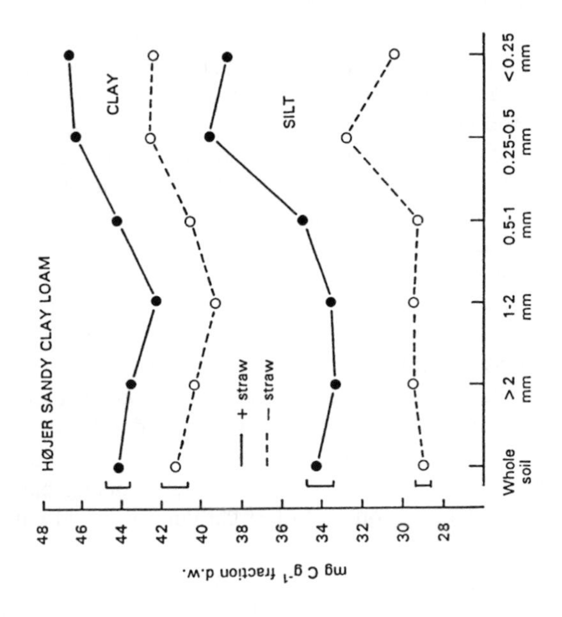

Figure 17. Carbon content in clay (<2 µm) and silt (2-20 µm) isolated from whole soil and dry-sieved aggregate fractions; soil originated from long-term field experiments on Studsgaard loamy sand (left) and Højer sandy clay loam (right); straw was either removed (open symbols) or incorporated (closed symbols). (Data from Christensen, 1986.)

In the sandy clay loam with straw incorporation, silt from the 0.25 to 0.5 mm and <0.25 mm fractions was somewhat higher in carbon. Comparing size separates obtained from wet sieved and from ultrasonically dispersed samples of an Alfisol, Water and Oades (1991) found that clay (<2 μm) from completely dispersed whole soil was somewhat higher in carbon than clay isolated from wet sieved soil. The opposite was true for silt sized separates (2-20 μm). In contrast, clay from ultrasonically dispersed samples of Mollisol and Oxisol was substantially lower in carbon than clay dispersed during wet sieving.

For a cultivated silty loam exposed to slaking and wet sieving, Elliott (1986) found microaggregates (0.05 to 0.30 mm) to be considerably lower in carbon than macroaggregates. The carbon level of the <0.05 mm fraction was in line with that of macroaggregates. When prewetted by misting, all macro- and microaggregate classes held similar carbon contents except for the 0.05 to 0.09 mm microaggregates which were comparatively low in carbon. Carbon in aggregates from slaked samples generally declined with decreasing size of aggregate classes. This pattern was maintained even when carbon contents in aggregate classes were expressed on a clay + silt content basis. It was concluded that microaggregates contained less carbon associated with clay plus silt than macroaggregates.

A subsequent study (Cambardella and Elliott, 1993b) essentially confirmed these observations, although misted samples in this study showed consistently lower carbon contents in the <0.05 mm fraction than larger fractions. Compared to capillary wetted samples, the concentration of carbon on a sand free basis was increased in small macroaggregates following slaking, suggesting that slaking breaks less-stable aggregates leaving behind more stable aggregates enriched in carbon. For Alfisol under pasture and wheat rotations, Smetten et al. (1992) also observed that >1 mm aggregates surviving slaking and wet sieving were substantially higher in carbon than aggregates broken down to <250 μm. However, when aggregates were corrected for sand and for particulate organic matter trapped within the aggregates, Cambardella and Elliott (1993b) found slaked and capillary-wetted small aggregates to be similar in carbon content. Thus clay + silt (<53 μm particles) associated with aggregates >0.05 mm did not differ in carbon concentrations regardless of aggregate size and prewetting method, whereas slightly less carbon was present in the <0.05 mm fraction.

Similar conclusions were reached in a study of a cultivation chronosequence on tropical sandy loam Ultisols (Elliott et al., 1991), although the 0.02 to 0.05 mm microaggregates were higher in carbon when corrected for sand content.

Elliott et al. (1991) found that light fractions (<1.8 g ml^{-1}) comprised 24 and 28% of the carbon found in 1 to 8 and 0.2 to 1 mm wet stable macroaggregates, respectively. For the 0.05-0.2 mm microaggregates, 14% of the carbon was in light fractions while corresponding values for 0.02-0.05 and <0.02 mm fractions were 5 and 2% only. A high proportion of the light fraction was however charcoal considered to be biologically inert and to have insignificant aggregate forming and stabilizing potential.

Dormaar (1984) found that the proportion of carbon in sugars declined with decreasing size of macroaggregates from cultivated Haploboroll. Resin extractable carbon showed no consistent trend across macroaggregate classes, but was higher for micro- than for macroaggregates (Dormaar, 1983). All aggregates contained however the same suite of monosaccharides and aliphatic carboxylic acids. Skjemstad et al. (1990) examined wet stable aggregates from a previous rainforest soil (C_3 species) which had been under C_4 grasses for up to 83 years. After 35 and 83 years of pasture, microaggregates from 0 to 7.5 cm depth contained 59 and 33% of rainforest carbon, while macroaggregates held 51 and 25%, respectively. Since macroaggregates were considered to be composed of microaggregates containing higher proportions of old carbon, the SOM responsible for stabilization of macroaggregates was taken to be younger and more transient.

Cambardella and Elliott (1993a) observed that 18-20% of the carbon in wet stable macroaggregates was in silt (2 to 20 μm) particles having a density of 2.1 to 2.2 g ml^{-1}. This fraction was termed ELF (enriched labile fraction) because it showed a higher nitrogen mineralization rate than intact macroaggregates. Skjemstad et al. (1993) used high energy ultraviolet photo-oxidation to study physically protected SOM in clay and silt sized particles. For carbon in clay (<2 μm) and silt (2-20 μm) up to 23 and 36%, respectively, was considered to be physically protected within the particles. Photo-oxidation resulted in a reduction in particle size as the organic cementing agents were oxidized. Although clay and silt fractions were found to contain essentially modern carbon, the carbon in SOM resisting photo-oxidation was older.

Available evidence suggests that macroaggregates and whole soil hold similar proportions of clay, silt and sand sized primary particles, provided that aggregate classes are corrected for loose sand accumulating during the sieving procedure. Differences in carbon content of clay and silt isolated from dry macroaggregates of different size are relatively small. For soils low in macro-OM, the carbon content of true macroaggregates is highly correlated with their clay + silt content. It is indicated that SOM in microaggregates is more stable than the SOM responsible for binding microaggregates into macroaggregates, but more research is needed to substantiate the concepts of aggregate hierarchy and inter- and intraaggregate macro-OM.

C. Carbon Mineralization

For a cultivated loam, Elliott (1986) found that the proportion of carbon evolved as CO_2 during a 20 day incubation was smaller for intact wet-stable macroaggregates (>0.3 mm) than for microaggregates and macroaggregates crushed to <0.3 mm, the last two fractions releasing almost equal proportions of CO_2. Microaggregates from a native grassland mineralized less carbon than intact macroaggregates which in turn released a smaller proportion of carbon than crushed macroaggregates. However, differences in the percent of carbon mineralized from the various fractions were generally small, the most significant trend being an increased CO_2 evolution early in the incubation period when macroaggregates were crushed to the size of microaggregates. Gupta and Germida (1988) found for both cultivated and native grass soils that macroaggregates released significantly more CO_2 during a 2 week incubation than did microaggregates. Crushing macroaggregates to <0.25 mm increased the carbon mineralization. For dry-sieved macroaggregates of a cultivated silt loam, Seech and Beauchamp (1988) observed that CO_2 production during a 2 week incubation was greatest for microaggregates (<0.25 mm), and that CO_2 evolution decreased with increasing size of macroaggregates. Substantially lower carbon mineralization was seen for aggregates >5 mm, while differences within the four aggregate classes ranging from 0.25 to 5 mm was less than 7%. During the first 10 days, however, CO_2 production in 0.25 to 1 mm aggregate classes was almost twice as great as in 1 to 20 mm aggregates. Crushing of aggregates to powder increased significantly the proportion of carbon released as CO_2 during incubation.

Gregorich et al. (1989) examined the CO_2 evolution from 1 to 2 mm aggregates subjected to increasing intensity of ultrasonic dispersion. Microaggregate disruption was considered to peak at an energy level of 300 to 500 J ml^{-1} soil suspension. Table 10 shows the CO_2 evolution from sand, silt and clay sized material during a 20 day incubation. Increased disruption of 1 to 2 mm aggregate samples significantly increased the CO_2 release from the sand size fraction while silt sized material showed almost similar specific carbon mineralization across disruption treatments. The proportion of carbon released from clay decreased dramatically when passing from the shake to the 100 to 300 J ml^{-1} ultrasonic treatment, but at the same time clay was accumulating carbon released from breakdown of microaggregates. Thus the intra-aggregate carbon released during dispersion and accumulating in the >50 μm fraction was significantly more decomposable than the carbon left in clay sized organomineral particles after microaggregate disruption.

Freeze/thaw cycles were found to double the bioavailability of SOM in whole soil samples and wet stable 1 to 2 mm aggregates from a sandy loam under anaerobic conditions (Christensen and Christensen, 1991). Two successive freeze/thaw events with an intervening incubation period was found to make similar amounts of SOM bioavailable. It was calculated that the decrease in microbial biomass carbon accounted for only half the freeze/thaw derived increase in bioavailable carbon.

Examining the bioavailability of oily waste organics stabilized in 1 to 2 mm soil aggregates from a clay loam, Rasiah et al. (1992) found that the carbon mineralization rate increased when aggregates were exposed to stepwise increased dispersion by ultrasonic treatment. Compared to no ultrasonic treatment, severe disruption of aggregates increased the carbon mineralization by 48%. The specific carbon mineralization rate did not however change significantly with increased dispersion level. It was therefore concluded that substantial amounts of potentially biodegradable carbon from oily wastes were physically

Table 10. Cumulated CO_2 release during a 20 day incubation of clay, silt, and sand size fractions isolated from 1 to 2 mm aggregates of a cultivated silt loam; aggregates were exposed to A) gently shaking for 20 minutes; B) ultrasonic treatment at 100-300 J ml^{-1} soil suspension, or C) ultrasonic treatment at 500-1500 J ml^{-1}; treatment A and B were considered to leave microaggregates intact while partly (A) or completely (B) breaking down macroaggregates; at treatment C, microaggregates were disrupted

	mg CO_2-C/g fraction-C		
Aggregate treatment	Clay (<2 μm)	Silt (2-50 μm)	Sand (50-2000 μm)
A. Shake for 20 min., mAG[a] intact	26	13	31
B. 100-300 J ml^{-1}, mAG intact	11	12	27
C. 500-1500 J ml^{-1}, mAG disrupted	8	15	47

[a] mAG = microaggregates. (Data from Gregorich et al., 1989.)

protected in soil aggregates but also that the oily waste carbon exposed by dispersion at different intensity was chemically similar with regard to association with primary organomineral complexes.

The method of separating soil into aggregate size classes most likely influences the results obtained on carbon mineralization. As noted by Seech and Beauchamp (1988), water soluble SOM considered to be readily metabolizable is probably lost from aggregates during wet sieving. Further, the drying of field moist samples and rewetting procedures employed in wet sieving techniques may well increase the availability of SOM in aggregates. To relate more directly the availability of SOM in different aggregate classes to conditions *in situ*, it is advocated to employ field moist soils avoiding drying/rewetting cycles. Dry sieving of field moist samples may be advantageous when availability of carbon in macroaggregates is assessed.

IV. Carbon Storage in Soil

Less than 10% of the carbon isolated with sand sized primary particles is bound into organomineral complexes. The major part occurs as light fraction SOM or macro-OM, encompassing mainly particulate plant and animal residues undergoing decomposition, but also biologically inert SOM such as charcoal. A number of studies suggest that macro-OM isolated from sand sized separates and from whole soil samples behave similarly during SOM turnover and may be identical SOM pools (Christensen, 1992). Macro-OM from sand is considered to be an important pool of bioavailable carbon. At least part of the sand sized macro-OM is involved in formation and stabilization of macroaggregates, e.g. as intra-aggregate particulate organic matter (Cambardella and Elliott, 1992, 1993b). The amount of carbon in macro-OM is sensitive to fluctuations in plant litter inputs and may show significant seasonal and spatial variability in forest soils (Spycher et al., 1983). Seasonal variations are less prominent in arable soils (Adams, 1980), although macro-OM has been found to reflect differences in tillage system (Cambardella and Elliott, 1992). Accumulation of macro-OM is favoured in cold and dry soils, in soils low in pH and in permanently vegetated soils with significant returns of plant litter, e.g. forests and permanent grasslands. In these soils, sand sized macro-OM may account for more than 40% of whole soil carbon whereas macro-OM in long-term arable soils often holds less than 10% of the soil carbon. In accordance with this it is generally observed that carbon in macro-OM is readily depleted when native soils are brought under cultivation and when permanent grasslands are converted to all arable rotation. Often the drop in macro-OM accounts for a significant part of the initial loss in soil carbon.

The bulk of available information suggests that sand sized macro-OM represents a dynamic property of potential use in assessing shorter-term changes in soil carbon storage induced by changes in soil management, land use and vegetation regimes. It may be hypothesized that macroaggregation and physical protection of SOM is more closely linked to the abundance and turnover of sand sized macro-OM than to whole soil carbon levels. Clearly, sand sized macro-OM must be considered to encompass an active SOM pool. It is indicated that macro-OM is quantitatively more important in sandy than in clayey soils (Greenland and Ford, 1964; Richter et al., 1975; Leuschner et al., 1981).

Biologically based and mathematically formulated SOM turnover models generally recognize the importance of soil texture by incorporating adjustment factors or functions related primarily to soil clay content. Response functions are typically based on experimental evidence from different soil types using laboratory incubations with sieved samples. Additions of pure, mined clay species to study the stabilization and subsequent turnover of decomposition products as a function of clay type and clay content are not feasible, the effect of pure clays most likely being different from that of 'dirty', naturally occurring mixtures of clay minerals. Studies involving different soil types differing in clay content suffer from uncertainties related to degree of similarity in soil mineralogy, land use history and other factors related to soil origin. Thus effects considered to be related to clay contents may be significantly influenced by differences in other soil parameters of importance to SOM behaviour. SOM turnover models may gain in realism and also in performance if texture/structure adjustment factors were derived from experiments involving "natural" organomineral complexes and texture/structure gradients established with mineralogically similar soils.

For temperate arable soils, whole soil carbon storage has previously been shown to be positively correlated with soil clay content (e.g. Heinonen, 1974; Körschens, 1980). In line with this, the high carbon enrichment of the clay and silt sized organomineral complexes in soils low in these separates suggest that the carbon storage potentials of clay and silt is lower in sandy than in clayey soils. Active sites for stabilization of newly formed SOM are probably reduced as particles become more saturated with SOM. Sørensen (1967, 1972) showed that addition of mined clay minerals to sandy soils increased the retention of microbial products derived from ^{14}C labelled substrates. Mineralogical species and particle size influenced the stabilization potential of added clay minerals. Further, an increased retention of ^{14}C with increased content of natural soil clays has been demonstrated for a variety of soils and added substrates (Jenkinson, 1977; Sørensen, 1981, 1983; Amato et al., 1987). The lower carbon enrichment for silt than for clay at similar yields of these separates illustrates that clay is more effective in storing carbon. Silt sized primary organomineral complexes do however exhibit a significant potential for carbon storage which should be accounted for when simulating soil texture effects on carbon turnover.

More than 80% of the carbon in temperate arable soils is associated with primary <20 μm organomineral separates. The proportion allocated to clay is positively related to soil clay content at least in the range 5 to 20% clay, and coarse clay may retain soil carbon in more stable organomineral interactions than medium clay. Proportions of whole soil carbon found in sand sized separates could not be related to the abundance of sand in soil. Carbon enrichments of silt was related to silt yields. The higher proportion of carbon found in clayey soils may reflect a more efficient interaction between microbial metabolites and abundant binding sites being available on clay particles less embedded in native SOM. Assuming relatively low and similar annual inputs of organic residues to arable soils, the stabilizing effect of clay on metabolites formed during the decomposition of organic materials will therefore be higher in soils rich in clay. Hence a larger proportion of the organic matter can be found in the clay fraction of these soils because differences in the decomposability of stabilized organic matter has been shown to be largely independent of soil clay + silt content (Sørensen, 1975).

The decay of plant materials is accompanied by a simultaneous synthesis of a range of microbial products that are exposed to further decomposition or stabilized in soil. In this process the synthesis of aromatics appears less prominent, the increased content of aromatic carbon in silt probably being due to a selective stabilization of residual plant derived compounds. In contrast, accumulation of alkyl carbon in clay appear to result both from stabilization of plant derived substances and microbial synthesis *in situ*. Evidence from ^{13}C-NMR studies suggest a higher microbially derived SOM input to clay than to silt, a concept supported by the results on the distribution of soil carbohydrates and diaminopimelic acid.

Natural ^{13}C abundance studies involving more than one sampling after change over of vegetation (Balesdent et al., 1988; Vitorello et al., 1989; Bonde et al., 1992) clearly demonstrate that individual size separates regardless of size class limits encompass pools of carbon that differ in turnover rates. However, available results suggest that in tropical soils the most stable carbon is in clay sized organomineral complexes while in temperate soils silt holds the most resistant carbon.

While Christensen (1987) found good agreement between CO_2 evolved from whole soil and from size separates, other studies show either lower (Gregorich et al., 1989) or higher (Amato and Ladd, 1980; Christensen and Christensen, 1991) pooled contributions from size separates. Accordance between whole soil CO_2 and pooled CO_2 from primary size separates suggests that SOM protected physically within aggregates is quantitatively less important. Higher cumulated CO_2 evolution from size separates may indicate that dispersion has relieved bioavailable carbon either from physical protection within secondary organo-mineral complexes or from microbial biomass killed during soil dispersion. SOM rendered soluble by the dispersion treatment and subsequently included in the finest size separate may also add to an increased CO_2 evolution.

A different mineralization rate of SOM associated with differently sized organomineral complexes may reflect differences in the bioavailability of particle bound SOM. However, differently sized primary organomineral complexes may also affect differently the microenvironment in the immediate vicinity of particle surfaces. Clay minerals probably create a more favourable environment for the decomposer organisms than relatively inert minerals such as quartz. Clay minerals may enhance yields of biomass and microbial metabolites (Filip, 1977) and possess significant catalytic potentials (Wang et al., 1983). Differences between size separates in CEC, specific surface area, carbon load, pH etc. have also been demonstrated.

The distribution pattern of new SOM formed during decomposition of a variety of organic substrates appears similar but remains different from that of native SOM for up to several years. When a pulse of readily available substrate is added, the characteristic distribution of new SOM is established within a few days. For complex substrates the distribution pattern takes longer to be manifest because of interference from residual substrate. Assuming that the sites where new SOM is fixed are determined by the location of the initial decomposers, successive microbial turnover does not induce a transfer of SOM between size separates unless the microbes are adhering to particulate plant remains which are subsequently exposed to comminution by fauna, soil management or abiotic factors such as freeze/thaw or wet/dry cycles. Any substrate derived soluble organics (e.g. exoenzymes or cell contents leaked from lysed organisms) are probably fixed *in situ* through exploitation by neighbouring organisms or chemical/physical stabilization in the organomineral complex. Once established, the distribution pattern of new SOM across organomineral complexes can be traced for extensive periods, indicating that differences in the stability of substrate derived and native SOM associated with differently sized separates is operative over time spans up to decades.

Although individual soil particles are not envisaged to have a finite capacity for binding SOM, it is hypothesized that the strength with which recently formed SOM initially becomes attached to organomineral complexes will be impeded on mineral surfaces already covered in SOM. Although it has been shown that sorption of hydrophobic ring compounds on mineral surfaces is enhanced when minerals are coated with a layer of humic substances (Murphy et al., 1990) and that CEC is positively related to SOM, clay and silt particles already embedded in a matrix of stabilized and chemically less reactive SOM probably exhibit surfaces which are less capable of binding new SOM strongly enough for it to resist further biological attack. Indeed, adsorption and other reversible binding mechanisms may increase the decomposability of various organic compounds.

Substrate derived carbon may become stabilized temporarily by being incorporated into the soil microbial biomass. However, organomineral complexes highly enriched in relatively inert SOM are probably less attractive for microbial adhesion and colony formation. This could result from a reduced ability of the organomineral complex to favourably modify the physico-chemical environment of the microorganisms. Such particles may therefore receive reduced inputs of microbial metabolites formed during turnover of substrate derived SOM.

Conversion of permanently vegetated soils to arable rotations induces a general loss in whole soil carbon. Davidson and Ackerman (1993) estimated an initial average loss of 30% of the carbon in the entire soil profile. A major part of the initial more rapid decline may be ascribed to decomposition and redistribution of sand sized macro-OM, some of which is associated with macroaggregates. As cultivation proceeds and the rate of carbon loss decreases exponentially, residual carbon accumulates mainly in microaggregates and in silt and clay sized primary organomineral complexes. This carbon is relatively stable, causing SOM turnover in long-term arable soils to be limited by the availability of carbon. Therefore the macro-OM is considered to be the most important reservoir of short-term biologically available carbon. Physical protection of macro-OM may well be a key issue in regulating carbon turnover in arable soils. Also bioavailable macro-OM with high C/N ratios may represent a significant nitrogen immobilization potential.

Feller (1993) demonstrated that soil texture influenced cultivation induced changes in the carbon distribution across size separates isolated from three tropical oxisols. For coarse textured cultivated soils, the increase in soil carbon following a 6 year fallow period was mainly in the sand sized macro-OM. In medium textured soil, all size separates showed similar increases. In clay rich cultivated soil converted to permanent pasture, accumulation in clay ($<2\ \mu m$) accounted for most of the increase in whole soil carbon. Similarly, decreases in soil carbon following a forest clearing-cultivation sequence could be ascribed to losses of carbon from sand in coarse textured soils and from clay sized organomineral complexes in the clayey soils.

It appears that for a particular soil different land use strategies may have only moderate consequences for the chemical composition of SOM associated with primary organomineral complexes $<20\ \mu m$. Differences are recorded, however, particularly when very contrasting land use are compared. Further, quantitative changes in carbon return to soil and changes in the temporal and spatial distribution pattern of plant litters and macro-OM induced by change-over of land use may be very significant. Obviously both qualitative and quantitative changes, including distribution patterns, should be considered when evaluating the impact of land use on soil carbon storage and turnover.

To stimulate an increased experimental activity on the particle composition of secondary organomineral complexes, it is ventured that macro- and micro-aggregates contain primary particles in proportions similar to the textural composition of the corresponding fraction of the whole soil. This suggestion does not necessarily imply a fortuitous spatial organization of all primary particles engaged in an aggregate or that aggregates are simply random agglomerates of primary organomineral particles. A restricted selective transport in soil of primary particles during formation of $>2\ \mu m$ aggregates is however implied.

Contempory concepts of aggregates emphasize the role of particulate plant remains in aggregate formation and stabilization (Cambardella and Elliott, 1992, 1993a, 1993b; Oades, 1993; Oades and Waters, 1991; Tisdall, 1991; Waters and Oades, 1991). In many soils, encrustation of particulate residues of plants and fungi is considered an important feature of stable aggregates and a major mechanism in the accumulation of physically protected but otherwise bioavailable SOM. Only a few studies have however attempted to quantify experimentally the pool of intraaggregate particulate organic matter.

The attempts to describe the character of carbon involved in soil aggregation clearly indicate that aggregates contain different pools of carbon, and that novel approaches when further developed may provide new insight into the nature of intra-aggregate carbon. The prevailing models of aggregate formation and stabilization implies that macroaggregates should be enriched in recently formed SOM responsible for cementing microaggregates, primary organomineral complexes, sand and particulate SOM into macroaggregates. Disintegration of macroaggregates should therefore expose bioavailable carbon with a relative high turnover rate.

While sand sized macro-OM holds a carbon pool with short-term turnover and of importance in macroaggregation, carbon in microaggregates is generally thought to be more stable. Evidence from studies where CO_2 release from micro- and macroaggregates have been compared suggest that carbon in microaggregates represents a less labile carbon reservoir. Similarly incubated samples of primary organomineral complexes have shown carbon in sand to be considerable more mineralizable than SOM in clay and silt which are the dominant constituents of microaggregates.

Carbon storage potentials in soil are thus made up of a more readily responding component encompassing sand sized macro-OM of which part is involved in macroaggregation, and a more slowly reacting and less labile pool encompassing clay and silt associated SOM engaged in microaggregates. For most soils the more persisting carbon storage is determined by the proportion of clay and silt in soil, while carbon storage in sand does not appear to depend on the amount of sand in a soil. The potential of soils in storing carbon in sand size macro-OM is not known.

The amount and distribution of carbon stored in soil can be considered a key ecosystem characteristic, reflecting the balance between inputs from primary producers and the herbivore based system, and outputs due to the activity of microbial saprovores and invertebrate saprovores and microbivores (Swift et al.,

1979). The activity of the decomposer system is influenced by edaphic factors and the physico-chemical soil environment including soil moisture, temperature and aeration status. Ruling out extreme environments (e.g. water logged soils, deserts, strongly acidified soils and tundras) and assuming that the diversity and the multiplication potential of the decomposer system is sufficient to allow a full exploitation of available substrate, substrate availability becomes decisive for the decomposer activity in soil. Availability of substrate is largely determined by initial substrate quality (chemical and physical) and by the interaction of substrate and decomposition products with the soil matrix. Thus for agricultural soils, the tertiary structure is important for their carbon storage potentials (Oades, 1984, 1993; van Veen and Kuikman, 1990; Ladd et al., 1993).

In this chapter, the tertiary soil structure has been broken down to primary and secondary structural elements in order to examine their possible contribution to the soil carbon storage potential. Integrating the information on primary and secondary organomineral complexes will of course not produce a complete picture, but may pinpoint important elements to be included in carbon turnover models and provide explanations on cause/effect behind phenomena expressed on the tertiary structural level.

V. Summary

This paper reviews the distribution, chemical composition and turnover of carbon in primary and secondary organomineral complexes obtained by physical fractionation procedures. Secondary complexes result from aggregation of primary complexes which in this context are the basic units of soil structure. Primary complexes represent a functional equivalent to soil texture, and are generally isolated from soil dispersed by ultrasonic vibrations in water. Secondary complexes are obtained following limited dispersion and dry or wet sieving procedures.

Complete soil dispersion is essential for isolation of primary organomineral complexes. The highest carbon concentrations are observed in <5 μm primary organomineral separates. For a number of soils, carbon enrichment of clay and silt has been found to decline following a power function as the proportion of these separates increases in whole soil. Clay generally accounts for >50% of the carbon in soils low in macro-OM, clay and silt (<20 μm) together may account for >90%. The proportion of carbon in carbohydrates generally increases with decreasing particle size, and a greater microbial input to clay than to silt is indicated by hexose to pentose ratios. Analyses of CuO oxidation products reveal that lignin degradation is more advanced in clay than in silt organomineral separates. Clay associated SOM is more aliphatic in nature than SOM in coarser separates while aromatic structures peak in silt. O-alkyl carbon decrease and alkyl carbon increase with decreasing particle size.

For arable soils, the CEC, the adsorption capacity and the specific surface area of primary organomineral complexes decrease with increasing particle size.

While the CEC of particles $>2\ \mu m$ depends on their carbon content, the mineral and the organic part both contribute substantially to clay CEC. Partial removal of SOM increased the specific surface area of clay sized particles while coarser fractions were slightly affected.

Generally the carbon mineralization rate decreases in the order: sand, clay, whole soil and silt. Following addition of simple substrates, new SOM is found in most separates although clay sized organomineral complexes most often show greater accumulations. In the longer term, the proportion of substrate derived SOM in clay declines more rapidly than that in silt indicating a higher stability. It is indicated that newly formed clay SOM is less stable in sandy than in clayey soils.

Light fraction carbon associated with coarse organomineral complexes is much reduced upon cultivation of virgin soils, forests and grasslands. The loss of carbon is due to mineralization and transfer of SOM to silt and clay sized separates. In tropical soils, clay SOM may contain the most stable carbon while in temperate soils SOM in silt appears more stable than clay SOM. Available data suggest that the chemical composition of SOM in $<20\ \mu m$ separates is not drastically influenced by changes in plant residue inputs although permanently vegetated soils and arable rotation may differ somewhat in terms of SOM quality. The sand associated SOM responds more rapidly to changes in land use and soil management than separates $<20\ \mu m$.

Macroaggregates and whole soil appear to hold similar proportions of clay, silt and sand sized primary particles, provided that aggregate classes are corrected for loose sand accumulating during the sieving procedure. Differences in carbon content of clay and silt isolated from dry macroaggregates of different size are relatively small. For soils low in inter- and intraaggregate macro-OM, the carbon content of true macroaggregates is highly correlated with their clay + silt content. It is indicated that SOM in microaggregates is more stable than the SOM responsible for binding microaggregates into macroaggregates, but more research is needed to substantiate the concepts of aggregate hierarchy and inter- and intraaggregate SOM.

The mineralization rate of carbon tends to differ between various aggregate size classes. Crushing of macroaggregates generally increases carbon mineralization.

Sand sized macro-OM represents a dynamic property of potential use in assessing shorter-term changes in soil carbon storage induced by changes in soil management, land use and vegetation regimes. It may be hypothesized that macroaggregation and physical protection of SOM is more closely linked to the abundance and turnover of sand sized macro-OM than to whole soil carbon levels.

Natural ^{13}C abundance studies involving more than one sampling after change over of vegetation demonstrates that individual size separates regardless of size class limits encompass pools of carbon that differ in turnover rates. However, available results suggest that in tropical soils the most stable carbon is in clay sized organomineral complexes while in temperate soils silt holds the most resistant carbon.

For a particular soil different land use strategies may have only moderate consequences for the chemical composition of SOM associated with <20 μm primary complexes. Differences are recorded, however, particularly when very contrasting land use are compared. Both qualitative and quantitative changes, including distribution patterns should be considered, however, when evaluating the impact of different land use on soil carbon storage and turnover.

It is ventured that macro- and microaggregates contain primary particles in proportions similar to the textural composition of the corresponding fraction of the whole soil. This does not necessarily imply a fortuitous spatial organization of all primary particles engaged in an aggregate.

Contempory concepts of aggregates emphasize the role of particulate plant remains in aggregate formation and stabilization. Encrustation of particulate residues is considered an important feature of stable aggregates and a major mechanism in the accumulation of physically protected but otherwise bioavailable SOM. Integration of information on primary and secondary organomineral complexes may pinpoint important elements to be included in carbon turnover models, and provide explanations on cause/effect behind phenomena expressed on the tertiary structural level.

Acknowledgements

I wish to thank Ms. Anne Sehested Jensen for her excellent technical assistance in preparing the manuscript text and figures, and Dr. Per Schjønning for helpful discussions on aggregates. This work was financially supported by the Danish Environmental Research Programme and by the Ministry of Agriculture.

References

Adams, T.McM. 1980. Macro organic matter content of some Northern Ireland soils. *Record Agric. Res.* 28: 1-11.

Amato, M. and J.N. Ladd. 1980. Studies of nitrogen immobilization and mineralization in calcareous soils-V. Formation and distribution of isotope-labelled biomass during decomposition of ^{14}C- and ^{15}N-labelled plant material. *Soil Biol. Biochem.* 12: 405-411.

Amato, M., J.N. Ladd, A. Ellington, G. Ford, J.E. Mahoney, A.C. Taylor, and D. Walsgott. 1987. Decomposition of plant material in Australian soils. IV. Decomposition *in situ* of ^{14}C- and ^{15}N-labelled legume and wheat materials in a range of Southern Australian soils. *Aust. J. Soil Res.* 25: 95-105.

Anderson, D.W., S. Saggar, J.R. Bettany, and J.W.B. Stewart. 1981. Particle size fractions and their use studies of soil organic matter: I. The nature and distribution of forms of carbon, nitrogen and sulfur. *Soil Sci. Soc. Am. J.* 45: 767-772.

Anderson, T.-H. 1991. Bedeutung der Mikroorganismen für die Bildung von Aggregaten im Boden. *Z. Pflanzenernähr. Bodenk.* 154: 409-416.

Angers, D.A. and G.R. Mehuys. 1990. Barley and alfalfa cropping effects on carbohydrate contents of a clay soil and its size fractions. *Soil Biol. Biochem.* 22: 285-288.

Angers, D.A. and A. N'Dayegamiye. 1991. Effects of manure application on carbon, nitrogen, and carbohydrate contents of a silt loam and its particle-size fractions. *Biol. Fertil. Soils* 11: 79-82.

Baldock, J.A., J.M. Oades, A.M. Vassallo, and M.A. Wilson. 1989. Incorporation of uniformly labelled ^{13}C-glucose carbon into the organic fraction of a soil. Carbon balance and CP/MAS ^{13}C NMR measurements. *Aust. J. Soil Res.* 27: 725-746.

Baldock, J.A., J.M. Oades, A.M. Vassallo, and M.A. Wilson. 1990. Solid state CP/MAS ^{13}C N.M.R. analysis of particle size and density fractions of a soil incubated with uniformly labelled ^{13}C-glucose. *Aust. J. Soil Res.* 28: 193-212.

Baldock, J.A., G.J. Currie, and J.M. Oades. 1991. Organic matter as seen by solid state ^{13}C NMR and pyrolysis tandem mass spectrometry. pp. 45-60. In: W.S. Wilson (ed.) *Advances in Soil Organic Matter Research: The Impact on Agriculture and the Environment.* The Royal Society of Chemistry, Cambridge, UK.

Baldock, J.A., J.M. Oades, A.G. Waters, X. Peng, A.M. Vassallo, and M.A. Wilson. 1992. Aspects of the chemical structure of soil organic materials as revealed by solid-state ^{13}C NMR spectroscopy. *Biogeochemistry* 16: 1-42.

Balesdent, J., A. Mariotti, and B. Guillet. 1987. Natural ^{13}C abundance as a tracer for studies of soil organic matter dynamics. *Soil Biol. Biochem.* 19: 25-30.

Balesdent, J., G.H. Wagner, and A. Mariotti. 1988. Soil organic matter turnover in long-term field experiments as revealed by carbon-13 natural abundance. *Soil Sci. Soc. Am. J.* 52: 118-124.

Balesdent, J., J.-P. Pétraud, and C. Feller. 1991. Effets des ultrasons sur la distribution granulométrique des matières organiques des sols. *Science du Sol* 29: 95-106.

Beare, M.H. and R.R. Bruce. 1993. A comparison of methods for measuring water-stable aggregates: implications for determining environmental effects on soil structure. *Geoderma* 56: 87-104.

Black, J.M.W. and D.S. Chanasyk. 1989. The wind erodibility of some Alberta soils after seeding: aggregation in relation to field parameters. *Can. J. Soil Sci.* 69: 835-847.

Bonde, T.A., B.T. Christensen, and C.C. Cerri. 1992. Dynamics of soil organic matter as reflected by natural ^{13}C abundance in particle size fractions of forested and cultivated Oxisols. *Soil Biol. Biochem.* 24: 275-277.

Braunack, M.V. and A.R. Dexter. 1989. Soil aggregation in the seedbed: A review. I. Properties of aggregates and beds of aggregates. *Soil Tillage Res.* 14: 259-279.

Bremner, J.M. and D.A. Genrich. 1990. Characterisation of the sand, silt, and clay fractions of some Mollisols. pp. 423-438 In: M.F. De Boodt, M.H.B. Hayes and A. Herbillon (eds.) *Soil Colloids and Their Associations in Aggregates*. Plenum Press, New York, USA.

Broersma, K. and L.M. Lavkulich. 1980. Organic matter distribution with particle-size in surface horizons of some sombric soils in Vancouver Island. *Can. J. Soil Sci.* 60: 583-586.

Burford, J.R., T.L. Deshpande, D.J. Greenland, and J.P. Quirk. 1964. Influence of organic materials on the determination of the specific surface areas of soils. *J. Soil Sci.* 15: 192-201.

Cambardella, C.A. and E.T. Elliott. 1992. Particulate soil organic-matter changes across a grassland cultivation sequence. *Soil Sci. Soc. Am. J.* 56: 777-783.

Cambardella, C.A. and E.T. Elliott. 1993a. Methods for physical separation and characterization of soil organic matter fractions. *Geoderma* 56: 449-457.

Cambardella, C.A. and E.T. Elliott. 1993b. Carbon and nitrogen distribution in aggregates from cultivated and native grassland soils. *Soil Sci. Soc. Am. J.* 57: 1071-1076.

Carter, M.R. 1992. Influence of reduced tillage systems on organic matter, microbial biomass, macro-aggregate distribution and structural stability of the surface soil in a humid climate. *Soil Tillage Res.* 23: 361-372.

Catroux, G. and M. Schnitzer. 1987. Chemical, spectroscopic, and biological characteristics of the organic matter in particle size fractions separated from an Aquoll. *Soil Sci. Soc. Am. J.* 51: 1200-1207.

Cerri, C., C. Feller, J. Balesdent, R. Victoria, and A. Plenecassagne. 1985. Application du tracage isotopique naturel en ^{13}C, a létude de la dynamique de la matiere organique dans les sols. *C. R. Acad. Sc. Paris, Serie II,* 300: 423-428.

Cheshire, M.V. and C.M. Mundie. 1981. The distribution of labelled sugars in soil particle size fractions as a means of distinguishing plant and microbial carbohydrate residues. *J. Soil Sci.* 32: 605-618.

Cheshire, M.V., B.T. Christensen, and L.H. Sørensen. 1990. Labelled and native sugars in particle-size fractions from soils incubated with ^{14}C straw for 6 to 18 years. *J. Soil Sci.* 41: 29-39.

Chiou, C.T., J-F. Lee, and S.A. Boyd. 1990. The surface area of soil organic matter. *Environ. Sci. Technol.* 24: 1164-1166.

Christensen, B.T. 1985. Carbon and nitrogen in particle size fractions isolated from Danish arable soils by ultrasonic dispersion and gravity-sedimentation. *Acta Agric. Scand.* 35: 175-187.

Christensen, B.T. 1986. Straw incorporation and soil organic matter in macro-aggregates and particle size separates. *J. Soil Sci.* 37: 125-135.

Christensen, B.T. 1987. Decomposability of organic matter in particle size fractions from field soils with straw incorporation. *Soil Biol. Biochem.* 19: 429-435.

Christensen, B.T. 1988. Effects of animal manure and mineral fertilizer on the total carbon and nitrogen contents of soil size fractions. *Biol. Fertil. Soils* 5: 304-307.

Christensen, B.T. 1992. Physical fractionation of soil and organic matter in primary particle size and density separates. *Adv. Soil Sci.* 20: 1-90.

Christensen, B.T. and S. Bech-Andersen. 1989. Influence of straw disposal on distribution of amino acids in soil particle size fractions. *Soil Biol. Biochem.* 21: 35-40.

Christensen, B.T. and L.H. Sørensen. 1985. The distribution of native and labelled carbon between soil particle size fractions isolated from long-term incubation experiments. *J. Soil Sci.* 36: 219-229.

Christensen, B.T. and L.H. Sørensen. 1986. Nitrogen in particle size fractions of soils incubated for five years with ^{15}N-ammonium and ^{14}C-hemicellulose. *J. Soil Sci.* 37: 241-247.

Christensen, B.T., F. Bertelsen, and G. Gissel-Nielsen. 1989. Selenite fixation by soil particle-size separates. *J. Soil Sci.* 40: 641-647.

Christensen, S. and B.T. Christensen. 1991. Organic matter available for denitrification in different soil fractions: effect of freeze/thaw cycles and straw disposal. *J. Soil Sci.* 42: 637-647.

Coleman, D.C., J.M. Oades, and G. Uehara (eds.). 1989. *Dynamics of Soil Organic Matter in Tropical Ecosystems.* NifTAL Project, University of Hawaii at Manoa, USA.

Cortez, J. 1989. Effect of drying and rewetting on mineralization and distribution of bacterial constituents in soil fractions. *Biol. Fertil. Soils* 7: 142-151.

Cortez, J. and A. Cherqui. 1991. Plant growth and the mineralization of adsorbed ^{14}C- and ^{15}N-labelled organic compounds. *Soil Biol. Biochem.* 23: 261-267.

Coughlan, K.J., G.D. Smith, and W.E. Fox. 1978. Measurement of the size of natural aggregates. pp. 205-210. In: W.W. Emerson et al. (eds.) *Modifications of Soil Structure.* Wiley Interscience, Chichester, UK.

Curtin, D., P.M. Huang, and H.P.W. Rostad. 1987. Components and particle size distribution of soil titratable acidity. *Soil Sci. Soc. Am. J.* 51: 332-336.

Dalal, R.C. and R.J. Henry. 1988. Cultivation effects on carbohydrate contents of soil and soil fractions. *Soil Sci. Soc. Am. J.* 52: 1361-1365.

Dalal, R.C. and R.J. Mayer. 1986. Long-term trends in fertility of soils under continuous cultivation and cereal cropping in Southern Queensland. III. Distribution and kinetics of soil organic carbon in particle-size fractions. *Aust. J. Soil Res.* 24: 293-300.

Davidson, E.A. and I.L. Ackerman. 1993. Changes in soil carbon inventories following cultivation of previously untilled soils. *Biogeochemistry* 20: 161-193.

Dormaar, J.F. 1983. Chemical properties of soil and water-stable aggregates after sixty-seven years of cropping to spring wheat. *Plant Soil* 75: 51-61.

Dormaar, J.F. 1984. Monosaccharides in hydrolysates of water-stable aggregates after 67 years of cropping to spring wheat as determined by capillary gas chromatography. *Can. J. Soil Sci.* 64: 647-656.

Dudas, M.J. and S. Pawluk. 1970. Naturally occurring organo-clay complexes of orthic black chernozems. *Geoderma* 3: 5-17.

Edwards, A.P. and J.M. Bremner. 1967. Microaggregates in soils. *J. Soil Sci.* 18: 64-73.

Elliott, E.T. 1986. Aggregate structure and carbon, nitrogen, and phosphorous in native and cultivated soils. *Soil Sci. Soc. Am. J.* 50: 627-633.

Elliott, E.T. and C.A. Cambardella. 1991. Physical separation of soil organic matter. *Agric. Ecosyst. Environ.* 34: 407-419.

Elliott, E.T., C.A. Palm, D.E. Reuss, and C.A. Monz. 1991. Organic matter contained in soil aggregates from a tropical chronosequence: correction for sand and light fraction. *Agric. Ecosyst. Environ.* 34: 443-451.

Elustondo, J., D.A. Angers, M.R. Laverdiere, and A. N'Dayegamiye. 1990. Étude comparative de l'ágrégation et de la matiére organique associée aux fractions granulométriques de sept sols sous culture de maïs ou en prairie. *Can. J. Soil Sci.* 70: 395-402.

Emerson, W.W., R.C. Forster, and J.M. Oades. 1986. Organo-mineral complexes in relation to soil aggregation and structure. pp. 521-548. In: P.M. Huang and M. Schnitzer (eds.) *Interactions of Soil Minerals with Natural Organics and Microbes*. SSSA, Wisconsin, USA.

Feller, C. 1993. Organic inputs, soil organic matter and functional soil organic compartments in low-activity clay soils in tropical zones. pp. 77-88. In: K. Mulongoy and R. Merckx (eds.) *Soil Organic Matter Dynamics and Sustainability of Tropical Agriculture.* John Wiley & Sons, Chichester, UK.

Feller, C., G. Burtin, B. Gérard, and J. Balesdent. 1991. Utilisation des résines sodiques et des ultrasons dans le fractionnement granulométrique de la matière organique des sols. Intérêt et limites. *Science du Sol* 29: 77-93.

Feller, C., E. Schouller, F. Thomas, J. Rouiller, and A.J. Herbillon. 1992. N_2-BET specific surface areas of some low activity clay soils and their relationships with secondary constituents and organic matter contents. *Soil Sci.* 153: 293-299.

Filip, Z. 1977. Einfluss von Tonmineralen auf die mikrobielle Ausnutzung der kohlenstoffhaltigen Substanzen und Bildung der Biomasse. *Ecol. Bull. (Stockh.)* 25: 173-179.

Gee, G.W. and J.W. Bauder. 1986. Particle-size analysis. pp. 383-441. In: A. Klute (ed.), *Methods of Soil Analysis, Part 1, 2nd Edition.* ASA and SSSA Publ., Madison, WI, USA.

Greenland, D.J. and G.W. Ford. 1964. Separation of partially humified organic materials from soils by ultrasonic dispersion. *Trans. 8th Int. Congr. Soil Sci., Bucharest,* 3: 137-148.

Gregorich, E.G. 1989. *The effects of texture on the stabilization and physical protection of organic matter in soil.* Ph. D. thesis, The Faculty of Graduate Studies, University of Guelph, Ontario, Canada.
Gregorich, E.G., R.G. Kachanoski, and R.P. Voroney. 1988. Ultrasonic dispersion of aggregates: Distribution of organic matter in size fractions. *Can J. Soil. Sci.* 68: 395-403.
Gregorich, E.G., R.G. Kachanoski, and R.P. Voroney. 1989. Carbon mineralization in soil size fractions after various amounts of aggregate disruption. *J. Soil Sci.* 40: 649-659.
Gregorich, E.G., R.P. Voroney, and R.G. Kachanoski. 1991. Turnover of carbon through the microbial biomass in soils with different textures. *Soil Biol. Biochem.* 23: 799-805.
Guggenberger, G. and B.T. Christensen. 1993. Organische Substanz in Korngrössenfraktionen unterschiedlich genutzter Böden. *Mitteilgn. Dtsch. Bodenkundl. Gesellsch.* 72: 357-360.
Guggenberger, G., B.T. Christensen, and W. Zech. 1994. Land use effects on the composition of organic matter in soil particle size separates I. Lignin and carbohydrate signature. *Eur. J. Soil Sci.* 45:449-458.
Guggenberger, G., W. Zech, L. Haumaier, and B.T. Christensen. 1995. Land use effects on the composition of organic matter in soil particle size separates II. CPMAS and solution ^{13}C NMR analysis. *Eur. J. Soil Sci.* 46: (in press).
Gupta, V.V.S.R. and J.J. Germida. 1988. Distribution of microbial biomass and its activity in different soil aggregate size classes as affected by cultivation. *Soil Biol. Biochem.* 20: 777-786.
Hamdy, A.A. and G. Gissel-Nielsen. 1977. Fixation of selenium by clay minerals and iron oxides. *Z. Pflanzenernaehr. Bodenkd.* 140: 63-70.
Hassink, J., L.A. Bouwman, K.B. Zwart, J. Bloem, and L. Brussaard. 1993. Relationships between soil texture, physical protection of organic matter, soil biota, and C and N mineralization in grassland soils. *Geoderma* 57: 105-128.
Haynes, R.J. 1993. Effect of sample pretreatment on aggregate stability measured by wet sieving or turbidimetry on soils of different cropping history. *J. Soil Sci.* 44: 261-270.
Heinonen, R. 1974. Humusversorgung, Bodenstruktur und Wasserhaushalt. *Landwirtsch. Forsch. Sonderheft* 30/II: 123-124.
Huang, P.M. and M. Schnitzer (eds.) 1986. *Interactions of Soil Minerals with Natural Organics and Microbes.* SSSA Publ. Inc., Madison, WI, USA.
Huang, P.M., R. Grover, and R.B. McKercher. 1984. Components and particle size fractions involved in atrazine adsorption by soils. *Soil Sci.* 138: 20-24.
Jastrow, J.D. and R.M. Miller. 1991. Methods for assessing the effects of biota on soil structure. *Agric. Ecosyst. Environ.* 34: 279-303.
Jenkinson, D.S. 1977. Studies on the decomposition of plant material in soil. V. The effects of plant cover and soil type on the loss of carbon from ^{14}C labelled ryegrass decomposing under field conditions. *J. Soil Sci.* 28: 424-434.

Jenkinson, D.S., D.E. Adams, and A. Wild. 1991. Model estimates of CO_2 emissions from soil in response to global warming. *Nature* 351: 304-306.

Jocteur-Monrozier, L., J.N. Ladd, R.W. Fitzpatrick, R.C. Foster, and M. Raupach. 1991. Components and microbial biomass content of size fractions in soils of contrasting aggregation. *Geoderma* 49: 37-62.

Kanazawa, S. and Z. Filip. 1986. Distribution of microorganisms, total biomass, and enzyme activities in different particles of Brown soil. *Microb. Ecol.* 12: 205-215.

Kemper, W.D. 1965. Aggregate stability. pp. 511-519. In: C.A. Black et al. (eds.) *Methods of Soil Analyses, Part 1,* SSSA Publ. Madison, WI, USA.

Kemper, W.D. and R.C. Rosenau. 1986. Aggregate stability and size distribution. pp. 425-442. In: A. Klute (ed.) *Methods of Soil Analysis, Part 1, 2nd Edition.* ASA and SSSA Publ., Madison, WI, USA.

Körschens, M. 1980. Beziehungen zwischen Feinanteil, C_t- and N_t-Gehalt des Bodens. *Arch. Acker- Pflanzenbau Bodenkd.* 24: 585-592.

Ladd, J.N., J.W. Parsons, and M. Amato. 1977a. Studies of nitrogen immobilization and mineralization in calcareous soils-I. Distribution of immobilized nitrogen amongst soil fractions of different particle size and density. *Soil Biol. Biochem.* 9: 309-318.

Ladd, J.N., J.W. Parsons, and M. Amato. 1977b. Studies of nitrogen immobilization and mineralization in calcareous soils-II. Mineralization of immobilized nitrogen from soil fractions of different particle size and density. *Soil Biol. Biochem.* 9: 319-325.

Ladd, J.N, M. Amato, and J.W. Parsons. 1977c. Studies of nitrogen immobilization and mineralization in calcareous soils. III. Concentration and distribution of nitrogen derived from the soil biomass. pp. 301-311. In: *Soil Organic Matter Studies, Vol. 1.* IAEA, Vienna, Austria.

Ladd, J.N., M. Amato, and J.M. Oades. 1985. Decomposition of plant material in Australian soils. III. Residual organic and microbial biomass C and N from isotope-labelled legume material and soil organic matter, decomposing under field conditions. *Aust. J. Soil Res.* 23: 603-611.

Ladd, J.N., R.C. Foster, and J.O. Skjemstad. 1993. Soil structure: carbon and nitrogen metabolism. *Geoderma* 56: 401-434.

Leinweber, P. 1988. *Erfassung und Charakterisierung organischmineralischer Komplexe (OMK) und ihrer Differenzierung in Böden von Dauerveldversuchen der DDR.* Ph. D. thesis, Diss. A., Wilhelm-Pieck-Universität, Rostock, Germany.

Leinweber, P. and G. Reuter. 1988. Menge und Qualität organisch-mineralischer Komplexe in Böden unterschiedlicher Standorte. *Tag.-Ber. Akad. Landwirtsch.-Wiss. DDR.* 269: 223-235.

Leinweber, P. and G. Reuter. 1990. Zum Einfluss unterschiedlicher Düngung auf Kohlenstoff und Stickstoff in organisch-mineralischen Komplexen. *Tag.-Ber., Akad. Landwirtsch.-Wiss.* 295: 169-178.

Leinweber, P. and H.-R. Schulten. 1992. Differential thermal analysis, thermogravimetry and in-source pyrolysis-mass spectrometry studies on the formation of soil organic matter. *Thermochim. Acta* 200: 151-167.

Leinweber, P., G. Reuter, E. Schnieder, M. Smukalski, and F. Asmus. 1990. Einfluss unterschiedlicher Düngungsmassnahmen auf organisch-mineralische Komplexe in Ap-Horizonten von Dauerfeldversuchen auf weichselglazialen Grundmoränen der DDR. *Arch. Acker- Pflanzenbau Bodenkd.* 34: 411-417.

Leinweber, P., G. Reuter, M. Körschens, and J. Garz. 1991. Veränderungen der organischen Komponenten organisch-mineralischer Komplexe (OMK) in Dauerfeldversuchen auf Löss und lössbeeinflussten Substraten. *Arch. Acker-Pflanzenbau Bodenkd.* 35: 85-93.

Leinweber, P., G. Reuter, and K. Brozio. 1993. Cation exchange capacities of organo-mineral particle-size fractions in soils from long-term experiments. *J. Soil Sci.* 44: 111-119.

Leuschner, H.H., R. Aldag, and B. Meyer. 1981. Dichte-Fraktionierung des Humus in A_p-Horizonten von Sandböden mit unterschiedlicher Körnung und Nutzungs-Vorgeschichte. *Mitteilgn. Dtsch. Bodenkundl. Gesellsch.* 32: 583-592.

Martin, A., A. Mariotti, J. Balesdent, P. Lavelle, and R. Vuattoux. 1990. Estimate of organic matter turnover rate in a savanna soil by ^{13}C natural abundance measurements. *Soil Biol. Biochem* 22: 517-523.

Morra, M.J., R.R. Blank, L.L. Freeborn, and B. Shafii. 1991. Size fractionation of soil organo-mineral complexes using ultrasonic dispersion. *Soil Sci.* 152: 294-303.

Murphy, E.M., J.M. Zachara, and S.T. Smith. 1990. Influence of mineral-bound humic substances on the sorption of hydrophobic organic compounds. *Environ. Sci. Technol.* 24: 1507-1516.

Nkedi-Kizza, P., P.S.C. Rao, and J.W. Johnson. 1983. Adsorption of diuron and 2,4,5-T on soil particle-size separates. *J. Environ. Qual.* 12: 195-197.

Oades, J.M. 1984. Soil organic matter and structural stability: mechanisms and implications for management. *Plant Soil* 76: 319-337.

Oades, J.M. 1989. An introduction to organic matter in mineral soil. pp. 89-159. In: J.B. Dixon and S.B. Weed (eds.) *Minerals in Soil Environments, Second Edition.* SSSA Publ. Inc. Madison, WI, USA.

Oades, J.M. 1990. Associations of colloids in soil aggregates. pp. 463-483. In: M.F. De Boodt, M.H.B. Hayes, and A. Herbillon (eds.) *Soil Colloids and Their Associations in Aggregates.* Plenum Press, New York, USA.

Oades, J.M. 1993. The role of biology in the formation, stabilization and degradation of soil structure. *Geoderma* 56: 377-400.

Oades, J.M. and A.G. Waters. 1991. Aggregate hierarchy in soils. *Aust. J. Soil Res.* 29: 815-828.

Oades, J.M., A.M. Vassallo, A.G. Waters, and M.A. Wilson. 1987. Characterization of organic matter in particle size and density fractions from a red-brown earth by solid-state ^{13}C N.M.R. *Aust. J. Soil Res.* 25: 71-82.

Oades, J.M., A.G. Waters, A.M. Vassallo, M.A. Wilson, and G.P. Jones. 1988. Influence of management on the composition of organic matter in a red-brown earth as shown by ^{13}C nuclear magnetic resonance. *Aust. J. Soil Res.* 26: 289-299.

Raine, S.R. and H.B. So. 1993. An energy based parameter for the assessment of aggregate bond energy. *J. Soil Sci.* 44: 249-259.

Rasiah, V., R.P. Voroney, and R.G. Kachanoski. 1992. Bioavailability of stabilized oily waste organics in ultrasonifield soil aggregates. *Water, Air, Soil Pollut.* 63: 179-186.

Richter, M., I. Mizuno, S. Aranguez, and S. Uriarte. 1975. Densimetric fractionation of soil organo-mineral complexes. *J. Soil. Sci.* 26: 112-123.

Schnitzer, M. and P. Schuppli. 1989a. The extraction of organic matter from selected soils and particle size fractions with 0.5 M NaOH and 0.1 M $Na_4P_2O_7$ solutions. *Can. J. Soil Sci.* 69: 253-262.

Schnitzer, M. and P. Schuppli. 1989b. Method for the sequential extraction of organic matter from soils and soil fractions. *Soil Sci. Soc. Am. J.* 53: 1418-1424.

Schnitzer, M., J.A. Ripmeester, and H. Kodama. 1988. Characterization of the organic matter associated with a soil clay. *Soil Sci.* 145: 448-454.

Schulten, H.-R. and P. Leinweber. 1991. Influence of long-term fertilization with farmyard manure on soil organic matter: Characteristics of particle-size fractions. *Biol. Fertil. Soils* 12: 81-88.

Schulten, H.R. and M. Schnitzer. 1990. Aliphatics in soil organic matter in fine-clay fractions. *Soil Sci. Soc. Am. J.* 54: 98-105.

Seech, A.G. and E.G. Beauchamp. 1988. Denitrification in soil aggregates of different sizes. *Soil Sci. Soc. Am. J.* 52: 1616-1621.

Skjemstad, J.O., R.C. Dalal, and P.F. Barron. 1986. Spectroscopic investigations of cultivation effects on organic matter of Vertisols. *Soil Sci. Soc. Am. J.* 50: 354-359.

Skjemstad, J.O., R.P. Le Feuvre, and R.E. Prebble. 1990. Turnover of soil organic matter under pasture as determined by ^{13}C natural abundance. *Aust. J. Soil Res.* 28: 267-276.

Skjemstad, J.O., L.J. Janik, M.J. Head, and S.G. McClure. 1993. High energy ultraviolet photo-oxidation: a novel technique for studying physically protected organic matter in clay- and silt-sized aggregates. *J. Soil Sci.* 44: 485-499.

Smetten, K.R.J., A.D. Rovira, S.A. Wace, B.R. Wilson, and A. Simon. 1992. Effect of tillage and crop rotation on the surface stability and chemical properties of a red-brown earth (Alfisol) under wheat. *Soil Tillage Res.* 22: 27-40.

Spycher, G., P. Sollins, and S. Rose. 1983. Carbon and nitrogen in the light fraction of a forest soil: Vertical distribution and seasonal patterns. *Soil Sci.* 135: 79-87.

Swift, M.J., O.W. Heal, and J.M. Anderson. 1979. *Decomposition in Terrestrial Ecosystems.* Blackwell Sci. Publ., Oxford, UK.

Sørensen, L.H. 1967. Duration of amino acid metabolites formed in soils during decomposition of carbohydrates. *Soil Sci.* 104: 234-241.

Sørensen, L.H. 1972. Stabilization of newly formed amino acid metabolites in soil by clay minerals. *Soil Sci.* 114: 5-11.

Sørensen, L.H. 1975. The influence of clay on the rate of decay of amino acid metabolites synthesized in soils during decomposition of cellulose. *Soil Biol. Biochem.* 7: 171-177.

Sørensen, L.H. 1981. Carbon-nitrogen relationships during the humification of cellulose in soils containing different amounts of clay. *Soil Biol. Biochem.* 13: 313-321.

Sørensen, L.H. 1983. Size and persistence of the microbial biomass formed during the humification of glucose, hemicellulose, cellulose, and straw in soils containing different amounts of clay. *Plant Soil* 75: 121-130.

Theng, B.K.G, G.J. Churchman, and R.H. Newman. 1986. The occurrence of interlayer clay-organic complexes in two New Zealand soils. *Soil Sci.* 142: 262-266.

Tiessen, H. and J.W.B. Stewart. 1983. Particle-size fractions and their use in studies of soil organic matter: II. Cultivation effects on organic matter composition in size fractions. *Soil Sci. Soc. Am. J.* 47: 509-514.

Tisdall, J.M. 1991. Fungal hyphae and structural stability of soil. *Aust. J. Soil Res.* 29: 729-743.

Tisdall, J.M and J.M. Oades. 1982. Organic matter and water-stable aggregates in soils. *J. Soil Sci.* 33: 141-163.

Turchenek, L.W. and J.M. Oades. 1978. Organo-mineral particles in soils. pp. 138-144. In: W.W. Emerson et al. (eds.) *Modification of Soil Structure.* Wiley, Chichester, UK.

Turchenek, L.W. and J.M. Oades. 1979. Fractionation of organo-mineral complexes by sedimentation and density techniques. *Geoderma* 21: 311-343.

Van Veen, J.A. and P.J. Kuikman. 1990. Soil structural aspects of decomposition of organic matter by micro-organisms. *Biogeochemistry* 11: 213-234.

Vitorello, V.A., C.C. Cerri, F. Andreux, C. Feller, and R.L. Victoria. 1989. Organic matter and natural carbon-13 distribution in forested and cultivated oxisols. *Soil Sci. Soc. Am. J.* 53: 773-778.

Wang, T.S.C., S.W. Li, and Y.L. Ferng. 1978. Catalytic polymerization of phenolic compounds by clay minerals. *Soil Sci.* 126: 15-21.

Wang, T.S.C., M.C. Wang, and Y.L. Ferng. 1983. Catalytic synthesis of humic substances by natural clays, silts and soils. *Soil Sci.* 135: 350-360.

Wang, T.S.C., P.M. Huang, C.-H. Chou, and J.-H. Chen. 1986. The role of soil minerals in the abiotic polymerization of phenolic compounds and formation of humic substances. pp. 251-281. In: P.M. Huang and M. Schnitzer (eds.) *Interactions of Soil Minerals with Natural Organics and Microbes.* SSSA, Madison, Wisconsin, USA.

Waters, A.G. and J.M. Oades. 1991. Organic Matter in Water-stable Aggregates. pp. 163-174. In: W.S. Wilson (ed.) *Advances in Soil Organic Matter Research: The Impact on Agriculture and the Environment*. The Royal Society of Chemistry, Cambridge, UK.

Watson, J.R. and J.W. Parsons. 1974a. Studies of soil organo-mineral fractions. I. Isolation by ultrasonic dispersion. *J. Soil Sci.* 25: 1-8.

Watson, J.R. and J.W. Parsons. 1974b. Studies of soil organo-mineral fractions. II. Extraction and characterization of organic nitrogen compounds. *J. Soil Sci.* 25: 9-15.

Williams, R.J.B. 1971. Relationships between the composition of soils and physical measurements made on them. pp. 5-35. *Rep. Rothamsted Exp. St. 1970, Part 2.* Harpenden, UK.

Zech, W. and I. Kögel-Knabner. 1994. Patterns and regulation of organic matter transformation in soils: litter decomposition and humification. pp. 303-334. In: *Flux Control in Biological Systems*. Academic Press Inc., San Diego, CA, USA.

Zeikus, J.G. 1981. Lignin metabolism and the carbon cycle. Polymer biosynthesis, biodegradation, and environmental recalcitrance. *Adv. Microb. Ecol.* 5: 211-243.

Zhang, H., M.L. Thompson, and J.A. Sandor. 1988. Compositional differences in organic matter among cultivated and uncultivated Argiudolls and Hapludalfs derived from loess. *Soil Sci. Soc. Am. J.* 52: 216-222.

Storage of Soil Carbon in the Light Fraction and Macroorganic Matter

E.G. Gregorich and H.H. Janzen

I. Introduction

Plant residues are the largest source of carbon entering the soil. These residues enter the soil as dead and decaying above-ground biomass, senescent root tissue, sloughed root cells, and root exudates. The carbon in these residues is present in a wide range of substrates, from readily decomposable cytoplasmic materials to more resistant cell wall components. Upon entering the soil, these residues

1-56670-033-7/96/$0.00+$.50

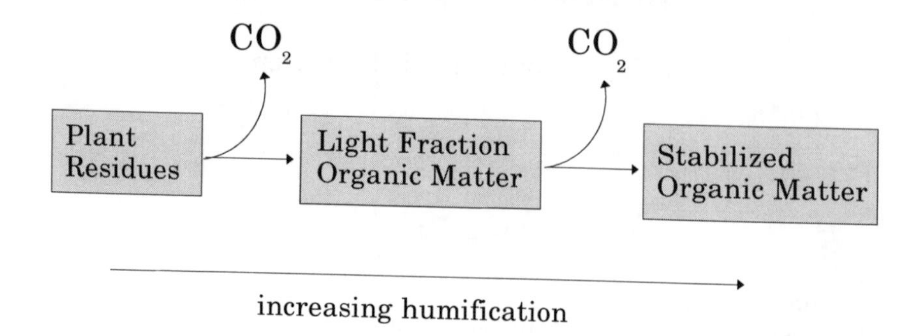

Figure 1. Conceptual view of light fraction organic matter as a transitory intermediate between plant litter and stable organic matter.

undergo physical and chemical transformations, and the organic carbon is eventually released from the soil as CO_2 or stabilized by humification or association with mineral components.

Knowledge about the nature, distribution, and turnover of organic matter in soils is critical to understanding its storage of carbon and nutrients. This chapter will consider the *light fraction* and *macroorganic matter* as a transitory pool of organic matter between fresh residues and humified, stable organic matter (Figure 1). The 'free' or uncomplexed organic matter in soil is that which has not undergone significant transformations and can be separated by density using heavy liquids, or by size using sieving.

Free organic matter plays an important role in determining the structure and function of the soil ecosystem by acting as an energy source for heterotrophic organisms and as a reservoir of relatively labile C and plant nutrients. This pool of organic matter is one of several indicators that may be used to evaluate the quality of organic matter (Gregorich et al., 1994) and permits assessment of the effects of decomposition on soil fertility, residue persistence, and organisms.

II. Definitions and Methods of Separation

A. Light Fraction

The light fraction (LF) of soil is isolated from soils by flotation on dense liquids. The density of the heavy liquids varies among investigations but is typically between 1.5 and 2.0 g cm^{-3}, a density which yields fractions with the highest proportions of undecomposed plant material (Turchenek and Oades, 1979). Because the density of soil minerals is usually > 2.0 g cm^{-3}, flotation on dense liquids yields LFs composed mainly of free organic material. Small deviations in liquid density, particularly between 1.9 and 2.4 g cm^{-3} (Richter et al., 1975), can result in large differences in carbon concentration of the LF. While there is no definitive density for separation of the LF, we will define this fraction as organic material with a specific density < 2.0 g cm^{-3}.

Various liquids have been used to separate the LF from the heavier mineral components (Gregorich and Ellert 1993; Christensen 1992). The organic solvents initially used to fractionate soil presented problems of toxicity, carbon contamination, and coagulation of suspended particles. These problems have been largely alleviated by using inorganic media.

Some of the LF is entrapped within aggregates, and plant fragments, such as roots, are loci for aggregate formation (Dormaar and Foster, 1991; Oades, 1993). As a result, recovery of all of the LF requires disruption of aggregates. Oades (1972) reported that substances resembling plant fragments were released from the heavy fraction of a soil after treatment with acid and alkali and further density fractionation. Golchin et al. (1994) isolated LF material before and after aggregate disruption, and found that the organic matter occluded within aggregates comprised 0.69 to 1.81% of the soil mass and 9.2 to 17.5% of the total soil C.

The organic matter in the LF is comprised primarily of plant debris, as evident by its cellular structure, but may also contain fungal hyphae, spores, seeds, charcoal, and animal remains (Greenland and Ford, 1964). The LF contains a portion of the microbial biomass (Kanazawa and Filip, 1986) and may also contain some humified organic matter (Greenland and Ford, 1964; Skjemstad et al., 1986).

B. Macroorganic Matter

Large particles of organic matter, termed 'macroorganic matter', can be separated by sieving or by flotation and sieving, using methods similar to those used for removing roots (Al-Khafaf et al., 1977). Winnowing techniques have also been used; for example, Theodorou (1990) sieved air-dried soils into different size fractions and then separated the heavier and lighter components of each fraction by winnowing in a gentle stream of air. The size of macroorganic matter varies among investigations (e.g., Oades, 1989), but is usually equated with the sand-size fraction (50 to 2000 μm). The sieving procedures used to isolate macroorganic matter are usually the same as those used in particle-size analysis of soil, but without pretreatment for removing organic matter, carbonates, and iron oxides (Gregorich and Ellert, 1993).

Microscopic examination shows that macroorganic matter may include fungal hyphae, seeds, spores, and faunal skeletons but is dominated by undecomposed plant residues with a recognizable cellular structure. The organic matter in the sand-size fraction has also been shown to contain a large proportion of extractable humic substances (Catroux and Schnitzer, 1987).

C. Differences between LF and Macroorganic Matter

Both the LF and macroorganic matter contain mostly plant residues, but their chemical and biological properties are apparently not identical. Macroorganic matter may contain organic matter > 2.0 g cm^{-3} and, therefore, may be chemically different from LF. Gregorich et al. (1995) observed large differences between the $\delta^{13}C$ of the LF and the sand-size fractions from a soil under maize (*Zea mays* L.) for 25 years. More than 70% of the C in the LF was derived from maize, compared to only 45% of the C in the sand-size fraction. In a survey of 20 forest and cropped soils, Ellert and Gregorich (1994) found that the sand-size fraction contained more organic C than the LF. The C:N ratio of the organic matter was lower in the sand-size fraction, suggesting that it was more decomposed than that of the LF. The sand-size fraction may also include organic matter as coatings on the sand grains; material that is different in composition from that floated off the sand-size fraction (Zhang et al., 1988; Baldock et al., 1992).

Dalal and Mayer (1987) concluded that density fractionation may be more effective than particle-size fractionation in separating labile from non-labile organic matter fractions. Because particles of macroorganic matter may be disintegrated into finer particles as a result of cultivation or by the mechanical action of sieving, the LF may more accurately reflect the changes from decomposition and humification. As well, the macroorganic matter may include non-labile materials associated with mineral particles.

III. Comparison of LF and Macroorganic Matter among Soils and Ecosystems

The diversity in the sources and composition of plant residues produces extreme heterogeneity of the LF and macroorganic matter among soils. Other factors such as environmental conditions and management practices may also influence the amount and composition of soil LF and macroorganic matter.

A. Climosequences and Soil Depth

Molloy and Speir (1977) evaluated the relationship between LF and environmental factors in a climosequence of soils. They concluded that the contribution of the LF to whole soil organic C and total N was highest at the extreme ends of the climosequence, where factors such as aridity and cold temperatures limited decomposition. At various sites in Saskatchewan, Canada, the large differences in the C content of the LF found under similarly managed soils were also attributed partly to differences in environmental characteristics that influence decomposition rate (Janzen et al., 1992). Under relatively arid conditions, LF material probably decomposes at a slower rate and accumulates at higher levels than under more humid conditions.

Data from a series of forested sites show that the proportion of C and N in the LF tends to increase with increasing latitude (Christensen, 1992). This trend reflects the combined effect of vegetation and climate on the chemical composition of the LF.

The distribution of LF in the soil profile is related to the placement of recent residues. Studies of the LF in grassland (Garwood et al., 1972) and forest (Spycher et al., 1983) soil profiles have shown that the amount of LF and the proportion of organic C in LF decline abruptly below the surface 2 to 3 cm and then diminish more gradually with depth. Biederbeck et al. (1994) found that the LF C content of the 7.5 to 15 cm layer of cultivated soils was less than half that in the 0 to 7.5 cm layer. In intensively tilled soils the distribution of LF within the profile is likely more uniform.

B. Chemical Composition

The composition of the LF is controlled primarily by the chemical composition of the carbon inputs (i.e., vegetation type) and the nature and extent of soil decomposition processes (i.e., type of organisms and soil environmental conditions).

The LF is enriched in organic carbon and plant nutrients, such as nitrogen (Table 1). For cultivated soils the carbon concentration in the LF typically ranges from 20 to 30% and the nitrogen concentration from 0.5 to 2.0% (Janzen et al., 1992; Greenland and Ford, 1964; Whitehead et al., 1975; Dalal and Mayer, 1986d). The carbon concentration of the LF is usually lower than that of plant tissue (Table 2), probably because of the presence of small amounts of mineral matter.

Although the LF occupies a small portion of the soil mass, it can contain a substantial portion of the total carbon in soil because of its high carbon concentration. Janzen et al. (1992) reported that the LF accounts for 2 to18% of the total soil carbon and 1 to 12% of the total nitrogen. A study of German soils found that the LF contained 7 to 12% of the soil carbon (Leuschner et al., 1981). The LF in cultivated soils under cereals in Greenland and Ford's (1964) study contained slightly higher proportions (from 21 to 25%) of total soil carbon.

Table 1. Proportion of soil mass and whole soil carbon and nitrogen in the light fraction of soils under different managements

			Light fraction		
Reference[a]	Fraction density or size, location	Management	% Dry weight of whole soil	Carbon ----------% of total	Nitrogen -----------
1	< 1.7 g cm^{-3}	Continuous wheat (+)[b]	0.49	5.4	2.8
	Indian Head, Sask., Canada	Wheat/fallow (+)	0.25	2.9	1.4
1	< 1.7 g cm^{-3}	Continuous wheat (-)	1.23	6.4	3.5
	Melfort, Sask., Canada	Wheat fallow (-)	0.72	3.9	2.3
1	< 1.7 g cm^{-3}	Continuous wheat	2.39	17.5	11.5
	Scott, Sask., Canada	Canola/fallow	1.01	7.4	4.9
2	< 1.8 g cm^{-3}	Forest	1.38	8.8	5.6[d]
	Plainfield, Ont., Canada	Pasture	0.91	5.6	3.1[d]
		Maize cropping	0.42	5.4	2.8[d]
3	< 1.8 g cm^{-3}	Forest	1.15	5.0	2.6
	Highgate, Ont., Canada	Maize cropping	0.95	4.5	2.4
4	< 2.0 g cm^{-3}	Continuous Pasture	2.81	28.5	28.7
	Australia	Continuous Wheat	1.27	20.6	11.5
		Wheat/fallow	0.78	25.3	10.8
5	< 2.0 g cm^{-3}	Forest	2.8	27.1	15.0
	Langlands-Logie, Qld., Australia	Cereal cropping	n.d.[c]	10.4	4.2
5	< 2.0 g cm^{-3}	Forest	2.2	20.7	10.0
	Cecilvale, Qld., Australia	Cereal cropping	n.d.	6.8	2.8
5	< 2.0 g cm^{-3}	Forest	2.4	30.0	14.0
	Riverview, Qld., Australia	Cereal cropping	n.d.	22.5	9.1

Table 1. continued--

5	< 2.0 g cm^{-3}	Forest	1.8	14.6	8.0
	Thallon, Qld., Australia	Cereal cropping	n.d.	7.2	4.4
5	< 2.0 g cm^{-3}	Grassland	3.2	30.8	15.0
	Billa Billa, Qld., Australia	Cereal cropping	n.d.	13.9	7.1
5	< 2.0 g cm^{-3}	Grassland	2.6	21.7	8.0
	Waco, Qld., Australia	Cereal cropping	n.d.	9.6	3.8
6	< 2.06 g cm^{-3}	Continuous pasture	5.36	23.0	17.8
	Berkshire, England	Ryegrass/clover	1.46	20.2	11.4
		Barley	0.36	8.5	4.1
7	> 53 μm	Native grassland	n.d.	39	29
	Nebraska, U.S.A.	Wheat/fallow (no-till)	n.d.	25	12
		Wheat/fallow (stubble mulch)	n.d.	19	16
		Wheat/fallow (bare fallow)	n.d.	18	16
8	< 1.7 g cm^{-3}	Continuous wheat (+)	1.95	14.6	8.2
	Swifit Current, Sask., Canada	Continuous wheat (-)	1.19	10.2	5.9
		Wheat/fallow	0.70	6.8	3.4
9	< 1.7 g cm^{-3}	Grassland	2.92	24.1	16.8
	Lethbridge, Alta., Canada	Continuous wheat	1.85	17.5	12.4
		Wheat/fallow	0.84	10.1	6.5

[a] 1, Janzen et al., 1992; 2, Ellert and Gregorich, 1994; 3 and [d], Ellert and Gregorich, 1994 (unpublished data); 4, Greenland and Ford, 1964; 5, Dalal and Mayer, 1986a, 1986d, 1987; 6, Whitehead et al., 1975; 7, Cambardella and Elliott, 1992; 8, Biederbeck et al., 1994; 9, Bremer et al., 1994.
[b] (+) = fertilized; (-) = unfertilized; [c] n.d. = data not given in reference.

Table 2. Organic C, total N and C:N ratio of plant tissue, light fraction and whole soils under forest and corn

Fraction	Organic C ($g\ kg^{-1}$)	Total N ($g\ kg^{-1}$)	C:N ratio
Forest			
Leaf tissue	475	13.7	35
Light fraction	333	15.2	22
Whole soil (0-5 cm)	194	12.2	16
Corn			
Leaf tissue	440	7.8	56
Light fraction	250	11.2	22
Whole soil (0-5 cm)	20	2.0	10

(Gregorich, 1994 unpublished data.)

The C:N ratio of the LF is usually intermediate between that of whole soil and plant tissue (Table 2), in part because LF organic matter is less humified than the rest of the whole soil organic matter. Greenland and Ford (1964) found that only a small proportion (4 to 14%) of the LF was humic material. As well, during the early stages of decomposition of plant material in soil, nitrogen is conserved and C mineralization is relatively high for materials with wide C:N ratios. Adams (1980) observed an increase in CO_2 production and a decrease in N mineralization during an incubation of soil with added macroorganic matter. He concluded that the added macroorganic matter stimulated microbial activity, because the microorganisms metabolizing the macroorganic matter of high C:N ratio immobilized inorganic N to form new microbial tissue of low C:N ratio. Theodorou (1990) attributed the increased N in LF partly to immobilization and partly to the nitrogen composition of the decomposer organisms.

Management practices may alter the chemical composition of the LF and macroorganic matter. Garwood et al. (1972) reported that the C:N ratio of macroorganic matter was narrower in soils receiving higher rates of nitrogen, because the fertilization had increased the nitrogen content of the macroorganic matter.

The LF is more enriched than soil C in carbohydrates (Dalal and Henry, 1988) apparently because of the influence of plant residues. The monosaccharide composition of different soils is usually very similar, and most of the differences in relative amounts of monosaccharides are associated with those present in plant fragments (Oades, 1972). Different sugars predominate in the LF (Oades, 1972; Whitehead et al., 1975) and macroorganic matter (Dalal and Henry, 1988; Angers and Mehuys, 1990), probably reflecting the differences in residue monosaccharide composition.

Based on ^{13}C NMR analyses, the largest proportion of C in macroorganic matter and LF is present as carbohydrates (Baldock et al., 1992; Golchin et al., 1994). Baldock et al. (1992) observed that the LF in the largest particle-size fractions from a wide range of soils was composed of 39 to 57% O-alkyl, 16 to 26% alkyl, 16 to 28% aromatic and 6 to 12% carboxyl carbon. Golchin et al. (1994) observed comparable values for free LF but found that the chemical composition of LF occluded within aggregates had lower O-alkyl C and higher alkyl C than free LF. This difference was attributed to the nature and extent of decomposition processes in each soil fraction.

Some research has been conducted to determine the origin of carbohydrates in the LF and macroorganic matter. The ratio of galactose plus mannose to arabinose plus xylose has been used to separate plant- and microbe-derived carbohydrates (Oades, 1984). This technique is based on the premise that pentoses such as arabinose and xylose are major constituents of plant residues but are not synthesized to a great extent by soil microorganisms. Angers and Mehuys (1990) used this ratio, the C:N ratio, and the sugar content to show that the macroorganic matter was not composed solely of plant residues. Dalal and Henry (1988), however, concluded that no clear distinction can be made between light and heavy fractions with regard to the origin of carbohydrates.

The LF is composed mainly of plant debris, but other materials may be present in sufficient amounts to affect its chemical composition. For example, charcoal may be an important constituent of the LF in some soils (Molloy and Speir, 1977; Spycher et al., 1983; Skjemstad et al., 1990). Elliott et al. (1991) estimated that charcoal, resulting from slash-and-burn agriculture, accounted for about half of the LF in a tropical soil.

C. Biological Characterization

On the basis of density, microorganisms and their products are included in the LF, unless they are adsorbed to minerals. Furthermore, the LF is a likely habitat for microorganisms and the site of intense decomposer activity because of its enrichment with C and N. Microscopic examinations of LF have verified fungal hyphae and faunal debris as significant constituents of LF (Spycher et al., 1983; Molloy and Speir, 1977). Kanazawa and Filip (1986) reported that 34 to 42% of the bacteria, 27% of the actinomycetes and 33% of the fungi in soil were associated with organic particles (< 1.0 g cm^{-3}). Another study, however, found insignificant amounts of ATP and microbial biomass associated with the LF (Ahmed and Oades, 1984).

Ladd et al. (1977) used fractionation of soils before and after fumigation to show that destruction of soil biomass was accompanied by significant decreases in the organic N of the LF in straw-amended soils. This finding implies that the microbial biomass represents a significant proportion of the LF.

Separation into light and heavy fractions by density may be useful to isolate fractions that are enriched or depleted in C, but the distribution of these fractions within the soil matrix is not fully understood. Young and Spycher (1979) proposed that LF materials are located at the surface of aggregates and therefore are easily accessible to soil organisms that are confined to aggregate surfaces and spaces between aggregates. The heavy fraction materials are confined to regions within aggregates, thereby protected from degradation by soil organisms and enzymes. According to this view, the LF would be more readily mineralized than the heavy fraction.

IV. Dynamics of LF and Macroorganic Matter in Soil

The complex processes of decomposition, mineralization, and humification can be evaluated using simulation models. Such models also provide insight into the role of LF and macroorganic matter in soil organic matter dynamics and carbon storage.

A. Carbon Pools and Cycling

Several simulation models express a number of pools of organic matter in terms of chemical or physical stabilization (Jenkinson and Raynor, 1977; van Veen and Paul, 1981; Parton et al., 1987; McGill et al., 1981). The pools of organic matter are largely conceptual and are considered to be regulated by a number of physical, chemical, and biological factors, which in turn can be modified by management practices.

In the models, plant residues added to soil are considered to consist of two components, metabolic and structural, to account for the unequal accessibility and decomposability of these components in the soil environment (McGill et al., 1981). The metabolic component is more decomposable than the structural material (Jenkinson and Raynor, 1977; Parton et al., 1987). It provides a readily available energy source for decomposers and is therefore most influential during the initial stages of decomposition. Usually there is a rapid loss of soluble metabolic material due to microbial decomposition and leaching.

In the CENTURY model (Parton et al., 1987), residues are partitioned into structural and metabolic components based on lignin-to-nitrogen ratios. Metabolic components are assimilated into the active C pool and the remaining C is respired. The structural components are assimilated into the active and slow pools of soil organic matter, with the remaining C evolved as CO_2 as a function of lignin content.

Cambardella and Elliott (1992) suggested that particulate organic matter, a fraction analogous to macroorganic matter, accounts for most of the loss of organic matter upon cultivation. They equated this fraction to the intermediate pool, described as the "slow" (Parton et al., 1987) or "decomposable" (van Veen and Paul, 1981) pool in simulation models. They further hypothesized that after long-term cultivation, this fraction is less important in providing mineralized N than microbial decomposition products that are physically protected within aggregates.

Models which simulate decomposition and stabilization of soil organic matter adjust decay rates to account for physical protection of substrates within aggregates at sites inaccessible to microorganisms (van Veen and Paul, 1981). Golchin et al. (1994) reported that the chemical composition of LF material within aggregates was different from that of free LF material and that the amount of occluded LF was higher in soils with more clay. They hypothesized that the occluded organic matter was an old pool of C that had been accreted within aggregates over decades of root growth, and that it was this pool that was lost due to cultivation.

B. Turnover and Mineralization

The LF and macroorganic matter are generally free of mineral particles and therefore lack the protection from decomposition by microorganisms and enzymes that these particles impart. Consequently, these materials decompose more quickly than whole soil or mineral particle fractions (Sollins et al., 1984; Bonde et al., 1992; Christensen, 1987; Gregorich et al., 1989). A large portion of the C lost following the initial cultivation of native grassland or forest soils can be attributed to C loss from the LF and macroorganic matter (Adams, 1980; Dalal and Mayer, 1986b,c; Tiessen and Stewart, 1983). Conversely, this fraction is increased rapidly when a degraded soil is put under a continuous forage crop (Angers and Mehuys, 1990).

Some understanding of the C turnover in LF can be inferred from field studies of changes in LF C. For example, Spycher et al. (1983) found that LF material increased by 50 to 100% from early spring to summer. They inferred that the LF was an important reservoir of labile C. Conti et al. (1992) also observed appreciable seasonal fluctuations in LF. Studies of LF C in different phases of various cropping systems are inconclusive. Janzen et al. (1992) observed some significant fluctuations in response to factors such as fallow, whereas Bremer et al. (1994) found only minor changes in LF among phases, indicating that the turnover time of the LF was longer than the duration of the crop rotation.

Adams (1980) measured the CO_2 production from soil with and without macroorganic matter added. Addition of macroorganic matter stimulated CO_2 evolution, indicating the relatively rapid turnover of this material. The importance of LF organic matter as a source of labile C is confirmed by studies showing highly significant positive relationships between LF C and C mineralization (Janzen et al., 1992; Bremer et al., 1994)

Although the LF generally exhibits high rates of C mineralization, it is not always an important source of mineral N. Adams (1980) observed that soil with added macroorganic matter released less mineral N than an unamended soil, apparently because N was immobilized during the decomposition of the macroorganic matter. Similarly, Sollins et al. (1984) found that N mineralization of heavy-fraction organic matter was higher than that of a whole soil. They suggested that N mineralized from the heavy fraction in unfractionated soil was immobilized by microorganisms decomposing the LF. This phenomenon is supported by studies that have found that the correlation of LF with C mineralization is much stronger than that of LF with N mineralization (Janzen et al., 1992; Biederbeck et al., 1994).

Natural abundance of ^{13}C has been used to determine the turnover of soil organic matter (Balesdent et al., 1987; 1988) in agroecosystems where there has been a shift in species of different photosynthetic pathways (C3 or C4 plants). In tropical soils in Brazil, 83 to 93% of the initial forest-derived C in the macroorganic matter (63 to 2000 μm) decomposed after 12 years of cropping to sugar cane (*Saccharum spp.* L.) (Bonde et al., 1992). The calculated turnover time of 4 years for the macroorganic matter was comparable to that of the active fraction (1 to 5 years) in the CENTURY model (Parton et al., 1987). Further losses of only 1 to 7% occurred with an additional 38 years of cropping, suggesting that an equilibrium had been reached after 50 years of cropping.

The rapid turnover of organic matter in the LF was also observed by Gregorich et al. (1995), who used natural ^{13}C abundance to measure the turnover of soil organic matter and storage of maize residue C in eastern Canadian soils that had been under maize for 25 years. Mineralization of the LF was faster than that of organic matter associated with any particle-size fraction. More than 70% of the LF C had turned over since the start of cropping to maize, whereas only about 45% of the C in the macroorganic matter had turned over during this time. About 27% of the organic C in the LF was derived from C3 vegetation, and the estimated half-life of this C was 8 years. These results differ from those of Skjemstad et al. (1990), who found that 35 years after rainforest in Australia had been replaced by pasture, about 70% of the original LF soil organic matter remained. However, the LF was contaminated with charcoal, which may compromise the results from soils that have undergone regular burning of vegetation.

C. Indicator of Organic Matter Dynamics

Several studies have suggested that the LF be used as an indicator of soil quality, specifically as a predictor of organic matter changes (Shaymukhametov et al., 1984; Dalal and Mayer, 1987; Skjemstad et al., 1988; Janzen et al., 1992). This application stems, in part, from the observation that LF is much more responsive than total organic matter to changes in management. For example, in a study of long-term wheat (*Triticum aestivum* L.) systems, Biederbeck et al. (1994) observed that the range of LF C among treatments was about 15 times greater than the range in total organic C concentration. Similar findings have been reported by others (Janzen et al., 1992; Bremer et al., 1994). As a result, LF allows for greater sensitivity and resolution in the measurement of management effects on soil organic matter.

Although the sensitivity of LF to management effects is well established, its usefulness as a predictor of organic matter change has not been verified. An increase in the LF concentration generally signals enhanced levels of labile organic matter, but the eventual conversion of that transient C to stable organic matter has not been adequately described. As a result, it is conceivable that an increased LF may only be proportional and may not indicate an increase in organic matter beyond the accumulation of the LF itself.

V. Management Effects on LF and Macroorganic Matter

Since the LF is a transitory pool of organic matter between fresh residues and humified, stable organic matter (Skjemstad et al., 1986; Christensen, 1992), any factor which differentially affects the accretion or decomposition of this substrate will determine the LF concentration. Thus LF content is influenced by the amount, chemical composition, and accessibility of added residues, as well as by environmental factors, such as moisture, temperature, and nutritional status, that constrain microbial activity. Agronomic practices directly affect most of these variables and therefore strongly influence the amount of LF in soil.

A. Conversion to Arable Agriculture

The initial cultivation of virgin lands typically results in a large loss of LF organic matter (Table 1). Greenland and Ford (1964) observed that LF C concentrations in a lateritic red earth were about four times higher under native vegetation than under arable agriculture. In a series of Saskatchewan soils, cultivation of grassland soils resulted in losses of floatable C from the A horizon ranging from 67 to 76% (Tiessen and Stewart, 1983). Balesdent et al. (1988) found that a native grassland soil had more than four times as much LF C as a soil used for production of wheat or timothy (*Phleum pratense* L.). Similar results were reported by Cambardella and Elliott (1992) for particulate organic matter. Skjemstad et al. (1986) found that cultivation of an Australian site originally under *Acacia harpophylla* vegetation reduced LF C to almost negligible levels.

The LF usually represents a much smaller proportion of total organic C in cultivated soils than in virgin soils (Table 1). Skjemstad et al. (1986) reported that the LF accounted for 9% of organic C in a virgin soil, but only 1% of organic C in a cultivated soil. Tiessen and Stewart (1983) observed that floatable organic C accounted for 20-38% of organic C in virgin soil but for only 8 to 14% of organic C in cultivated soils. Comparable values were reported by Cambardella and Elliott (1992) and Dalal and Mayer (1986a). These studies consistently demonstrate that cultivation results in the disproportionate loss of LF organic matter. Indeed, the loss of accumulated LF may account for much of the organic matter loss typically observed after cultivation (Greenland and Ford, 1964). For example, Tiessen and Stewart (1983) reported that loss of floatable organic material accounted for 40% of the organic matter lost upon cultivation of a heavy clay soil.

Much of the LF loss occurs shortly after cultivation, an observation consistent with the concept of the LF as a highly decomposable, transitory substrate. Adams (1980) found that the largest decrease in macroorganic matter content occurred during the first 3 years of arable cropping, with a more gradual decrease with subsequent use (Figure 2). Similarly, Tiessen and Stewart (1983) observed that much of the loss of floatable C occurred within 4 years of initial cultivation, and Balesdent et al. (1988) reported that LF losses occurred within the first several decades of cultivation.

Dalal and Mayer (1986d; 1987) proposed that loss of LF followed first order kinetics. They calculated the amount of C in the LF initially and at equilibrium in soils under continuous cereal-cropping in Australia. Their calculated rates of LF C loss (k), ranging from 0.12 to 0.83 $year^{-1}$, were significantly correlated to the reciprocal of the clay content and the frequency of stubble retention. Between 49 and 85% of the organic C in LF would be lost at equilibrium. In fine-textured soils, the losses from the LF were 2 to 11 times faster than the losses of organic matter associated with inorganic particles. The estimated decay rate constants of the LF compared favorably with those estimated using stable isotopes for the macroorganic matter in temperate soils (Balesdent and Mariotti, 1990).

The rapid loss of LF that consistently occurs when a virgin soil is initially cultivated is probably attributable to a number of factors. The input of residues in agricultural soils may be lower than that in native ecosystems, particularly if a large proportion of the plant biomass is exported. As well, decomposition rates may be enhanced in arable agriculture because of more favorable hydrothermal conditions, disturbance of soil by tillage, and a greater supply of decomposable residues.

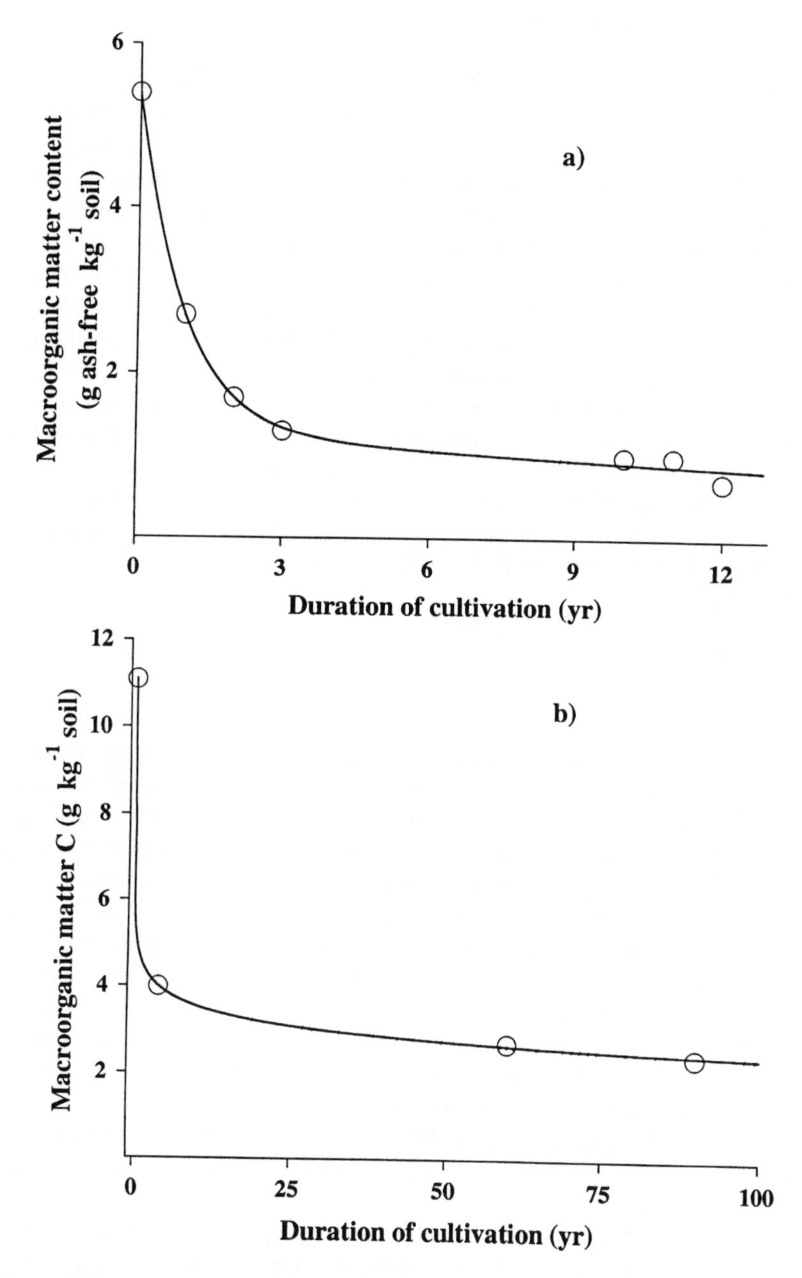

Figure 2. Loss of macroorganic matter under continuous cultivation in a) Northern Ireland, and b) Canada. (From Table 1 in Adams, 1980; and Figure 1 and Table 1 in Tiessen and Stewart, 1983.)

B. Frequency of Summerfallow

In soils used for arable agriculture, the frequency of summerfallow has a dominant effect on LF concentrations. Greenland and Ford (1964) and Ford and Greenland (1968) found higher LF C and N concentrations in a continuous wheat system than in a fallow-wheat system. Janzen (1987) similarly observed a strong inverse relationship between LF N and frequency of fallow in a series of spring wheat-based crop rotations. This observation has been consistently confirmed at five sites in western Canada (Janzen et al., 1992; Bremer et al., 1994; Biederbeck et al., 1994). On average, the LF C content of soils under frequent summerfallow was about half that in soils that had been cropped every year. For example, mean LF C concentrations in the surface soil of a long-term rotation study at Lethbridge were 3.3, 2.2, and 1.6 g C kg^{-1} for continuous wheat, fallow-wheat-wheat, and fallow-wheat, respectively (Bremer et al. 1994). The lower LF in soils under fallow is attributable, in part, to the absence of appreciable primary production during the fallow phase. As well, fallowing may reduce LF content by creating soil moisture and temperature conditions conducive to the breakdown of free organic matter (Janzen et al., 1992; Douglas and Rickman, 1992).

C. Tillage Practices

Tillage has long been recognized as detrimental to soil organic matter content, but few studies have specifically measured its effect on LF material. Cambardella and Elliott (1992) compared particulate organic matter in wheat-fallow systems that had used various tillage intensities for 20 years. Soils under no-till treatment had significantly higher concentrations of particulate organic matter than those under stubble mulch and bare-fallow treatments (Figure 3), even though inputs of residue were comparable among the three treatments. The higher concentrations of particulate organic matter in the no-till soil are attributed to slower decomposition of residues because of their placement on the soil surface. This observation was confirmed by ^{13}C natural abundance measurements, which demonstrated that only 13% of wheat straw C remained in the bare fallow treatment, compared with 31% in the stubble mulch and no-till treatments. Roberson et al. (1991) observed no effect of various soil management practices on LF carbohydrates in an orchard soil, but LF C was not directly measured.

Angers et al. (1993) separated soils by particle size to assess the impact of 11 years of reduced tillage on soil organic matter. The organic C of the macro-organic matter accounted for 7% of the total C under moldboard plowing and up to 19% under reduced tillage. Little difference was found in the organic C content of the silt- and clay-size fractions, indicating that reducing tillage

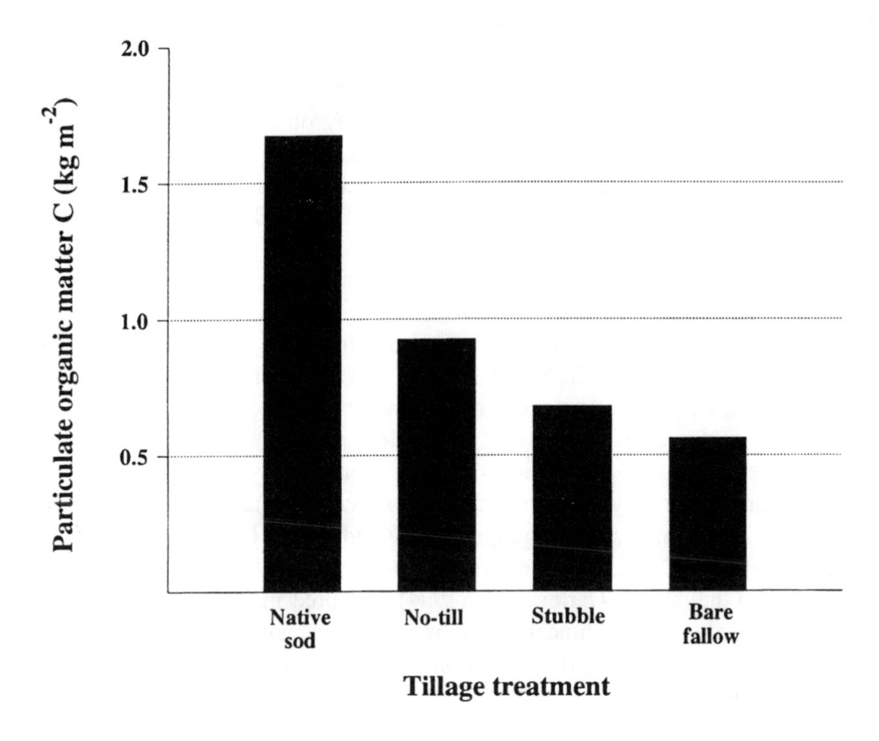

Figure 3. Particulate organic matter in a wheat-fallow production system with various tillage intensities and in a native sod. (Adapted from Cambardella and Elliott, 1992.)

intensity can result in the maintenance or increase of the labile macroorganic matter.

D. Ley Crops

Seeding arable land to grasses or legumes typically enhances LF organic matter, particularly if the forages are maintained for a sufficient duration (Greenland, 1971; Garwood et al., 1972). Ford and Greenland (1968) observed that a 4-year rotation with 2 years of pasture maintained higher LF-N concentrations than wheat or wheat-fallow systems. Continuous pasture, in turn, had appreciably higher LF N than the four-year rotation. Whitehead et al. (1975) reported that LF-C in an 'old pasture' was higher than that in a soil that had been under a ryegrass- (*Lolium perenne* L.) white clover (*Trifolium repens* L.) sward for 17 years after prolonged arable agriculture. The LF in the latter treatment was higher than that in an arable soil used for barley (*Hordeum vulgare* L.) production. The composition of the LF was generally comparable among treatments, though some differences were observed in concentration of N and several organic acids. Conti et al. (1992) generally observed higher LF under various grassland systems than under maize in a series of sites in Argentina.

Introduced grasses may not maintain as much LF as native grasses. For example, Balesdent et al. (1988) found that LF in soil under a 98-year stand of timothy was appreciably lower than that under native prairie. In part, this observation may be attributed to greater diversion of C below ground in native species (Smoliak and Dormaar, 1985).

The management of the forage crop also affects the accumulation of LF material. For example, Garwood et al. (1972) observed slightly higher macroorganic matter levels under grazed than under cut perennial ryegrass.

The effect of grasses on LF may be evident within several years of stand establishment. Angers and Mehuys (1990) found that short-term effects of cropping on the carbohydrate content of soils were attributed to differences in the macroorganic matter. Cropping to barley and alfalfa (*Medicago sativa* L.) resulted in larger C, N, and carbohydrate contents in the macroorganic matter compared to the fallow. Carter et al. (1994) measured macroorganic matter in plots 4 years after establishment of temperate grasses and found macroorganic matter concentrations comprised about 40% of the total organic C: Differences in soil structural stability induced by the different grass species were not related to macroorganic matter.

Bremer et al. (1994) measured LF C in plots 8 years after seeding to various native grasses and found that LF C concentrations were already significantly higher than those in any of several spring wheat systems. Short-term leguminous hay crops in rotation with annual crops, such as wheat, tend to maintain LF concentrations higher than those in fallow-wheat systems and roughly equivalent to those in continuously cropped systems (Janzen et al., 1992; Bremer et al., 1994).

E. Crop Sequence

Apart from the effects of fallow and forages, little is known about the relative contribution of various crop rotations to LF organic matter. Studies in western Canada found no consistent effects of oilseed crops, such as flax (*Linum usitatissum* L.) and canola (*Brassica campestris* L.), on LF, relative to comparable wheat rotations (Janzen et al., 1992; Biederbeck et al., 1994). Rotation of wheat with lentils (*Lens culinaris* L.) tended to reduce LF relative to that in continuous wheat (Biederbeck et al., 1994), perhaps because of the higher decomposability of the annual legume residue. Apparently for the same reason legume green manures did not contribute appreciably to soil LF (Janzen et al. 1992).

F. Nutritive Amendments

Application of fertilizers often enhances LF organic matter. For example, Biederbeck et al. (1994) and Janzen et al. (1992) observed that application of N fertilizer significantly increased LF C in a continuous cropping system at some sites (Table 1), probably because of the increased residue production associated with the yield response to N. As a result, fertilizer may have no effect on LF in instances where yield response to the added nutrient is minimal (Bremer et al., 1994).

Manure application may also appreciably enhance LF organic matter. In soil under grass, Adams (1980) observed that 200 m^3 ha^{-1} $year^{-1}$ pig slurry dressings resulted in higher macroorganic matter contents than in unamended soil. Bi-annual applications of solid cattle manure over 10 years to soils under maize have been shown to increase macroorganic matter levels (Angers and N'dayegamiye, 1991). Bremer et al. (1994) observed that long-term application of low rates of animal manure resulted in a nearly two-fold increase in LF-C in a fallow-wheat-wheat system. The increases observed in these studies are probably attributable both to the direct application of decomposable organic matter in the manure and to the yield response arising from its application.

VI. Summary

The LF and macroorganic matter are a transitory pool of organic matter between fresh plant residues and humified soil organic matter. This pool occupies a small portion of the soil mass, but is highly enriched with C and may contain a substantial portion of the total carbon in soil. Moreover, its short turnover time gives the LF a prominent role as a C substrate and source of nutrients.

By virtue of its transitory nature, the LF concentration of soils is highly variable, depending on the amount and characteristics of C inputs, and environmental factors that affect rates of decomposition. For the same reason, the LF is very responsive to management practices in agroecosystems. This sensitivity suggests a possible role for LF as an indicator of soil quality and productivity.

Characterization of LF organic matter has been hindered by the variation in methodology. Standardization of the methods used to isolate this organic matter has been proposed (Gregorich and Ellert, 1993), but more extensive testing of the methods is warranted. More research is required to identify the importance of the role of LF and macroorganic matter in soil organic matter dynamics. One strategy may be to use actual measurements of this fraction in simulation models.

References

Adams, T. McM. 1980. Macro organic matter content of some Northern Ireland soils. *Record Agric. Res.* 28:1-11.

Ahmed, M. and J.M. Oades. 1984. Distribution of organic matter and adenosine triphos-phate after fractionation of soils by physical procedures. *Soil Biol. Biochem.* 5:465-470.

Al-Khafaf, S., P.J. Wierenga, and B.C. Williams. 1977. A flotation method for determining root mass in soil. *Agron. J.* 69:1025-1026.

Angers, D.A. and G.R. Mehuys. 1990. Barley and alfalfa cropping effects on carbohydrate contents of a clay soil and its size fractions. *Soil Biol. Biochem.* 22:285-288.

Angers, D.A. and A. N'dayegamiye. 1991. Effects of manure application on carbon, nitrogen, and carbohydrate contents of a silt loam and its particle size fractions. *Biol. Fertil. Soils* 11:79-82.

Angers, D.A., A. N'dayegamiye, and D. Côté. 1993. Tillage-induced differences in organic matter of particle size fractions and microbial biomass. *Soil Sci. Soc. Am. J.* 57:512-516.

Baldock, J.A., J.M. Oades, A.G. Waters, X. Peng, A.M. Vassallo, and M.A. Wilson. 1992. Aspects of the chemical structure of soil organic materials as revealed by solid-state ^{13}C NMR spectroscopy. *Biogeochemistry* 16:1-42.

Balesdent, J., A. Mariotti, and B. Guillet. 1987. Natural ^{13}C abundance as a tracer for studies of soil organic matter dynamics. *Soil Biol. Biochem.* 19:25-30.

Balesdent, J., G.H. Wagner, and A. Mariotti. 1988. Soil organic matter turnover in long-term field experiments as revealed by carbon-13 natural abundance. *Soil Sci. Soc. Am. J.* 52:118-124.

Balesdent, J. and A. Mariotti. 1990. The turnover of total organic carbon and of soil organic fractions estimated from ^{13}C abundance in corn fields. *Trans. 14th Int. Cong. Soil Science.* Kyoto, Japan, Vol. II:411-412.

Biederbeck, V.O., H.H. Janzen, C.A. Campbell, and R.P. Zentner. 1994. Labile soil organic matter as influenced by cropping practices in an arid environment. *Soil Biol. Biochem.* 26:1647-1656.

Bonde, T.A., B.T. Christensen, and C.C. Cerri. 1992. Dynamics of soil organic matter as reflected by natural ^{13}C abundance in particle size fractions of forested and cultivated oxisols. *Soil Biol. Biochem.* 24:275-277.

Bremer, E., H.H. Janzen, and A.M. Johnston. 1994. Sensitivity of total, light fraction and mineralizable organic matter to management practices in a Lethbridge soil. *Can. J. Soil Sci.* 74:131-138.

Cambardella, C.A. and E.T. Elliott. 1992. Particulate soil organic-matter changes across a grassland cultivation sequence. *Soil Sci. Soc. Am. J.* 56:777-783.

Carter, M.R., D.A. Angers, and H.T. Kunelius. 1994. Soil structural form and stability, and organic matter under cool-season perennial grasses. *Soil Sci. Soc. Am. J.* 58:1194-1199.

Catroux, G. and M. Schnitzer. 1987. Chemical, spectroscopic, and biological characteristics of the organic matter in particle size fractions separated from an Aquoll. *Soil Sci. Soc. Am. J.* 51:1200-1207.

Christensen, B.T. 1987. Decomposability of organic matter in particle size fractions from field soils with straw incorporation. *Soil Biol. Biochem.* 19:429-435.

Christensen, B.T. 1992. Physical fractionation of soil and organic matter in primary particle size and density separates. *Adv. Soil Sci.* 20:1-90.

Conti, M.E., R.M. Palma, N. Arrigo, and E. Giardino. 1992. Seasonal variations of the light organic fractions in soils under different agricultural management systems. *Comm. Soil Sci. Plant Anal.* 23:1693-1704.

Dalal, R.C. and R.J. Henry. 1988. Cultivation effects on carbohydrate contents of soil and soil fractions. *Soil Sci. Soc. Am. J.* 52:1361-1365.

Dalal, R.C. and R.J. Mayer. 1986a. Long-term trends in fertility of soils under continuous cultivation and cereal cropping in Southern Queensland. I. Overall changes in soil properties and trends in winter cereal yields. *Aust. J. Soil Res.* 24:265-279.

Dalal, R.C. and R.J. Mayer. 1986b. Long-term trends in fertility of soils under continuous cultivation and cereal cropping in Southern Queensland. II. Total organic carbon and its rate of loss from the soil profile. *Aust. J. Soil Res.* 24:281-292.

Dalal, R.C. and R.J. Mayer. 1986c. Long-term trends in fertility of soils under continuous cultivation and cereal cropping in Southern Queensland. III. Distribution and kinetics of soil organic carbon in particle-size fractions. *Aust. J. Soil Res.* 24:293-300.

Dalal, R.C. and R.J. Mayer. 1986d. Long-term trends in fertility of soils under continuous cultivation and cereal cropping in Southern Queensland. IV. Loss of organic carbon from different density fractions. *Aust. J. Soil Res.* 24:301-309.

Dalal, R.C. and R.J. Mayer. 1987. Long-term trends in fertility of soils under continuous cultivation and cereal cropping in Southern Queensland. VI. Loss of total nitrogen from different particle-size and density fractions. *Aust. J. Soil Res.* 25:83-93.

Dormaar, J.F. and R.C. Foster. 1991. Nascent aggregates in the rhizosphere of perennial ryegrass (*Lolium perenne* L.). *Can. J. Soil Sci.* 71:465-474.

Douglas, C.L. and R.W. Rickman. 1992. Estimating crop residue decomposition from air temperature, initial nitrogen content, and residue placement. *Soil Sci. Soc. Am. J.* 56:272-278.

Ellert, B.H. and Gregorich, E.G. 1994. Management-induced changes in the actively cycling fractions of soil organic matter. In: W.W. McFee and J.M. Kelly (eds.), Proceedings of the 8th North American Forest Soils Conference. Gainsville, FL. American Society of Agronomy, Madison, WI (in press).

Elliott, E.T., C.A. Palm, D.E. Reuss, and C.A. Monz. 1991. Organic matter contained in soil aggregates from a tropical chronosequence: correction for sand and light fraction. *Agric. Ecosyst. Environ.* 34:443-451.

Ford, G.W. and D.J. Greenland. 1968. The dynamics of partly humified organic matter in some arable soils. *Trans. 9th Int. Cong. Soil Sci.* II:403-410.

Garwood, E.A., C.R. Clement, and T.E. Williams. 1972. Leys and organic matter. III. The accumulation of macro-organic matter in the soil under different swards. *J. Agric. Sci. Camb.* 78:333-341.

Golchin, A., J.M. Oades, J.O. Skjemstad, and P. Clarke. 1994. Study of free and occluded particulate organic matter in soils by solid state ^{13}C CP/MAS NMR spectroscopy and scanning electron microscopy. *Aust. J. Soil Res.* 32:285-309.

Greenland, D.J. 1971. Changes in the nitrogen status and physical condition of soils under pastures, with special reference to the maintenance of the fertility of Australian soils used for growing wheat. *Soils Fert.* 34:237-251.

Greenland, D.J. and G.W. Ford. 1964. Separation of partially humified organic materials from soils by ultrasonic dispersion. *Trans. 8th Int. Cong. Soil Sci.* II:137-147.

Gregorich, E.G. and B.H. Ellert. 1993. Light fraction and macroorganic matter in mineral soils. pp. 397-407. In: M.R. Carter (ed.), Soil Sampling and Methods of Analysis. Canadian Society of Soil Science. Lewis Publishers, Boca Raton FL.

Gregorich, E.G., R.G. Kachanoski, R.P. Voroney. 1989. Carbon mineralization in soil size fractions after various amounts of aggregate disruption. *J. Soil Sci.* 40:649-659.

Gregorich, E.G., M.R. Carter, D.A. Angers, C.M. Monreal, and B.H. Ellert. 1994. Towards a minimum data set to assess soil organic matter quality. *Can. J. Soil Sci.* 74:367-385.

Gregorich, E.G., B.H. Ellert, and C.M. Monreal. 1995. Turnover of soil organic matter and storage of corn residue carbon estimated from natural ^{13}C abundance. *Can. J. Soil Sci.* 75:161-167.

Janzen, H.H. 1987. Soil organic matter characteristics after long-term cropping to various spring wheat rotations. *Can. J. Soil Sci.* 67:845-856.

Janzen, H.H., C.A. Campbell, S.A. Brandt, G.P. Lafond, and L. Townley-Smith. 1992. Light-fraction organic matter in soils from long-term crop rotations. *Soil Sci. Sci. Am. J.* 56:1799-1806.

Jenkinson, D.S. and J.H. Raynor. 1977. The turnover of soil organic matter in some of the Rothamsted classical experiments. *Soil Sci.* 123:298-305.

Kanazawa, S. and Z. Filip. 1986. Distribution of microorganisms, total biomass, and enzyme activities in different particles of brown soil. *Microb. Ecol.* 12:205-215.

Ladd, J.N., M. Amato, and J.W. Parsons. 1977. Studies of nitrogen immobilization and mineralization in calcareous soils. III. Concentration and distribution of nitrogen derived from soil biomass. pp. 301-311. In: Soil Organic Matter Studies. Vol. I. International Atomic Energy Agency, Vienna.

Leuschner, H.H., R. Aldag, and B. Meyer. 1981. Dichte-Fraktionierung des Humus in Ap-Horizonten von Sandboden mit unterschiedlichter Kornung und Nutzungs-Vorgeschichte. *Mitteilgn. Dtsch. Bodenkundl. Gesellsch.* 32:583-592.

McGill, W.B., H.W. Hunt, R.G. Woodmansee, and J.O. Reuss. 1981. Phoenix, A model of the dynamics of carbon and nitrogen in grassland soils. In: F.E. Clark and T. Rosswall (eds.), Terrestial Nitrogen Cycles, Processes, Ecosystem Strategies and Management Inputs. *Ecol. Bull. (Stockholm)* 33:45-115.

Molloy, L.F. and T.W. Speir. 1977. Studies on a climosequence of soils in tussock grasslands. 12. Constituents of the soil light fraction. *N.Z. J. Science* 20:167-177.

Oades, J.M. 1972. Studies on soil polysaccharides III. Composition of polysaccharides in some Australian Soils. *Aust. J. Soil Res.* 10:113-126.

Oades, J.M. 1984. Soil organic matter and structural stability: mechanisms and implications for management. *Plant Soil* 76:319-337.

Oades, J.M. 1989. An introduction to organic matter in mineral soils. pp. 89-159. In: J.B. Dixon and S.B. Weed (eds.), Minerals in soil environments. Second Edition. Soil Sci. Soc. Am. Pub., Madison WI.

Oades, J.M. 1993. The role of biology in the formation, stabilization and degradation of soil structure. *Geoderma* 51:377-400.

Parton, W.J., D.S. Schimel, C.V. Cole, and D.S. Ojima. 1987. Analysis of factors controlling soil organic matter levels in Great Plains grasslands. *Soil Sci. Soc. Am. J.* 51:1173-1179.

Richter, M., I. Mizuno, S. Aranguez, and S. Uriarte. 1975. Densimetric fractionation of soil organo-mineral complexes. *J. Soil Sci.* 26:112-123.

Roberson, E.B., S. Sarig, and M.K. Firestone. 1991. Cover crop management of polysaccharide-mediated aggregation in an orchard soil. *Soil Sci. Soc. Am. J.* 55:734-739.

Shaymukhametov, M.S., N.A. Titova, L.S. Travnikova, and Y.M. Labenets. 1984. Use of physical fractionation methods to characterize soil organic matter. *Soviet Soil Sci.* 16:117-128.

Skjemstad, J.O., R.C. Dalal, and P.F. Barron. 1986. Spectroscopic investigations of cultivation effects on organic matter of Vertisols. *Soil Sci. Soc. Am. J.* 50:354-359.

Skjemstad, J.O., I. Vallis, and R.J.K. Myers. 1988. Decomposition of soil organic nitrogen. pp. 134-144. In: E.F. Henzell (ed.), Advances in nitrogen cycling in agricultural ecosystems. CAB International, Wallingford, England.

Skjemstad, J.O., R.P. Le Feuvr, and R.E. Prebble. 1990. Turnover of soil organic matter under pasture as determined by ^{13}C natural abundance. *Aust. J. Soil Res.* 28:267-276.

Smoliak, S. and J.F. Dormaar. 1985. Productivity of Russian wildrye and crested wheat grass and their effect on prairie soils. *J. Range Manage.* 38:403-405.

Sollins, P., G. Spycher, and C.A. Glassman. 1984. Net mineralization from light- and heavy-fraction forest soil organic matter. *Soil Biol. Biochem.* 16:31-37.

Spycher, G., P. Sollins, and S. Rose. 1983. Carbon and nitrogen in the light fraction of a forest soil: vertical distribution and seasonal fluctuations. *Soil Sci.* 135:79-87.

Theodorou, C. 1990. Nitrogen transformations in particle size fractions from a second rotation pine forest soil. Comm. Soil Sci. *Plant Anal.* 21:407-413.

Tiessen, H. and J.W.B. Stewart. 1983. Particle-size fractions and their use in studies of soil organic matter: II. Cultivation effects on organic matter composition in size fractions. *Soil Sci. Soc. Am. J.* 47:509-514.

Turchenek, L.W. and J.M. Oades. 1979. Fractionation of organo-mineral complexes by sedimentation and density techniques. *Geoderma* 21:311-343.

van Veen, J.A. and E.A. Paul. 1981. Organic carbon dynamics in grassland soils. 1. Background information and computer simulation. *Can. J. Soil. Sci.* 61:185-201.

Whitehead, D.C., H. Buchan, and R.D. Hartley. 1975. Components of soil organic matter under grass and arable cropping. *Soil Biol. Biochem.* 7:65-71.

Young, J.L. and G. Spycher. 1979. Water-dispersible soil organic-mineral particles: I. Carbon and nitrogen distribution. *Soil Sci. Soc. Am. J.* 43:324-328.

Zhang, H., M.L. Thompson, and J.A. Sandor. 1988. Compositional differences in organic matter among cultivated and uncultivated Argiudolls and Hapludalfs derived from loess. *Soil Sci. Soc. Am. J.* 52:216-222.

Impact of Climate, Soil Type, and Management

Aggregation and Organic Matter Storage in Cool, Humid Agricultural Soils

D.A. Angers and M.R. Carter

I. Introduction

Cool to cold humid climates characterised by long cold winters are mainly found in the northern hemisphere between 40°C to 60°N (Money, 1974). Native vegetation for this climatic region is primarily boreal forest which is dominantly coniferous. In southern parts, the conifers grade into forests of broad-leaved deciduous trees. Soil types developed in the above glaciated ecosystems are predominantly Podzols, Luvisols and Gleysols (Canada Expert Committee on Soil Survey, 1978).

In the native state (usually forest) soils of the cool, humid climatic regions usually show large accumulation of organic matter. Post et al. (1982) and Zinke et al. (1984) estimated the C density of cool temperate forest soils to be about 13 kg C m^{-2}, which contrasts with a value of approximately 5 kg C m^{-2} for dry soils (Eswaran et al., 1993). Generally, the influence of climate on soil C content is reflected in the balance between C inputs from vegetation and C losses via decomposition (Mele and Carter, 1993). This balance can be characterized by the quotient of mean annual temperature and annual precipitation, where a decreasing ratio (i.e. cool wet climate) is associated with slow soil C turnover

1-56670-033-7/96/$0.00+$.50

(Tate, 1992). However, type of vegetation and agricultural system will also greatly influence the content and storage of soil C (Schlesinger, 1990). This underlines the importance of characterizing the organic carbon storage potential for agricultural soils in cool humid regions.

Cultivation of virgin soils (native grassland or forest) has usually resulted in a decline in soil organic matter and net release of CO_2 to the atmosphere (Schlesinger, 1990; Davidson and Ackerman, 1993). For forest soils, this decline in organic matter is related first of all to a change in type of vegetation and subsequent change in C inputs and magnitude of various C pools (e.g. litter); secondly, to increases in soil temperature, associated with a reduction in soil shading and greater incidence of net radiation at the soil surface, which can influence organic C decomposition; and thirdly to changes associated with cultivation such as modification of soil structure, incorporation of organic C into the soil, and increased potential for soil erosion. However, although the above changes are operative, the actual net change in total soil organic C mass can be minor. Martel and Deschênes (1976) reported an average loss of C of 30% upon long-term (> 30 years) cultivation of three Québec forest soils. More recently, in eastern Canada, the comparison of 22 boreal forest soils with adjacent agricultural soils not subject to significant soil redistribution via erosion showed that the mass of soil organic C under the latter system was depleted by an average of 22% (Gregorich et al., 1993). This is generally comparable or somewhat lower than the reported losses of organic matter inventory from North American prairie soils (Mollisols) subject to cultivation (Davidson and Ackerman, 1993).

The role of soil structure modification as expressed by aggregate degradation (breakdown) in the loss of soil organic C upon cultivation is still unclear. The introduction of tillage may cause a decline in organic C through disruption of aggregates and subsequent release of $C0_2$. In contrast, a change in plant productivity associated with arable farming may result in reduced organic C inputs and organic aggregate binding agents, causing a consequent decrease in aggregation.

The main objective of this chapter is to evaluate the potential for managing soil aggregation for the storage and sequestration of organic matter under cool, humid climatic conditions. Theng et al. (1989) emphasized the advantage of characterizing organic matter within well-defined climatic regions. To this end, the chapter will concentrate mainly on studies conducted in Podzolic, Luvisolic and Gleysolic soils of eastern Canada to illustrate short-term dynamics of soil aggregation and organic C storage under various cropping systems.

II. Soil Structure and Aggregation Models

Soil structure can be defined in terms of form and stability (Kay et al., 1988). Structural form refers to the heterogeneous arrangement of solid and void space that exists at a given time, whereas the stability of a soil's structure is its ability to retain this arrangement when exposed to different stresses *(ibid.)*. Soil aggregates are normally not a random arrangement of primary particles. Primary particles and aggregates of different size are usually arranged in a hierarchical fashion. Tisdall and Oades (1982) presented an aggregation model for Australian grassland soils, which appears to have universal application for soils when organic matter is the main aggregate stabilizing agent (Oades and Waters, 1991). The model suggests that, as proposed by Edwards and Bremner (1967), the building blocks or elementary units are stable microaggregates (<250 μm) which are bound together to form stable macroaggregates (>250 μm). The cementing or binding agent between the microaggregates is relatively labile organic matter which Tisdall and Oades (1982) called transient and temporary binding agents. These were identified to be fungal hyphae and fine roots (Tisdall and Oades, 1982), although other binding agents are polysaccharides (Angers and Mehuys, 1989; Haynes and Swift, 1990) and hydrophobic aliphatics (Capriel et al., 1990). Binding material within microaggregates is recalcitrant organic matter and inorganic constituents (Tisdall and Oades, 1982).

The hierarchical model appears to be valid for soils in cool, humid climates. Baldock and Kay (1987) found that conventional corn (*Zea mays* L.) production decreased the proportion of macroaggregates >1 mm of a silty loam relative to a bromegrass (*Bromus inermis* Leyss) stand. The decrease in macroaggregates was associated with an increase in stable microaggregates <250 μm. For a clay soil, Angers and Mehuys (1988) found that, under an alfalfa (*Medicago sativa* L.) stand, aggregates >2 mm were formed mostly at the expense of stable aggregates <1 mm. Ultrasonic energy was necessary to disrupt these aggregates ($<$ 1 mm) into smaller units (Angers and Mehuys, 1990). These results confirm that the hierarchical model holds for these soils but that the size of the building blocks (microaggregates) and the resulting macroaggregates may vary. The model is also very consistent with the fact that rapid management-induced changes in soil aggregation are most often observed in the large macroaggregate size fraction (>1 mm) (Baldock and Kay, 1987; Angers and Mehuys, 1988; Carter, 1992) which reinforces the hypothesis that labile organic matter is responsible for binding microaggregates into macroaggregates.

III. Organic Matter Protection and Macroaggregation

The hierarchical model described above would suggest a differentiation in the concentration and composition of the soil organic matter between the aggregate size classes. However, experimental results show that the variations in total organic C and N contents with aggregate size are not always clear and consistent. For grassland soils, Dormaar (1984), Elliott (1986), and Cambardella and Elliott (1993) generally found that organic C and N contents decreased with decreasing aggregate size. Conversely, Baldock et al. (1987) working on an Ontario Luvisol observed that both organic C and total carbohydrate contents increased with decreasing aggregate size. Gupta and Germida (1988) and Elliott et al. (1991) found no clear trend in C concentration between aggregate size fractions. It should be pointed out that in most of these studies, the observed trends were not always strong and that the size classes as well as the methods used to separate these size classes varied. The procedure used to separate the aggregates may impact on the results obtained. Generally, when slaking is allowed to take place, C concentration increases with aggregate size whereas when slaking is minimized (dry sieving or aggregates pre-wetted under tension or vacuum), no clear trends are observed (Elliott, 1986; Cambardella and Elliott, 1993). Elliott (1986) and Baldock et al. (1987) also emphasized that the nutrient concentrations in aggregate size classes should be corrected for the presence of sands and primary particles as these may preferentially accumulate in size fractions during wet sieving. Waters and Oades (1991) concluded that the bulk composition of soil aggregates is generally similar to that of whole soil and questioned the appropriateness of this approach to elucidate mechanisms of aggregate stabilization. Clearly, more studies are needed to better understand the interactions between aggregate size and organic matter content.

On the other hand, studies on soil organic matter composition have consistently shown that the C:N ratio decreases with decreasing aggregate size, whereas mineralization experiments suggested that, in line with the hierarchical model, macroaggregates contain more labile organic matter than microaggregates (Elliott, 1986; Gupta and Germida, 1988). From the hierarchical model, it can also be hypothesized that the organic matter contained in the macroaggregates should be younger than that in the microaggregates. This was verified by Skjemstad et al. (1990) who showed, using ^{13}C natural abundance, that the microaggregates of an Australian soil contained more old C than the macroaggregates. Recent studies in France and eastern Canada further confirmed this hypothesis for temperate soils (Table 1). There is a clear trend towards increasing proportion of young corn-derived-C with increasing aggregate size.

The distribution of organic C in aggregate size factions is a function of the quantity of soil and the C concentration in the size fractions (Cambardella and Elliott, 1993). As organic C concentration appears to be relatively uniform across aggregate size fractions, the level of water-stable macroaggregation should provide an estimation of the amount of organic C stored and physically protected in the soil, especially the labile fractions. This is demonstrated by the proportion of whole-soil C present in water-stable aggregates from Brunisolic and Podzolic soils (Table 2). The proportion of organic C stored in water-stable aggregates varies with management; generally, aggrading systems show the largest values. This storage, however would be temporary or transient since these large aggregates can be degraded rapidly upon cultivation (Kay, 1990; Carter, 1992; Angers et al., 1992).

Evidence that aggregation physically protects organic matter from decomposition has been provided by various soil macroaggregate disruption studies (Rovira and Greacen, 1957; Powlson, 1980; Elliott, 1986; Gupta and Germida, 1988; Gregorich et al., 1989; Hassink, 1992). Measurement of the resistance to disruption (i.e. aggregate stability) can therefore be interpreted as an indication

Table 1. Percent of young C (C_4 -derived C) in water-stable aggregates of two soils

Aggregate size[1]	France[2]	Eastern Canada[3]
mm	---------%	---------
>2	32	19
1-2	21	18
0.5-1	20	10
0.2-0.5	14	7
0.05-0.2	8	4
<0.05	3	2

[1]Aggregate size classes obtained by slaking (wet-sieving of air-dry soil).
[2]Silty soil which had been under C_4 (grain corn) for 6 years (From P. Puget et al., 1995).
[3]Silty loam which had been under C_4 (silage corn) for 15 years (From D.A. Angers and M. Giroux, unpublished results).

Table 2. Proportion of whole-soil C present in macroaggregates under two cropping systems on a Brunisolic silt loam in southern Ontario (Recalculated from Baldock and Kay, 1987) and a Podzolic fine sandy loam in Prince Edward Island (Recalculated from Carter, 1992)

Cropping system	Water-stable[a] aggregates (>1mm)	C content of whole soil	C content of aggregates	Whole-soil C in macro-aggregates
	(%)	(g C kg^{-1} soil)	(g C kg^{-1} aggregates)	(%)
Ontario				
Grass (15 years)	60	20	19	57
Corn (15 years)	30	19	19	28
Prince Edward Isl.				
Grass (10-40 years)	89	44	49	99
Cereal (5 years)	37	20	22	40

[a]Wet sieving conducted on whole soil samples.

of organic C protection. However, a large proportion of the organic C in macroaggregates would be not sequestered in microaggregates, but exist as inter-microaggregate 'free' or particulate C and thus easily subject to mineralization upon macroaggregate disruption.

IV. Potential for Increasing Organic Matter Retention

The previous section emphasized that water-stable aggregation and aggregate stability are associated with the protection and/or storage of soil organic matter, especially labile organic C compounds. The following sections present the results of recent studies, mostly from eastern Canada, which demonstrate the influence of agricultural cropping practices on soil C storage and the dynamics of water-stable aggregation and aggregate stability in cool, humid soils. Emphases are placed on short-term (<10 years) management effects on water-stable macroaggregation, and the potential of agricultural practices to increase soil organic C storage.

A. Perennial Forages

Beneficial effects of perennial forages on soil macroaggregation are well recognized. Several studies have shown that water-stable macroaggregation increases rapidly when arable or degraded lands are put into continuous perennial forages. Maximum stability is often achieved after 3 to 5 years (Low, 1955; Perfect et al., 1990; Angers, 1992).

The involvement of grass roots and fungal hyphae in stabilizing macro-aggregates has been shown under grassland soils (Tisdall and Oades, 1982; Jastrow, 1987) and suggested for cool, humid soils (Carter et al., 1994). Caron et al. (1992), on the other hand, attributed the early improvement in aggregate stability under bromegrass to ill-defined tetraborate-sensitive organic materials from the grass. Perennial forage legumes and grass-legume mixtures have also been shown to increase aggregate stability. Angers (1992) showed that the improvement in stability under alfalfa (Figure 1a) took place more rapidly than the increase in total organic C content (Figure 1b). Early improvement (after only two years) in water-stable aggregation was partly attributed to production of binding carbohydrates (soluble to dilute acid) (Angers and Mehuys, 1989). Similar early improvement under mixed pastures and perennial crops was related to soluble carbohydrates produced by the large microbial biomass present in the pasture rhizosphere (Haynes and Swift, 1990; Haynes et al., 1991; Haynes and Francis, 1993).

Choice of perennial forage crop species and cultivar will also influence the extent of aggregate formation and stabilization. Alfalfa was as efficient as bromegrass but slightly less than timothy (*Phleum pratense L.)* and reed canarygrass (*Phalaris arundinacea* L.) in improving the water-stable aggregation of two Québec soils (Chantigny et al., 1995). Carter et al. (1994) showed that the potential for increasing the stability of soils from Atlantic Canada varied with grass species but also with cultivar. Management of perennial forages should also influence aggregation through factors which influence crop C production. However, Perfect et al. (1990) did not see any effect of cutting

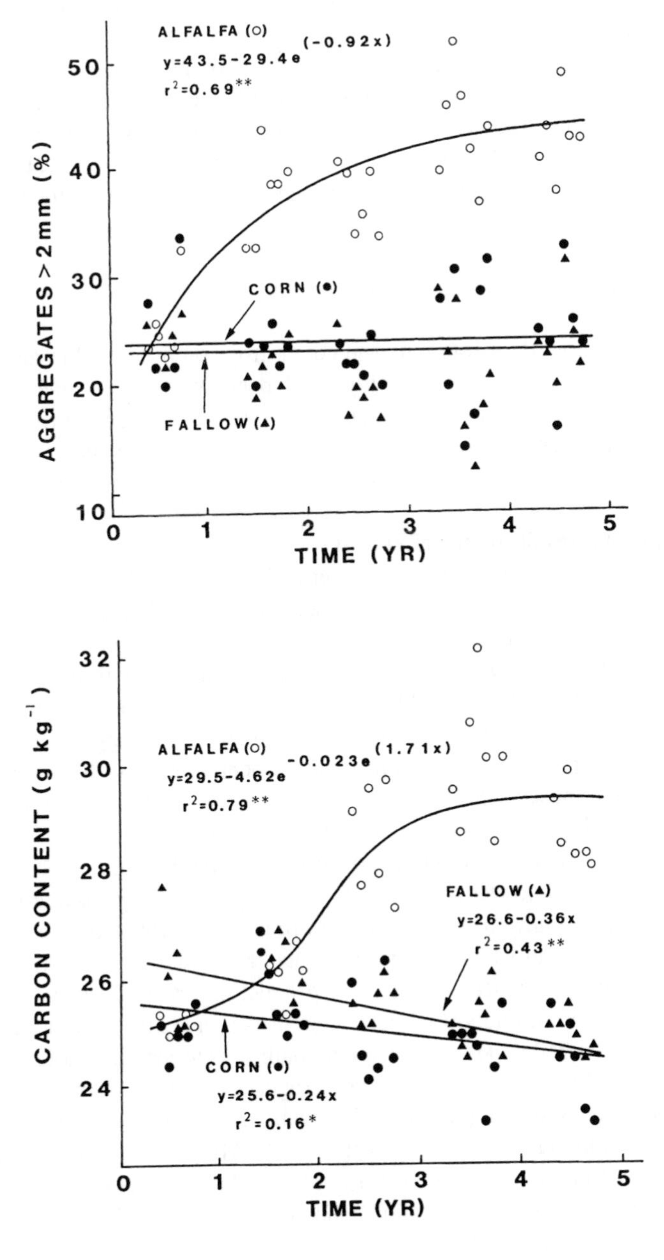

Figure 1. Changes in a) water-stable macroaggregation and b) organic C content under alfalfa, corn and a bare soil in a Kamouraska clay (Humic Gleysol). (From Angers, 1992.)

regime of red clover (*Trifolium pratense* L.) or N fertilization of bromegrass on aggregation.

These studies show that the use of perennial forages represent an excellent means of storing and protecting C in cool-humid soils. Within 2 to 3 years after establishment, measurable changes in labile organic C and water-stable macroaggregation can be detected. These can be used as indicators of the potential of a given forage crop to encourage soil C storage over the long-term.

B. Short-Term Rotations and Cover Crops

The effects of short-term rotations and cover crops on water-stable aggregation and soil organic C is obviously highly dependent on crop species and in particular the amount of residues that each crop of the rotation returns to the soil in the form of either roots or above-ground residues. Management components such as fertilizer and tillage practices are also important factors which will be treated separately.

Experimental results on the effects of short-term rotations and cover crops on soil C content and aggregation in eastern Canada and elsewhere have been inconclusive. In some cases cover crops and rotation with a legume or a grass-legume mixture significantly improved soil macroaggregation (Webber, 1965; Raimbault and Vyn, 1991; Latif et al., 1992) but in others little or no effects were observed (MacRae and Mehuys, 1987; Carter and Kunelius, 1993). Clearly, the results of such studies will depend on the initial aggregation and soil C levels and on previous and current management practices. It should also be pointed out that most of these studies were concerned only with the top few (7 to 20) cm of soil. A better appreciation of their effects on soil C storage will be achieved by considering the whole soil profile or Ap horizon.

An appropriate knowledge of the C inputs provided by each crop in the rotation would certainly be the best way to predict the effects of these rotations on soil aggregation and soil C. Carbon inputs vary widely among crop species, and good estimates of below-ground C production are still lacking for most agricultural crops. Simulation models suggest that an annual input of 2 to 3 Mg C ha^{-1} is required under eastern Canadian conditions to maintain soil C levels at approximately 20 g C kg^{-1} (Voroney and Angers, 1994). With the exception of some row crops such as potatoes (*Solanum tuberosum* L), silage corn and soybean (*Glycine max* (L.) Merr.) (residues <1 Mg C ha^{-1}), most crops grown under cool and humid climates will provide residues in excess of 2 Mg C ha^{-1} and may potentially ensure the maintenance of soil C levels (Carter et al., 1989; Bolinder and Angers, 1993).

Another approach to monitor changes in C inputs is to use early indicators of the change in soil organic C. In a recent review, Gregorich et al. (1994) suggested that microbial biomass, light fraction and labile carbohydrates could be appropriate early indicators of management-induced changes in soil C. Labile carbohydrates may be particularly relevant when soil structure is of interest.

C. Fertilization and Organic Additions

Mineral fertilizer, in particular N, can have both positive and negative effects on soil aggregation and C storage. In the short term, N fertilization can accelerate mineralization of organic binding agents (e.g. Acton et al., 1963) but, in the longer-term, N fertilization seems to have limited effects on C turnover (E.G. Gregorich and C.F. Drury, unpublished results). On the other hand, N fertilization can increase C content by improving crop yield and crop residue inputs to the soil (Liang and MacKenzie, 1992; E.G. Gregorich and C.F. Drury, unpublished results). Kay (1990) suggested that the potential improvement in soil aggregation associated with increased production of roots due to fertilization may not be fully realized if fertilization also causes increased rates of mineralization of the binding material.

There is a multitude of organic materials applied to agricultural soils. The most widespread is farmyard manure. The beneficial effects of farmyard manure on soil structure and organic matter is well documented under various soil and climatic conditions (e.g. Ketcheson and Beauchamp, 1978; Magdoff and Amadon, 1980; Sommerfeldt and Chang, 1985; Dormaar et al., 1988). N'dayegamyie and Angers (1990) and N'dayegamyie (1990) observed a linear increase in water-stable macroaggregation and soil C content with cattle manure application rates in a silage corn production system. An application rate of approx. 40 Mg ha^{-1} every 2 years (2 Mg C ha^{-1} yr^{-1}) was adequate to maintain the C and aggregation levels of the silty loam studied. The application of other organic materials such as compost, sewage sludge and ligno-cellulosic materials on soil aggregation and organic C are also probably beneficial to soil aggregation and organic C storage. However, there is wide variation in the composition of these materials and more research is required to quantify the decomposition rates and effects of these organic amendments on soil organic C and aggregation under contrasting climatic conditions. Paustian et al. (1992) reported that materials with high-lignin content (sawdust, manure) resulted in greater C accumulations per unit C input than low-lignin amendments (straw, green manure). However, the effects of lignin-rich organic amendments on water-stable macroaggregation can be limited (N'Dayegamyie and Angers, 1993).

D. Conservation Tillage

Several studies in cool, humid climatic regions have shown an accumulation of organic C at the surface depth (upper 5 to 10 cm) of no-till soils relative to

Table 3. Percent of organic C present as microbial biomass (MBC), hot water-soluble carbohydrates (WSC), and acid-hydrolysable carbohydrates (AHC) under different tillage systems after 3 years in a Humic Gleysol (clay) in the 0-7.5 cm soil layer

Tillage	MBC	WSC[a]	AHC[a]
		-------------%-------------	
MP[b]	1.2a	0.52a	8.8a
CP	1.5b	0.58b	9.6b
NT	1.7c	0.66c	10.6c

[a]Based on 40% C in the carbohydrate fractions.
[b]MP=Moldboard plow, CP=Chisel plow, NT=No-till. Means followed by the same letter are not statistically different at 0.05 probability level.
(Adapted from Angers et al., 1993a.)

moldboard plowed soils (Carter, 1986, 1992; Angers et al., 1993a). This has importance for improved soil physical quality (Carter, 1992; Angers et al., 1993b). In addition to increases in total organic C, several detailed studies have shown that conservation tillage can also increase labile C fractions. Greater contents of microbial biomass (Carter, 1986; Angers et al., 1993a,c; Simard et al., 1994), labile carbohydrates (Angers et al., 1993a), macroorganic matter (Angers et al., 1993c) and potentially mineralizable N (Simard et al., 1994) have been found under no-till and reduced tillage relative to moldboard plowing at the soil surface layer, although little or no measurable differences in total organic C occurred between tillage treatments. As illustrated in Table 3, greater proportions of organic C present as labile organic matter can indicate longer-term trends of organic C accumulation under reduced tillage systems in cool, humid climates (Mele and Carter, 1993).

However, in studies where the whole Ap horizon is considered (Carter, 1986; Angers et al., 1993c; O'Hallaran, 1993), differences in total soil organic C are not so obvious, especially when calculations are based on equivalent depth to compensate for differences in soil density as illustrated in Table 4. Natural ^{13}C abundance can be used to study the effects of tillage on organic matter turnover (Balesdent et al., 1990). Angers et al. (1995) used this approach on a cool, poorly drained soil in eastern Canada and found that reduced tillage did not significantly alter the decomposition rate of either the original meadow-derived C or the 'young' corn-derived C (Table 5). These results suggest that conservation tillage may have limited impact in reducing organic C turnover under cool-humid climates. Furthermore, studies on organic C distribution in particle size fractions (Dalal and Mayer, 1986; Angers et al., 1993c) suggest that under moldboard plowing the mineral-associated organic matter ($< 50\ \mu m$) increases relative to conservation tillage practices. The mixing effect of plowing

Table 4. Comparison of organic C in the profile of a fine sandy loam (Orthic Podzol) at two sites (1 and 2) and a silty loam (site 3) under conservation tillage systems

	C in sampled profile			C in equivalent soil mass		
Tillage system[a]	Depth (mm)	Soil mass (Mg ha^{-1})	C mass (Mg ha^{-1})	Depth (mm)	Soil mass (Mg ha^{-1})	C mass (Mg ha^{-1})
Site 1						
MP	240	2688	55.4	276	3048	63.2
ST	240	2976	58.9	246	3048	60.2
DD	240	3048	61.5	240	3048	61.5
Site 2						
MP	240	3072	57.1	261	3344	62.1
DD	240	3344	65.1	240	3344	65.1
Site 3						
MP	240	2976	76.6	252	3120	80.4
RT	240	3120	82.6	240	3120	82.6
ST	240	2952	79.2	254	3120	83.8

[a]MP=Moldboard plowing, RT=Ridge tillage, ST=Shallow tillage, DD=direct drilling.
(Recalculated from Carter et al., 1988 and Carter, 1991; Angers et al., 1993c.)

Table 5. Total amounts of organic C in the upper 2880 Mg soil ha^{-1} in a Gleyed Humo-Ferric Podzol (Silty loam)

Treatment	Equivalent depth	Total C	Meadow-C	Corn-C
	cm	----------------Mg ha^{-1}	----------------	
MD[a]	24.0	83.2	83.2	0
ST	24.8	70.9a	66.4a	4.5a
RT	25.9	71.0a	65.6a	5.4a
MP	24.4	66.2a	61.2a	5.0a

[a]MD=Meadow, MP=Moldboard plow, RT=Ridge tillage, ST=Shallow tillage. Means followed by the same letter are not statistically different at the 0.05 probability level.
(Adapted from Angers et al., 1995.)

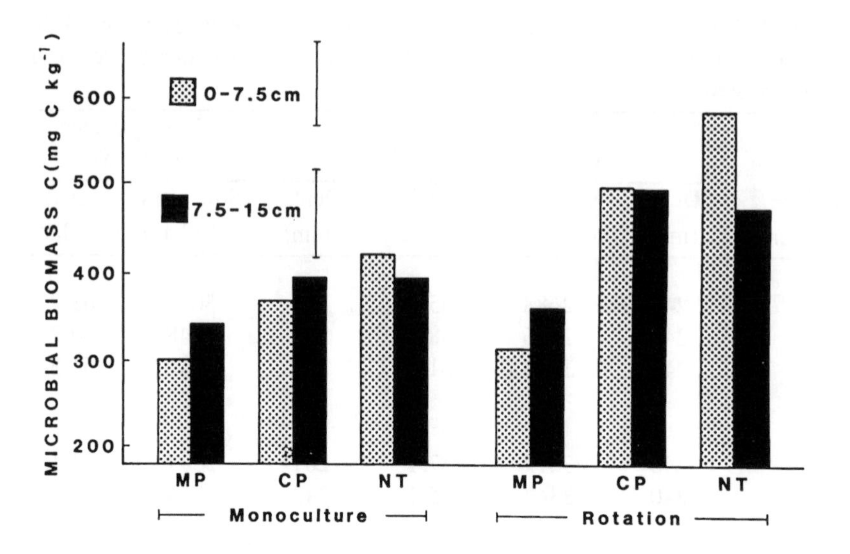

Figure 2. Effects of rotation and tillage treatment combinations on microbial biomass C in two soil layers averaged across sampling dates. MP=Moldboard Plow, CP=Chisel Plow, and NT=No-Till. The bars represent the LSD (0.05 level) for treatment comparisons (tillage x rotation interaction) within each soil layer. Kamouraska clay (Humic Gleysol) (From Angers et al., 1993a.)

increases the association of mineral and organic particles resulting in the formation of organo-mineral complexes. This suggests that tillage may play a positive role in long-term C sequestration. Although in many cropping practices (especially no-till) the mixing activity of soil macrofauna may complement, to some degree, the role of tillage. Clearly, more studies considering the whole soil profile and using isotopes are required to determine the real impact of conservation tillage on soil C storage and turnover.

Generally, unless conservation tillage increases crop C production and inputs to the soil, the major influence of conservation tillage will be on organic C placement and distribution within the soil profile (Carter, 1986; Angers et al., 1992). Thus, the whole soil profile or at least the whole tillage layer has to be considered when C storage is discussed. Furthermore, the effect of reducing tillage on organic C storage may be better achieved if forage-based rotations or cover crops are used. No-till and reduced-till barley in rotation with red clover resulted in greater microbial biomass C than barley monoculture using any tillage treatment (Figure 2). These rapid changes in microbial biomass may be indicative of longer-term trends in C storage under these treatments. Recently, Lee et al. (1993) suggested, using simulation studies, that a combination of no-

Table 6. Various comparisons of C and N storage under forested (native), pasture and arable management in a range of soil types in Prince Edward Island; based on 3500 Mg soil ha^{-1} mass

		C and N in equivalent soil mass			
Management	Soil texture	Depth	C	N	C/N
		cm	-------------Mg ha^{-1}	-------------	
Forest (mixed)	Sandy loam	32.5	87.0	6.06	14.4
Arable		27.3	76.5	7.79	9.8
Forest (coniferous)	Sandy loam	28.9	79.9	8.56	14.4
Pasture		25.2	90.8	6.50	14.0
Arable		30.7	63.5	5.76	11.0
Forest (deciduous)	Sandy loam	38.6	76.0	5.03	15.1
Pasture		24.8	70.7	8.40	8.4
Arable		24.5	54.4	5.16	10.5
Pasture	Sandy loam	24.1	61.6	5.55	11.1
Arable		26.2	42.0	4.60	9.1
Pasture	Sandy loam	25.4	85.5	6.68	12.8
Arable		23.3	49.8	4.72	10.6
Pasture	Loam	25.2	68.8	5.42	12.7
Arable		26.1	44.1	4.69	9.4
Pasture	Loamy sand	29.8	54.4	5.25	10.4
Arable		27.8	38.8	3.17	12.3

till corn with winter wheat cover crops has the potential to provide a net increase in soil organic C, in contrast to corn alone.

E. Type of Vegetation

In cool and humid regions, type of vegetation has the main influence on storage of organic C within the soil profile. Several studies conducted on Podzolic soils in eastern Canada have compared the mass of organic C and N on an equivalent soil mass basis (Table 6). These illustrate that vegetation changes have the main influence on organic matter storage. For example, forested areas in Prince Edward Island provide a mean of 81 Mg C ha^{-1} compared to 53 Mg C ha^{-1} under long-term arable conditions when calculated on an equivalent soil mass basis. Use of forages, however, restored the organic C stored to 72 Mg C ha^{-1}. These comparisons should also take into consideration the relatively large pool of C stored in the above-ground biomass in forested areas.

V. Conclusions

In soils typical of cool, humid climates storage of organic C is associated with water-stable aggregation. The development of the latter agrees with the hierarchical model, in which labile organic fractions bind microaggregates (which contain sequestered organic matter) into macroaggregates. Thus, levels of macroaggregation, organic C and forms of labile C are usually positively related.

Agricultural management practices, such as use of perennial forages and organic amendments, can significantly increase soil macroaggregation and C storage. Cover crops and short-term forage-based rotations may also be beneficial if soil C inputs increase. Conservation tillage may significantly influence organic C placement and distribution within the soil profile, but will have marginal effects on organic C storage unless soil C inputs are enhanced.

Generally, the C storage potential of cool, humid agricultural soils in eastern Canada is mainly associated with type of vegetation. Under perennial forage grasses or legumes, organic C is stored predominantly in macroaggregates which are formed by root-microbe-organic matter interactions. Storage of inter-aggregate organic C in macroaggregates, however, should not be equated with sequestration as the C is relatively labile, easily available to heterotrophic microbes, and consequently subject to potential decomposition. Storage or maintenance of soil organic C, under these conditions is temporary and dependent on providing a continuous C source and preserving the soil structure.

Acknowledgements

The authors would like to thank Drs. H.H. Janzen and R.R. Simard for reviewing an early draft of this chapter.

References

Acton, C.J., D.A. Rennie, and E.A. Paul. 1963. Dynamics of soil aggregation. *Can. J. Soil Sci.* 43:201-209.

Angers, D.A. 1992. Changes in soil aggregation and organic carbon under corn and alfalfa. *Soil Sci. Soc. Am. J.* 56:1244-1249.

Angers, D.A. and G.R. Mehuys. 1988. Effects of cropping on macro-aggregation of a marine clay soil. *Can. J. Soil Sci.* 68:723-732.

Angers, D.A. and G.R. Mehuys. 1989. Effects of cropping on carbohydrate content and water-stable aggregation of a clay soil. *Can. J. Soil Sci.* 69:373-380.

Angers, D.A., and G.R. Mehuys. 1990. Barley and alfalfa cropping effects on carbohydrate contents of a clay soil and its particle size fractions. *Soil Biol. Biochem.* 22:285-288.

Angers, D.A., A. Pesant, and J. Vigneux. 1992. Early cropping-induced changes in soil aggregation, organic matter, and microbial biomass. *Soil Sci. Soc. Am. J.* 56:115-119.

Angers, D.A., N. Bissonnette, A. Légère, and N. Samson. 1993a. Microbial and biochemical changes induced by rotation and tillage in a soil under barley production. *Can. J. Soil Sci.* 73:39-50.

Angers, D.A., N. Samson, and A. Légère. 1993b. Early changes in water-stable aggregation induced by rotation and tillage in a soil under barley production. *Can. J. Soil Sci.* 73:51-59.

Angers, D.A., A. N'dayegamyie, and D. Côté. 1993c. Tillage-induced differences in organic matter of particle size fractions and microbial biomass. *Soil Sci. Soc. Am. J.* 57:512-516.

Angers, D.A., R.P. Voroney, and D. Côté. 1995. Dynamics of soil organic matter and corn residues as affected by tillage practices. *Soil Sci. Soc. Am. J.* (in press).

Balesdent, J., A. Mariotti, and D. Boigontier. 1990. Effect of tillage on soil organic carbon mineralization estimated from ^{13}C abundance in maize fields. *J. Soil Sci.* 41:587-596.

Baldock, J.A. and B.D. Kay. 1987. Influence of cropping history and chemical treatments on the water-stable aggregation of a silt loam soil. *Can. J. Soil Sci.* 67:501-511.

Baldock, J.A., B.D. Kay, and M. Schnitzer. 1987. Influence of cropping treatments on the monosaccharide content of the hydrolysates of a soil and its aggregate fractions. *Can. J. Soil Sci.* 67:489-499.

Bolinder, M.A. and D.A. Angers. 1993. Evaluation of the potential for carbon sequestration for selected agro-ecosystems in Quebec using simulation modelling. Research Report, Agriculture Canada, Research Station, Sainte-Foy, Quebec.

Cambardella, C.A. and E.T. Elliott. 1993. Carbon and nitrogen in aggregates from cultivated and native grassland soils. *Soil Sci. Soc. Am. J.* 57:1071-1076.

Canada Expert Committee on Soil Survey. 1978. *The Canadian System of Soil Classification*. Agriculture Canada Publ. No. 1646. Canadian Gov. Pub. Centre, Ottawa, Ontario.

Capriel, P., T. Beck, H. Borchert, and P. Harter. 1990. Relationship between soil aliphatic fraction extracted with supercritical hexane, soil microbial biomass, and soil aggregate stability. *Soil Sci. Soc. Am. J.* 54:415-420.

Caron, J., B.D. Kay, and J.A. Stone. 1992. Improvement of structural stability of a clay loam with drying. *Soil Sci. Soc. Am. J.* 56:1583-1590.

Carter, M.R. 1986. Microbial biomass as an index for tillage-induced changes in soil biological properties. *Soil Tillage Res.* 7:29-40.

Carter, M.R. 1991. Evaluation of shallow tillage for spring cereals on a fine sandy loam. 2. Soil physical, chemical and biological properties. *Soil Tillage Res.* 21:37-52.

Carter, M.R. 1992. Influence of reduced tillage systems on organic matter, microbial biomass, macro-aggregate distribution and structural stability of the surface soil in a humid climate. *Soil Tillage Res.* 23:361-372.

Carter, M.R. and H.T. Kunelius. 1993. Effect of undersowing barley with annual ryegrasses or red clover on soil structure in a barley-soybean rotation. *Agric. Ecosyst. Environ.* 43:245-254.

Carter, M.R., H.W. Johnston, and J. Kimpinski. 1988. Direct drilling and soil loosening for spring cereals on a fine sandy loam in Atlantic Canada. *Soil Tillage Res.* 12:365-384.

Carter, M.R., H.T. Kunelius, and L.M. Edwards. 1989. The importance of crop rotation for soil structure and conservation. p. 45-54. In: M.R. Carter and H.W. Johnston (eds) *The importance of crop rotations in maritime agriculture.* Agriculture Canada, Charlottetown Research Station, P.E.I.

Carter, M.R., D.A. Angers, and H.T. Kunelius. 1994. Soil structural form and stability, and organic matter under cool-season perennial grasses. *Soil Sci. Soc. Am. J.* 58:1194-1199.

Chantigny, M.H., D.A. Angers, D. Prévost, L.P. Vezina, and F.P. Chalifour. 1995. Plant-induced changes in soil aggregate stability and organic matter in two arable soils. *Soil Sci. Soc. Am. J.* (submitted).

Dalal, R.C. and R.J. Mayer. 1986. Long-term trends in fertility of soils under continuous cultivation and cereal cropping in southern Queensland. III. Distribution and kinetics of soil organic carbon in particle-size fractions. *Aust. J. Soil Res.* 24:293-300.

Davidson, E.A. and I.L. Ackerman. 1993. Changes in soil carbon inventories following cultivation of previously untilled soils. *Biogeochemistry* 20:161-193.

Dormaar, J.F. 1984. Monosaccharides in hydrolysates of water-stable aggregates after 67 years of cropping to spring wheat as determined by capillary gas chromatography. *Can. J. Soil Sci.* 64:647-656.

Dormaar, J.F., C.W. Lindwall, and G.C. Kosub. 1988. Effectiveness of manure and commercial fertilizer in restoring productivity of an artificially eroded dark brown chernozemic soil under dryland conditions. *Can. J. Soil Sci.* 68:669-679.

Edwards, A.P. and J.M. Bremner. 1967. Domains and quasicrystalline regions in clay systems. *Soil Sci. Soc. Am. Proc.* 35:650-654.

Elliott, E.T. 1986. Aggregate structure and carbon, nitrogen, and phosphorus in native and cultivated soils. *Soil Sci. Soc. Am. J.* 50:627-633.

Elliott, E.T., C.A. Palm, D.E. Reuss, and C.A. Monz. 1991. Organic matter contained in soil aggregates from a tropical chronosequence: correction for sand and light fraction. *Agric. Ecosyst. Environ.* 34:443-451.

Eswaran, H., E. van den Berg, and P. Reich. 1993. Organic carbon in soils of the world. *Soil Sci. Soc. Am. J.* 57:192-194.

Gregorich, E.G., R.G. Kachanoski, and R.P. Voroney. 1989. Carbon mineralization in soil size fractions after various amounts of aggregate disruption. *J. Soil Sci.* 40:649-659.

Gregorich, E.G., B.H. Ellert, D.A. Angers, C.M. Monreal, M.R. Carter, and R.G. Donald. 1993. Impact of management on the quantity and composition of organic matter in soils of eastern Canada. Can. Soc. Soil Sci. Annual meeting. St. John's Newfoundland. August 18-22, 1993.

Gregorich , E.G., M.R. Carter, D.A. Angers, C.M. Monreal, and B.H. Ellert. 1994. Towards a minimum data set to assess soil organic matter quality in agricultural soils. *Can. J. Soil Sci.* 74:367-385.

Gupta, V.V.S.R. and J.J. Germida. 1988. Distribution of microbial biomass and its activity in different soil aggregate size classes as affected by cultivation. *Soil Biol. Biochem.* 20:777-786.

Hassink, J. 1992. Effects of soil texture and structure on carbon and nitrogen mineralization in grassland soils. *Biol. Fertil. Soils* 14:126-134.

Haynes, R.J. and R.S. Swift. 1990. Stability of soil aggregates in relation to organic constitutents and soil water. *J. Soil Sci.* 41:73-83.

Haynes, R.J. and G.S. Francis. 1993. Changes in microbial biomass C, soil carbohydrates and aggregate stability induced by growth of selected crop and forage species under field conditions. *J. Soil Sci.* 44:665-675.

Haynes, R.J., R.S. Swift, and R.C. Stephen. 1991. Influence of mixed cropping rotations (pasture-arable) on organic matter content, water stable aggregation and clod porosity in a group of soils. *Soil Tillage Res.* 19:77-87.

Jastrow, J.D. 1987. Changes in soil aggregation associated with tallgrass prairie restoration. *Am. J. Bot.* 74:1656-1664.

Kay, B.D. 1990. Rates of change of soil structure under different cropping systems. *Adv. Soil Sci.* 12:1-52.

Kay, B.D., D.A. Angers, P.H. Groenevelt, and J.A. Baldock. 1988. Quantifying the influence of cropping history on soil structure. *Can. J. Soil Sci.* 68:359-368.

Ketcheson, J.W. and E.G. Beauchamp. 1978. Effects of corn stover, manure, and nitrogen on soil properties and crop yield. *Agron. J.* 70:792-797.

Lee, J.J., D.L. Phillips, and R. Liu. 1993. The effect of trends in tillage practices on erosion and carbon content of soils in the US corn belt. *Water, Air, Soil Poll.* 70:389-401.

Latif, M.A., G.R. Mehuys, A.F. MacKenzie, I. Alli, and M.A. Farris. 1992. Effects of legumes on soil physical quality in a maize crop. *Plant Soil* 140:15-23.

Liang, B.C. and A.F. MacKenzie. 1992. Changes in soil organic carbon and nitrogen after six years of corn production. *Soil Sci.* 153:307-313.

Low, A.J. 1955. Improvements in the structural state of soil under leys. *J. Soil Sci.* 6:177-199.

Magdoff, F.R. and J.F. Amadon. 1980. Yield trends and soil chemical changes resulting from N and manure application to continuous corn. *Agron. J.* 72:161-164.

MacRae, R.J. and G.R. Mehuys. 1987. Effects of green manuring in rotation with corn on the physical properties of two Quebec soils. *Biol. Agric. Hortic.* 4:257-270.

Martel, Y.A. and J.M. Deschênes. 1976. Les effets de la mise en culture et de la prairie prolongée sur le carbone, l'azote et al structure de quelques sols du Québec. *Can. J. Soil Sci.* 56:373-383.

Mele, P.M. and M.R. Carter. 1993. Effect of climatic factors on the use of microbial biomass as an indicator of changes in soil organic matter. p. 57-63. In: K. Mulongoy and R. Merckx (eds), *Soil Organic Matter Dynamics and Sustainability of Tropical Agriculture*. J. Wiley & Sons, NY

Money, D.C. 1974. *Climate, soils and vegetation*. University Tutorial Press, London, U.K.

N'dayegamyie, A. 1990. Effets à long terme d'apports de fumier solide de bovins sur l'évolution des caractéristiques chimiques et de la production de maïs ensilage. *Can. J.Plant Sci.* 70:767-775.

N'dayegamyie, A. and D.A. Angers. 1990. Effets de l'apport prolongé de fumier de bovins sur quelques propriétés physiques et biologiques d'un loam limoneux Neubois sous culture de maïs. *Can. J. Soil Sci.* 70:259-262..

N'dayegamyie, A. and D.A. Angers. 1993. Organic matter characteristics and water-stable aggregation after 9 years of wood residue application. *Can. J. Soil Sci.* 73:115-122.

Oades, J.M. and A.G. Waters. 1991. Aggregate hierarchy in soils. *Aust. J. Soil Res.* 29:815-828.

O'Hallaran, I.P. 1993. Effect of tillage and fertilization on inorganic and organic soil phosphorus. *Can. J. Soil Sci.* 73:359-369.

Paustian, K., W.J. Parton, and J. Persson. 1992. Modeling soil organic matter in organic-amended and nitrogen-fertilized long-term plots. *Soil Sci. Soc. Am. J.* 56:476-488.

Perfect, E., B.D. Kay, W.K.P. van Loon, R.W. Sheard, and T. Pojasok. 1990. Rates of change in soil structural stability under forages and corn. *Soil Sci. Soc. Am. J.* 54:179-186.

Post, W.M., W.R. Emmanuel, P.J. Zinke, and A.G. Stangenberger. 1982. Soil carbon pools and world life zones. *Nature* 298:156-159.

Powlson, D.S. 1980. The effects of grinding on microbial and non-microbial organic matter in soil. *J. Soil Sci.* 31:77-85.

Puget, P., C. Chenu, and J. Balesdent. 1995. Total and young organic matter distributions in aggregates of silty cultivated soils. *Eur. J. Soil Sci.* (submitted).

Raimbault, B.A. and T.J. Vyn. 1991. Crop rotation and tillage effects on corn growth and soil structural stability. *Agron. J.* 83:979-985.

Rovira, A.D. and E.L. Greacen. 1957. The effect of aggregate disruption on enzyme activity of microorganisms in the soil. *Aust. J. Agric. Res.* 8:659-673.

Schlesinger, W.H. 1990. Evidence from chronosequence studies for a low carbon-storage potential of soils. *Nature* 348:232-234.

Simard, R.S., D.A. Angers, and C. Lapierre. 1994. Soil organic matter quality as influenced by tillage, lime and phosphorus. *Biol. Fertil. Soils* 18:13-18.

Skjemstad, J.O., R.P. Le Feuvre, and R.E. Preble. 1990. Turnover of soil organic matter under pasture as determined by ^{13}C natural abundance. *Aust. J. Soil Res.* 28:267-276.

Sommerfeldt, T.G. and C. Chang. 1985. Changes in soil properties under annual applications of feedlot manure and different tillage practices. *Soil Sci. Soc. Am. J.* 49:983-987.

Tate, K.R. 1992. Assessment, based on a climosequence of soils in tussock grasslands, of soil carbon storage and release in response to global warming. *J. Soil Sci.* 43:697-707.

Theng, B.K.G., K.R. Tate, and P. Sollins. 1989. Constituents of organic matter in temperate and tropical soils. *In:* D.C. Coleman, J.M. Oades and G. Vehara (eds.), *Dynamics of soil organic matter in tropical ecosystems.* Univ. Hawaii Press, Honolulu.

Tisdall, J.M. and J.M. Oades. 1982. Organic matter and water stable aggregates in soils. *J. Soil Sci.* 33:141-163.

Voroney, R.P. and D.A. Angers. 1994. Analysis of the short-term effects of management on soil organic matter using the CENTURY model. p. 113-120. In: R. Lal, J. Kimble, E. Levine, and B. Stewart (eds.), *Soil Management and Greenhouse Effect.* Adv. Soil Sci., CRC Press, Inc., Boca Raton, Fl.

Waters, A.G. and J.M. Oades. 1991. Organic matter in water-stable aggregates. In: Wilson W.S. (Ed.) *Advances in soil organic matter research. The impact on agriculture and the environment.* Royal. Soc. Chem., Cambridge.

Webber, L.R. 1965. Soil polysaccharides and aggregation in crop sequences. *Soil Sci. Soc. Am. Proc.* 29:39-42.

Zinke, P.J., A.G. Strangenberger, W.M. Post, W.R. Emmanuel, and J.S. Olsen. 1984. Worldwide organic soil carbon and nitrogen data. Report ORNL/TM-8857. Oak Ridge National Laboratory, TN.

Aggregation and Organic Matter Storage in Meso-Thermal, Humid Soils

R.J. Haynes and M.H. Beare

I. Introduction

Meso-thermal humid soils are those in temperate regions not subject to extreme freezing or long periods of snow cover. Agriculture in these regions is often centred around culture of arable crops such as cereals (e.g. wheat, barley), greenfeed crops (e.g. turnips, mangolds), pulses (e.g. beans, lentils, peas), oilseed crops (e.g. rape, sunflowers) small seed crops (e.g. grass and clover seed crops), green manures (e.g. lupins, mustard) and other miscellaneous crops

1-56670-033-7/96/$0.00+$.50

(e.g. potatoes). Grassland farming, either with all-grass or grass-legume mixes, is another common land use. Grassland may be grown for hay or grazed with sheep or cattle some or all of the year. In some cases grassland and arable farming are integrated by growing pasture leys in rotation with arable crops. Forestry and tree crop production can also be important land uses but these are beyond the scope of this review.

Maintenance and improvement in soil organic matter content is generally accepted as being an important aim for any sustainable system of agriculture. In temperate agricultural soils, organic C content often ranges from three to six percent in pastoral soils but is usually in the range of one to two percent or even lower under continuous arable management. Soil organic matter contributes to soil fertility in a number of ways. It is important as a reservoir of nutrients such as N, S and P and upon microbial oxidation, plant-available mineral forms of these nutrients are released. It also increases the cation exchange capacity and water-holding capacity of soils. In addition, soil organic matter contributes significantly to the formation and stabilization of soil structure so that maintenance of soil organic matter and soil structure often go hand-in-hand.

Soil structure can be defined in terms of form and stability. Structural form refers to the arrangement of solid soil particles and the void spaces between them. A desirable range of pore sizes for a tilled soil occurs when most of the clay fraction is flocculated into microaggregates, defined as <250 μm diam., and secondly these microaggregates and other particles are bound together into macroaggregates >250 μm diam. (Tisdall and Oades, 1982; Oades, 1984). The majority of macroaggregates should be in the range 1 to 10 mm diam. (Edwards and Bremner, 1967). Stability of soil structure refers to the ability of the soil fabric to withstand the disrupting action of external forces, particularly drying and rewetting. Both microaggregates and macroaggregates depend on organic matter for their integrity and stability. The breakdown of macro aggregates upon rewetting (slaking) is caused primarily by pressures exerted by entrapped air inside aggregates and also rapid swelling of clays (Kemper and Rosenau, 1986). Slaking results in the release of microaggregates from which clay particles may or may not be released (dispersed).

Slaking at the soil surface can result in fine soil particles moving into interaggregate pores in the surface layer. This can reduce the infiltration rate of rainfall or irrigation water and reduce hydraulic conductivity. Upon drying, a surface crust can form that can impede emergence and growth of seedlings. Extensive slaking and downward movement of dispersed soil particles into the surface horizon can result in the formation of a dense, massive plough layer. Macroaggregate stability is therefore of central importance to maintaining a desirable soil condition for crop growth.

Soil organic matter content is greatly influenced by soil management practice, for example, increasing greatly under pastoral management and decreasing rapidly under arable cropping. The organic matter that accumulates under pasture and declines under arable is mainly that which is involved in binding microaggregates into stable macroaggregates. This organic matter can become incorporated into macroaggregates in ways such that much of it is apparently inaccessible to microbial degradation unless aggregates are physically disturbed by factors such as cultivation or drying and rewetting (Haynes, 1986). The nature of this organic matter and the way that it is deposited and bound within macroaggregates is therefore of great importance in relation to understanding of how management practice affects both soil organic matter content and aggregate stability.

This paper reviews the mechanisms by which soil organic matter is incorporated into soil aggregates and the role of specific components of organic matter as aggregating agents in warm-humid temperate soils. The importance of soil microflora and macrofauna in depositing and mixing organic binding agents throughout the soil volume is also examined. Based on this discussion, the various soil properties that have been found to be closely correlated with aggregate stability are considered and the effects of soil management practice on organic matter content and aggregate stability are discussed. For the latter discussion, particular emphasis is placed on results obtained in New Zealand and Western Europe. Europe has a history of long-term experiments which chronicle the effects of various management practices and cropping rotations on soil organic matter levels. The classical experiments at Rothamsted are of particular note and are used as examples throughout the text.

II. The Aggregation Process

The processes of aggregation are discussed in detail in chapters 1 to 4 of this volume. A broad outline, with particular reference to temperate soils, is presented below in order to give a basis for the following discussion on the effects of management practice on organic matter content and aggregate stability. A summary of the major binding and aggregating agents and their role in temperate soils is given in Table 1.

The major organic binding agents in soils are humic molecules and polysaccharides. These molecules can strongly bind to the mineral components of soils and to each other and are therefore of central importance in relation to binding aggregates together. The generally accepted model of aggregation involves two physical units, microaggregates < 250 μm diam. and macroaggregates > 250 μm diam. (Edwards and Bremner, 1967; Tisdall and Oades, 1982). Humic molecules become associated with clay minerals and amorphous Al and Fe oxides with the formation of microaggregates. Microaggregates are the building blocks of soil structure. Through the cementing actions of polysaccharides and humic substances, they can become united with one another and with other components such as fragments of decomposing organic material and sand particles to form macroaggregates. Physical enmeshment by roots, associated mycorrhizal hyphae and other fungal hyphae help in this macroaggregation process. So too do the mixing actions of macrofauna, such as earthworms,

Table 1. Summary of the major binding and aggregating agents and their role in soils of warm humid climates

Aggregating agents	Aggregation process	Major scale of aggregation
Humic substances	Form strong bonds with soil mineral components	Basis of microaggregate formation
Polysaccharides	Act as gelatinous glueing agents Form organo-mineral associations	Involved in stabilization of both micro- and macroaggregates
Plant roots	Enmesh soil aggregates Exude polysaccharides	Agents of macroaggregate formation and short-term binding
Fungal hyphae	Enmesh soil aggregates Exude polysaccharides	Agents of macroaggregate formation and short-term binding
Earthworms	Mix organic matter and clay colloids together Mix decaying detritus with the bulk soil	Agents of macroaggregate formation

which ingest and mix soil and decomposing organic material and excrete the mixtures in the form of excretal pellets.

A. Organic Binding Agents

1. Humic Substances

Soil humic substances are complex systems of high molecular weight organic molecules made up of phenolic polymers produced from the products of biological degradation of plant and animal residues and the synthetic activity of microorganisms. They exist in soils as heterogenous, complex three dimensional amorphous structures. Humic substances usually make up 70 to 80% of the soil organic matter content in most mineral soils. Since soils in temperate regions have a generally high organic matter content, humic substances play a major role in aggregation in such soils.

A hypothetical model of a humic acid molecule is shown in Figure 1. The basic structure is an aromatic ring of di- or trihydroxyphenol type, bridged by -0-, $-CH_2-$, -NH-, -N=, -S- and other groups containing both free hydroxyl

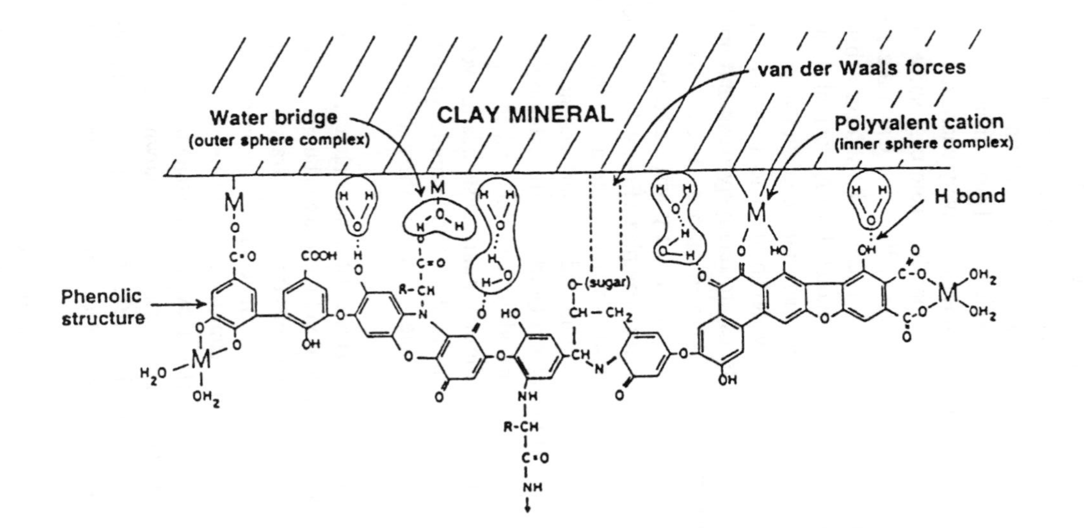

Figure 1. Schematic diagram of a humic acid molecule bound to a phyllosilicate clay. (Adapted from Stevenson and Ardakani, 1972.)

groups and double linkages of quinones. The molecule contains attached proteinaceous and carbohydrate residues.

Soil humic substances characteristically interact with the mineral component of soils to form water-soluble and water-insoluble associations of widely differing chemical and biological stabilities (Schnitzer, 1986). Soils in the temperate regions that are used for intensive agricultural production are usually derived from alluvium, loess or volcanic ash. They can contain a wide range of clay minerals including phyllosilicate clays, aluminum and iron oxides and hydroxides and amorphous aluminosilicates. Thus, many types or organo-mineral associations can be formed. Since negatively charged humic molecules are repelled from negatively charged phyllosilicate clay surfaces, adsorption occurs when polyvalent cations (e.g. Al^{3+}, Fe^{3+} and Ca^{2+}) are present on the cation exchange surface of the clays. These cations form a bridge between negatively charged clays and humic molecules. An anionic (carboxylate or hydroxide) group of the humic polymer interacts with a polyvalent cation through a water molecule (i.e. a water bridge) to form an "outer sphere" complex (Figure 1). These complexes are relatively weak and predominate in aqueous solution or with strongly hydrated cations. A second type of bridging occurs when a humic anionic group is directly associated with a polyvalent cation in the form of an "inner sphere" complex (Figure 1). This occurs under dehydrating conditions or with weakly hydrated cations. Other bonding forces may also operate between organic anions and clay surfaces including H-bonding and van der Waals forces (Murray and Quirk, 1990). Although the strength of these bonds is small their effects are additive so that total adsorption energy can be appreciable particularly where the clay-humic system is allowed to dehydrate, bringing clay particles closer together. Drying soil samples presumably induces inner sphere complex formation and enhances these other bonding forces (Greenland, 1971).

Hydrous oxides and aluminosilicates exist in soils alone and as coatings over phyllosilicate clay mineral surfaces. Bonding between humic molecules and these materials can occur through simple electrostatic attraction (anion exchange) between an anionic group of a humic molecule and a positively charged site. Alternatively these anionic groups can displace a water molecule (coordinated to aluminium exposed at the surface) by ligand exchange and as a result the surface becomes more negative. This mechanism is known as specific adsorption and a very strong bond is formed if more than one group on the humic molecule participates. Specific adsorption can also occur on the crystal edge surfaces of 1:1-type clay minerals.

The processes by which humic molecules become sorbed to clay colloids are poorly understood. The highly soluble (fulvic acid) component of humus could sorb directly from solution but most humic substances in soils are in solid or gel states (Hayes and Swift, 1990). Mechanical mixing by soil microfauna would be one way to bring the inorganic colloids into close contact with humic molecules in the solid and gel state. Soil moisture will be required to ensure humic molecules attain an expanded gel state in which interactions with adjacent inorganic colloids can take place. As noted above, it appears that drying is necessary subsequently in order to 'cement' the adhesion.

2. Polysaccharides

Soil carbohydrate is derived from a mixture of plant and microbial polysaccharides (Cheshire, 1979). Carbohydrates are introduced to the soil principally in decaying plant material (which contains about 75% polysaccharide). These compounds are, however, rapidly degraded by the heterotrophic microbial biomass during decomposition of the plant material. Nonetheless, most soil microorganisms characteristically produce extracellular gelatinous polysaccharide materials in which they become encapsulated (Cheshire and Hayes, 1990). Thus, when readily decomposable organic material is added to soil there is a rapid increase in the size and activity of the microbial biomass and therefore an increase in the production of microbial polysaccharide mucigel material. In addition, living roots exude mucilaginous material which is dominantly polysaccharide and this can form a layer several microns thick around the roots (Oades, 1978; Morel et al., 1991). This material is readily degraded by soil microorganisms which, as already noted, produce extracellular polysaccharides themselves. Microbial polysaccharides can, in turn, be biodegraded rapidly by other microbes.

Polysaccharide macromolecules have high molecular weights, long chains and are often charged so that they take up random conformations in soil and can sorb to surfaces at many points (Cheshire and Hayes, 1990). Many types of interaction may take place between polysaccharides and clays. For neutral polymers H-bond formation may be most important whilst cationic polymers can participate in normal exchange reactions. Negatively charged polymers are involved in interactions similar to those outlined previously for humic substances. That is, they become linked to clay minerals through polyvalent cations whilst H-bonding and van der Waals forces may also be important. They may also link to hydrous oxides through anion exchange and specific adsorption. The extent of sorption of polysaccharides by clay depends on the molecular weight, conformation and configuration of the molecules. As polymer molecular weight increases so too does the extent of its sorption since the greater is its ability to form bridges between soil components (Pagliai et al., 1979; Cheshire and Hayes, 1990).

Polysaccharides are thought to be transitory binding agents (Tisdall and Oades, 1982) since they can be rapidly metabolized by microorganisms. If their role in aggregate stabilisation is more than transitory then they must be continually replenished and/or become physically or chemically protected from microbial decomposition in some way. In the rhizosphere region, particularly of grasses, a large active microbial biomass proliferates. Thus, there is production of a large amount of microbial mucigel and a balance develops between microbial synthesis and degradation of polysaccharide binding agents (Burns and

Davies, 1986). Protection of polysaccharides occurs through their association with inorganic or organic moieties and/or their inaccessibility. Adsorption of polysaccharides to clay surfaces tends to reduce their biodegradation (Cheshire and Hayes, 1990). In addition, electron microscopy has demonstrated that microbial mucilages, either alone or attached to microorganisms, become coated by clay platelets effectively restricting their degradation by microbes (Foster, 1988). Soil polysaccharides are also known to form complexes with metal ions and humic molecules (Lynch and Bragg, 1985). Both these interactions result in an increase in the persistence of polysaccharides (Martin et al., 1966, 1978). Because of the above associations with other soil components, polysaccharides can become incorporated into aggregates in ways that make them physically inaccessible to microorganisms and enzymes (Burns and Davies, 1986) and as a result about 10% of soil organic matter in surface soils occurs as polysaccharides. It is noted here that much of this carbohydrate is probably linked to, or associated with soil humic substances and may not be directly involved in aggregation. A more labile, "easily extractable" fraction of soil carbohydrate, however, appears to be closely related to aggregation in temperate soils (Haynes and Swift, 1990; Haynes and Francis, 1993).

B. Biological Aggregating Agents

1. Roots and Fungal Hyphae

Roots and root hairs of plants, particularly grasses, can act as aggregating and binding agents. They have been shown to enmesh fine particles of soil into aggregates even after their death (Clarke et al. 1967; Coughlan et al., 1973; Forster, 1979). In the rhizosphere, vesicular arbuscular mycorrhizae (VAM) associated with the roots also have an enmeshing effect. VAM produce extensive hyphae in soil which can extend 10 to 30 mm from the surface of the root (Tisdall and Oades, 1979; Miller, 1986) and there can be up to 50 m hyphae per gram of stable aggregate (Miller, 1986). The stability of macroaggregates of several soils has been related to the length of these external hyphae in soil (Tisdall and Oades, 1980a; Miller and Jastrow, 1990). Hyphae can persist in soils for several months after the plant dies (Tisdall and Oades, 1980b). Whilst individual hyphae are not strong, the combined strength of all hyphae and fine roots, especially in a 3-dimensional network, holds particles so that aggregates do not slake when wetted rapidly. Such binding agents build up in the soil within a few weeks or months as the root systems and associated hyphae grow. As discussed previously, the aggregating effect of roots and fungal hyphae not only occurs through physical entrapment since they both exude polysaccharides that have an aggregating effect. Fungi also produce phenolic humic materials that participate in aggregate binding.

Stabilisation of aggregates in bulk soil by filamentous fungi is likely to occur mainly at sites around readily decomposable organic material and will be favoured by additions of organic matter to the soil (Burns and Davies, 1986). By ramifying through the soil, fungal hyphae may bring soil particles in contact and enhance their contact with binding agents. Filamentous fungi differ in their stabilising abilities, the most effective being vigorous growing species with woolly hyphae (Lynch and Bragg, 1985). Saprophytic fungal populations often decline in soils as readily available substrates are depleted so their stabilisation of aggregates can sometimes last for only a few weeks (Low and Stuart, 1974; Molope et al., 1987). Where plant litter material is left to slowly decompose at the soil surface (e.g. under zero tillage), fungal-mediated soil aggregation may, however, be much more persistent (Beare et al., 1995).

The contribution of roots, root hairs and fungal hyphae to aggregate stabilisation probably differs in different situations. However, as noted by Tisdall (1991), root hairs and fungal hyphae can enter smaller pores than fine roots and hence can distribute organic material more evenly throughout the soil. The surface area per unit volume of hyphae is about 100 times that of a root (Harley, 1989) so that hyphae produce less material to develop the same amount of surface to which soil particles may become attached. It should be noted here that roots, root hairs and hyphae absorb water from the surrounding soil and thus cause localized drying. As noted in following sections, this can have both stabilizing and disrupting effects on soil macroaggregates.

2. Role of Earthworms

Soil macrofauna also play an important role in macroaggregation. The major group of macrofauna in temperate soils are earthworms and in temperate pastures their populations may reach 1000-2000 m^{-2} (Lee, 1985). Earthworms ingest large amounts of organic litter material as well as soil and their casts frequently have a higher organic matter content than bulk soil. In addition, casts have often been found to have a greater aggregate stability than the surrounding soil (Shaw and Pawluk, 1986; Brussard et al., 1990). Earthworms deposit casts both at the soil surface and in the surface soil horizon. For many populations, the majority of casts appear to be deposited within the surface soil horizon. For example, Graff (1971) estimated that up to 1000 t ha^{-1} or 25% of the Ah horizon of a pasture soil in Germany was turned over each year by earthworms yet only about 40 to 50 t ha^{-1} yr^{-1} are normally deposited at the surface of temperate pastures (Lee, 1985). Similarly, in savanna soils Lavelle (1978) estimated that 1200 t ha^{-1} yr^{-1} of soil was ingested and cast by earthworms but only 2.5 t ha^{-1} yr^{-1} was deposited at the soil surface.

The importance of earthworm activity is illustrated by the fact that several workers have suggested that more than 50% of the structural aggregates in the upper 10 to 20 cm of soils under temperate pastures are recognisable as earthworm casts (van de Westeringh, 1972; Lee and Foster, 1991). Similarly, in mull-type forest soils of Europe, practically all aggregates in the surface horizon have been identified as earthworm casts or residues of casts (Lee and

Foster, 1991). Earthworm casts can be of two basic types (Lee, 1985). These are (1) ovoidal to spherical pellets ranging from < 1 mm to > 1 cm in diam. and (2) paste-like slurries that form more-or-less rounded shapes. Casts of the common earthworms of agricultural lands are mainly in the range 2 to 10 mm diam. with many of them being composites of the two basic forms (Lee and Foster, 1991).

Ultrastructural studies on casts have shown that organic material is present as fresh root material, partially decomposed organic remnants and as humified material (Lee and Foster, 1991). Litter is fragmented by the grazing activity of earthworms and further fragmentation occurs in the gizzard which grinds food with the aid of mineral particles (Edwards and Lofty, 1977). Physical breakdown of the tissue combined with microbial activity and the action of digestive enzymes and secretions produced in the digestive tract results in the decomposition of some organic material as well as the formation and release of organic binding agents (Shipitalo and Protz, 1989). Ingested microaggregates appear intact and hardly modified by passage through the gut (Lee and Foster, 1991). There is much evidence for the stimulatory effect of passage of earth through the earthworm gut on microbial activity (Lee, 1985; Daniel and Anderson, 1992). Most microorganisms in casts appear to be in protected sites either in plant debris within soil microaggregates or as clay-coated microcolonies (Lee and Foster, 1991).

The feeding habit of earthworms may influence their effect on aggregate formation and stabilization. For example, *Lumbricus terrestris* is an anecic species that removes large quantities of litter from the soil surface into its permanent burrows that extend down below the plough layer. The walls of the permanent burrows become coated with cutaneous polysaccharide mucus material from the outside of the earthworm (Zhang and Schrader, 1993). In addition, pieces of litter material line the walls of the burrows which further promotes microbiological activity. As a result, the stability of the soil near the burrow walls is greatly increased but the remainder of the soil is little affected (Shaw and Pawluk, 1986; Zhang and Schrader, 1993). On the other hand, endogenic species ingest large amounts of soil (but select organic materials within the soil) and make extensive systems of burrows that ramify through the soil. They deposit their casts mainly in their burrows and in other soil voids. Such earthworms have a large effect on aggregate stability in the surface 20 cm of soil (Lee and Foster, 1991).

C. Structural Organisation

Diverse organic and inorganic constituents participate in the binding of soil particles into water-stable aggregates and the relative importance of each varies in differing situations. Indeed, the structural organisation of soils is heterogeneous over short distances and can also change over short time scales by processes such as wetting and drying, penetration by roots and burrowing by

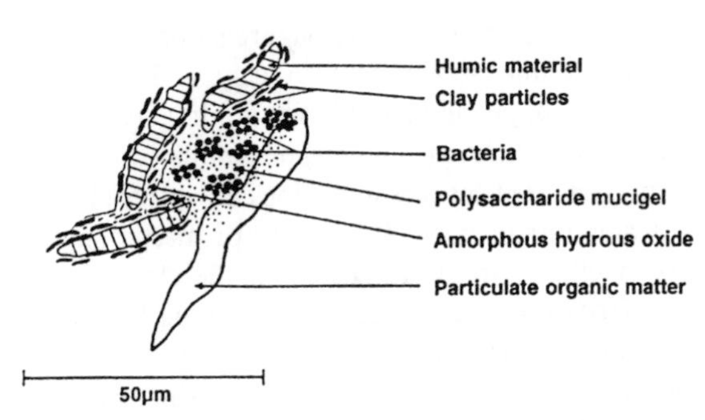

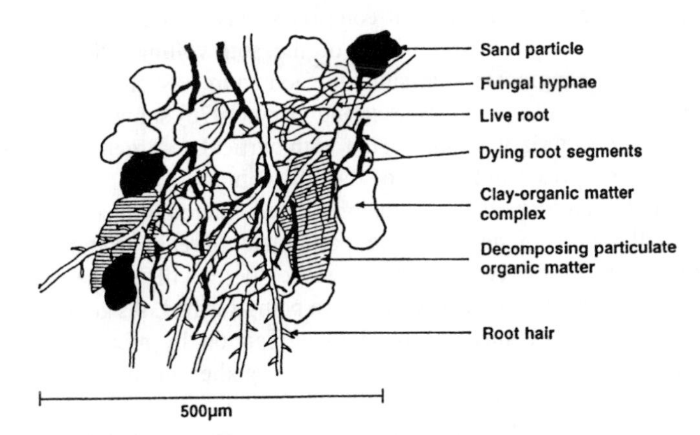

Figure 2. Schematic diagram of the structure of a micro- and a macroaggregate.

processes such as wetting and drying, penetration by roots and burrowing by soil animals (Young and Crawford, 1991). As a result, only generalized models of soil microaggregate and macroaggregate structure and formation are practical. Even then, considerable controversy surrounds the relative importance of various processes involved in the formation of both microaggregates and macroaggregates and there are no universally accepted models of their structures. Below, an overview of microaggregation and macroaggregation is presented (schematic structures and shown in Figure 2) in order to give a basis for future discussion. Particular attention is paid to macroaggregate formation and stabilization since it is the organic matter involved in these processes that is greatly affected by soil management practice in temperate soils.

1. Microaggregation

As already noted, the driving force of microaggregation is the sorption of organic colloids (mainly humic substances) to clay and Al and Fe oxide materials. Humic substances in gel and solid states are largely immobile in soil and so too are polysaccharide macromolecules. The mobile component involved in microaggregate formation often appears to be fine clay material. For example, clay can move with water in soil pores by gravity or matric suction towards roots until it contacts mucilage. In the rhizosphere, the clay plates then align themselves with the surface of the mucilages, encapsulating bacteria, surrounding fungal hyphae or aligning parallel to roots (Oades, 1984).

Since humic molecules and polysaccharides are principally products of microbial metabolism it follows that microaggregate formation occurs mainly at sites in soil of high microbial activity. The microbial biomass can constitute up to 50 kg dry weight ha^{-1} although it comprises only 1 to 3% of soil organic C and may occupy as little as 0.001% of the soil volume (Sparling, 1985). Microorganisms congregate near suitable food sources such as decaying plant and animal debris. As a result, Hissett and Gray (1976) showed that in a sandy soil 64% of the bacteria were associated with particulate organic particles even though these comprised only 15% of the soil volume.

It is likely therefore, that the major site of microaggregate formation is around small pieces of decomposing particulate organic matter (Elliott and Coleman, 1988; Beare et al., 1994a), since these will be the predominant microsites in soil where humic polymers are being formed. As noted previously, soil fauna could also contribute to microaggregation by mixing fragments of detritus and inorganic and organic colloids together during passage of soil through their gut.

Particles of slowly decomposing particulate organic matter (100-200 μm diam.) encrusted with microbial mucilage and clay make up larger microaggregates (Oades and Waters, 1991). As this organic material slowly decomposes, smaller organo-mineral complexes can be formed. These latter microaggregates consist of clay particles, clay domains and hydrous oxides of Al and Fe (either alone or as amorphous coatings over clays) sorbed to organic colloids (mainly humic materials and some polysaccharides) (Figure 2). Small pores (1μm diameter or less) exist within microaggregates, and bacteria tend to occur within these pores and within cell wall debris (Foster, 1988).

2. Macroaggregation

The relative importance of various binding agents in the formation of macroaggregates is subject to some debate. On the semi-arid soils studied by Oades and coworkers (Tisdall and Oades, 1982; Oades 1984; Oades and Waters, 1991;

Tisdall, 1991) the enmeshing effect of roots and VAM hyphae has been emphasised. Soils from mesothermal humid regions generally have a considerably higher organic matter contents that those from semi-arid regions. Thus, polysaccharides and particularly humic substances may have a greater influence in the former soils. For this reason, the roles of humic and polysaccharide binding agents are stressed throughout the following sections.

Macroaggregate formation may occur through entrapment of groups of microaggregates and other particles (e.g. sand particles) by ramifying roots, root hairs, VAM and other fungal hyphae (Figure 2). The excretal pellets of soil fauna (e.g. earthworms) may also act as precursors for stable macroaggregates. Not only does the network of roots and hyphae physically enmesh particles but it also produces organic binding agents which help cement them together. A large microbial population (dominated by bacteria) proliferates in the rhizosphere of plant roots (see Section III D). Organic binding agents (humic substances and polysaccharides) are deposited by the fungal and bacterial biomass and polysaccharides are also exuded by growing roots. Furthermore, as parts of the hyphal and root network die organic substrate is deposited throughout the macroaggregate structure. Microflora produce more binding agents as they degrade this deposited particulate organic matter thus further contributing to macroaggregate integrity. As discussed above, the decomposing organic particles can, in turn, become the centre for microaggregate formation. At the same time that some root segments, root hairs and VAM hyphae are senescing, new tissue is being produced and is ramifying through and around macroaggregates. Thus, the processes involved in the formation and stabilization of macroaggregates are dynamic in nature.

Ingestion of soil and litter by soil macrofauna (especially earthworms) represents a mechanism whereby clay is brought into intimate association with decomposing mucilage-coated organic fragments. Large amounts of watery mucus are added to the comminuted ingested material in the anterior part of the gut (Barois, 1987) and the soil undergoes a thorough kneading. A liquified soil-debris mixture is excreted from the animal. Most of the coherence of fresh earthworm casts appears to be attributable to interlocking of various components together with some short-range binding by microbial polysaccharides (Lee and Foster, 1991). Fresh casts tend to have a low density so that not all the particles are in close association (McKenzie and Dexter, 1987). As a result, the aggregate stability of fresh casts can be less than that of the surrounding soil (Shipitalo and Protz, 1988; Marinissen and Dexter, 1990; Zhang and Schrader, 1993). In excreted pellets bonds between organic and mineral components are greatly strengthened when these materials are brought into close association. This is encouraged by drying which causes a marked increase in the stability of cast material (Marinissen and Dexter, 1990). If casts remain moist their stability increases greatly with aging. This is principally related to the colonization of casts by fungal hyphae from the surrounding soil (Marinissen and Dexter, 1990;

Lee and Foster, 1991). The hyphal network helps bind particles together and increases stability. Within the casts undigested, mucilage-covered organic fragments probably become foci for further aggregation. Moreover, microbial decomposition of organic fragments would result in continued replenishment of polysaccharide binding agents.

It is sometimes suggested that polysaccharides and humic molecules play a minor role in stabilising macroaggregates. For example, polysaccharide macromolecules have a chain length of a few hundred angstroms and act only over short (sub-micron) distances (Tisdall and Oades, 1982; Foster, 1988). However, as outlined previously, these substances are being produced throughout macroaggregates. Both polysaccharides and phenolic substances are produced by bacterial colonies which may be occluded within and between microaggregates or be present in the rhizosphere region of growing roots or in and around decomposing fragments of particulate organic material (e.g. dying roots). Similarly, they are being produced by fungal communities associated with decaying particulate organic matter and around hyphae of VAM. Thus, although these binding agents act over relatively short distances they are acting at a large number of points within a macroaggregate. The polysaccharides and humic substances may form strong associations with each other as well as with mineral surfaces.

It is, therefore, not surprising that when microbial polysaccharides isolated from cultures of soil microorganisms are added to soils they are effective agents for improving macroaggregate stability (Martin, 1971; Moavad et al., 1976; Chaney and Swift, 1986a). The improvement in aggregation is, however, usually short-lived since soil microflora soon break down the polysaccharide material. In the rhizosphere of grasses, where large amounts of microbial polysaccharides are being continually produced, polysaccharide-mediated stability of macroaggregates may be of particular importance (Haynes and Francis, 1993).

The interaction between polysaccharides and humic molecules helps long-term stabilisation of macroaggregates. Where polysaccharides were added to soils to improve macroaggregate stability Griffiths and Burns (1972) found that subsequent addition of phenolic substances prolonged stabilisation. Chaney and Swift (1986b) found that when humic acid was adsorbed onto soil minerals there was a significant increase in macroaggregate stability. The stability was even greater when the samples were incubated with glucose in order to stimulate microbial production of extracellular polysaccharides.

Drying and rewetting cycles in soil can greatly influence macroaggregate stability. Roots can cause localised drying of soil due to water uptake by the plant. Drying causes dehydration and shrinkage of soil organic materials. As a result, additional intermolecular associations are formed between organic macromolecules (humic substances and polysaccharides) which are acting as binding agents and between these organic molecules and mineral surfaces (Haynes and Swift, 1990). As a result, the stability of macroaggregates can be greatly increased by drying (Churchman and Tate, 1987; Haynes, 1993). The effects of drying on increasing the strength of bonding of microaggregates were noted in section IIA.

D. Aggregate Structure and Organic Matter Availability

1. Physical Inaccessibility and Organic Matter Turnover

The way in which organic matter becomes bound into soil aggregates will influence its availability to microbial degradation. In general, it is thought that humified organic matter bound to inorganic colloids in the form of microaggregates is rather stable (Edwards and Bremner, 1967; Tisdall and Oades, 1982). However, organic materials involved in binding microaggregates together to form macroaggregates are more easily mineralizable but are often physically inaccessible to soil microflora and extracellular enzymes. Much evidence regarding inaccessibility of organic matter to microbial attack comes from studies of the effects of grinding or crushing macroaggregates into smaller particles on subsequent mineralisation in laboratory incubations. Generally, grinding greatly increases mineralisation of C and N and this is attributed to release of organic matter previously inaccessible to attack (Edwards and Bremner, 1967; Craswell and Waring, 1972; Elliott, 1986). Calculations by Adu and Oades (1978) showed that only a very small fraction of organic matter in soils is likely to be in close proximity to soil organisms at any one time. They calculated that in a clay loam with a microbial population of 10^8 bacteria g^{-1} soil, the organisms would occupy about 0.1% of the pore volume and cover about 0.01% of the soil surface. Assuming that enzyme diffusion in soil is limited by adsorption, they concluded that at any one time at least 90% of surface soils are not accessible to microorganisms or their enzymes. In a laboratory experiment, Adu and Oades (1978) found that the artificial distribution of ^{14}C-labelled substrate into macro- and micropores of soil aggregates rendered a portion inaccessible to microbial attack. Disruption of aggregates, either by mechanical disturbance or a drying and rewetting cycle during incubation, resulted in a flush of $^{14}CO_2$ evolution.

Little is known about organic matter turnover in relation to its location within the matrix of soil aggregates. Nevertheless, it is the structural relationships between soil particles and pores which provides variation in habitat for biological (faunal, microbial and enzymatic) attack of organic substrates. For example different microbial populations are associated with different sized soil pores. Most soil protozoa and fungi are predominantly located in large pores (i.e. > 6 μm diam.) whereas most soil bacteria lie in smaller pores (i.e. < 6 μm diam.) (Elliott et al., 1980; Foster, 1988). The turnover of added ^{14}C-labelled substrate in soil aggregates, and evolution of $^{14}CO_2$, has been shown to be affected by both its location in terms of pore size (i.e. $< 6\ \mu m$ or 6-30 μm diam.) and the soil matric potential under which turnover takes place (Killham et al., 1993). Turnover was found by Killham et al. (1993) to be greater in larger, more accessible, pores particularly at soil matric potentials which ensured the larger pores were water-filled. The larger, water-filled pores may well have allowed greater access of protozoan grazers to the ^{14}C-labelled microbial biomass which proliferated in the pores and this would have increased ^{14}C evolution. Even so, location of potentially decomposable organic matter in relation to pore size is only one factor leading to physical inaccessibility. For instance, as already noted, fragments of particulate organic matter may become encrusted with microbial mucilage and clay particles during decomposition and this renders them partially inaccessible to further microbial attack (Oades, 1984; Beare et al., 1994a).

Despite the inaccessibility of organic matter in macroaggregates, degradation of organic matter occurs continuously as shown by the fact that an equilibrium soil organic matter level is reached under any given soil management practice. Thus, under a long-term pasture sizable amounts of organic material are being constantly deposited within macroaggregates in the form of dying and decaying roots and hyphae of VAM and other fungi, polysaccharide gels formed by fungi, bacteria and growing roots and polyphenolic macromolecules formed by the activity of microorganisms. At the same time, similar amounts of organic matter are being decomposed. For example, although humic material is often considered to be rather intransigent in soils, humic molecules are being synthesized and degraded at similar rates. Drying and rewetting cycles contribute greatly to the breakdown of intra-macroaggregate organic matter in temperate soils.

Drying causes shrinkage of soil aggregates and fracture planes and cracks can develop along lines of natural weakness within and between macroaggregates (Kay, 1990). As noted in section II C 2, however, drying can also help stabilize intact macroaggregates because upon shrinkage additional intermolecular associations can be formed. The pattern of drying is often spatially variable with localized drying of soil occurring around growing plant roots. Rewetting can also induce cracking as a consequence of differential swelling of the soil matrix and the pressure of entrapped air inside macroaggregates. The formation of fracture planes and cracks caused by drying and rewetting cycles exposes organic matter previously physically inaccessible to microbial attack (i.e. physically protected through being bound within macroaggregates) and thus there is a flush of organic matter decomposition. In soils that freeze, freezing and thawing cycles can have a similar effect in stimulating organic matter breakdown. The above discussion further emphasizes the dynamic nature of aggregation processes in soils.

2. Cultivation and Organic Matter Turnover

The effect of physical disruption of macroaggregates on mineralization of C and N in laboratory incubations is generally much greater from grassland (Elliott, 1986; Gupta and Germida, 1988) and at the surface of zero-tilled (Beare et al., 1994b) than long-term cultivated soils. Such a phenomenon is not altogether surprising since, as discussed in Section IV, pastoral and zero-tilled soils generally have considerably higher soil organic matter contents than continually cultivated soils. The additional organic matter held in these soils is mainly inter-macroaggregate material which has been deposited within macroaggregates and is involved in maintaining their integrity and stability. It consists of a wide range of organic material including humic substances, polysaccharide gels and decomposing particulate organic matter and is bound within macroaggregates in ways that render a substantial amount of it inaccessible to microbial attack. When macroaggregates are crushed much of this organic matter is exposed to microbial attack and can consequently be mineralized in laboratory incubations.

Similarly, when a grassland soil is converted to conventional arable cultivation, the major source of organic matter that is mineralized is the inter-macroaggregate organic material (Elliott, 1986; Gupta and Germida, 1988). As a result there is a decline in macroaggregation and macroaggregate stability (Elliott, 1986; Haynes and Swift, 1990) whilst microaggregate integrity remains virtually unaffected (Elliott, 1986). As well as disrupting aggregates and promoting organic matter breakdown, conversion of grassland to arable management also results in a marked reduction in the input of organic material to the soil particularly in the form of plant roots (see sections IV B and C). Thus, extractable particulate organic matter, which is comprised primarily of partially decomposed root segments, can account for a significant portion of organic matter that is initially lost from macroaggregates as a result of cultivation of grassland (Cambardella and Elliott, 1992).

In warm-humid climates, the loss of organic matter from grassland soils following continual cultivation may well alter the relative importance of various binding and aggregating agents to macroaggregate stability. The high organic matter content (and hence high content of humic material) in grassland macroaggregates results in them being very stable (Haynes and Swift, 1990; Haynes et al., 1991). Physical enmeshment by roots and hyphae and production of microbial polysaccharides provide additional stability to macroaggregates. As humic material is degraded under continual cultivation, macroaggregation in arable soils becomes more dependent on the actions of crop roots, fungal hyphae and microbial polysaccharides. As a result, there can be pronounced seasonal fluctuations in macroaggregation and macroaggregate stability in long-term arable soils (Blackman, 1992; Mulla et al., 1992). During spring and summer there is a large increase in crop root growth and turnover, a resulting increase in microbial activity and increased production of microbial polysaccharides.

Macroaggregate stability therefore increases. During winter, root turnover and microbial activity are lower and aggregate stability declines. Such seasonal variation in macroaggregate stability is very much less pronounced in the very stable grassland aggregates (Blackman, 1992).

III. Indicators of Aggregate Stability

There are a number of different methods for measuring the stability of macroaggregates to the degrading action of water. The two most common are wet sieving and turbidimetry (Haynes, 1993). These methods rely on arbitrary degrees of mechanical disruption due to rate of rewetting and the shattering and abrasive actions of sieving or shaking of macroaggregates in water. For wet sieving, the proportion of macroaggregates greater than a certain diameter remaining is normally measured whilst for turbidimetry the amount of clay plus silt-sized particles produced is quantified. In the following discussions, the term aggregate stability is used to describe the ability of macroaggregates to withstand the disrupting action of water irrespective of the method used in its measurement.

An important component of research on the effects of soil management on soil structure and aggregate stability has been attempting to relate changes in various soil properties to measured changes in aggregate stability. Such studies have aimed to improve our understanding of the mechanisms responsible for changes in aggregate stability and/or to find an easily-measurable predictor of changes in stability that occur due to changing soil management practice. Properties investigated include total soil organic matter content, size of the microbial biomass, quantity of an extractable 'active' binding component of soil organic matter, crop rooting density and previous cropping history. The basis and merits of these properties are discussed below for warm-humid soil types.

A. Soil Organic Matter Content

Since soil organic matter (particularly humic substances and polysaccharides) is central to the formation of stable soil macroaggregates there is likely to be a close relationship between soil organic matter content and water stable aggregation. Indeed, a strong positive relationship usually exists in agricultural soils (Hamblin and Davies, 1977; Douglas and Goss, 1982; Chaney and Swift, 1984; Haynes et al., 1991). Such a relationship is shown in Figure 3 for a group of agricultural soils from England and Scotland. Total N content of soils is also strongly correlated with aggregate stability (Williams, 1970; Chaney and Swift, 1984; Haynes et al., 1991). Indeed, in a range of arable soils from England and Ireland, Williams (1970) found that total N content was more closely correlated

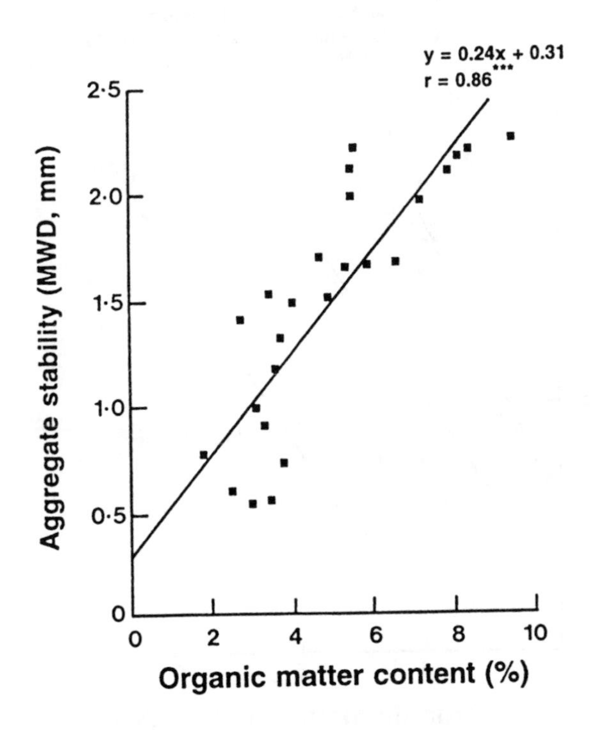

Figure 3. Relationship between aggregate stability (mean weight diameter) and soil organic matter content for a group of soils. (Redrawn from Chaney and Swift, 1984.)

with aggregate stability than was organic C content. It was suggested by Williams (1970) that total organic C values include "inactive" materials, such as coarse lignified particles of low N content. Thus, total N would be a better indicator of the active organic components which are more likely to be associated with mineral surfaces and involved in the formation and stabilisation of aggregates. However, other workers have found that total N is no more closely correlated with aggregate stability than organic C content (Chaney and Swift, 1984; Haynes et al., 1991).

With increasing clay content, soils require a higher organic C content in order to maintain a given aggregate stability value. Such a relationship is shown for a group of soils from southern England in Figure 4. For a wet sieving index of

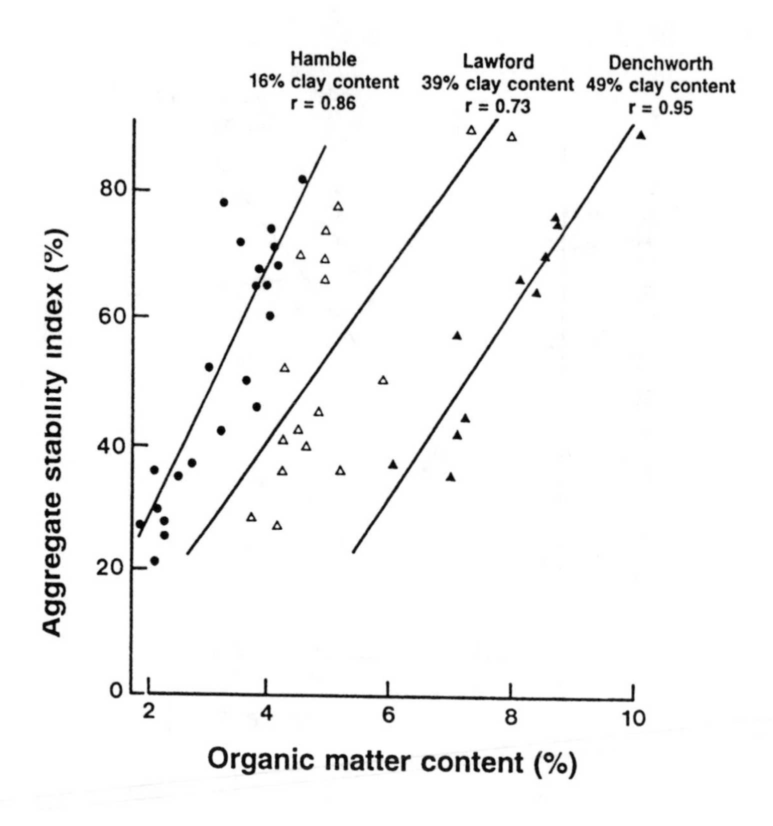

Figure 4. Relationship between aggregate stability and organic matter content for samples from the Hamble, Lawford and Denchworth soil series. (Redrawn from Douglas and Goss, 1982.)

50%, the contents of organic matter for the Hamble, Lawford and Denchworth series were 3.0, 4.6 and 7.2% respectively. A similar relationship was observed by Haynes et al. (1991) when determining the relationship between organic C content and aggregate stability for a sandy loam, a silt loam and a clay loam. The reason for this is presumably that in the absence of organic coatings, clay particles readily disperse in water. Thus, soils with a higher clay content require a higher organic matter content in order to stabilise aggregate structure. It is interesting to note that the slopes of the three regression lines in Figure 4 did not differ significantly indicating that for each soil a given change in organic matter content produced a similar increase in the stability index.

B. Microbial Biomass

In studies of the short term (e.g. 1 to 4 years) effects of soil management, total soil organic C content is often not measurably affected yet large changes in microbial biomass C and aggregate stability can occur. In studies of mixed cropping rotations (e.g. 4 years pasture followed by 4 years arable) no appreciable changes in soil organic C were detected yet microbial biomass C and aggregate stability increased greatly under the pasture phase and declined rapidly under arable (Haynes and Swift, 1990; Haynes et al., 1991). Similarly, Drury et al. (1991) found that biomass C was significantly correlated with aggregate stability being highest under 3 years grass or lucerne (*Medicago sativa* L.) and lowest under 3 years of soybean [*Glycine max* (L.) Merr.] or maize (*Zea mays* L.) yet organic C content remained relatively unchanged. Robertson et al. (1991) observed that aggregate stability and biomass C were increased by 2 years of grass cover although organic C remained unchanged. Furthermore, a rapid increase in microbial biomass C and aggregate stability has been observed in the rhizosphere of grass species without any changes in organic C being detected (Haynes and Francis, 1993). In longer term (i.e. 5 to 8 years) experiments investigating the effects of converting pastoral land to continuous arable management aggregate stability has also been observed to be more closely associated with changes in biomass C than total organic C contents (Sparling and Shepherd, 1986; Hart et al., 1988; Sparling et al., 1992).

Hart et al. (1988) modelled the decline in organic C and microbial biomass C when a soil under permanent pasture was plowed and continuously cropped (Figure 5). It is evident that during the first 5 years, microbial biomass C declined much more rapidly than did total organic C. Similarly, so too did aggregate stability. The rapid initial decline in aggregate stability following plowing in pasture is discussed more fully in section IV C.

It is not surprising that there is a close relationship between microbial biomass C and aggregate stability since as discussed in previous sections, the microbial population plays a central role in aggregation through the production of fungal hyphae which can enmesh aggregates and extracellular polysaccharide gels which act as glues. When long-term treatments are compared, microbial biomass C is often directly related to the soil organic C content (Sparling, 1985). However, in the shorter-term, soil management practice influences microbial biomass C and aggregate stability before changes in total organic C are detected. Indeed the large background level of relatively stable organic matter makes it very difficult to measure small changes that occur over relatively short time periods (e.g. 1 to 4 years). By contrast, the living component of soil organic matter, the microbial biomass comprises only a small proportion (1 to 4%) of total organic C, but has a rapid rate of turnover and responds rapidly to changes in C availability (Jenkinson and Ladd, 1981). Generally, microbial biomass C is a more sensitive indicator of soil organic matter dynamics than total organic C.

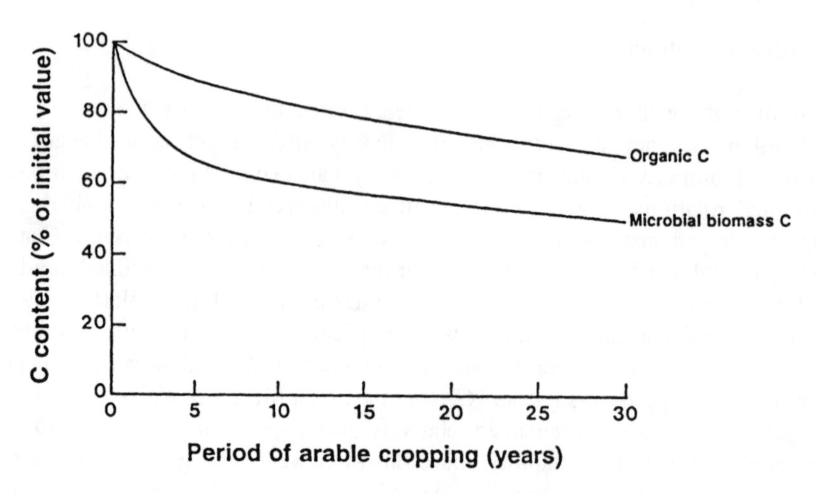

Figure 5. Modelled changes in organic C and microbial biomass C under continuous cropping for 30 years on a silt loam which was previously under permanent pasture. (Redrawn from Hart et al., 1988.)

As a result, in temperate soils microbial biomass C can be used as an indicator of early changes in soil organic matter brought about by management practices such as tillage and straw incorporation (Ross et al., 1982; Carter, 1986; Powlson et al., 1987; Hart et al., 1988). It is apparent that it can also be used as an early indicator of changes in aggregate stability.

C. Extractable Carbohydrates

Total carbohydrate content of soils is frequently no more strongly correlated with aggregate stability than total organic C content (e.g., Chaney and Swift, 1984; Haynes et al., 1991). This is because when total carbohydrates are measured, no differentiation between active and inactive carbohydrate binding agents is made.

A promising approach is to attempt to extract an active fraction of soil carbohydrate that is involved in binding aggregates together. In a study of the effects of previous cropping history on aggregate stability in three soils from the South Island of New Zealand, Haynes et al. (1991) fractionated soil carbohydrate sequentially with cold water, hot water, 1M HCl and 0.5M NaOH. They observed that HCl (which extracts carbohydrates without extracting humic materials) and NaOH (which extracts carbohydrates associated with humic substances) extractable carbohydrates did not appear to be active binding carbohydrates since they were not closely correlated with aggregate stability.

However, hot-water extractable carbohydrates (which constitute about 6 to 8% of total carbohydrate content) were closely correlated with aggregate stability. Hot water was used as a mild agent that would extract a fraction involved in short-term binding of aggregates. In addition it has been observed that both hot water extractable carbohydrates and aggregate stability increase under short-term pasture and decline under short-term arable management yet total organic C and total carbohydrate content remain relatively unaffected (Haynes and Swift, 1990; Haynes et al., 1991). Such a relationship is illustrated in Table 2 for a soil from the South Island of New Zealand under an 8-year rotation (4 year pasture followed by 4 year arable crops). Similarly, in a microthermal climate Angers et al. (1993) studied changes in water stable aggregation induced by 4 years of different rotations and tillage practices. They found that among different organic matter fractions measured, the proportion of hot water-extractable carbohydrate was best correlated with aggregate stability. Studying aggregation of soils amended with sewage sludge, both Metzger et al. (1987) and Kinsbursky et al. (1989) also observed that the hot water-extractable carbohydrate fraction was closely correlated with aggregate stability. In the rhizosphere of grass species, Haynes and Francis (1993) recently found that both aggregate stability and hot water-extractable carbohydrate content were higher than in bulk soil yet total carbohydrate content was unchanged.

In all the above studies, increases in hot water extractable carbohydrates and aggregate stability were associated with concomitant increases in the size of the microbial biomass (see Table 2). Haynes and Swift (1990) suggested that the hot water extractable carbohydrate fraction represents extracellular polysaccharides mainly of microbial origin that are involved in the short-term stabilisation of soil aggregates. Other workers have also concluded that hot water-extractable substances in soils represent mainly extracellular metabolic products of the microbial biomass (Redl et al., 1990).

The soil microbial population synthesizes dominantly galactose (G), glucose and mannose (M) and only small amounts of arabinose (A) and xylose (X) (Martens and Frankenberger, 1991). Plant polysaccharides, on the other hand, contain substantial quantities of arabinose and xylose. As a result, the [(G+M):(A+X)] ratio is typically low (<0.5) for plant polysaccharides and high (>2.0) for microbial polysaccharides (Oades, 1984). Haynes and Francis (1993) showed that the [(G+M):(A+X)] ratio of hot water extracts of a soil was 2.1 confirming it was mainly of microbial origin. As is commonly the case, the [(G+M):(A+X)] ratio of the total carbohydrate fraction was 1.3 to 1.4 confirming that total soil carbohydrates are a mixture of those of plant and microbial origin.

Table 2. Effect of previous cropping history on aggregate stability, organic C, acid hydrolysable and hot water-extractable carbohydrate and biomass C content of a soil from the South Island of New Zealand

Previous cropping history	Aggregate stability (MWD, mm)	Organic C (%)	Acid-hydrolysable carbohydrate (%C)	Hot water-extractable carbohydrate ($\mu g\ C^{-1}$)	Microbial biomass C ($\mu g\ C\ g^{-1}$)
18 yr pasture	2.7	3.2	0.35	208	1018
4 yr pasture[1]	2.5	2.5	0.26	169	890
1 yr pasture	2.0	2.4	0.25	152	801
1 yr arable	1.3	2.4	0.23	140	738
4 yr arable	1.2	2.4	0.23	134	712
10 yr arable	1.0	2.0	0.19	127	610

[1] The 1 year and 4 year pasture and 1 year and 4 year arable soils come from a cropping rotation of 4 years arable followed by 4 years pasture. (Data from Haynes et al., 1991.)

D. Crop Root Density

Large quantities of organic material are supplied to soils from roots, especially in warm-humid climates. This process is known as rhizodeposition. The major source of rhizodeposited material is through the normal growth and senescence of root segments and root hairs but roots also exude a range of organic substances. Newman (1985) estimated that for every 100 g dry weight of senescent roots, soluble exudates from living roots commonly contribute only 1 to 10 g and root cap plus mucigel from living roots perhaps 2.5 g. The magnitude of rhizodeposition of organic C is closely correlated with total root mass (Figure 6). Since the root mass and root length density in the surface soil is considerably greater for grass species than for most arable annual arable crops (Perfect et al., 1990; Haynes and Francis, 1993) rhizodeposition is also greater for grasses. Calculated rhizodeposition for annual crops in temperate regions is in the range 0.24-1.6 t C ha^{-1} yr^{-1} but for perrenial grassland systems it is believed to be 2.0 t C ha^{-1} yr^{-1} or greater (Whipps, 1990). On a field scale, Davenport and Thomas (1988) calculated that the total annual rhizodeposition in the surface 15 cm of soil would be nearly five times higher with bromegrass (*Bromus inermis* Leyss.) than with maize.

As a result of rhizodeposition of C, a large active microbial biomass develops in the rhizosphere and therefore large amounts of organic binding agents are produced. The greater the root density the larger the volume of rhizosphere soil and the greater are the amounts of binding agents produced. In addition, the greater the root density, the larger are the effects of roots, root hairs and VAM hyphae ramifying through and around macroaggregates.

It is therefore not surprising that crops with the greatest root mass show the greatest improvement in aggregation (Stone and Buttery, 1989). When crops are compared under field conditions with respect to their ability to improve aggregate stability the same sequence is observed as that for root mass or root length density (Perfect et al., 1990; Haynes and Francis, 1993). The root mass and length density reflects a number of factors including the normal sowing rate of the crop, the survival rate of seedlings, the rooting characteristics of individual plants and their spreading characteristics (e.g. tillering and/or stolon forming ability). Thus, Kay (1990) observed that root density in the surface 15 cm of soil generally increases in the following order: row crops (e.g. maize, soybeans) < cereals [wheat (*Triticum aestivum* L.), barley (*Hordeum vulgare* L.) < grasses and a similar sequence applies for increases in aggregate stability. Forage grasses are particularly effective at stabilising aggregates because of their characteristically high ratio of below- to above-ground biomass, their dense fibrous root system and the fact that they generally spread rapidly (thus increasing their root mass) by tillering (Haynes and Francis, 1993).

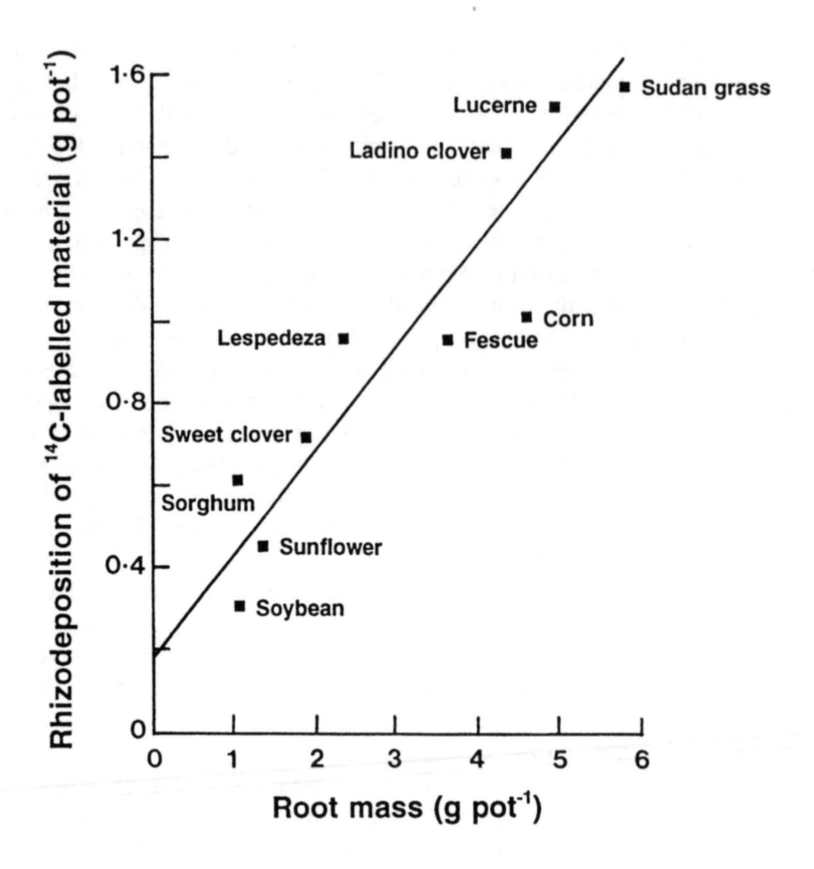

Figure 6. Relationship between root mass produced by a range of crops and the amount of ^{14}C-labelled organic debris deposited in the soil (rhizodeposition). (Redrawn from Shamoot et al., 1968.)

E. Previous Cropping History

In view of the preceding discussion, it is clear that previous cropping history will greatly influence soil organic matter content and aggregate stability. Increasing years under pasture generally result in greater aggregate stability whilst increasing periods under arable cropping have the opposite effect. This relationship is demonstrated for a soil from southern New Zealand in Figure 7 where an index of previous cropping history was more closely correlated with

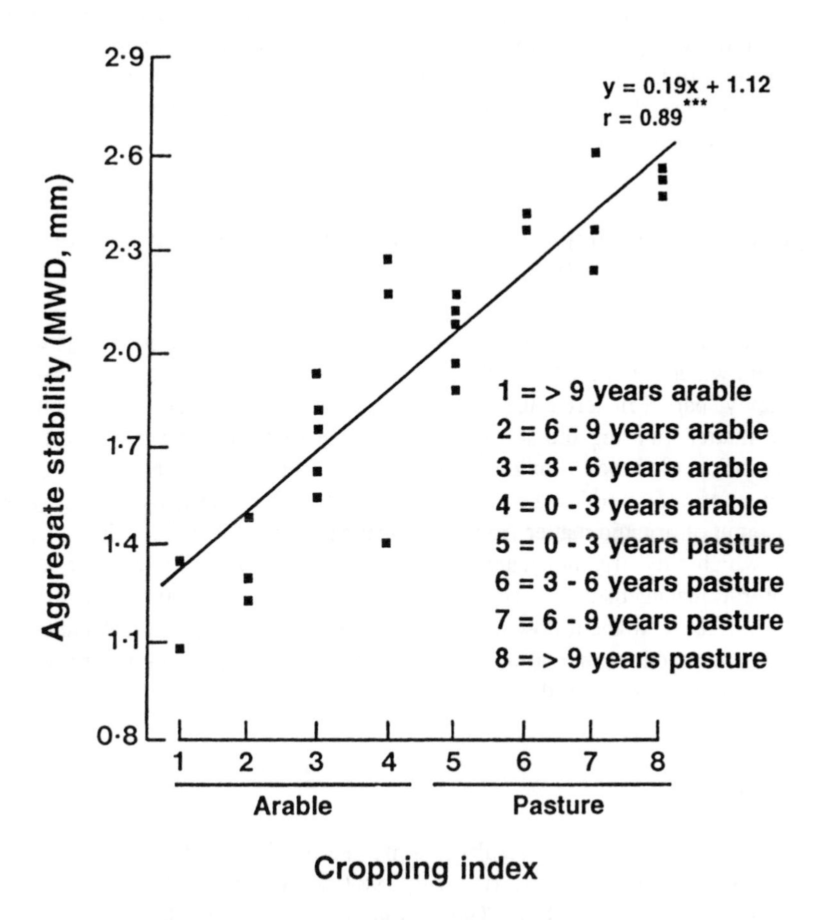

Figure 7. Relationship between previous cropping history and aggregate stability (mean weight diameter) for a silt loam. (Redrawn from Haynes et al., 1991.)

aggregate stability than was total N, organic C or total carbohydrate content of soils (Haynes et al., 1991).

Cropping history reflects changes in a number of factors that influence aggregate stability. For example, recent work (Haynes et al., 1994) using the same index as used in Figure 7 has shown that with increasing time under pasture there was a progressive increase in soil organic C and microbial biomass C content and a rapid increase in earthworm density. The reverse was true for increasing time under arable cropping.

As discussed later in section IV B and C, the type of arable crop or grass grown and its management is also likely to influence organic matter inputs and microbial activity. Nevertheless, use of a simple cropping index can give a good indication of aggregate stability. In many parts of the world fertilizer N rates to arable crops are made based on an index of previous cropping history (Goh and Haynes, 1986) so there is possible scope to use such indices as an indicator of aggregate stability as well.

IV. Effects of Management on Soil Organic Matter Content and Aggregate Stability

An equilibrium soil organic matter content is reached within a mature natural ecosystem which is dependent on the interaction of soil forming factors (climate, topography, parent material and time). At this equilibrium level, the amount of organic matter entering the soil each year is the same as the amount lost by mineralization to carbon dioxide. In agricultural soils changes in soil management practice affect soil organic matter content in two ways, by altering the annual input of organic matter from dead plants and animals and by altering the rate at which this organic material decays. Under any particular long-term agricultural management, at any given site, soil organic matter reaches an equilibrium level. In the following section, the effects of important soil and crop management practices on soil organic matter levels and aggregation for warm-humid regions are discussed.

A. Fallow

Leaving a soil fallow usually results in a marked decline in soil organic matter content and aggregate stability (Kofoed, 1982; Sauerbeck, 1982; Tisdall and Oades, 1982; Johnston, 1986). The reason for this is that plant returns are insignificant and, in addition, the soil is often tilled several times per year to maintain it weed-free. Tillage breaks up soil clods and exposes previously inaccessible organic matter to microbial attack. Thus mineralization of native soil organic matter is stimulated. Less frequent tillage, leaving some weeds on the surface (i.e. green fallow) is not quite as damaging to soil organic matter content as bare fallow (Sauerbeck, 1982) since tillage is less frequent and returns of plant material are greater.

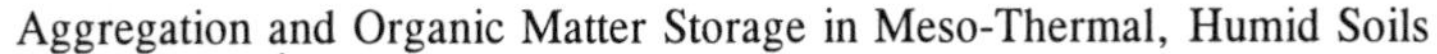

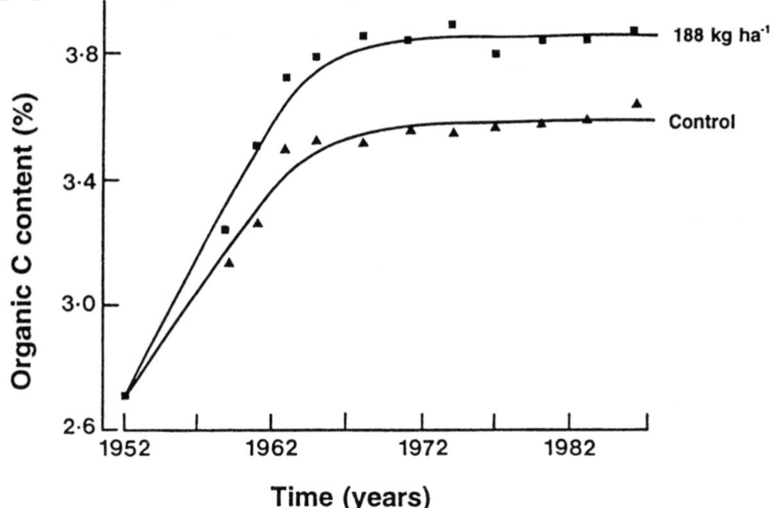

Figure 8. Changes in soil organic C contents (0-7.5 cm) over a 34-year period of pasture development with no fertilizer applied (control) or 188 kg superphosphate ha^{-1} applied annually. (Adapted and recalculated from Nguyen and Goh, 1990.)

B. Pasture

Long-term pastoral management generally results in the attainment of a high soil organic matter content (Haynes and Williams, 1993). When land under arable management is converted to pasture there is often an appreciable increase in soil organic matter. The rate of organic matter accumulation and the time taken to reach an equilibrium, where organic matter additions are balanced by mineralization and losses, varies considerably with initial organic matter level, soil type, climate and management.

The accumulation of organic C over a 38-year period under improved pasture is shown in Figure 8. The site in the South Island of New Zealand had been cultivated and arable crops grown for several years prior to it being sown down in pasture (Haynes and Williams, 1992a). Soil organic matter accumulation was particularly rapid during the initial 10 years. Organic matter inputs under pasture arise from senescing plant tops and roots, exudation of organic compounds from pasture roots, turnover of the large microbial biomass in the pasture rhizosphere and return of ingested organic material by grazing animals in the form of dung. Annual additions of superphosphate at a rate of 188 kg ha^{-1} increased mean

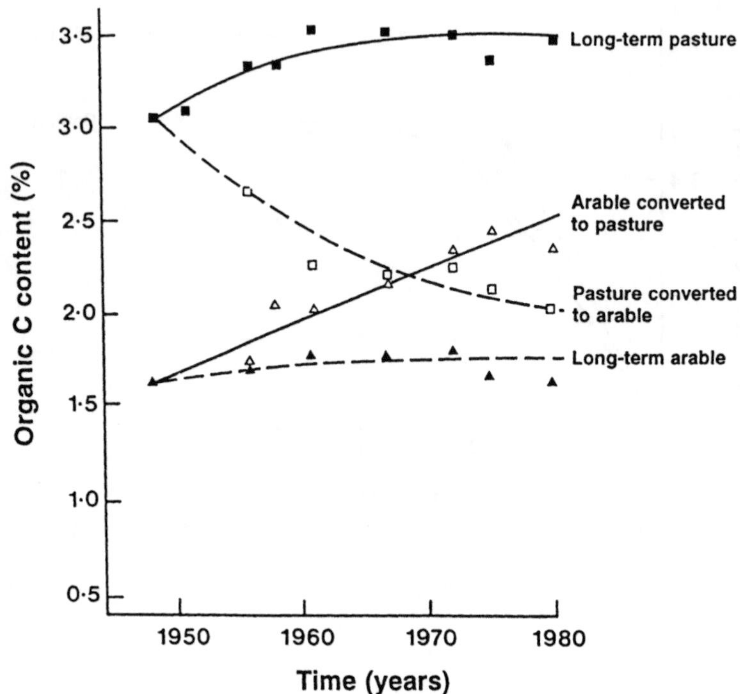

Figure 9. Organic C content of soils from the Rothamsted ley-arable experiment. Treatments consist of: long-term grassland, long-term grassland soil converted to arable, long-term arable and long-term arable soil converted to grassland. (Redrawn from Johnston, 1986.)

pasture dry matter yields over the 37-year period from 4 to 12 t ha^{-1} (Haynes and Williams, 1992b). This increased dry matter production resulted in greater organic matter inputs to the soil and thus the equilibrium organic matter reached was higher with superphosphate applications.

Results of a long-term experiment at Rothamsted in Hertfordshire on a flinty silt loam (Batcombe series) initiated in 1949 also demonstrate the positive effect of long-term grassland on organic matter content (Figure 9). One site had been under grass for at least 100 years and contained 3.1% C. Soil C increased to 3.4% during the first 15 years under continuous grass probably because of improved management and increased dry matter production (and increased organic matter inputs). On an old arable soil initially containing 1.6% organic C the organic matter content increased steadily under grass management.

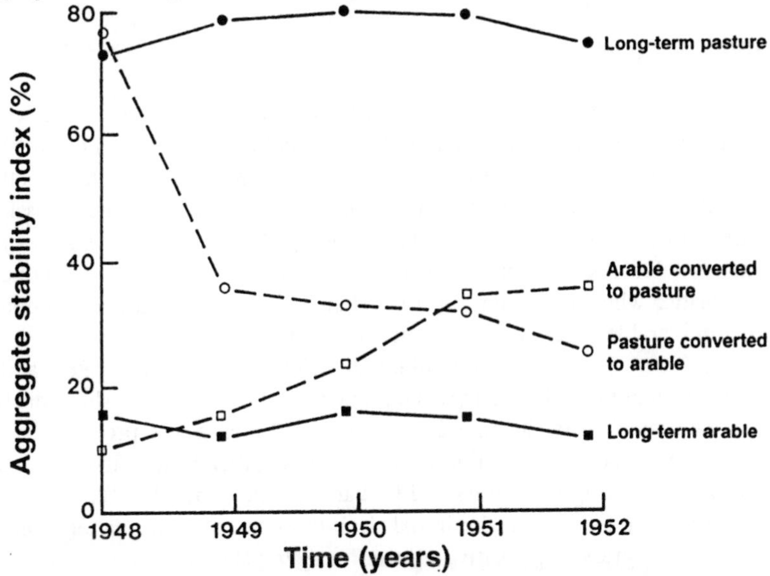

Figure 10. Changes in aggregate stability of a soil under contrasting long-term treatments. Treatments consist of long-term grassland, long-term grassland soil converted to arable, long-term arable and long-term arable soil converted to grassland. (Data from Low, 1955, 1972.)

Since pastoral management increases soil organic matter content, it also increases aggregate stability. Although data is from separate sites, results presented in Figure 10 for long-term trends in aggregate stability on soils developed on chalky boulder clay in southern England (Hanslope series) mirror those for organic matter shown in Figure 9. On an old grassland site aggregate stability was high whilst on an old arable site that had been converted to grassland, aggregate stability steadily increased. When a long-term arable site is sown to pasture it may take up to 50 years for aggregate stability to return to that of old grassland (Low, 1955). The rate of change in aggregate stability can be influenced by the grass species present. In grass-white clover (*Trifolium repens* L.) pastures Clement and Williams (1958) observed that the positive effect on aggregate stability followed the order perennial ryegrass (*Lolium perenne* L.) > timothy (*Phleum pratense* L.) = meadow fescue (*Festuca pratensis* Huds.) > cocksfoot (*Dactylis glomerata* L.). This difference was attributed to perennial ryegrass developing a greater root mass than other species.

C. Arable Cropping

Under arable cropping, the amount of organic material returned to the soil is considerably lower than that under pasture. In addition, the soil is often tilled several times a year which favours decomposition of native soil organic matter. As a result, soil organic matter content characteristically declines when grassland soils are converted to continuous arable cropping (Johnston, 1986; Haynes and Williams, 1992b). Results presented in Figure 9 demonstrate such a decline in organic matter when a long-term grassland site was put under arable cropping. On the old arable site organic matter had equilibrated at about 1.6% C. Similarly, Figure 10 shows that an old arable site had a low aggregate stability whilst when an old grassland site was put under arable there was a sharp decline in aggregate stability in the first 2 to 3 years and then a less pronounced but steady decline. The sharp decline in aggregate stability in the first few years after plowing in a long-term grassland is characteristic (Low, 1972) and appears to be more pronounced than the initial decline in soil organic matter content. This is because grass has positive effects on aggregate stability in addition to those associated with increased organic matter content. When grass is removed these effects are lost. The additional effects have been discussed earlier and include production of large quantities of polysaccharide binding agents by the large microbial biomass in the pasture rhizosphere and the enmeshing effects of fine grass roots and associated mycorrhizal hyphae (Haynes et al., 1991).

Crop management practice and the type of arable crop grown can also influence soil organic matter content. Fertilizer additions normally tend to increase organic matter accumulation because they promote greater crop growth and thus increase returns of organic matter as roots and stubble to the soil. This is demonstrated by the results of a long-term manuring experiment at Rothamsted on a flinty silt loam (Batcombe series) (Figure 11) where the plot which has received annual N, P, K fertilizers has equilibrated to an organic matter content that is 15% higher than the unfertilized plot. In cereal production if straw is plouwed-back instead of being removed or burned more organic material is returned to the soil. Thus, the equilibrium soil organic matter level attained under continuous cereal production is slightly higher with straw incorporation (Johnston, 1979; Christensen, 1986). Organic matter returns are generally greater for cereal crops than root crops (where below-ground matter is harvested). In addition, during harvest of root crops the soil is often greatly disturbed which encourages decomposition of native soil organic matter. As a result increasing proportions of root crops in arable rotation generally result in attainment of a lower equilibrium organic matter level (Kofoed, 1982; Sauerbeck, 1982). Sauerbeck (1982) observed a greater decline in soil organic matter under continuous potatoes (*Solanum tuberosum* L.) or sugar beet (*Beta*

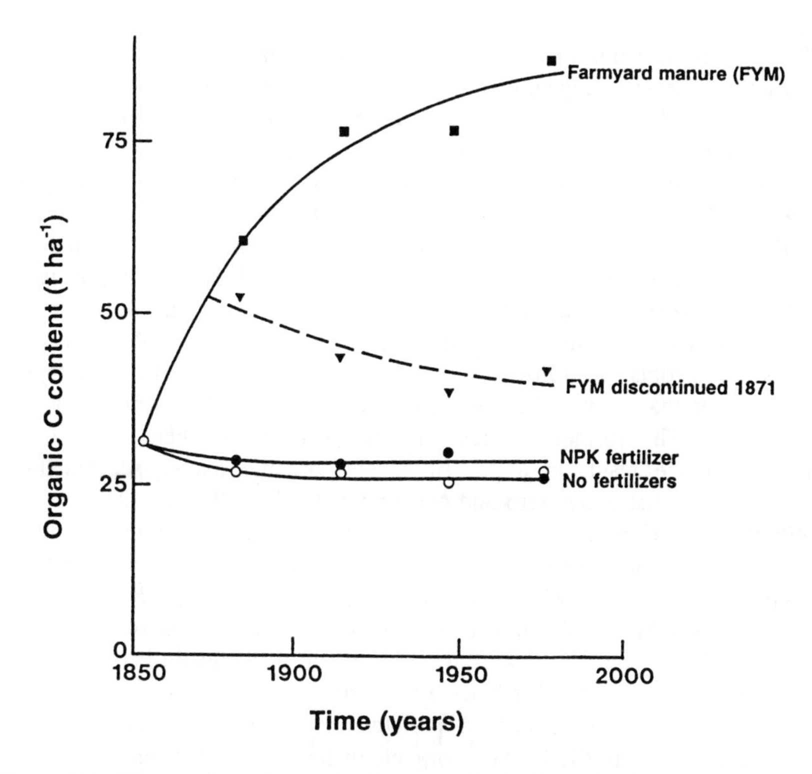

Figure 11. Changes in soil organic C content in the Hoosfield continuous barley experiment at Rothamsted. Treatments consist of: no fertilizers applied, NPK fertilizer applied annually, farmyard manure (FYM) applied annually at 35t ha^{-1} and FYM applied 1852-1871 and none since. (Redrawn from Johnston, 1986.)

vulgaris L.) than under continuous cereals. Further, the decline was greater under sugar beet with leafy tops removed than when leaves were plouwed in.

D. Tillage Method

In warm-humid climates, conventional tillage generally uses the mouldboard plough to bury plant residues and additional tillage for seedbed preparation. By contrast, minimum and no-tillage involve a system where weed control is accomplished mainly by herbicides. Following herbicide application, the soil can either be lightly worked (e.g. harrowed) before drilling (minimum tillage) or soil disturbance can be limited to opening a small slot for seed placement (zero tillage).

Tillage method can influence both the pattern of organic matter accumulation in the soil profile and the quantity of organic matter present. Under conventional tillage the plough layer is often inverted each year so there is a fairly uniform distribution of organic matter through the plough layer. Such a distribution is shown for a silty soil in Berkshire, England in Figure 12. After several years, conventionally tilled soils that have been converted to zero tillage develop an organic matter gradient with a large concentration of organic matter at the soil surface and a sharp decrease with depth (Figure 12). This is mainly because weeds and crop residues are not incorporated but left to decompose at the soil surface. Thus organic matter inputs originate close to the soil surface. In addition, crop roots can sometimes be more abundant in the surface soil layers of zero rather than conventionally tilled soils (Cheng et al., 1990; Drew and Saker, 1978). This further enhances the organic matter gradient. Methods of minimum tillage result in a pattern of soil organic matter accumulation intermediate to that under zero and conventional tillage (Figure 12).

Where a soil has been under continuous arable for many years, conversion to zero-tillage often has little effect on total soil organic matter content of the soil (Gowman et al., 1979; Powlson and Jenkinson, 1981; Haynes and Knight, 1989; Francis and Knight, 1993). For example, on four old arable sites from Scotland and England Powlson and Jenkinson (1981) found that organic matter content of soils (0 - 25 cm layer) after 5 to 10 years of zero tillage was similar to that under conventional tillage. Since crop yields are usually similar under the two farming systems (Cannell, 1985) the organic matter input to the soils in the form of stubble, roots and root exudates are also similar. Presumably, the annual rate of decomposition of organic matter inputs is similar and as a result soil organic matter content is unchanged by tillage method. As noted earlier, the distribution of organic matter in the profile is, however, different.

When soils with an initially high soil organic matter content (i.e. previously under long-term pasture) are converted to arable cropping the decline in native soil organic matter over time is typically more pronounced under conventional than zero tillage (Blevins et al., 1983; Dick, 1983; Francis and Knight, 1993). It was found by Francis and Knight (1993) that when a five year-old grass/clover pasture was converted to arable crops the annual rate of decline in organic C content over a nine-year period was 1.8 t C ha^{-1} under zero tillage but 2.6 t C ha^{-1} under conventional tillage. A number of factors could contribute to this. Aeration is generally better in plowed soil particularly during winter and this leads to a more oxidative microbial metabolism under plouwing than zero tillage. Ploughing may promote microbial decomposition by exposing previously inaccessible soil organic matter to microbial attack. It has also been shown that in comparison with zero tillage ploughing promotes decomposition of plant residues by increasing contact between soil and the residues (Hendrix et al.,

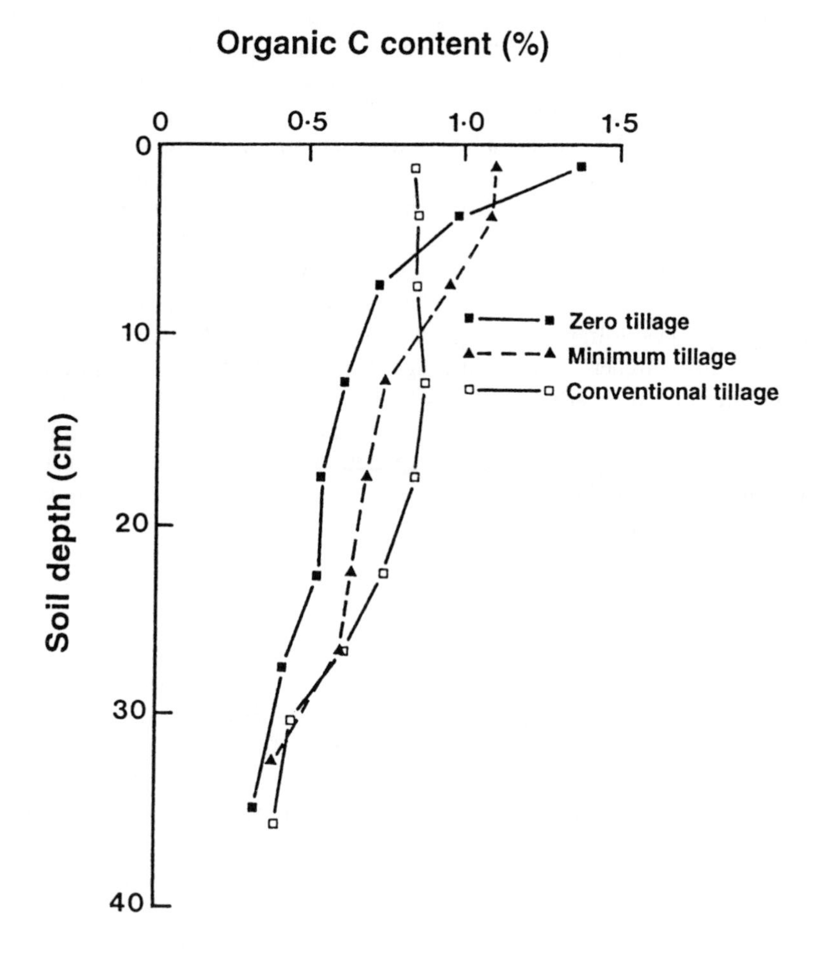

Figure 12. Profiles of organic C in a soil after 10 years of zero tillage, minimum tillage, conventional ploughing or continuous grass. (Redrawn from Douglas et al., 1986.)

1986; Beare et al., 1992). This effect will be of particular importance where large amounts of crop residues are being returned to the soil.

Because of the higher organic matter content, aggregate stability in the surface 10 cm of soil is characteristically higher under zero than conventional tillage (Douglas and Goss, 1982; Douglas et al., 1986; Haynes and Knight, 1989;

Horne et al., 1992; Beare et al., 1994a). Even though soil organic matter content can tend to be greater under conventional than zero tillage at lower soil depths (e.g. 10 to 20 cm) aggregate stability is often equal or higher under zero tillage (Haynes and Knight, 1989; Beare et al., 1994a). This suggests that within the soil profile the distribution of specific binding agents (e.g. carbohydrates) involved in aggregate stability may be different to that of total soil organic matter. Indeed, it is known that a higher total soil organic matter content is not the only factor causing increased aggregate stability under zero tillage. In comparison with conventional cultivation, zero tillage favours the development of a fungal-dominated microflora in the surface soil (Hendrix et al., 1986; Beare et al., 1992, 1993). This fungal biomass is of particular importance in aggregation. Beare et al. (1995) found that the greater aggregate stability under zero tillage for a well-drained sandy loam from Georgia, USA, corresponds with 1.3 to 1.5 times higher densities of fungal hyphae and nearly two-fold higher concentrations of total carbohydrates (which also contained a higher proportion of microbial-derived monosaccharides) compared with conventional tillage. Fungal inhibition by application of a fungicide to the soil resulted in a marked reduction in aggregate stability under zero tillage but had little effect under conventional cultivation.

E. Ley-Arable Farming

In Britain in the 1930s concern regarding loss of soil organic matter led to the widespread adoption of ley-arable farming. Pasture leys (3 to 6 years) were alternated with 3 to 4 years of arable crops instead of having some fields in permanent grass and some in continuous arable. Over the last 30 years this practice has declined and many farmers now grow only arable crops. However, in other parts of the world, such as New Zealand, arable crops are still commonly grown in rotation with pasture. The inclusion of grass species in these rotations means that the organic matter input is greater than that under continuous arable. As a result, the long-term equilibrium level of soil organic matter reached is greater than under continuous arable but considerably less than that reached under permanent pasture (Johnston, 1982, 1986; Kofoed, 1982; Sauerbeck, 1982).

Results presented in Table 3 demonstrate the surprisingly small long-term effects of a 3 year arable-3 year ley rotation on organic C content of soils at two sites at Rothamsted on a flinty silt loam previously described in relation to Figure 9. On the old grassland site the decline in organic C content under a grass/clover-arable or grass-arable rotation was slightly less than that under continuous arable. At the old arable site, these two rotations resulted in a small increase in organic C content compared with continuous arable. A similar ley experiment was initiated on an old arable field at Woburn in 1938 (Johnston,

Table 3. Effects of continuous pasture, continuous arable, and 3-year leys on organic C contents (0-23 cm) on two sites on the Rothamsted ley-arable experiment

Previous history and measurement	Initial	Cropping history			
		Continuous pasture	Continuous arable	3 yr grass/ clover - 3 yr arable	3 yr grass - 3 yr arable
Old grassland soil					
Organic C content (%)	3.1	3.4	2.0	2.3	2.2
Percent increase due to ley				15	10
Old arable soil					
Organic C content (%)	1.6	2.4	1.6	1.8	1.8
Percent increase due to ley				13	13

(Data from Johnston, 1986.)

1986). After 33 years the organic C content had decreased from 1.0 to 0.9% under continuous arable. Under a rotation of 3 years grass followed by 2 years arable the organic C content had increased a little to 1.3%. In agreement with the above results, when a number of old arable sites were sown to grass for 3 to 4 years (Clement and Williams, 1958) appreciable increases in aggregate stability were found but the degree of improvement was relatively small compared with conditions found under permanent grassland.

Under rotations of 3 to 4 years arable followed by 3 to 4 years pasture, soil organic matter content equilibrates to a certain value and no appreciable changes in organic C content are discernable during the rotation (Haynes et al., 1991). Nonetheless, as illustrated in Table 2 aggregate stability values fluctuate during the rotation generally increasing greatly under the pasture phase and declining rapidly under the arable phase (Low, 1972; Haynes and Swift, 1990; Haynes et al., 1991). It appears that under a short term pasture ley the increase in aggregate stability is primarily mediated by the very large microbial biomass that accumulates in the pasture rhizosphere (Haynes et al., 1991). The biomass produces binding polysaccharides which increase stability and in addition, fine grass roots and associated mycorrhizal hyphae enmesh soil particles. When the pasture is plowed-under, the dense pasture root mass in the surface soil is replaced by a much more sparse arable crop root system and the microbial biomass declines greatly as does aggregation.

F. Organic Manures

Appreciable increases in soil organic matter can be achieved by adding large quantities of organic material to soil. As shown in Figure 11, when farmyard manure (FYM) was applied to an arable soil at Rothamsted (cropped continuously with cereals) each year at 35 t ha^{-1}, soil organic C content increased exponentially over a 140-year period. Organic C content is now approaching an equilibrium at an amount nearly three and one third times that of the NPK-treated soil.

On a market garden site at Woburn in Hertfordshire on a sandy loam with an initially low organic matter content (0.87% C) soil organic C content was increased to 2.26 and 2.29% when 75t ha^{-1} yr^{-1} of FYM or sewage sludge were applied for 25 and 20 years respectively (Johnston, 1975). However only 25 and 35% of the organic C added as FYM or sewage sludge respectively was still in the soil when applications ceased. As soon as manure applications ceased the organic C content began to decline (Johnston, 1975). Similarly on a plot on the Hoosfield continuous barley experiment (Figure 11), soil organic C content began to decline as soon as annual additions of FYM ceased in 1871. Even so, 104 years after the last addition, the plot contained more organic C than did plots not receiving organic manures. This demonstrates the great stability of soil humic material.

The amount of organic matter accumulated in the soil per tonne of organic material applied can vary greatly depending mainly on its ease of decomposition. Sauerbeck (1982), for example, showed that when a range of organic materials were added to a soil in central Germany, accumulation of soil organic C increased in the order green manure < straw < fresh FYM < composted FYM. Similarly Johnston (1975) showed that the increase in soil organic C per tonne of organic matter added was greater for composted than fresh material. During composting there is much breakdown of easily decomposable organic material and loss of $C0_2$ so that composted material added to the soil is relatively more resistant to further breakdown than fresh material.

Since amendment of soils with organic manures increases soil organic matter content it is not surprising that it also improves aggregate stability (Williams and Cooke, 1961; Hafez, 1974; Tiarks et al., 1974). Many workers have demonstrated the value of adding easily decomposable C sources (e.g. green manures) to soils or artificial aggregates (Low, 1954; Rennie et al., 1954; Griffiths and Jones, 1965; Monnier, 1965). After a brief lag phase there is a flush of microbial growth with a concomitant increase in physical entanglement by fungal hyphae and the production of extracellular polysaccharides capable of linking soil particles together. There is therefore a rapid rise in aggregate stability. Usually, this improved stabilisation is only temporary since the microbial population rapidly declines and hyphae and polysaccharides decompose so that the newly formed interparticle links become severed. By contrast, the addition of well-decomposed farmyard manure induces a slow, more steady increase in aggregate stability since the organic matter consists mainly of humic substances which are relatively stable binding agents. Presumably, regular additions of decomposable manures will cause a slow increase in aggregate stability in the long term as soil organic matter content steadily increases.

Long-term additions of organic amendments to soils seem to be less effective at increasing aggregate stability than a grass ley even where soil organic matter is higher in amended soils. Ekwue (1992), for example, studied the long-term changes in manure plots on soils of the Cottenham series at Woburn. He found that annual applications of FYM increased soil organic matter content to a higher level than a grass ley yet aggregate stability was significantly less than that under the ley. Similarly, straw incorporation was more effective in increasing the organic matter content of soil than aggregate stability. Such results probably reflect the fact that grass ley has additional effects in increasing aggregate stability apart from increasing soil organic matter content. As discussed previously these are likely to include maintenance for a large active microbial biomass in the pasture rhizosphere and the enmeshing effect of fine roots and associated mycorrhizal hyphae.

V. Summary and Conclusions

The residues of decomposition of plant detritus contribute to the formation of soil organic matter. The humic component (which usually constitutes 70 to 80 % of total soil organic matter content) consists of a mixture of large, complex polymeric molecules that are synthesized by the decomposer microflora during the decomposition process. Although plant material contains about 70% carbohydrate, this is rapidly degraded during decomposition and carbohydrates constitute only about 10% of total soil organic matter content. Much of this soil carbohydrate actually originates from extracellular polysaccharide mucigels produced by soil microflora. Both humic substances and polysaccharides can become strongly bound to phyllosilicate clay minerals, amorphous aluminosilicates and metal hydrous oxides as well as to each other. All of these interactions can be important in relation to aggregation in temperate soils.

Soil microaggregates (< 250 μm diam.) begin as small pieces of decaying particulate organic matter which become encrusted with microbial mucilage and clay. As the organic material decomposes organo-mineral associations (e.g. clay-humic complexes) are formed. Microaggregates are bound together to form macroaggregates (> 250 μm diam.) partly by the enmeshing action of ramifying roots, root hairs, VAM and other fungal hyphae. Their effect is not only physical since they also produce binding agents (polysaccharide gels and/or humic molecules). These ramifying agents die and decay depositing organic material throughout and around macroaggregates and providing a substrate for microbes which in turn, produce additional binding agents. Earthworms, which are the dominant macrofauna in temperate soils, can have a large influence on macroaggregation since they ingest substantial amounts of soil and organic debris, mix them together and their casts may form the foundation for many macroaggregates particularly in pasture soils.

In temperate regions, the major agricultural land uses are pastoral and arable crop production. When soil organic matter increases in a soil, such as under pasture or in the surface layer under zero tillage, it mainly accumulates as intermicroaggregate binding material which results in the formation of very stable macroaggregates. Much of this organic matter is bound into macroaggregates in ways such that it is physically inaccessible to microbial attack. When a pasture soil is converted to arable cultivation, macroaggregates are disrupted exposing the previously inaccessible organic matter to microbial breakdown. As a result, much of the organic matter involved in stabilising macroaggregates is lost and there is a decline in the stability of macroaggregates.

Over the long-term, changes in total soil organic matter content are closely correlated with changes in aggregate stability. Long-term agricultural practices that increase soil organic matter content (e.g. long-term pasture or regular use of organic manures) also result in increases in aggregate stability. Conversely, long-term management practices that decrease soil organic matter content (e.g.

conventional cultivation or fallow) also cause a decline in aggregate stability. By contrast, short-term (e.g. 1 to 4 years) effects of agricultural treatments can have relatively large effects on aggregate stability without having a measurable effect on soil organic matter content. The large background levels of relatively stable organic matter make it difficult to measure small changes that occur over comparatively short time periods. Other measures such as microbial biomass C and hot water-extractable carbohydrate content (a measure of the amount of binding agent produced by the microbial biomass) change more rapidly in response to alterations in C availability and can be strongly correlated with changes in aggregate stability that occur over short time periods.

Under agricultural cropping systems, changes in aggregate stability caused by specific crops are often closely related to the size of the root mass of the crops. A large root mass means that there is a large input of C to the soil, a large microbial biomass in the rhizosphere producing binding agents and a significant physical enmeshing effect of roots, root hairs and VAM hyphae. Forage grasses have a dense root mass and are particularly effective at improving macroaggregation and in the long-term in increasing soil organic matter content. Such grasses will be an important component of cropping rotations that are designed to maintain and improve soil organic matter and/or soil physical conditions.

References

Angers, D.A., N. Samson, and A. Legégè. 1993. Early changes in water-stable aggregation induced by rotation and tillage in a soil under barley production. *Can. J. Soil. Sci.* 73:51-59.

Barois, I. 1987. Interactions entre les vers de terre (Oligochaeta) tropicaux géophages et la microflore pour l'exploration de materielles organiques du sol. *Trav. Chercheurs Sation Lamto* 7:1-15.

Beare, M.H., R.W. Parmelee, P.F. Hendrix, W. Cheng, D.C. Coleman and D.A. Crossley. 1992. Microbial and faunal interactions and effects on litter nitrogen and decomposition in agroecosystems. *Ecol. Mono.* 62:569-591.

Beare, M.H., B.R. Pohland, D.H. Wright, and D.C. Coleman. 1993. Residue placement and fungicide effects on fungal communities in conventional and no-tillage soils. *Soil Sci. Soc. Am. J.* 57:392-399.

Beare, M.H., P.F. Hendrix, and D.C. Coleman. 1994a. Water-stable aggregates and organic matter fractions in conventional and no-tillage soils. *Soil Sci. Soc. Am. J.* 58:777-786.

Beare, M.H., M.L. Cabrera, P.F. Hendrix, and D.C. Coleman. 1994b. Aggregate-protected and unprotected pools of organic matter in conventional and no-tillage soils. *Soil Sci. Soc. Am. J.* 58:787-795.

Beare, M.H., S. Hu, D.C. Coleman, and P.F. Hendrix. 1995. Influences of mycelial fungi on soil aggregation and soil organic matter retention in conventional and no-tillage soils. *Soil Sci. Soc. Am. J.* (in press).

Blackman, J.D. 1992. Seasonal variation in the aggregate stability of downland soils. *Soil Use Manag.* 8:142-150.

Blevins, R.L., G.W. Thomas, M.S. Smith, W.W. Frye, and P.L. Cornelius. 1983. Changes in soil properties after 10 years continuous non-tilled and conventionally tilled corn. *Soil Tillage Res.* 3:135-145.

Brussard, L., D.C. Coleman, D.A. Crossley, W.A.M. Didden, P.F. Hendrix, and J.C.Y. Marinseen. 1990. Impacts of earthworms on soil aggregate stability. *Trans 14th Int. Cong. Soil Sci. Vol. 3, Commission* 3: 100-103.

Burns, R.G. and J.A. Davies. 1986. The microbiology of soil structure. p. 9-27. In: J.M. Lopez-Real and R.D. Hodges (eds.), *The role of microorganisms in a sustainable agriculture.* A.B. Academic Publishers, Berkhamstead.

Cambardella, C.A. and E.T. Elliott. 1992. Particulate soil organic matter changes across a grassland cultivation sequence. *Soil Sci. Soc. Am. J.* 56:777-783.

Cannell, R.Q. 1985. Reduced tillage in North-West Europe - A Review. *Soil Tillage Res.* 5:129-177.

Carter, M.R. 1986. Microbial biomass as an index for tillage-induced changes in soil biological properties. *Soil Tillage Res.* 7:29-40.

Chaney, K. and R.S. Swift. 1984. The influence of organic matter on aggregate stability in some British soils. *J. Soil Sci.* 35:223-230.

Chaney, K. and R.S. Swift. 1986a. Studies on aggregate stability. I. Re-formation of soil aggregates. *J. Soil Sci.* 37:329-335.

Chaney, K. and R.S. Swift. 1986b. Studies on aggregate stability II. The effect of humic substances on the stability of re-formed soil aggregates. *J. Soil Sci.* 37:337-343.

Cheng, W., D.C. Coleman and J.E. Box. 1990. Root dynamics, production and distribution in agroecosystems on the Georgia Piedmont using minirhizotrons. *J. Appl. Ecol.* 27:592-604.

Cheshire, M.V. 1979. *Nature and origin of carbohydrates in soils.* Academic Press, London.

Cheshire, M.V. and M.H.B. Hayes. 1990. Composition, origins, structures, and reactivities of soil polysaccharides. p. 307-336. In: M.F. De Boodt, M.H.B. Hayes, and A. Herbillon (eds.), *Soil colloids and their associations in aggregates.* Plenum, New York.

Christensen, B.T. 1986. Straw incorporation and soil organic matter in macro-aggregates and particle size separates. *J. Soil Sci.* 37:125-135.

Churchman, G.J. and K.R. Tate. 1987. Stability of aggregates of different size grades in allophanic soils from volcanic ash in New Zealand. *J. Soil Sci.* 38:19-27.

Clarke, A.L., D.J. Greenland, and J.P. Quirk. 1967. Changes in some physical properties of the surface of an impoverished red-brown earth under pasture. *Aust. J. Soil. Res.* 5:59-68.

Clement, C.R. and T.E. Williams. 1958. An examination of the method of aggregate analysis by wet sieving in relation to the influence of diverse leys on arable soils. *J. Soil Sci.* 9:252-266.

Coughlan, K.J., W.E. Fox, and J.D. Hughes. 1973. A study of the mechanisms of aggregation in a krasnozem soil. *Aust. J. Soil Res.* 11:65-73.

Daniel, O. and J.M. Anderson. 1992. Microbial biomass and activity in contrasting soil materials after passage through the gut of the earthworm *Lumbricus rubellus* Hoffmeister. *Soil Biol. Biochem.* 24:465-470.

Davenport, J.R. and R.L. Thomas. 1988. Carbon partitioning and rhizodeposition in corn and bromegrass. *Can. J. Soil Sci.* 68:693-701.

Dick, W.A. 1983. Organic carbon, nitrogen and phosphorus concentrations and pH in soil profiles as affected by tillage intensity. *Soil Sci. Soc. Am. J.* 47:102-107.

Douglas, J.T. and M.J. Goss. 1982. Stability and organic matter content of surface soil aggregates under different methods of cultivation and in grassland. *Soil Tillage Res.* 2:155-175.

Douglas, J.T., M.G. Jarvis, K.R. Howse, and M.J. Goss. 1986. Structure of a silty soil in relation to management. *J. Soil Sci.* 37:137-151.

Drew, M.C. and L.R. Saker. 1978. Effects of direct drilling and ploughing on root distribution in spring barley, and on the concentrations of extractable phosphate and potassium in the upper horizons of a clay soil. *J. Sci. Food Agric.* 29:201-206.

Drury, C.F., J.A. Stone, and W.I. Findlay. 1991. Microbial biomass and soil structure associated with corn, grasses and legumes. *Soil Sci. Soc. Am. J.* 55:805-811.

Edwards, A.P. and J.M. Bremner. 1967. Microaggregates in soils. *J. Soil Sci.* 18:64-73.

Edwards, C.A. and J.R. Lofty. 1977. *Biology of earthworms.* Chapman and Hall, London.

Ekwue E.I. 1992. Effect of organic and fertilizer treatments on soil physical properties and erodibility. *Soil Tillage Res.* 22:199-209.

Elliott, E.T. 1986. Aggregate structure and carbon, nitrogen and phosphorus in native and cultivated soils. *Soil Sci. Soc. Am. J.* 50:627-633.

Elliott, E.T. and D.C. Coleman. 1988. Let the soil work for us. *Ecol. Bull.* 39:23-32.

Forster, S.M. 1979. Microbial aggregation of sand in an embryo dune system. *Soil Biol. Biochem.* 11:537-543.

Foster, R.C. 1988. Microenvironments of soil microorganisms. *Biol. Fertil. Soils* 6:189-203.

Francis, G.S. and T.L. Knight. 1993. Long-term effects of conventional and no-tillage on selected soil properties and crop yields in Canterbury, New Zealand. *Soil Tillage Res.* 26:193-210.

Goh, K.M. and R.J. Haynes. 1986. Nitrogen and agronomic practice. p. 379-468. In: R.J. Haynes (ed.), *Mineral nitrogen in the plant-soil system.* Academic Press, Orlando.

Gowman, M.A., J. Coutts, and D. Riley. 1979. Changes in soil structure and physical properties associated with continuous direct drilling of cereals in the UK. *Proc. 11th Inter. Congr. Soil Sci. Edmonton, Alberta, Canada,* Vol 1. p. 290.

Graff, O. 1971. Stickstoff Phosphor und Kalium in der Regenwurmlosung auf der Wiesenversuchsfläche des Solling projekts. In: J. d'Aguilar (ed.), *IV Colloquium Pedobiologiae.* Institut National des Rescherches Agronomiques Publ. 71-77, Paris.

Greenland, D.J. 1971. Interaction between humic and fulvic acids and clays. *Soil Sci.* 111:34-41.

Griffiths, E. and R.G. Burns. 1972. Interaction between phenolic substances and microbial polysaccharides in soil aggregation. *Plant Soil* 36:599-612.

Griffiths, E. and D. Jones. 1965. Microbiological aspects of soil structure. I. Relationships between organic amendments, microbial colonisation and changes in aggregate stability. *Plant Soil* 23:23-28.

Gupta, V.V.S.R. and J.J. Germida. 1988. Distribution of microbial biomass and its activity in different soil aggregate size classes as affected by cultivation. *Soil Biol. Biochem.* 20:777-786.

Hafez, A.A.A. 1974. Comparative changes in soil physical properties induced by admixtures of manures from various domestic animals. *Soil. Sci.* 118:53-59.

Hamblin, A.P. and D.B. Davies. 1977. Influence of organic matter on the soil physical properties of some East Anglian soils of high silt content. *J. Soil Sci.* 28:11-22.

Harley, J.L. 1989. The significance of mycorrhiza. *Mycol. Res.* 92:129-139.

Hart, P.B.S., J.A. August, C.W. Ross, and J.F. Julian. 1988. Some biochemical and physical properties of Tokomaru silt loam under pasture and after 10 years of cereal cropping. *N.Z. J. Agric. Res.* 31:77-86.

Hayes, M.H.B., and R.S. Swift. 1990. Genesis, isolation, composition and structures of soil humic substances. p. 245-305. In: M.F. DeBoodt, M.H.B. Hayes, and A. Herbillon (eds.), *Soil colloids and their associations in aggregates.* Plenum, New York.

Haynes, R.J. 1986. The decomposition process : Mineralization, immobilization, humus formation, and degradation. p. 52-126. In: R.J. Haynes (ed.), *Mineral nitrogen in the plant-soil system.* Academic Press, Orlando.

Haynes, R.J. 1993. Effect of sample pretreatment on aggregate stability measured by wet sieving or turbidimetry on soils of different cropping history. *J. Soil Sci.* 44:261-270.

Haynes, R.J. and G.S. Francis. 1993. Changes in microbial biomass C, soil carbohydrate composition and aggregate stability induced by growth of selected crop and forage species under field conditions. *J. Soil. Sci.* 44: 665-675.

Haynes, R.J. and T.L. Knight. 1989. Comparison of soil chemical properties, enzyme activities, levels, of biomass N and aggregate stability in the soil profile under conventional and no-tillage in Canterbury, New Zealand. *Soil Tillage Res.* 14:197-208.

Haynes, R.J. and R.S. Swift. 1990. Stability of soil aggregates in relation to organic constituents and soil water content. *J. Soil Sci.* 41:73-83.

Haynes, R.J. and P.H. Williams. 1992a. An overview of pasture response, nutrient turnover and nutrient accumulation on the grazed, long-term superphosphate trial of Winchmore, New Zealand. *Proc. XVII Intern. Grassl. Congr. Hamilton,* p.1430-1433.

Haynes, R.J. and P.H. Williams. 1992b. Accumulation of soil organic matter and the forms, mineralisation potential and plant-availability of accumulated organic sulphur: effects of pasture improvement and intensive cultivation. *Soil. Biol. Biochem.* 24:209-217.

Haynes, R.J. and P.J. Williams. 1993. Nutrient cycling and soil fertility in the grazed pasture ecosystem. *Adv. Agron.* 49:119-199.

Haynes, R.J., R.S. Swift, and R.C. Stephen. 1991. Influence of mixed cropping rotations (pasture-arable) on organic matter content, water stable aggregation and clod porosity in a group of soils. *Soil Tillage Res.* 19:77-87.

Haynes, R.J., P.M. Fraser, and P.H. Williams. 1994. Earthworm population size, composition and microbial biomass: effect of pastoral and arable management in Canterbury, New Zealand. *Functional significance and regulation of soil biodiversity,* Kluwer Academic, Dordrecht (in press).

Hendrix, P.F., R.W. Parmelee, D.A. Crossley, D.C. Coleman, E.P. Odum, and P.M. Groffman. 1986. Detritus food webs in conventional and no-tillage agroecosystems. *Bioscience* 36:374-380.

Hissett, R., and T.R.G. Gray. 1976. Microsites and time changes in soil microbial ecology. p. 23-39. In: *The role of terrestrial and aquatic organisms in decomposition processes.* Blackwell, Oxford.

Horne, D.J., C.W. Ross, and K.A. Hughes. 1992. Ten years of a maize/oats rotation under three tillage systems on a silt loam in New Zealand. I. A comparison of some soil properties. *Soil Tillage Res.* 22:131-143.

Jenkinson, D.S. and J.N. Ladd. 1981. Microbial biomass in soil: measurement and turnover. p. 415-471. In: E.A. Paul and J.N. Ladd (eds.), *Soil Biochemistry.* Vol. 5. Marcel Dekker, New York.

Johnston, A.E. 1975, The Woburn market garden experiment, 1942-69. II. Effects of the treatments on soil pH, soil carbon, nitrogen, phosphorus and potassium. *Rothamsted Exp. Sta. Rep.* 1974, Part 2. p. 103-131.

Johnston, A.E. 1979. Some beneficial effects on crop yields from incorporating straw into the plough layer. p. 42-44. In: *Proceedings of the 4th A.D.A.S. Straw Utilization Conference, Oxford,* 1978. Ministry of Agriculture, Fisheries and Food, HMSO, London.

Johnston, A.E. 1982. The effects of farming systems on the amount of soil organic matter and its effect on yield. p. 187-202. In: D. Boels, D.B. Davies and A.E. Johnson (eds.), *Soil Degradation. Proceedings of the EEC Seminar held in Wageningen, Netherlands,* 13-15 October 1980. A.A. Balkema, Rotterdam.

Johnston, A.E. 1986. Soil organic matter effects on soils and crops. *Soil Use Manag.* 2:97-105.

Kay, B.D. 1990. Rates of change of soil structure under different cropping systems. *Adv. Soil Sci.* 12:1-52.

Kemper, W.D. and R.C. Rosenau. 1986. Aggregate stability and size distribution. p. 425-442. In: *Methods of soil analysis part 1. Physical and mineralogical methods.* Amer. Soc. Agron., Madison, Wisconsin.

Kinsbursky, R.S., D. Levanon, and B. Yaron. 1989. Role of fungi in stabilising aggregates of sewage sludge amended soils. *Soil Sci. Soc. Am. J.* 53:1086-1091.

Kofoed, A.D. 1982. Humus in long term experiments in Denmark p. 241-258. In: D. Boels, D.B. Davies and A.E. Johnston (eds.), *Soil Degradation. Proceedings of the EEC Seminar held in Wageningen, Netherlands,* 13-15 October 1980. A.A. Balkema, Rotterdam.

Lavelle, P. 1978. Les vers de terre de la savane de Lamto (Côte d;lvoire). Peuplements, populations et fonctions de l'écosystème. *Publ. Lab Zool. E.N.S.* 12:1-301.

Lee, K.E. 1985. *Earthworms: their ecology and relationships with soils and land use.* Academic Press, Sydney.

Lee, K.E. and R.C. Foster. 1991. Soil fauna and soil structure. *Aust. J. Soil Res.* 29:745-775.

Low, A.J. 1954. The study of soil structure in the field and in the laboratory *J. Soil Sci.* 5:57-74.

Low, A.J. 1955. Improvements in the structural state of soils under leys. *J. Soil Sci* 6:179-199.

Low, A.J. 1972. The effect of cultivation on the structure and other physical characteristics of grassland and arable soils (1945-1970). *J. Soil Sci.* 23:363-380.

Low, A.J. and P.R. Stuart, 1974. Micro-structural differences between arable and old grassland soils as shown in the scanning electron microscope. *J. Soil Sci.* 25:135-137.

Lynch, J.M. and E. Bragg. 1985. Microorganisms and soil aggregate stability. *Adv. Soil. Sci.* 2:133-171.

Marinissen, J.C.Y. and A.R. Dexter. 1990. Mechanisms of stablization of earthworm casts and artificial casts. *Biol. Fertil. Soils* 9:163-167.

Martens, D.A. and Frankenberger, W.T. 1991. Saccharide composition of extracellular polymers produced by soil micro-organisms. *Soil Biol. Biochem.* 8:731-736.

Martin, J.P. 1971. Decomposition and binding action of polysaccharides in soil. *Soil Biol. Biochem.* 3:33-41.

Martin, J.P., J.O. Ervin, and R.A. Shepard. 1966. Decomposition of the iron, aluminium, zinc and copper salts or complexes of some microbial polysaccharides. *Soil Sci. Soc. Am. Proc.* 30:196-200.

Martin, J.P., A.A. Parsa, and K. Haider. 1978. The influence of intimate association with humic polymers on biodegradation of [14C] - labelled organic substrates. *Soil Biol. Biochem.* 10:483-486.

McKenzie, B.M. and A.R. Dexter. 1987. Physical properties of casts of the earthworm *Aporrectodea rosea. Biol. Fertil. Soils* 5:152-157.

Metzger, L., D. Levanon, and U. Mingelgrin. 1987. The effect of sewage sludge on soil structural stability: microbial aspects. *Soil Sci. Soc. Am. J.* 51:346-351.

Miller, R.M. 1986. The ecology of vesicular-arbuscular mycorrhizae in grass - land shrublands. p. 135-167. In: G. Safir (ed.), *Ecophysiology of VAmycorrhizal plants.* CRC Press, Florida.

Miller, R.W. and J.D. Jastrow. 1990. Hierarchy of root and mycorrhizal fungal interactions with soil aggregation. *Soil Biol. Biochem.* 22:579-584.

Moavad, K., I.P. Bab'Yeva, and S. Ye. Gorin. 1976. Soil aggregation under the effect of extracellular polysaccharides of *Lipomyces eipofer. Soviet Soil Sci.* 9:65-68.

Molope, M.B., I.C. Grieve, and E.R. Page. 1987. Contributions by fungi and bacteria to aggregate stability of cultivated soils. *J. Soil Sci.* 38:71-77.

Monnier, G. 1965. Action des matieres organiques sur la stabilité structurale des sols. *Ann. Agron.* 16:471-534.

Morel, J.L., L. Habib, S. Plantureux, and A. Guckert. 1991. Influence of maize root mucilage on soil aggregate stability. *Plant Soil* 136:111-119.

Mulla, D.J., L.M. Huyck, and J.P. Reganold. 1992. Temporal variation in aggregate stability on conventional and alternative farms. *Soil Sci. Soc. Am. J.* 56:1620-1624.

Murray, R.S. and J.P. Quirk. 1990. Interparticle forces in relation to the stability of soil aggregates. p. 439-461. In: M.F. DeBoodt, M.H. B. Hayes, and A. Herbillon (eds.), *Soil colloids and their associations in aggregates.* Plenum, New York.

Newman, E.I. 1985. The rhizosphere: carbon sources and microbial populations. p. 107-121. In: A.H. Fitter, D. Atkinson, D.J. Read and M.B. Usher (eds.), *Ecological Interactions in Soil: Plants, Microbes and Animals.* Blackwell, Oxford.

Nguyen, M.L. and K.M. Goh. 1990. Accumulation of soil sulphur fractions in grazed pastures receiving long-term superphosphate applications. *N.Z. J. Agric. Res.* 32:245-262.

Oades, J.M. 1978. Mucilages at the root surface. *J. Soil Sci.* 29:1-16.

Oades, J.M. 1984. Soil organic matter and structural stability: mechanisms and implications for management. *Plant Soil* 76:319-337.

Oades, J.M. and A.G. Waters. 1991. Aggregate hierarchy in soils. *Aust. J. Soil Res.* 29:815-828.

Pagliai, H., G. Guidi, and G. Petrizzelli. 1979. Effect of molecular weight on dextran-soil interactions. p. 175-180. In: W.W. Emerson, R.D. Bond, and A.R. Dexter (eds.), *Modification of soil structure.* John Wiley, New York.

Perfect, E., B.D. Kay, W.K.P. Van Loon, R.W. Sheard, and T. Pojasok. 1990. Factors influencing soil structural stability within a growing season. *Soil Sci. Soc. Am. J.* 54:173-179.

Powlson, D.S. and D.S. Jenkinson. 1981. A comparison of the organic matter, biomass, adenosine triphosphate and mineralizable nitrogen contents of ploughed and direct-drilled soils. *J. Agric. Sci. Camb.* 97:713-721.

Powlson, D.S., P.C.R. Brookes, and B.T. Christensen. 1987. Measurement of soil microbial biomass provides an early indication of changes in total organic matter due to straw incorporation. *Soil Biol. Biochem.* 19:159-164.

Redl, G., C. Hubner, and F. Wurst. 1990. Changes in hot water soil extracts brought about by nitrogen immobilization and mineralization processes during incubation of amended soils. *Soil Sci. Soc. Am. J.* 55:734-739.

Rennie, D.A., E. Truog, and O.N. Allen. 1954. Soil aggregation as influenced by microbial gums, level of fertility and kind of crop. *Soil Sci. Soc. Am. Proc.* 18:399-403.

Robertson, E.B., S. Sarig, and M.K. Firestone. 1991. Cover crop management of polysaccharide-mediated aggregation in an orchard soil. *Soil Sci. Soc. Am. J.* 55:734-739.

Ross, D.J., K.R. Tate, and A. Cairns. 1982. Biochemical changes in a yellow-brown loam and a central gley soil converted from pasture to maize in the Waikato area. *N.Z.J. Agric. Res.* 25:35-42.

Sauerbeck, D.R. 1982. Influence of crop rotation, manurial treatment and soil tillage on the organic matter content of German soils. p. 163-179. In: D. Boels, D.B. Davies, and A.E. Johnston (eds.), *Soil degradation. Proceedings of the EEC seminar held in Wageningen, Netherlands,* 13-15 October 1980. A.A. Balkema, Rotherdam.

Schnitzer, M. 1986. Binding of humic substances by soil mineral colloids. p. 77-101. In: P.M. Huang, and M. Schnitzer (eds.), *Interactions of soil minerals with natural organics and microbes,* SSSA Spec. Pub. 17. Soil Sci. Soc. Amer., Madison, WI.

Shamoot, S., L. McDonald, and W.V. Bartholomew. 1968. Rhizo-deposition of organic debris in soil. *Soil Sci. Soc. Am. Proc.* 32:817-820.

Shaw, C. and P. Pawluk. 1986. The development of soil structure by *Octolasion tyrtaeum, Aporrectodea turgida* and *Lumbricus terrestris* in parent materials belonging to different textural classes. *Pedobiologia* 29:327-339.

Shipitalo, M.J. and R. Protz. 1988. Factors influencing the dispersibility of clay in worm casts. *Soil Sci. Soc. Am. J.* 52:764-769.

Shipitalo, M.J. and R. Protz. 1989. Chemistry and micromorphology of aggregation in earthworm casts. *Geoderma* 45:357-374.

Sparling, G.P. 1985. The soil biomass. p. 223-262. In: D. Vaughan and R.E. Malcolm (eds.), *Soil organic matter and biological activity.* Martinus Nijhoff, The Hague.

Sparling, G.P. and T.G. Shepherd. 1986. Physical and biochemical changes in Manawatu soils under maize cultivation. *Proc. NZSSS/ASSS Joint Conference on Surface Soil Management, 20-23 November 1986,* Rotorua, NZ. Soil Sci. Soc. p. 32-36.

Sparling, G.P., T.G. Shepherd, and H.A. Kettles. 1992. Changes in soil organic C, microbial C and aggregate stability under continuous maize and cereal cropping, and after restoration to pasture in soils from the Manawatu region, New Zealand. *Soil Tillage Res.* 24:225-241.

Stevenson, F.J. and M.S. Ardakani. 1972. Organic matter reactions involving micronutrients in soils. p. 79-114. In: J.J. Mortvedt, P.M. Giordano and W.L. Lindsay (eds.), *Micronutrients in agriculture. Soil Sci. Soc. Am.*, Madison, Wisconsin.

Stone, J.A. and B.R. Buttery. 1989. Nine forages and the aggregation of a clay loam soil. *Can. J. Soil Sci.* 69:165-169.

Tiarks, A.E., A.P. Mazurak, and L. Chesnin. 1974. Physical and chemical properties of soil associated with heavy applications of manure from cattle feedlots. *Soil Sci. Soc. Am. Proc.* 38:826-830.

Tisdall, J.M. 1991. Fungal hyphae and structural stability of soil. *Aust. J. Soil Res.* 29:729-743.

Tisdall, J.M. and J.M. Oades. 1979. Stabilization of soil aggregates by the root systems of ryegrass. *Aust. J. Soil Res.* 17:429-441.

Tisdall, J.M. and J.M. Oades. 1980a. The effect of crop rotation on aggregation in a red-brown earth. *Aust. J. Soil Res.* 18:423-433.

Tisdall, J.M. and J.M. Oades. 1980b. The management of ryegrass to stabilise aggregates of a red-brown earth. *Aust. J. Soil Res.* 18:415-422.

Tisdall, J.M. and J.M. Oades. 1982. Organic matter and water-stable aggregates in soils. *J. Soil Sci.* 33:141-163.

van de Westeringh, W. 1972. Deterioration of soil structure in worm free orchards. *Pedobiologia* 12:6-15.

Whipps, J.M. 1990. Carbon economy. p. 59-97. In: J.M. Lynch (ed.), *The Rhizosphere.* John Wiley, Chichester.

Williams, R.J.B. 1970. Relationship between the composition of soils and physical measurement made on them. *Rothamsted Exp. Stn. Rep. 1970, Pt 2,* p. 5-35.

Williams, R.J.B. and G.W. Cook. 1961. Some effects of farmyard manure and of grass residues on soil structure. *Soil Sci.* 92:30-39.

Young, I.M. and J.W. Crawford. 1991. The fractal structure of soil aggregates: its measurement and interpretation. *J. Soil Sci.* 42:187-192.

Zhang, H. and S. Schrader. 1993. Earthworm effects on selected physical and chemical properties of soil aggregates. *Biol. Fertil. Soils* 15:229-234.

Aggregation and Organic Matter Storage in Sub-Humid and Semi-Arid Soils

R.C. Dalal and B.J. Bridge

I. Introduction

Here we define soil aggregation as a state of soil aggregates in addition to their processes of formation and stabilization. A soil aggregate is a naturally occurring cluster or group of soil particles in which the forces holding the particles together are much stronger than the forces between adjacent aggregates (Martin et al., 1955). Although the formation of soil aggregates occurs mainly as a result of physical forces, the stabilization of soil aggregates is brought about by a number of factors, in particular the quantity and quality of soil aggregate stabilizing agents. The physical forces include alternate wetting and drying (semi-arid and sub-humid regions), freezing and thawing (temperate regions),

1-56670-033-7/96$0.00+$.50

the compressive as well as drying action of roots, and the mechanical alimentary action of soil fauna (mainly earthworms, termites and ants). The stabilizing agents include clays, polyvalent metal cations such as Ca^{2+}, Fe^{3+} and Al^{3+}, oxides of Fe, Al, Mn and Si, calcium and magnesium carbonates, and organic materials. The organic materials include fresh organic debris of plant, microbial and animal origin; root exudates and products of microbial synthesis; plant roots, fungal hyphae and microorganisms; and decomposed soil organic matter.

The quantity and quality of soil aggregate stabilizing agents of organic origin vary considerably depending upon the environment, soil type and soil matrix, and soil and crop management practices. For example, continuous cultivation and cropping of a soil previously supporting native vegetation and pasture generally leads to a lower content of organic matter in most soils under a wide range of environments (Clarke and Marshall, 1947; Haas et al., 1957; Jenny and Raychaudhuri, 1960; Russell, 1981) although the loss rates of organic matter differ considerably (Dalal and Mayer, 1986a). Moreover, the rate of loss of labile fractions of soil organic matter (SOM), such as macroorganic matter and microbial biomass, exceeds that of total SOM (Greenland and Ford, 1964; Dalal and Mayer, 1986b, 1987; Christensen, 1992). Therefore, the proportion of labile fraction to total SOM may provide an earlier indication of the consequences of different soil and crop managements than the levels of total soil organic matter (Ford and Greenland, 1968; Jenkinson, 1988). However, the effect of the labile fraction of organic matter on soil aggregation requires further examination.

Since the interactions of organic matter with metallic cations and clay particles play an important part in soil aggregation and the stability of both aggregates and the organic C contained within them (Oades, 1988), the loss of organic matter is associated with decreasing amount and degree of aggregation. This is especially so in Entisols, Ultisols and Alfisols of low clay content and/or oxides of iron and aluminium (Baver, 1968), where organic matter, especially the labile fraction, is the main aggregate stabilizing agent.

In a well-aggregated soil, a good combination of size, arrangement and stability of aggregates can give a wide range of pore sizes within and between aggregates. This assures adequate drainage and retention of plant available water for crop growth, good aeration and root growth. The breakdown of soil aggregates results in soil surface instability which leads to sealing, crusting, dispersion and blocked pores, with associated slow water infiltration rates and low hydraulic conductivities. Aggregate breakdown also results in surface and subsurface compaction which leads to water-logging, anaerobiosis, poor nutrient use and hence poor crop growth. In turn, slow water infiltration through soil results in increased runoff, soil erosion, and evaporation loss. Thus, in semi-arid and sub-humid regions where water is a major limitation for agricultural production, maintenance of adequate soil aggregation is imperative for the proper management of soil resources and hence sustainable agriculture.

II. Aggregate Formation and Stabilization

Soil particles are brought together into aggregates by physical forces and stabilized by binding agents. A summary of relative importance of agents of aggregate formation and aggregate stabilization in various soil groups in sub-humid and semi-arid regions is given in Table 1. For example, alternate wetting and drying is most important as an agent of aggregate formation in Vertisols whereas it is least important in Oxisols. In the latter, plant roots and rhizosphere are the most important agents of aggregate formation. In Entisols and Ultisols, SOM predominates as an agent of aggregate stabilization.

The role of SOM in aggregate stability can be summarised in Figure 1. The extent and degree of aggregation and its duration of stability is affected by the source of organic materials added to the soil and consequently to the nature and amount of organic matter produced and the decomposition rates of SOM fractions. The SOM in turn is protected within aggregates from decomposition. Thus, Dalal and Mayer (1986b) observed that the rate of decomposition of SOM in Vertisols and Alfisols brought under cultivation were inversely related to the degree of initial soil aggregation.

III. Environment, Soil and Cropping Effects on Aggregation and Organic Carbon

A. Environment

The sub-humid to semi-arid environmental regions receive between 400 mm and 1500 mm of total annual precipitation (about three to eight wet months), and have ratios of precipitation to potential evapotranspiration (P/PET) usually between 0.25 and 0.75. Thus, sub-humid and semi-arid regions are dominated by ustic and udic soil moisture regimes with frequent alternate soil wetting and drying cycles, the dominant physical forces of soil aggregate formation in these regions. Within a similar region, SOM content, a soil aggregate stabilizing agent, increases as P/PET ratio increases (Post et al., 1985) and as soil moisture changes from ustic to udic regime.

It is apparent, therefore, that the extent of aggregation is related to the climate under which the soil has been formed (Baver, 1968). A generalised relationship between the precipitation and potential evapotranspiration ratio (P/PET) and soil aggregation (Figure 2) shows that as the P/PET ratio increases, the percentage of silt and clay aggregated also increases and reaches a maximum value between a P/PET ratio of 0.4 and 0.6 so that the extent of soil aggregation increases from Entisols to Vertisols and Mollisols. At higher values of P/PET ratios, however, soil aggregation may decrease as in Alfisols and especially in Ultisols or it may increase even further as in Oxisols. The nature of the relationship

Table 1. Summary of relative importance of agents of aggregate formation and aggregate stabilization in various soil groups in sub-humid and semi-arid regions (1 = least important, 10 = most important)

Agent	Soil group						
	Entisols	Mollisols	Vertisols	Alfisols	Inceptisols	Oxisols	Ultisols
	Agents of aggregate formation						
Alternate wetting and drying	3	8	10	2	1	1	2
Alternate freezing and thawing	4	9	1	3	2	1	2
Roots and rhizosphere activity	9	7	7	9	9	9	9
Soil fauna	7	5	4	8	8	6	8
	Agents of aggregate stabilization						
Soil organic matter	10	7	6	9	9	8	10
Clays	4	8	9	5	6	6	3
Hydroxides of Fe and Al	3	5	5	7	8	10	3
Polyvalent metal cations	5	8	10	6	5	4	3

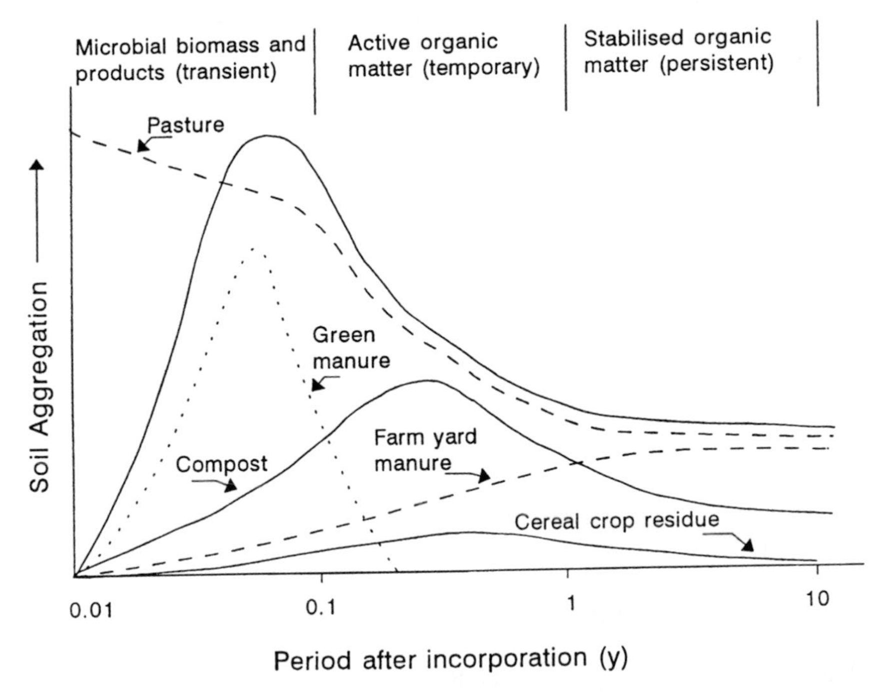

Figure 1. A generalized summary of soil aggregate stabilization by various sources of organic matter. (Adapted after Monnier, 1965; Tisdall and Oades, 1982.)

between climate and soil aggregation, shown in Figure 2, can be explained mainly in terms of aggregate forming and stabilizing agents (Table 1).

Figure 2 also emphasises the interaction between physical forces involved in the formation of soil aggregation. As P/PET ratio increases, clay content increases to a maximum at P/PET values where its formation (or alluviation) is in equilibrium with its eluviation (removed as soil suspension) or where clay eluviation is small as in Vertisols. Since SOM also increases with increasing P/PET ratio, soil aggregation reaches its maximum in Vertisols and Mollisols, the exception being Oxisols. As P/PET ratio increases further, clay eluviation also increases and in some Ultisols, even oxides of iron and aluminium and SOM also eluviate, thus leading to decreasing soil aggregation in Alfisols and Ultisols. However, in Oxisols, the hydroxides of iron and aluminium (along with increasing SOM contents) increase with P/PET, ensuring increasing soil aggregation.

Both organic matter and microbial biomass in soil increase with rainfall because of increased phytobiomass production. However, both decrease with

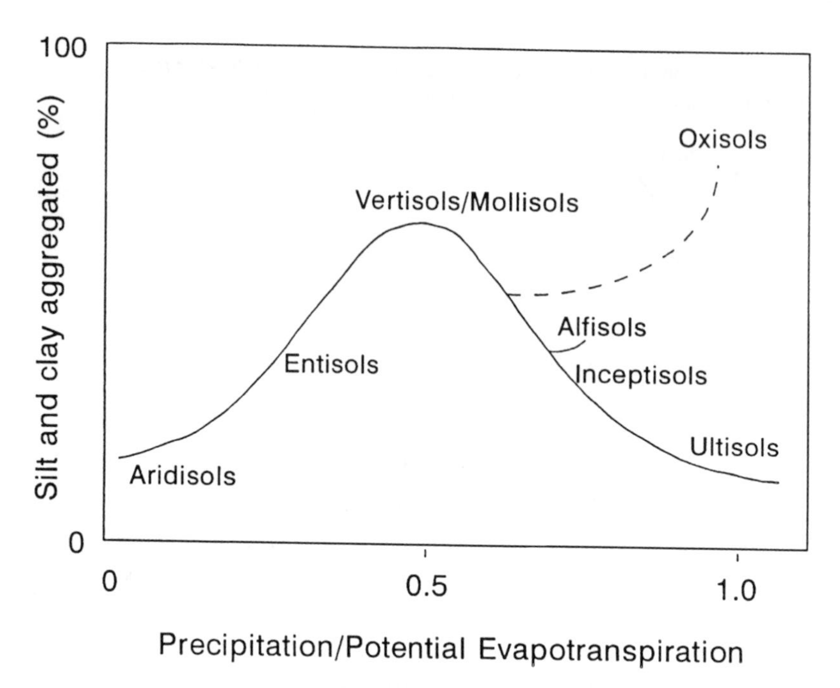

Figure 2. A generalized diagram of effect of climate on soil aggregation under natural ecosystems.

increasing temperature. Thus, Baver (1934) found that in the semi-arid regions, soil aggregation decreased with increasing mean annual temperature, while in the humid regions, soil aggregation actually increased with increasing temperature. In the warmer semi-arid regions, lower organic C inputs (i.e. lower yields) and higher SOM decomposition rate inhibit accumulation of SOM compared to the cooler regions (Mele and Carter, 1993).

In temperate regions, the proportion of microbial biomass C/organic C has been found to decrease exponentially with increasing ratio of annual P/PE (0.2 to 0.8) where PE is the mean annual pan evaporation (Insam et al., 1989). It is not known whether similar relationships hold for the warm regions. Sparling (1992) found that microbial biomass C/organic C ratio was useful as long as soil clay content, mineralogy and vegetation were also considered. Furthermore, Carter (1991) showed that the ratio must be assessed against a baseline (e.g., undisturbed grassland or forest soil) in the same soil type when it is used for comparison of cropping treatments. Moreover, almost nothing is known about the relationships between soil aggregation and the ratio of microbial biomass C/organic C although generally SOM has been found to be closely associated with soil aggregation in many soils (Tisdall and Oades, 1982; Acharya et al., 1988).

B. Soil Matrix and Soil Type

It is not only the degree but also the nature of soil aggregation that differs depending on soil matrix. Bodman and Constantin (1965) applied an ideal packing model for multisized spheres, as a simple binary model, to explain the arrangement of primary particles, sand, silt and clay in soil. The soil containing > 20% silt requires the application of ternary mixture model (Smith et al., 1978).

In a binary mixture model, if a soil contains < 30% clay and < 20% silt, it forms a coarse sand matrix (Smith et al., 1978). In this soil, clay orients around sand grains and pushes them apart; the spatial arrangement of the matrix is fixed by the distribution of sand grains. Organic matter is then oriented on the outer surfaces of the clay so that in coarse-textured soils, clay fraction is preferentially enriched in SOM (Sorensen, 1981; Dalal and Mayer, 1986c; Christensen, 1992). As the clay content increases from about 30 to 50% clay, the expansion of coarse matrix is complete and the available interstices are progressively filled with randomly oriented clay as tactoids; thus, both a coarse matrix and fine matrix coexist.

Above 50% clay, soil aggregates exist in a fine matrix; the nature of the clay distribution varies with the clay mineralogy and cations on the exchange sites. For example, in clay tactoids of Ca-montmorillonite, almost 80% of the surfaces are in close contact to each other compared to 20% in Ca-illite and < 10% in kaolinite; conversely, pore size within the clay tactoids increases from 1 to 4 nm in Ca-montmorillonite, about 10 nm in Ca-illite, to about 100 nm in kaolinite aggregates (Oades and Waters, 1991). Thus, in a fine matrix where clay tactoids dominate and SOM is oriented around the clay tactoids, clay fraction is likely to be less enriched in SOM than in the coarse matrix (Dalal and Mayer, 1986c).

The effect of differences in pore size distribution in different clay tactoids, and by implication aggregates formed from them, is that the smaller the pore size of clay tactoids the greater is the protection they offer to the aggregate stabilizing SOM against microbial and enzymic decomposition. Thus, Ca-montmorillonite protects the associated SOM most effectively whereas kaolinite is likely to be the least effective (Mortland, 1970), for although most microorganisms require >1 μm of pore size (Kilbertus, 1980), most extracellular degradative enzymes can still diffuse and be active within clay tactoids, with pore sizes exceeding 10 nm. The SOM existing in pore sizes < 10 nm is usually protected against the microbial and enzymic decomposition until the soil aggregates are broken down and that SOM fraction is exposed to microorganisms, for example, by cultivation.

Thus, the basic characteristics of the coarse matrix soil with respect to organic C and aggregation can be summarized as follows: (i) the soil organic matter would not build up to high levels, especially in cultivated soils, because of its relatively ready accessibility to decomposition by microorganisms; (ii) the stability of aggregates would depend on a continuous supply of fresh organic matter from plant growth and faunal activity so that organic bonding agents can be continually replaced as they are oxidised; and (iii) cultivation and clean fallowing these soils would result in a rapid decline of soil organic matter and aggregate stability. This behaviour is readily observed in Alfisols.

The fine matrix soils are likely to have the following properties with respect to organic matter and soil aggregation: (i) the soil organic matter can be maintained at relatively high levels as organic matter bonds with the clay tactoids which protect SOM from further decomposition by microorganisms by the presence of large numbers of very fine pore space; (ii) the degree and stability of the soil aggregation is likely to be high because of the protected bonding; (iii) cultivation and clean fallowing of these soils would result in only a slow decline of soil organic matter and aggregate stability. This behaviour is most readily observed in Vertisols which have smectitic clays and Oxisols with a clay fraction high in hydroxides of iron and aluminium.

C. Cultivation and Cropping

Cultivation and cropping of a soil can initiate physical forces for aggregate formation such as stress, shear and compression and dehydration through drying and mechanical pressure. Cropping can also provide the primary source of aggregate stabilizing agents and energy source for the diverse microbial biomass, which are aggregate stabilizing agents by themselves (such as having sticky and charged surfaces attracting clay particles) as well as producing a variety of aggregate stabilizing agents (Foster, 1988).

However, intensive cultivation for cropping, especially at an inappropriate soil moisture content can disrupt the soil aggregates. Under such conditions, cultivation disturbs the orientation of clay around coarse grains in the coarse matrix, and the distribution of the clay tactoids in the fine matrix. At the same time it physically breaks down macroorganic matter such as roots and hyphae. This not only leads to aggregate breakdown but also to increased SOM loss.

1. Cultivation

Intensive cultivation (tillage) for cropping has been associated with aggregate breakdown (Greenland, 1981) and decreasing SOM levels with increasing period of cultivation (Elliott, 1986; Dalal and Mayer, 1986b; Parton et al., 1987; Loch

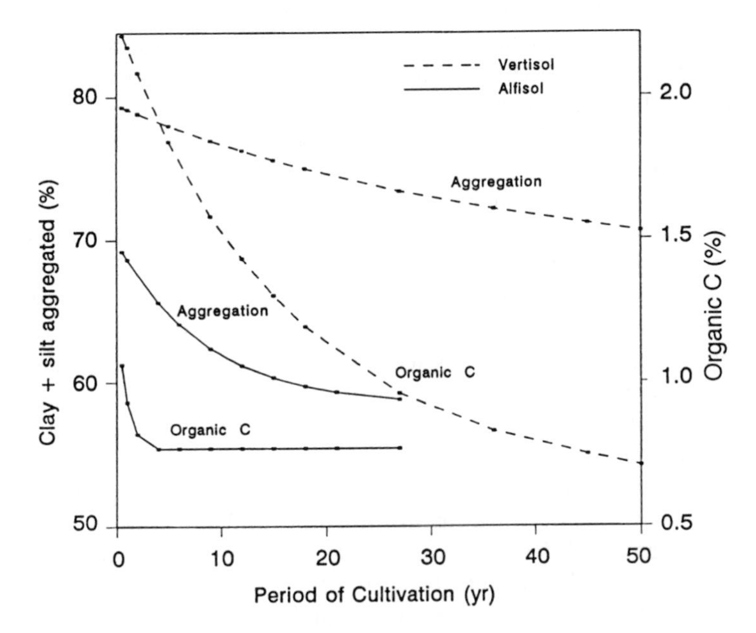

Figure 3. Decline in soil aggregation and organic C with increasing period of cultivation for cereal cropping in a Vertisol (Typic Paleustert) and an Alfisol (Rhodic Paleustalf):

Vertisol - Aggregation > 20 μm = 67.2 + 55.0 exp (-0.025 yr), $R^2 = 0.54$

Organic C (%) = 0.64 + 1.62 exp (-0.060 yr), $R^2 = 0.90$

Alfisol - Aggregation > 20 μm = 58.3 + 46.7 exp (-0.117 yr), $R^2 = 0.67$

Organic C (%) = 0.77 + 0.53 exp (-1.211 yr), $R^2 = 0.88$.

(Adapted from Dalal and Mayer, 1986b; and Dalal et al., 1991).

et al., 1987, Burke et al., 1989). In particular, tillage for cropping of well aggregated soils such as those under pastures, virgin grasslands and forests leads to considerable aggregate breakdown, especially in fallow-monoculture cropping systems (Figure 3). For example, Chan et al. (1988) found that a Vertisol which under native pasture had 60% aggregates > 500 μm, contained only 3% of this size aggregates after 45 years of tillage for cereal cropping.

The tillage of a soil brings about changes in temperature and moisture, aeration, exposure of new soil aggregate surfaces, increased contact between SOM and its decomposers and consequently aggregate breakdown and loss in SOM. Rovira and Greacen (1957) measured the effect of tillage on aggregate breakdown and microbial activity in two Ultisols, two Alfisols, an Oxisol and a Vertisol. They concluded that the main mechanism responsible for enhanced microbial activity (oxygen uptake) following tillage was the exposure of SOM previously present in inaccessible micropores when breakdown of aggregates occurred. Similar conclusions were reached by Edwards and Bremner (1967), Craswell and Waring (1972) and Powlson (1980). For example, Craswell and Waring (1972) found that by grinding the soil aggregates from 2 mm size to ~ 0.18 mm increased the microbial activity by 40 to 80%. The largest response to the grinding of aggregates was obtained in virgin soils, presumably because the greater exposure of SOM from the previously inaccessible micropores. Contribution to microbial activity from killed microbial biomass on grinding is also likely (Powlson, 1980). Admittedly, grinding of soil is too extreme to correspond to tillage, and the long-term SOM loss and decrease in agents of aggregate stabilization may largely be due to a decrease in organic C additions.

Dalal and Mayer (1986b) and Dalal et al. (1991) measured the rates of loss of total SOM and loss of soil aggregation > 20 μm when virgin Vertisols and Alfisols were mechanically cultivated for cereal cropping for up to 45 years (Figure 3). They found that the rates of loss in both the total SOM as well as aggregation were much faster in Alfisols than in Vertisols. Apparently, larger proportion of SOM in Alfisols remains unprotected compared to that in Vertisols. It is inferred that soil aggregation and SOM protection are interrelated since the rates of loss of SOM in a number of cultivated Alfisols and Vertisols were inversely related to their clay contents (Dalal and Mayer, 1986b).

The extent of aggregate breakdown due to tillage increases with the aggregate size. For example, the proportion of aggregates > 250 μm lost due to tillage of virgin soils is much larger than aggregates > 20 μm. This is because the former are largely dependent on live and decaying plant roots and fungal hyphae and probably casts of earthworms and termites, which are rapidly destroyed by tillage (Tisdall and Oades, 1982). Clean fallowing exacerbates this effect.

Unfortunately, the effect of tillage on the dynamics and the relationships between various sizes of aggregates and the nature of SOM associated with these aggregates is only qualitative at present. It is hoped that a better understanding of both the aggregates and SOM in cultivated soils will emerge from *in situ* quantification in the field. Increasing trends towards reduced and no-tillage practices would assist to elucidate the relationship between various size aggregates and SOM, their location in soil and the protection mechanism of SOM due to spatial distribution versus its presence in inaccessible pores in aggregates.

2. No-Tillage

Previously cultivated soils brought under no-tillage practice, especially when crop residues are retained, usually contain higher amounts of organic matter and increased soil aggregation (Hamblin, 1980; Dalal, 1989; Carter, 1992), often confined to the top soil layer. Thus, no-tillage practice enhances SOM stratification even though total amounts of SOM in soil may be similar to that in the conventional tillage practice.

Hamblin (1980) examined the effect of no-tillage and tillage on SOM and water-stable aggregates > 250 μm after 3 to 8 years of the tillage treatments (Table 2). In four of the five soils examined, there was an increase in SOM under no-tillage treatment. Soil aggregation also increased under no-tillage, increasing by 10 to 20% in Alfisols, 29% in a Vertisol and 92% in an Entisol. By selectively extracting SOM, sodium periodate sensitive carbohydrates appeared to be responsible for increases in soil aggregation in Entisol, while sodium pyrophosphate extractable organic C, and presumably associated polyvalent metal cations Fe and Al, were implicated in increased soil aggregation in Alfisols and a Vertisol.

From the analysis of SOM in different aggregate size fractions, SOM accumulated in 45-75 μm size fraction in Entisol and > 75 μm size fractions in Alfisol under no-tillage (Table 3). Conversely, when an Alfisol originally under forest was cultivated, the largest proportion of the organic C was lost from > 20 μm size fraction. In fact, cultivated Alfisols contained larger proportion of the organic C in < 2 μm fraction than either the soil under forest or no-tillage practice. In a Vertisol, on the other hand, no single aggregate-size dominated although large proportions of organic C were present in 2 to 20 μm and > 75μm fractions, presumably in aggregated clay and silt fractions.

Dalal (1989) measured a small but significant increase in soil aggregates > 20 μm in a Vertisol after 13 years of no-tillage treatment. Soil aggregation increased from 47% in the tilled treatment to 49% in the no-tillage treatment; the corresponding increases in SOM were 1.67% and 1.81% organic C, respectively. Thus the effects of no-tillage on SOM and soil aggregation are generally small in the semi-arid environment compared to the humid environments (Lal, 1989; Carter, 1992).

3. Plant Residues

The effect of plant residues returned or added to the soil on soil aggregation is schematically shown in Figure 1. Easily decomposable plant residue such as that from green manuring and legume-based pastures provides transient and temporary aggregate stabilizing agents while cereal crop residues provide persistent aggregate stabilizing agents, but usually in low amounts.

The effect of cereal crop residue retention on SOM as well as soil aggregation is enhanced when it is practiced along with no-tillage (Table 4). For example, R.J. Loch (unpublished data) found that although crop residue retention in a tilled soil increased soil aggregation only slightly in an Alfisol, a Vertisol and

Table 2. Effect of 3 to 8 years of no tillage (NT) and tillage (T) practices for continuous wheat (*Triticum aestivum* L.) on water-stable aggregates > 250 μm in different soils

Soil	Site	---Organic C (%)---		-----Clay (%)-----		--Aggregates > 250 μm (%)--		
		NT	T	NT	NT	NT	T	(NT-T)X100/T
Entisol	Yuna	1.54	1.75	12.7	12.0	16.5	8.6	92
Alfisol	Turretfield	1.63	1.37	19.7	19.0	15.1	12.5	21
Alfisol	Wagga Wagga	1.38	1.13	19.6	21.9	24.6	22.4	10
Vertisol	Hermitage	1.81[a]	1.62[a]	68.0[a]	68.0[a]	31.9	24.8	29

(Adapted from Hamblin, 1980; [a]from Dalal, 1989.)

Table 3. Distribution of organic C in different particle or aggregate-sized fractions from virgin (V), no-tillage (NT) and tillage (T) practices

Soil	Practice	Proportion of organic C (%)					
		<2μm	2-20 μm	>20 μm	20-45 μm	45-75 μm	>75 μm
Entisol[a]	NT	19.2	35.9		14.4	26.2	8.4
	T	19.1	45.8		10.2	17.3	7.5
Alfisol[a]	NT	27.3	22.3		12.7	9.6	28.0
	T	39.1	24.5		17.3	8.1	12.1
Alfisol[b]	V	47	20	33			
	T	69	27	6			
Vertisol[a]	NT	14.7	28.7		9.9	20.0	26.5
	T	16.1	24.5		17.8	18.4	22.3

[a]From Hamblin (1980).
[b]From Dalal and Mayer (1986c).

an Oxisol, crop residue retention in a no-tillage practice substantially increased the SOM and water-stable aggregates > 125 μm, more so in an Alfisol than in an Oxisol. In the former, with predominantly coarse matrix, additional organic matter provides effective aggregate-stabilizing agents for clay-enveloped coarse grains. Moreover, Carter and Mele (1992) showed that the effect of a combination of no-tillage and crop residue retention on soil aggregate depended on previous soil organic C and management regime. Further experimental evidence is required to elucidate the mechanism of organic matter protection against decomposition and enhanced soil aggregation following plant residue retention and no-tillage practice.

4. Organic Amendments

Organic amendments include compost, farm yard manure, plant residues including loppings from trees and shrubs grown as hedgerow, food processing wastes and sewage sludges. Since the nature and proportion of various aggregate stabilizing agents (gums, waxes, carbohydrates, uronides etc., or other transient, temporary and persistent agents) differ among these amendments, their effectiveness both in extent and stability in soil aggregation vary considerably. For example, green manuring increases soil aggregation much more rapidly but its effect lasts for a shorter period than compost or farm yard manure (Figure 1)

Table 4. Effect of crop residue retention and tillage practices on water stable aggregates > 125μm in three soils

Soil	Site	Crop	Treatment duration (y)	Aggregate > 125 μm %[a]			
				No-tillage		Tillage	
				Residue	No residue	Residue	No residue
Alfisol	Billa Billa	Wheat	33	43 (1.2)[b]	38 (0.9)	39 (0.9)	36 (0.8)
Vertisol	Hermitage	Wheat/Barley	19	67 (1.9)	61 (1.8)	64 (1.8)	62 (1.7)
Oxisol	Kairi	Maize	6	77 (1.6)	75 (1.4)	73 (1.1)	71 (1.1)

[a]After subtracting coarse sand content > 200 μm. In the Alfisol, aggregates > 125 μm are probably overestimated due to inclusion of some fine sand content. [b]Soil organic C (%). (Calculated from R.J. Loch, unpublished data.)

Although their benefits are universally recognised, the mechanism by which organic amendments improve soil aggregation and how the SOM produced from these amendments provides some protection against microorganisms and degradative enzymes in soil is not completely understood (Jenny, 1980; Boyle et al., 1989).

5. Cropping

The cropping of a soil provides organic matter directly and hence provides aggregate stabilizing agents. Moreover, plant roots are also involved in aggregate formation through compression and drying.

The contribution of annual crops to soil aggregation during their growing season is ephemeral and probably small in the semi-arid regions. For example, Chan and Heenan (1991) measured no significant difference in water-stable aggregates > 250 μm at maturity of barley (*Hordeum vulgare* L.), canola (*Brassica napus* L.), linseed (*Linum usitatissimum* L.), lentil (*Lens culinis* Medik), pea (*Pisum sativum* L. s lat.), and lupin (*Lupinus angustifolus* L.) crops, although there was a significant decrease in the dispersion of silt-size and clay-size particles by growing barley, canola, linseed and lentil but not pea and lupin crops (Table 5). Carter and Kunelius (1993) also observed significant differences between crops only in microaggregation. This may be due to root exudates and microbial debris which acted either as nuclei for aggregation of clay tactoids (Foster, 1988), or binding agents (Tisdall and Oades, 1982) including carbohydrates (Cheshire et al., 1983; Angers and Mehuys, 1989). However, cropping effects on soil aggregation disappeared on drying the moist soil (Chan and Heenan, 1991). Usually the effect of annual crops during one season on SOM content is small. Inclusion of forage legumes or grass-legume pastures in rotation significantly increased both organic C and soil aggregation in an Alfisol and a Vertisol (Figure 4). However, in an Alfisol the number of aggregates > 2 mm was increased although the stability of aggregates > 250 μm was decreased (Tisdall and Oades, 1982). On the other hand, there was a substantial increase in aggregates > 250 μm in a Vertisol, even with a moderate increase in organic C, albeit in a soil already low in organic matter (Tyagi et al., 1982). This is to be expected because Alfisols and Vertisols have different soil matrices and probably they also differ in aggregate hierarchy (Oades and Waters, 1991).

Another significant feature demonstrated in Figure 4 is the immediate adverse effect of clean fallowing on organic C as well as on soil aggregation > 2 mm in the Alfisol. This is to be expected in this soil where rate of SOM loss is very rapid (Dalal and Mayer, 1986b).

The effect of legume forage or grass-legume pasture on SOM as well as soil aggregation is well demonstrated in both semi-arid and sub-humid regions

Table 5. Effect of various crops on water-stable aggregation in an Alfisol

Crop	Aggregates > 250μm	Dispersible < 50 μm	Dispersible < 2μm
Barley	81.9	9.2	1.4
Canola	82.3	5.4	2.0
Linseed	86.0	5.3	1.0
Lentil	78.2	9.3	0.7
Pea	81.7	11.9	1.2
Lupin	74.0	13.4	1.8
LSD ($P < 0.05$)	N.S.[a]	3.4	0.7

[a]N.S., not significant. (Adapted from Chan and Heenan, 1991.)

(Tisdall and Oades, 1982; Payne, 1988). The plant roots and associated fungal hyphae (Tisdall, 1991), associated microorganisms and microbial debris (Lynch and Bragg, 1985) and soil fauna including earthworms (Brussaard et al., 1990), termites and ants (Lee and Foster, 1991) are implicated in improving soil aggregation and SOM.

In the sub-humid regions in Africa, Asia and Latin America, bush fallowing is practised to regenerate soil aggregation and build up SOM following exploitative cultivation of Entisols, Alfisols, Oxisols and Ultisols for a number of years (Nye and Greenland, 1960). The mechanisms of the improved soil aggregation and SOM are essentially similar to that brought about by pastures although the rate of improvement in the soil aggregation and SOM may be faster due to more dry matter production and greater return of plant residues to the soil by the bush vegetation.

Moreover, the basic metallic cations, especially Ca^{2+} build up in the top soil layers in these acidic soils, thus ensuring increased SOM-M^{n+}-clay bonding and hence improved aggregation.

IV. Aggregation and Organic Carbon in Soils of the Semi-Arid and Sub-Humid Regions

A. Entisols and Inceptisols

While Entisols are dominated by a coarse matrix and a generally ustic moisture regime, Inceptisols frequently are dominated by a fine matrix usually containing substantial amounts of amorphous materials and a generally udic moisture

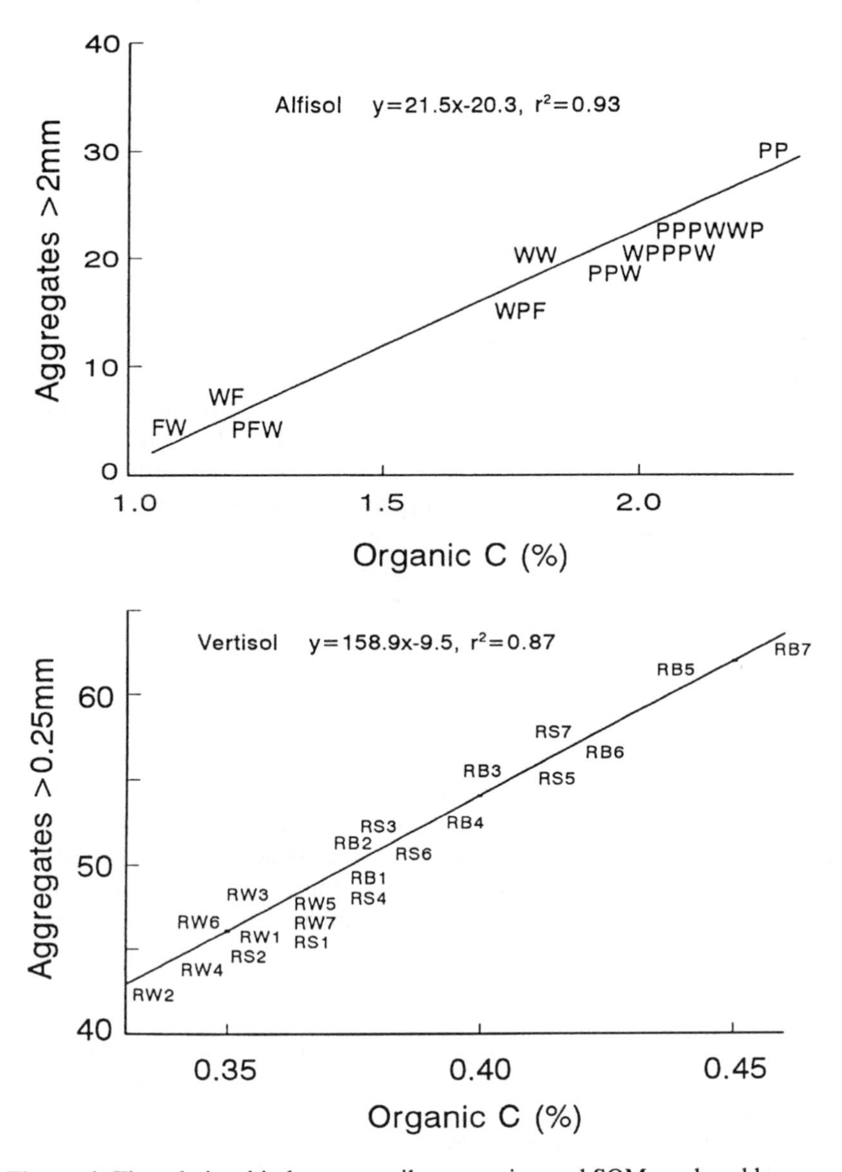

Figure 4. The relationship between soil aggregation and SOM produced by crop rotations after 50 years in an Alfisol, Australia (Tisdall and Oades, 1980) and after 14 crops were grown in a Vertisol, India (Tyagi et al., 1982) F, fallow; W, wheat; P, pasture; and R, rice (*Oryza sativa* L.); S, safflower (*Carthamus tinctorius* L.); B, berseem (*Trifolium alexandrinum* L.) and 1 to 7, number of rotations.

regime. In their natural ecosystems, although Entisols contain lower amount of organic matter than the Inceptisols, SOM plays a greater role in soil aggregation in the former (Table 1).

In an Entisol from Western Australia, Hamblin (1980) found that although waterstable aggregates > 250 μm increased from 8.6% in the tilled soil to 16.5% in the soil under no-tillage practice, organic C showed no significant differences. By using selective extracting reagents, it was inferred that carbohydrates were mainly responsible for aggregation in the Entisol studied. Rueda and Viqueira (1986) found that water-stable aggregates < 50 μm and to a lesser extent aggregates between 2 to 50 μm were principally stabilized by clay.

It is expected that in Entisols the primary mechanism of aggregation is clay (oriented around coarse grains) - polyvalent metal cations (mostly Ca^{2+}) - SOM (mainly carbohydrates) (Emerson et al., 1986). Thus, most of the SOM is likely to be exposed to microbial and enzymic degradation. It is not surprising, therefore, that many cultivated Entisols in the Gangetic Plains of India contain < 0.2% organic C in their 0 to 0.1 m depths (Jenny and Raychaudhuri, 1960).

Inceptisols usually occur in the sub-humid to humid zones and contain substantial amounts of amorphous material and poorly crystalline Al silicates. Andepts (or Andisols) are especially active in stabilizing organic C and commonly contain 4 to 30% organic matter (Zunino et al., 1982); most of the SOM is physically protected in water stable aggregates. The mechanism of soil aggregation is: amorphous Al silicates - Fe (Al) oxides - SOM - allophane (Wada and Higashi, 1976).

On the other hand, Inceptisols other than Andepts experience severe problems of compaction due to aggregate breakdown and the relatively rapid loss of SOM when they are cultivated. Sparling et al. (1992) found that in a Fluventic Eutrochrept the losses of SOM, microbial biomass and aggregates > 2 mm were 49%, 60% and 98% respectively, after cultivation of a pasture soil for maize (*Zea mays* L.) for 14 years. The relationships between soil aggregation, organic C and microbial biomass were close only during the cultivation phase following pasture when all three components declined rapidly.

It is postulated that in undisturbed pasture soils, a substantial proportion of the SOM (probably the light fraction) is not involved in soil aggregation and therefore, the correlation between the total SOM and aggregate stability, especially the macroaggregate stability is likely to be poor. Under such a soil management, this portion of the SOM is made spatially inaccessible to the microbial decomposition by separating substrates from decomposers (McGill and Myers, 1987).

Perfect et al. (1990) measured the changes in aggregate stability within a maize growing season on a fine-silty Inceptisol. They found that aggregate stability (measured using water-stable aggregates > 250 μm and dispersed clay) increased with decreasing soil water content and microbial biomass; that is,

Table 6. Effect of 3-year cropping treatments on turnover period of soil aggregates in an Inceptisol

Cropping treatment	T, Turnover period (yr)[a]	
	Dispersible clay	Aggregates > 250 μm
Lucerne (*Medicago sativa* L.)	3.6	7.6
Red clover (*Trifolium pratense* L.)	3.6	8.1
Maize - no tillage	6.3	∞
Maize - conventional tillage	∞	∞

[a]Using equation: Y_t = Ymax - (Ymax - Y_o) exp (Ct), and T = I/C, where Y_t and Y_0 and Ymax are dispersible clay or water-stable aggregates > 250 μm at t years, initially and after a long period (∞), respectively, and C is a constant (yr^{-1}). (Calculated from Perfect et al., 1990.)

when both aggregates forming (drying) and aggregate-stabilizing (microbial biomass and their products) agents were present simultaneously.

Stability of aggregates > 250 μm is also likely to be affected by the total lengths of roots and associated fungal hyphae. Miller and Jastrow (1990) and Thomas et al. (1993) demonstrated that the direct effect of vesicular-arbuscular mycorrhizae (VAM) soil hyphae on soil aggregation of an Inceptisol was at least equivalent to that of roots alone. This was attributed to drying action as well as root fragments and hyphae filaments as aggregate nuclei and excretion of extracellular polysaccharides (Foster, 1988).

Perfect et al. (1990) estimated the period required to attain steady-state soil aggregation in an Inceptisol under different cropping treatments. Under pastures, steady-state soil aggregation levels could be attained in about 8 years whereas soil under maize cropping may not attain a steady state in soil aggregation (Table 6). It appears that the soil particle sizes involved in aggregation (or reducing dispersible clay) are coarser than 50 μm and finer than 2 μm since aggregates of these particle sizes contained the youngest organic matter, as determined by natural ^{13}C abundance studies in an Inceptisol (Balesdent et al., 1988).

B. Mollisols

Mollisols are characterised by a mollic epipedon, that is a surface soil layer with dark, organic matter-rich, Ca-rich, stable aggregates of clay - Ca^{2+} - SOM. As much as 80% of the SOM is so intimately associated with the soil mineral fraction that it cannot be separated from it by sieving and sedimentation

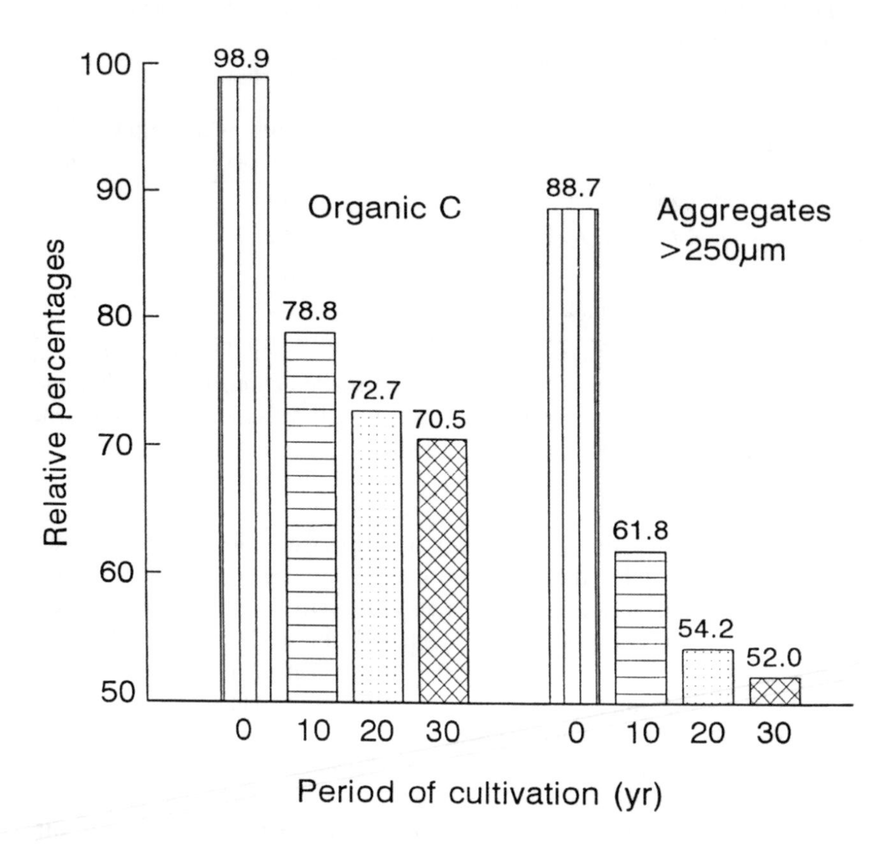

Figure 5. Decrease in soil organic C and water-stable aggregates > 250 μm with increasing period of cultivation of Mollisols. (Redrawn from Chen et al., 1982.)

(Greenland, 1965). Although a large proportion of the organic matter is associated with < 2 μm fraction, 2 to 20 μm size particles or aggregates may be equally rich in organic matter (Christensen, 1992).

Both organic C and water-stable aggregates > 250 μm decrease with increasing period of cultivation as mollisols originally under native grasses or pasture are brought under cultivation (Figure 5). Chen et al. (1982) showed that water-stable aggregates > 250 μm declined faster than the total organic C, presumably because the macroorganic matter and the light fraction organic matter was lost disproportionately more than the total SOM. Cambardella and Elliott (1992) also showed that the light fraction organic matter (< 1.85 Mg m^{-3}) loss exceeded 60% after 20 years of cultivation (bare-fallow and stubble

Table 7. Effect of 13 years of grain sorghum and wheat crops on wet aggregate stability and organic C content of a Mollisol

Crop	Wet aggregate stability (%)	Organic C[a] (%)
Wheat	39.6	0.93
Sorghum	52.5	1.04
LSD ($P < 0.01$)	7.1	0.05

[a]Organic C = 0.5 x Organic matter. (Adapted from Skidmore et al., 1986.)

mulch) of a grassland Mollisol. However, the loss of the heavy fraction (> 1.85 Mg m^{-3}) was less than 5%.

Elliott (1986) and Oades and Waters (1991) also found that the hierarchical conceptual model of soil aggregate structure (0.2 μm → 2 μm → 20 μm → 250 μm → > 2000 → μm) proposed by Tisdall and Oades (1982) for Alfisols also applied to the Mollisols studied. Elliott (1986) observed that the soil from the native grassland and the adjoining cultivation had essentially similar aggregate characteristics, although in the former the aggregates > 300 μm were more stable. Not only water-stable aggregates > 300 μm contained more organic C, N, and P but also more mineralizable C and N than the aggregates < 300 μm. Elliott (1986) suggested that the organic matter binding aggregates < 300 μm into aggregates > 300 μm is the primary source of nutrients released when the SOM is lost on cultivation of a virgin or pasture Mollisol.

Retention of crop residues may reduce the loss of soil aggregation in some cultivated Mollisols depending on the residue source. For example, Skidmore et al. (1986) found that sorghum (*Sorghum bicolor* L. Moench s. lat) crop residue incorporation for 13 years in an Ustoll resulted in a significantly higher percentage of water-stable aggregates and organic C than the wheat crop residue (Table 7). It appears that the greater amount of sorghum crop residue returned compared to that of wheat crop residue, accompanied by the differences in rates of residue decomposition and amounts of microbial biomass, were responsible for improved soil aggregation and organic C from the grain sorghum cropping.

Substantial improvement in soil aggregation is usually achieved by growing pastures. Even under pasture, a Mollisol in temperate ustic regime may require more than 50 years to attain soil aggregation levels similar to that in the virgin soil. For example, Jastrow (1987) obtained the following relationships between percentage aggregates > 0.2 mm (y) and years since last cultivation for maize or the numbers of years under pasture (x): $y = 95.8 - 56.2/x$, $r^2 = 0.93$; that is, 56 years are required to attain near steady-state level of soil aggregation. In tropical Mollisols, however, improvement in soil aggregation under pasture or

Table 8. Contribution of organic C and other soil properties to dispersible clay (%) in virgin Vertisols, Queensland

Soil property	Regression coefficient[a]	Partial r^2	Level of significance[b]
Organic C < 2 μm (%)	-0.47	0.58	< 0.01
Exchangeable Ca (c mol kg^{-1})	-0.09	0.18	4.06
Exchangeable Mg (c mol kg^{-1})	-0.23	0.46	0.03
Exchangeable Na (c mol kg^{-1})	3.89	0.82	< 0.01
Exchangeable K (c mol kg^{-1})	-1.30	0.47	0.02
CEC (c mol kg^{-1})	0.12	0.41	0.08
DTPA-Cu (mg kg^{-1})	-1.31	0.38	0.13
DTPA-Mn (mg kg^{-1})	0.12	0.63	< 0.01

[a]Using step-wise multiple regression (R. C. Dalal, unpublished data). [b]n = 36; r^2 = 0.95

forests (bush fallowing) of a previously intensively cultivated soil may be faster because of the rapid rates of organic matter accumulation (Lugo et al., 1986).

C. Vertisols

Vertisols usually have a fine or mixed clay matrix and contain > 50% of its SOM in the clay-size fractions (Dalal and Mayer, 1986d) although the 1 to 10 μm size fraction (silt and aggregated clay) is usually enriched in organic matter. This is presumably due to microbial biomass and microbial debris (Ladd et al., 1990).

Vertisols are easily identified by their expanding (and contracting) clay minerals, most frequently smectites. Aggregates are stabilized not only by SOM - clay complexes but also by metallic cations on the exchange sites of clays and total soil solution concentration. In Table 8, a comparative contribution of organic C and exchangeable cations towards soil aggregation in virgin Vertisols can be evaluated. As organic matter increased, dispersible clay (after 1 h end-over-end shaking in deionised water) decreased; significant relationships between the SOM and the dispersible clay being found with the clay-size C content ($r = -0.51$, $P < 0.01$, $n = 36$) and organic C content of the dispersible clay ($r = 0.53$, $P < 0.01$, $n = 36$). Although significant relationships between aggregation and organic C were obtained in a number of essentially similar Vertisols under cultivation (for example, in Figure 6), the dominant role of exchangeable cations other than Na^+ in the stabilization of aggregates, and that of Na^+ in the dispersion of aggregates in natural ecosystems of the diverse

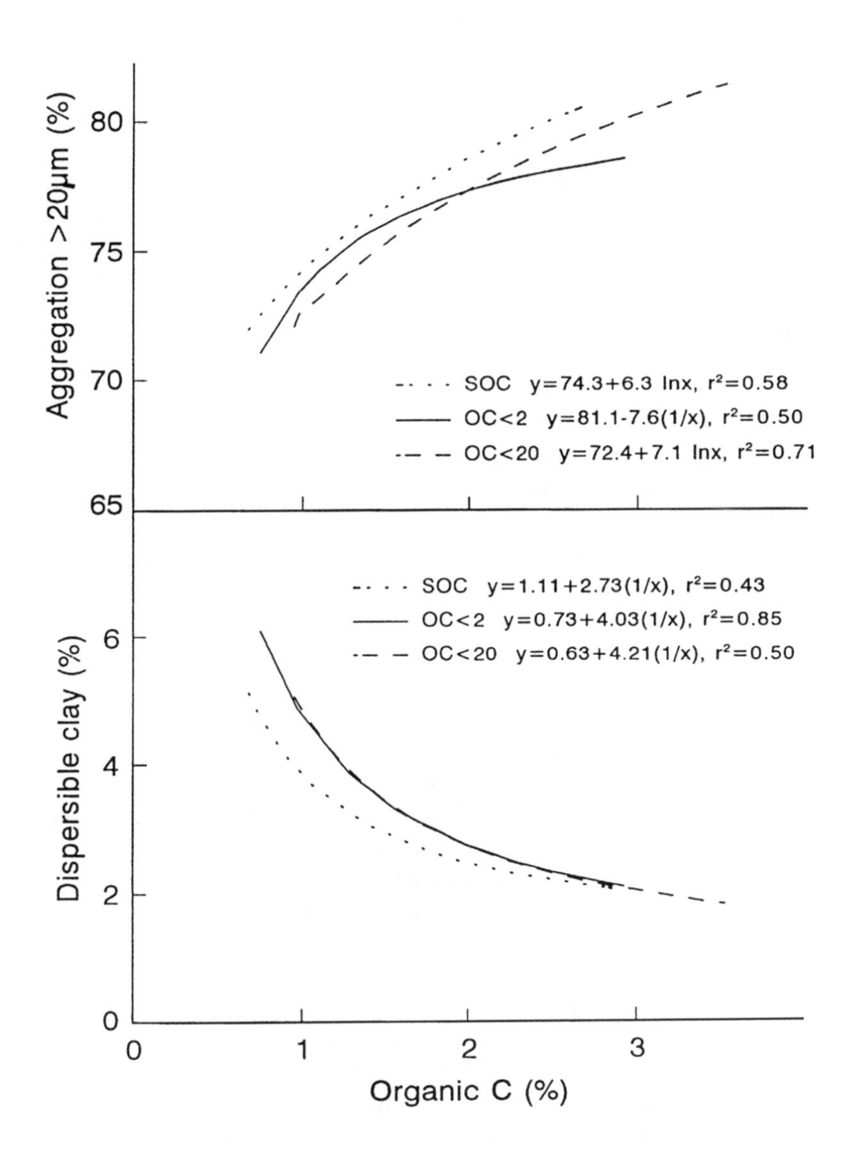

Figure 6. Relationship between dispersible clay and total organic C, organic C content of dispersible clay (OC < 2μm), and organic C content of water dispersed < 20 μm fraction. Similarly for the aggregated > 20 μm fraction in Vertisols (Typic Chromusterts, Queensland) subjected to long-term cereal cultivation. (From R.C. Dalal, unpublished data.)

Vertisols is apparent (Table 8). This corroborates the observations of Coughlan and Loch (1984), Dalal (1989) and Cook et al. (1992). They also found that the aggregate stability in many Vertisols was primarily associated with exchangeable cations, especially exchangeable Na^+ which is usually present in moderate amounts of < 10% of total exchangeable cations. High electrolyte concentration in soil solution reduces water-dispersible clay (Yerima et al., 1989).

Notwithstanding the predominant role of exchangeable cations in aggregate stability in Vertisols, the rates of loss of SOM and the extent of soil aggregation are interlinked. Dalal and Mayer (1986b) found that the rates of loss of SOM from the diverse virgin Vertisols on cultivation for up to 70 years decreased with the increasing initial percentages of aggregates > 20 μm initially as shown in Equation (1):

$$k_c = 0.319 + 0.442\, x_1 - 0.005 x_2 \qquad (1)$$

where k_c is the rate of loss of SOM (yr^{-1}), x_1 is the organic C/urease activity ratio (g organic C kg^{-1} soil/mg urea -N hydrolysed kg^{-1} soil h^{-1}) and x_2 is the percentage of aggregates > 20 μm. The increasing degree of aggregation in soil retards diffusion of organic substrates and/or products and reduces the accessibility to the organic substrate by the degradative enzymes and microorganisms (Ladd et al., 1990).

The ratio of organic C/urease activity in Vertisols is apparently a measure of relative accessibility of the SOM to proteolytic and other degradative enzymes since the urease activity of soils, in the absence of recently added organic materials, reflects their capacity for protection of the urease (Zantua and Bremner, 1977). Dalal and Mayer (1986b) inferred that the lower ratio of organic C/urease activity may, therefore, reflect greater protection afforded to the enzyme-organic matter complex entrapped within soil aggregates against biological degradation. They also observed that the ratio of organic C/urease activity in a Vertisol decreased when it was cultivated because relatively unprotected or easily accessible SOM was rapidly decomposed. Furthermore, the ratio of organic C/urease activity generally decreased with depth and so did the rate of loss of SOM. However, this ratio as an index of accessibility to or protection of SOM needs to be confirmed in different environments. The location of the urease in the urease-organic matter complex within aggregates can be ascertained by the cytochemical methods combined with back scatter scanning electron microscopy, transmission electron microscopy and electron probe microanalysis (Foster, 1988).

The relationships between water-stable aggregates > 20 μm, water dispersible clay and various soil organic fractions showed that the closest correlation was found between aggregate or dispersible clay and its organic C content (Figure 6). The soil aggregation increased as its organic C content increased. Conversely, the amount of water dispersible clay decreased as its organic C content increased, thus demonstrating the role of SOM or its fractions in aggregation of Vertisols. Hamblin (1980) and Chan (1989) demonstrated that organic matter is involved in aggregation of clay in Vertisols predominantly through polyvalent Fe and Al (Table 9).

Various functional groups of organic C have been implicated in soil aggregation (Theng, 1987). It appears from Table 10 that both alkyl C (including polymethylenes) and ethoxy C (alcohol, carbohydrates and ether) are involved in soil aggregation in Vertisols. However, the rapid loss of sand-size C (Table 10) and light fraction C (Dalal and Mayer, 1986d) may be less directly related to soil aggregation. Furthermore, the role of aromatic and carbonyl C as well as silt-size C (which may include most of the C groups) are probably small in soil aggregation of Vertisols.

The cessation of cultivation after changing to a no-tillage practice, especially when it is accompanied by crop residue retention, improves soil aggregation in Vertisols (Hamblin, 1980; Loch and Coughlan, 1984), especially that of large aggregates; although in some instances water-dispersible clay may actually increase. The no-tillage effects on aggregation are due to an increase in organic matter as well as lower exchangeable Na^+ or ESP (Dalal, 1989). Increase in dispersible clay in soil under no-tillage has been attributed to the increase in organic anions (Oades, 1987), although it may primarily be due to the decrease in electrolyte concentration frequently observed in Vertisols under no-tillage practice (Dalal, 1989).

It has frequently been observed that the Vertisols under grasslands are better aggregated than under forest (Prebble, 1987), horticultural crops (Albrecht, 1988), cereal grain crops (Chan et al., 1988; Dalal et al., 1991) and cotton (McKenzie et al., 1991). Also, grasses and forage legumes in crop rotation improve soil aggregation while continuous cereal cropping decreases it (Tyagi et al., 1982). The Vertisol under grasses and forage legumes not only receives additional organic matter (Tyagi et al., 1982) but also the protection is provided against rain drop damage to aggregates.

Addition of farm yard manure, composts and sewage sludge to Vertisols in small amounts may have only a marginal but longer-term effect on soil aggregation, whereas soil amendments such as gypsum improve aggregation where exchangeable Na^+ is high ($> 5\%$) but its effect is transitory. It follows, therefore, that a combination of organic residue additions and gypsum, especially when followed by minimum soil disturbance such as no-tillage practice provides a sound management option for improving soil aggregation. Good pasture management may be required, however, on Vertisols where severe breakdown of soil aggregates has already occurred.

Table 9. Water-stable aggregates and organic C in a pasture and adjoining cultivated Vertisol

Soil	Aggregates > 0.5 mm %	Dispersible clay %	Proportion of clay dispersed in		Organic C %	$Na_4P_2O_7$ - extract		
			$Na_4P_2O_7$ (%)	$NaIO_4/Na_2B_4O_7$ (%)		C %	Al/C x10^{-3}	Fe/C x10^{-3}
Native pasture	60	0.9	77	22	1.4	0.16	5.7	4.0
Cultivated	3	0.7	71	5	0.9	0.17	0.5	0.5

(Adapted from Chan et al., 1988, and Chan,1989.)

Table 10. Changes in soil aggregation and organic matter fractions in a Vertisol under cultivation

Period of cultivation (yr)	Aggregates[a] > 20 μm (%)	Dispersible[a] clay (%)	Total[a] organic C (%)	Carbohydrate[b] C (%)	^{13}C NMR analysis[a]			Particle-size[a]		
					Alkyl C (%)	Ethoxy C (%)	Aromatic Carbonyl C (%)	Clay C (%)	Silt C (%)	Sand C (%)
0	79.9	1.0	2.67	0.36	1.74	0.83	0.10	1.95	3.60	3.24
20	74.8	4.1	1.22	0.20	-	-	-	1.50	1.51	0.51
25	73.1	4.2	1.25	-	0.79	0.31	0.15	1.31	2.06	0.72
35	75.6	5.1	1.07	0.24	0.66	0.24	0.17	1.21	1.63	0.63
45	68.7	6.8	0.68	0.16	0.35	0.27	0.06	0.89	1.45	0.23

([a]R.C. Dalal, unpublished data; [b]Dalal and Henry, 1988; [c]calculated from Skjemsted et al., 1986; and Dalal and Mayer,1986b.)

D. Alfisols

Alfisols occur between the Mollisols and Vertisols of the ustic moisture regime and Spodosols and Inceptisols of the udic moisture regime in mesic or cooler climates, and they occur between Andisols of the xeric moisture regime and Inceptisols, Ultisols and Oxisols of the udic moisture regime in thermic or warmer climates. Alfisols usually possess a coarse matrix in 0 to 0.1 m depth of soil and follow a hierarchical arrangement of soil aggregates (Tisdall and Oades, 1982; Oades and Waters, 1991).

Tiessen and Stewart (1988) examined the role of organic matter and microbes in soil aggregation in an Alfisol (Paleustalf) using light and electron microscopy and Ruthenium/OsO_4 staining techniques. The majority of aggregates $> 50\ \mu m$ were formed on plant debris, into which soil particles have become entrapped. Older, more decomposed SOM appeared to be amorphous and intimately associated with soil particles. Soil aggregates were frequently invaded by fungi and actinomycetes whose hyphae interconnected several organic and mineral primary aggregates to form secondary organo-mineral aggregates. The attachments of microorganisms frequently bridged two or more mineral particles or aggregates, thus stabilizing microaggregates. Amorphous materials filled in the pores of larger aggregates, thus providing such SOM physical protection against microbial decomposition. The high affinity of extracellular polysaccharides for polyvalent cations, especially Fe, appeared to lead to the formation of relatively stable SOM-clay aggregation.

Oades and Waters (1991) also concluded that the organic materials were the dominant stabilizing agents in (larger) aggregates and fragments of roots acted as nuclei for smaller aggregates. The role of soil microorganisms in acting as nuclei for microaggregate formation has also been confirmed (Tiessen and Stewart, 1988; Foster, 1988).

R.C. Dalal (unpublished data) observed a close correlation between organic C content of clay particles and the amount of dispersible clay in a number of Alfisols subjected to continuous tillage for cereal cropping for up to 20 years. The amount of dispersible clay increased as the clay-size organic C content decreased (Figure 7). Cultivation reduced the clay-size organic C content as well as total organic C content although the total amount of remaining organic C increased proportionally from 47 to 72% in the clay-size organic matter after 20 years of cultivation (Dalal and Mayer, 1986c). Presumably this was due to protection or inaccessibility of the SOM provided by the clay fraction in microaggregates.

Continuous tillage for cereal cropping of Alfisols leads to reduced SOM, decreased aggregation (Figure 3 and Figure 8) and increased dispersible clay (Harte, 1984; Dalal and Mayer, 1986b; Loch et al., 1987; Lal, 1989; Dalal et al., 1991). No-tillage practice, on the other hand, usually results in increased SOM and soil aggregation (Aina, 1979; Hamblin, 1980; Lal, 1989; Smettem et

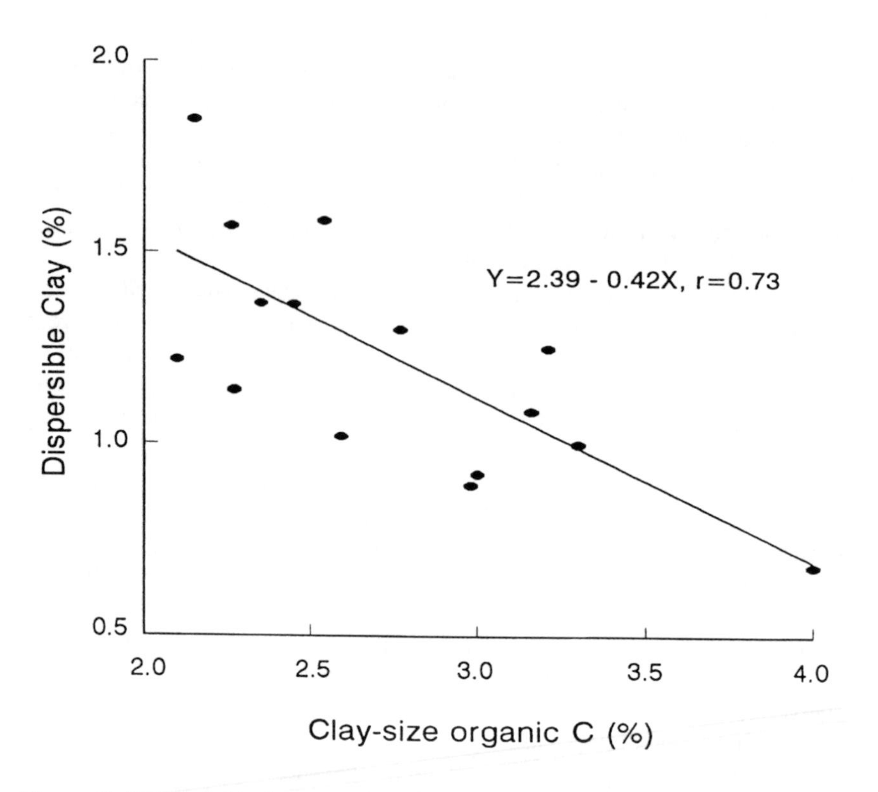

Figure 7. Relationship between clay-size organic C content and dispersible clay in Alfisols (Rhodic Paleustalfs). (From R.C. Dalal, unpublished data.)

al., 1992). Faster improvements in soil aggregation of Alfisols can be expected under pastures (Stoneman, 1973; Tisdall and Oades, 1980) although subsequent cropping results in rapid breakdown of aggregates > 2 mm (Figure 8), presumably due to a rapid loss of macroorganic matter. As a matter of fact, the rate of breakdown of aggregates > 2 mm (1.29 yr^{-1}), calculated from the data of Stoneman (1973) was similar to that of the SOM loss (1.22 yr^{-1}) from a number of Alfisols under continuous tillage (Dalal and Mayer, 1986b). Moreover, overgrazing of pastures can lead to aggregate breakdown (Stoneman, 1973; Bridge et al., 1983) and lower organic matter, resulting in bare areas in Alfisols. Therefore, optimum management of pastures, especially grass or grass-legume pastures in Alfisols is essential in building up the SOM and soil aggregation (Clarke et al., 1967).

Crops differ in their effects on soil aggregation in Alfisols. Even one season of lupin cropping produced significantly better aggregation than canola and

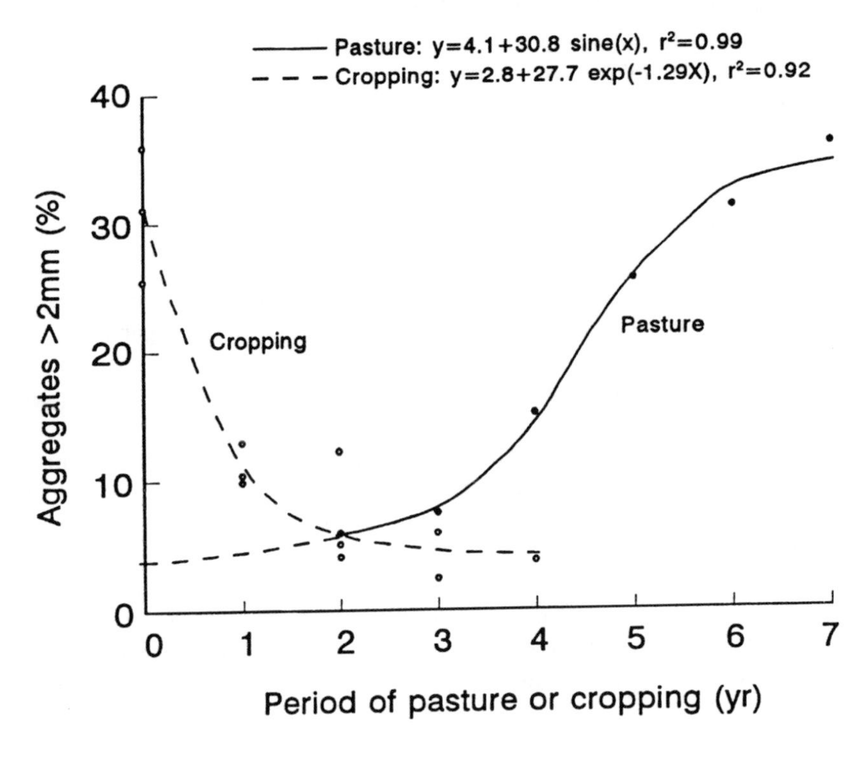

Figure 8. Dynamics of soil aggregation under pasture followed by tillage and cropping in an Alfisol. (Redrawn from Stoneman, 1973.)

linseed, presumably by increased extracellular polysaccharides resulting from roots, root exudates and microbial activity. For example, Oades (1987) demonstrated a substantial increase in soil aggregation > 1 mm following the addition of glucose which stimulated microbial activity.

Application of organic amendments to Alfisols, including farm yard manure and sewerage sludge improve soil aggregation and the SOM content (Acharya et al., 1988; Hamilton and Dindal, 1989). Both water-stable aggregates > 250 μm and mean weight diameter of aggregates and the SOM content increased significantly with the continuous application of farm yard manure for 13 years. However, the best improvement in soil aggregation and the SOM content in Alfisols was observed from the treatment receiving both farm yard manure and optimum rates of N, P and K fertilisers and the least from the N fertiliser alone treatment and the control treatment.

Application of sewerage sludge and the introduction of the earthworms, *Lumbricus terrestris* to an Alfisol significantly increased 4 mm diameter water-stable aggregates (Hamilton and Dindal, 1989). However, the introduction of the earthworms, *Eisenia fetida* had no significant effect on either organic matter content or water-stable aggregates.

As a soil amendment, gypsum appears to have a dual role in soil aggregation and soil organic matter. Gypsum application increases electrolyte concentration, thus decreasing dispersible clay. It also provides Ca^{2+} which not only provides polyvalent cationic bridges in soil aggregation but also retards organic matter decomposition in Alfisols (Oades, 1987), presumably through aggregate stabilization around microbial debris. Admittedly, the role of Ca^{2+} in retarding organic matter decomposition when acidic Alfisols are limed is equivocal since it is well established that, at least initially, liming increases organic matter decomposition (Allison, 1973).

The SOM turnover in long-term field experiments in Alfisols as revealed by ^{13}C-natural abundance (Balesdent et al., 1988) and ^{13}C NMR and FTIR spectroscopy (Capriel et al., 1992) have unequivocally shown that stabilization of SOM under various soil and cropping practices occurs primarily by interaction with the clay fraction and possibly the aggregated clay in the fine silt fraction. For example, Martin et al. (1990) measured from 43 to 48% loss of organic C from < 2 to 50 μm size fractions in an Ustalf after 16 years, whereas those from > 50 μm to 2000 μm varied from 89 to 97% during that period (Figure 9), thus again emphasizing the role of fine particle size fractions in protecting SOM against microbial decomposition.

E. Oxisols

Oxisols, which probably developed in a humid and perhumid climate (aqua moisture regime), occur widely in the tropical and sub-tropical regions under ustic and udic moisture regimes.

The mechanisms of aggregation in Oxisols apparently differ from that of other soils (Coughlan et al., 1973; Oades and Waters, 1991). Coughlan et al. (1973) observed that roots were the main aggregate binding agents of aggregates > 5 mm in a virgin Oxisol, but tillage operations resulted in clod formation in the cultivated Oxisol. The water stability of aggregates 0.5 to 5 mm diameter was due to the hydrophobic properties of organic matter, which along with free iron oxides, was the main aggregate stabilizing agent. The aggregates < 0.5 mm diameter were not affected by cultivation or organic matter content and reflected the characteristics of the basic soil particles.

Oades and Waters (1991) suggested that the aggregate hierarchy, proposed for Alfisols by Tisdall and Oades (1982), did not apply to the Oxisol they studied. Oades and Waters (1991) observed that soil aggregates > 250 μm were virtually stable (85%) to rapid wetting and had no dispersible clay. The disruption of these aggregates released clay-size particles and essentially no aggregates in 20

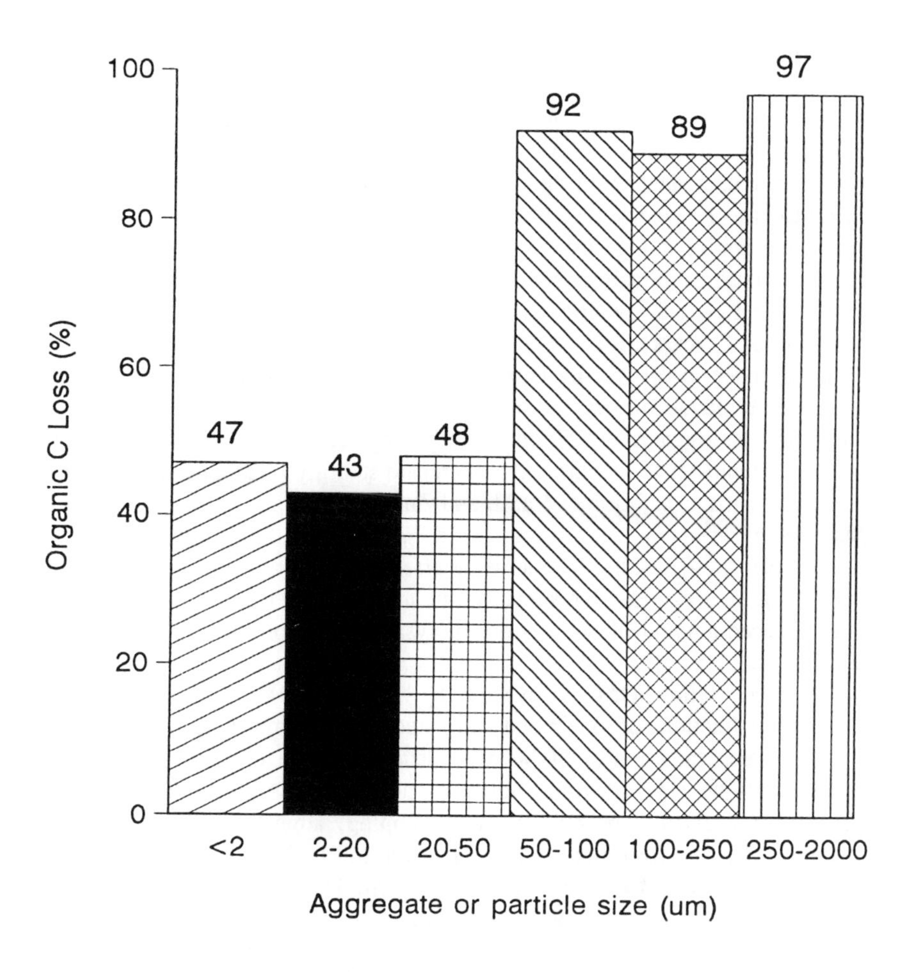

Figure 9. Loss of organic C from various aggregate or particle size fractions in 16 years from a savanna Ustalf (0-0.1 m) at Lamto, Ivory Coast, using ^{13}C natural abundance technique. (Drawn from Martin et al., 1990.)

- 250 μm size. Moreover, organic matter content and C:N ratios of aggregates in different size ranges were uniform throughout, unlike in Alfisols where organic C was concentrated in aggregates > 250 μm as plant and fungal debris and in aggregates < 20 μm; both of which were rich in organic C and N. Oades and Waters (1991) found no evidence of stepwise degradation of aggregates < 20 μm diameter, where close association of microorganisms and microbial debris, polyvalent metal cations and clay tactoids are likely to occur in Oxisols.

Continuous tillage of Oxisols decreases water-stable aggregates and organic C (Loch et al., 1987; Prove et al., 1990). Loch et al. (1987) found that continuous tillage for peanut (*Arachis hypogaea* L.) cropping for up to 60 years led to a substantial decrease in water-stable aggregates > 500 μm, which was significantly correlated with the organic C content.

Prove et al. (1990) observed that no-tillage and crop residue retention practice for maize cropping in Queensland significantly improved aggregate stability and organic C content. Small size aggregates < 125 μm decreased by 30% and organic C increased by 46% in Oxisol under no-tillage and crop residue retention for 7 years (Table 11), and thus large size aggregates > 125 μm increased under no-till practice compared with the conventional tillage practice. Similarly, removal of organic matter from an Oxisol in Bengal reduced water-stable aggregates > 250 μm, whereas the addition of organic matter and Ca, Mg and phosphate increased the aggregate stability (Adhikari et al., 1986). On the other hand, addition of $CuS0_4$ and $ZnS0_4$ (2.2 g kg^{-1} soil) had an adverse effect on water-stable aggregates > 250 μm although no explanation was provided for these results. The adverse effect of Cu^{2+} and Zn^{2+} additions on water-stable aggregates is presumably due to depressed microbial activity and hence reduced amount of extracellular polysaccharides produced for aggregate stabilization.

Dynamics of the SOM as revealed by ^{13}C natural abundance in various particle-size fractions of virgin and cultivated Oxisols from Brazil showed that the initial decomposition of SOM in the sand and silt-size fractions was more than that of the clay fraction (Bonde et al., 1992). After the 12 years of sugarcane (*Saccharum officinarum* L.) cropping, almost 90% of the initial organic C was lost from the sand-size fraction (63 to 200 μm), 63% from the silt fraction and only 40% from the clay fraction. Thus, the turnover periods of the initial organic C were 59, 6 and 4 years in the clay, silt and sand-size fractions, respectively, thereby emphasizing the role of the clay fraction in SOM stabilization through aggregation in Oxisols as well as other soils. Skjemstad et al. (1990) showed that organic matter within microaggregates < 200 μm in Oxisols from New South Wales contained up to 32% more old organic C than the remaining soil after 83 years of change in land use from forest to grassland. Thus, the presence of microaggregates is important in the stabilization of the SOM in Oxisols (Skjemstad et al., 1990).

F. Ultisols

Ultisols usually occur in the udic moisture regime in the sub-humid regions, both in the tropics and sub-tropics and even in temperate zones. The clay fraction

Table 11. Effect of tillage practice on water-stable aggregates and organic C in an Oxisol after 7 years of maize cropping

Tillage	Crop residue	Aggregates < 125 μm (%)	Organic C[a] (%)
Conventional till	Removed	19.7	1.06
Conventional till	Retained	18.0	1.13
No till	Removed	16.2	1.41
No till	Retained	13.8	1.55
LSD ($P < 0.05$)		3.2	0.10

([a]Adapted from R.J. Loch, unpublished data; Prove et al., 1990.)

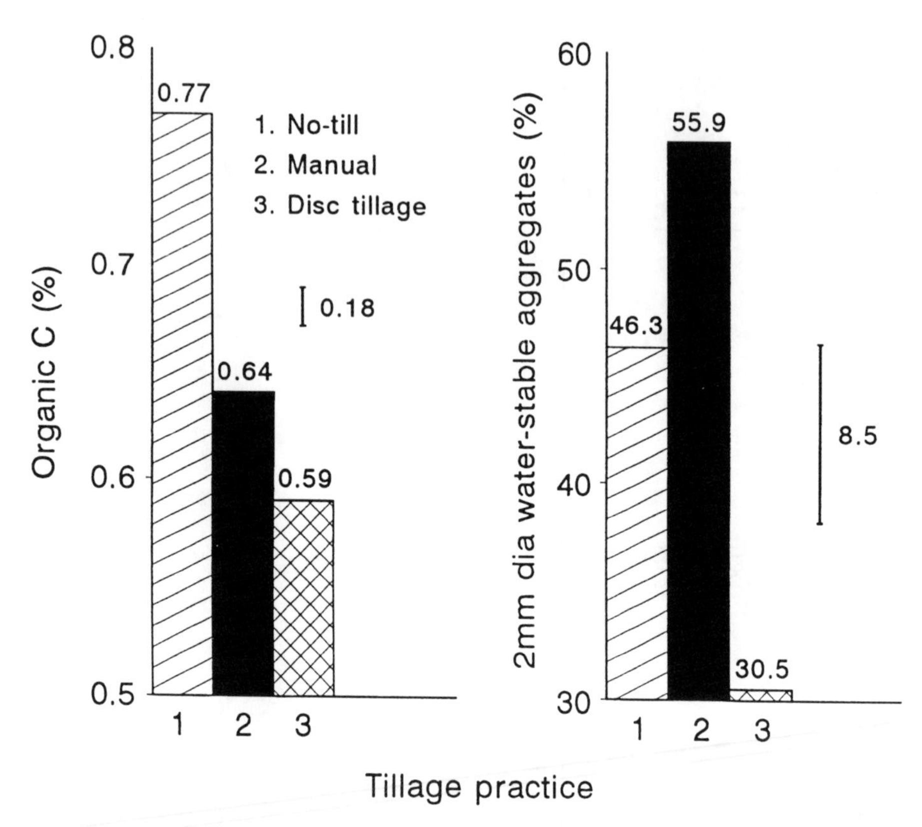

Figure 10. Effect of tillage practice on soil organic C and water-stable aggregates (2 mm dia) of 0 to 0.05 m layer of a Ultisol (Typic Haplustult) used for corn-cotton (*Gossypium arboreum* L.) cropping for 4 years. Bar represents $P < 0.05$. (Drawn from Ike, 1986.)

predominantly contains kaolinite, gibbsite and Al-interlayered minerals. Thus the soil aggregates are formed between these minerals and the SOM, bridged by polyvalent Al in the surface layers of Ultisols.

The SOM is essential for the soil aggregate stabilization in Ultisols because of the presence of low-activity clays and the absence of iron oxides. Cultural practices that lead to the decrease of SOM also reduce water-stable aggregates. For example, Ike (1986) found that the disc tillage of an Ultisol reduced the SOM by almost one-third compared to the no-till practice; water-stable aggregates were even more affected by the disc tillage (Figure 10), thus emphasizing the role of the SOM in aggregate stabilization in Ultisols.

Pastures managed on light-intensity grazing improve both soil organic matter and aggregate stability in Ultisols (Mbakaya et al., 1988). Earthworm activity also increases the mean diameter of water-stable aggregates. The effect is more pronounced in coarse-textured than in fine-textured soils. Brussaard et al. (1990) found that earthworm castings incorporated all grain sizes, with bridges of organic matter and fine material particles binding coarser grains together.

The distribution of the SOM in soil aggregates of the (tropical) Ultisols appears to differ from that of the (temperate) Alfisols and Mollisols. Contrary to the latter where macroaggregates have higher organic matter content than microaggregates (Sections IV B and D), various aggregate size classes may contain essentially similar amounts except the 0.02 to 0.053 mm size aggregates which may be enriched in the SOM. Elliott et al. (1991) believe that this size class of organic matter is due to microbial debris causing microsites of organic C enrichment. When forested Ultisol is cultivated, the distribution of aggregates shifts to smaller sizes, but the SOM content and the C/N ratio remain constant across most size classes. Upon reforestation, the SOM appears to shift from the smallest silt and clay size fractions to the aggregates > 1 mm diameter (Elliott et al., 1991), presumably due to the relatively spatial separation of the added organic substrates from the microbial decomposition.

V. Perspectives

Although the decomposition rates of various fractions of SOM vary widely, their resistance to decomposition is primarily due to their inaccessibility to microorganisms and degradative enzymes, and less due to their chemistry and structure (Skjemstad and Dalal, 1987). Thus, the SOM during aggregate formation could become incorporated or entrapped within aggregates, and depending on the pore size of aggregates becomes physically inaccessible to microorganisms, and probably to many proteolytic and polysaccharide hydrolysing enzymes. In some soils, however, Al by complexing with carboxyl groups of SOM could effectively protect SOM (Wada and Higashi, 1976). Unfortunately, information is lacking on spatial quantification of the distribution of organic matter, degradative enzymes, microorganisms and other larger organisms in the various aggregate pore sizes in soil, although the localizations of organic matter and some enzymes in soil's rhizosphere, mycorrhizas and mycelial strands has been intensively studied (Kilbertus, 1980; Foster, 1988; Van Veen and Kuikman, 1990).

Environment (precipitation, evapotranspiration, temperature and solar radiation) largely determines both the quality and quantity of agents for soil aggregate formation as well as soil aggregate stabilization (Figure 2). Soil aggregates are eventually broken down by both abiotic (mechanical disruption, e.g. tillage, slaking, dispersion) and biotic agents (decomposition of organic matter). Unfortunately, there is a paucity of information on the periodicity of aggregate turnover in various soil orders and in different environments. It is therefore important to examine the half-life of these aggregates, especially in the semi-arid environments where the supply of aggregate stabilizing agents, particularly the organic materials, tends to be periodic and insufficient under the various exploitative cropping systems prevalent in these regions. Therefore, one of the requirements of the so-called 'sustainable' agricultural systems must be that the rate of aggregate breakdown does not exceed that of aggregate formation. In many situations these conditions are satisfied where the levels of organic matter also meet the requirements of sustainable agriculture. Thus, these conditions are met in cropping systems that enhance the supply of transient as well as temporary aggregate binding agents (Figure 1) in soil management systems that reduce the turnover period of these agents by such practices as minimum soil disturbance (no-tillage), and by maintaining the optimum ionic environments (polyvalent metal cations, e.g. Ca^{2+}; and sufficient electrolyte concentration). Soil and crop management practices that leave plant residues on the soil surface not only reduce aggregate breakdown from the disruptive energy of raindrop impact but also are likely to enhance the SOM content at the soil surface by spatially and environmentally (e.g. moisture and temperature) isolating organic substrates from many decomposers and degradative enzymes. In most soils, although cropping system has the main effect rather than tillage, both interact in aggregate formation and stabilization in arable soils.

Among the soil types, Entisols ($P/PET < 0.5$) and Ultisols ($P/PET > 0.75$) and other soils of low clay contents (coarse matrix soils) depend primarily on organic matter as aggregate stabilizing agent. On the other hand, the role of organic matter in aggregate stabilization is less dominant in Vertisols (and Mollisols) and Oxisols. These differences are evident mainly in microaggregation $< 20\ \mu m$ because macroaggregation $> 250\ \mu m$ in all soils occurs mainly due to the activities of plant roots, fungal hyphae and meso- and macrofauna. Therefore, management practices such as cropping systems (pastures, legumes, arable cropping), plant residue retention, soil surface cover, reduced or no-tillage practices and addition of organic amendments are essential for the maintenance of macroaggregation and organic matter levels in soil.

Recent modelling of the dynamics of SOM invariably have included the protective effects of aggregate stabilizing agents, especially the clay fraction (Parton et al., 1987). No attempts have been made to model the relationships between the SOM turnover and aggregate hierarchy in soils, for such relationships have not been quantified. Moreover, appropriate methodology is lacking to quantify the relationship between soil aggregates (size, density as well as their pore size) and various biological processes in different soil types (Entisols, Inceptisols, Mollisols, Vertisols, Alfisols, Oxisols and Ultisols) and under different soil and crop management practices (tillage, pastures, organic and inorganic amendments).

VI. Conclusions

Allison (1968) stated that 'our understanding of soil aggregate formation and stabilization is probably as unsatisfactory at the present time as that of any other phase of soil science.' Since 1968, considerable progress has been made in elucidating to some extent the localization of agents of aggregate stabilization *in situ*, especially organic materials and some enzymes using staining techniques in conjunction with transmission and scanning electron microscopy and electron probe microanalysis. The ^{13}C-NMR spectroscopy of *in situ* characterization of the soil organic matter will further enhance the understanding of organic matter and aggregates in soil. Computer assisted tomography (CAT) scan in combination with other non-destructive techniques of soil matrix will lead to the better understanding of the dynamics of aggregate formation and stabilization in soils.

Soils differ considerably in the nature of their aggregate hierarchy, the predominance of the agents of aggregate formation and aggregate stabilization. What is common among these soils is that soil aggregation and the SOM are optimum when the soil is disturbed minimally, soil surface is covered with plant residues or litter, and, above all, optimum conditions exist for maximum phytobiomass production commensurate with the environment.

In spite of the considerable progress made in understanding the SOM and soil aggregation in the last 25 years, the question that still needs to be answered is: why can such large amounts of easily decomposable organic materials be found in the vicinity of largely starving microorganisms, especially in relatively undisturbed ecosystems?

Acknowledgments

We thank Kath Fowler for literature search and Dr R.J. Loch for permission to use his unpublished data, and for his constructive suggestions.

References

Acharya, C.L., S.K. Bishnoi, and H.S. Yaduvanshi. 1988. Effect of long-term application of fertilisers, and organic and inorganic amendments under continuous cropping on soil physical and chemical properties in an Alfisol. *Indian J. Agric. Sci.* 58:509-516.

Adhikari, M., S.K. Gupta, and M.K. Majumdar. 1986. Effect of organic matter and inorganic salts on some physical properties of a lateritic soil. *Indian Agric.* 30:171-174.

Aina, P.O. 1979. Soil changes resulting from long-term management practices in western Nigeria. *Soil Sci. Soc. Am. Proc.* 43:173-177.

Albrecht, A. 1988. Effect of cropping system on aggregation in a Vertisol and a ferrallitic soil, French West Indies. *Cahiers - ORSTOM - Serie - Pedologie.* 24:351-353.

Allison, F.E. 1973. *Soil organic matter and its role in crop production*. Elsevier Science Publishing Co., New York.

Allison, F.E. 1968. Soil aggregation - some facts and fallacies as seen by a microbiologist. *Soil Sci.* 106:136-143.

Angers, D.A., and G.R. Mehuys. 1989. Effects of cropping on carbohydrate content and water-stable aggregation of a clay soil: role of carbohydrates. *Can. J. Soil Sci.* 69:373-380.

Balesdent, J., G.H. Wagner, and A. Mariotti. 1988. Soil organic matter tumover in longterm field experiments as revealed by carbon-13 natural abundance. *Soil Sci. Soc. Am. J.* 52:118-124.

Baver, L.D. 1934. A classification of soil structure and its relation to the main soil groups. *Am. Soil Survey Assoc. Bull.* 15:107-109.

Baver, L.D. 1968. The effect of organic matter on soil structure. *Pontificia Accademia Scientiarum Scripta Varia.* 32:383-403.

Bodman, G.B., and G.K. Constantin. 1965. Influence of particle size distribution in soil compaction. *Hilgardia* 36:567-591.

Bonde, T.A., B.T. Christensen, and C.C. Cerri. 1992. Dynamics of soil organic matter as reflected by natural ^{13}C abundance in particle size fractions of forested and cultivated Oxisols. *Soil Biol. Biochem.* 24:275-277.

Boyle, M., W.T. Frankenberger Jr., and L.H. Stolzy. 1989. The influence of organic matter on soil aggregation and water infiltration. *J. Prod. Agric.* 2:290-299.

Bridge, B.J., J.J. Mott, and R.J. Hartigan. 1983. The formation of degraded areas in the dry savanna woodlands of northern Australia. *Aust. J. Soil Res.* 21:91-104.

Brussaard, L., D.C. Coleman, D.A. Crossley Jr., W.A.M. Didden, P.F. Hendrix, and J.C.Y. Marinissen. 1990. Impacts of earthworms on soil aggregate stability. *Proc. 14th Intern. Cong. Soil Sci.* 3:100-105.

Burke, I.C., C.M. Yonker, W.J. Parton, C.V. Cole, K. Flach, and D.S. Schimel. 1989. Texture, climate and cultivation effects on soil organic matter content in U.S. grassland soils. *Soil Sci. Soc. Am. J.* 53:800-805.

Cambardella, C.A. and E.T. Elliott. 1992. Particulate soil organic matter changes across a grassland cultivation sequence. *Soil Sci. Soc. Am. J.* 56:777-783.

Capriel, P., P. Harter, and D. Stephenson. 1992. Influence of management on the organic matter of a mineral soil. *Soil Sci.* 153:122-128.

Carter, M.R. 1991. The influence of tillage on the proportion of organic carbon and nitrogen in the microbial biomass of medium textured soils in a humid climate. *Biol. Fert. Soils* 11:135-139.

Carter, M.R. 1992. Influence of reduced tillage systems on organic matter, microbial biomass, macro-aggregate distribution and structural stability of the surface soil in a humid climate. *Soil Tillage Res.* 23:361-372.

Carter, M.R. and H.T. Kunelius. 1993. Effect of undersowing barley with annual ryegrasses or red clover on soil structure in a barley-soybean rotation. *Agric., Ecosyst. Environ.* 43:245-254.

Carter, M.R. and P.M. Mele. 1992. Changes in microbial biomass and structural stability at the surface of a duplex soil under direct drilling and stubble retention in North-eastern Victoria. *Aust. J. Soil Res.* 30:493-503.

Chan, K.Y. 1989. Effect of tillage on aggregate strength and aggregation of Vertisols. *Soil Tillage Res.* 13:163-175.

Chan, K.Y. and D.P. Heenan. 1991. Differences in surface soil aggregation under six different crops. *Aust. J. Exp. Agric.* 31:683-686.

Chan, K.Y., W.D. Bellotti, and W.P. Roberts. 1988. Changes in surface soil properties of Vertisols under dryland cropping in a semi-arid environment. *Aust. J. Soil Res.* 26:509-518.

Chen, E., L. Zhou, F. Qiu, C. Yan, and Z. Gao. 1982. An approach to the essence of soil fertility. *Z. Pflanzenernaehr. Bodenk.* 145:207-220.

Cheshire, M.V., G.P. Sparling, and C.M. Mundie. 1983. Effect of periodate treatment of soil on carbohydrate constituents and soil aggregation. *J. Soil Sci.* 34:105-112.

Christensen, B.T. 1992. Physical fractionation of soil and organic matter in primary particle size and density separates. *Adv. Soil Sci.* 20:1-90.

Clarke, G.B. and T.J. Marshall. 1947. Influence of cultivation on soil structure and its assessment in soils of variable mechanical composition. *J. Council Sci. Ind. Res. Aust.* 20:162-175.

Clarke, A.L., D.J. Greenland, and J.P. Quirk. 1967. Changes in some physical properties of the surface of an impoverished red brown earth under pasture. *Aust. J. Soil Res.* 5:59-68.

Cook, G.D., H.B. So, and R.C. Dalal. 1992. Structural degradation of Vertisols under continuous cultivation. *Soil Tillage Res.* 22:47-64.

Coughlan, K.J., W.E. Fox, and J.D. Hughes. 1973. A study of the mechanisms of aggregation. *Aust. J. Soil Res.* 11:65-73.

Coughlan, K.J. and R.J. Loch. 1984. The relationship between aggregation and other soil properties in cracking clay soils. *Aust. J. Soil Res.* 22:59-69.

Craswell, E.T. and S.A. Waring. 1972. Effect of grinding on the decomposition of soil organic matter. II. Oxygen uptake and nitrogen mineralization in virgin and cultivated cracking clay soils. *Soil Biol. Biochem.* 4:435-442.

Dalal, R.C. 1989. Long-term effects of no-tillage, crop residue and nitrogen application on properties of a vertisol. *Soil Sci. Soc. Am. J.* 53:1511-1515.

Dalal, R.C., and R.J. Henry. 1988. Cultivation effects on carbohydrate contents of soil and soil fractions. *Soil Sci. Soc. Am. J.* 52:1361-1365.

Dalal, R.C. and R.J. Mayer. 1986a. Long-term trends in fertility of soils under continuous cultivation and cereal cropping in Southern Queensland. I. Overall changes in soil properties and trends in winter cereal yields. *Aust. J. Soil Res.* 24:265-279.

Dalal, R.C. and R.J. Mayer. 1986b. Long-term trends in fertility of soils under continuous cultivation and cereal cropping in Southern Queensland. II. Total organic carbon and its rate of loss from the soil profile. *Aust. J. Soil Res.* 24:281-292.

Dalal, R.C. and R.J. Mayer. 1986c. Long-term trends in fertility of soils under continuous cultivation and cereal cropping in Southern Queensland. III. Distribution and kinetics of soil organic carbon in particle-size fractions. *Aust. J. Soil Res.* 24:293-300.

Dalal, R.C. and R.J. Mayer. 1986d. Long-term trends in fertility of soils under continuous cultivation and cereal cropping in Southern Queensland. IV. Loss of organic carbon from different density fractions. *Aust. J. Soil Res.* 24:301-309.

Dalal, R.C. and R.J. Mayer. 1987. Long-term trends in fertility of soils under continuous cultivation and cereal cropping in Southern Queensland. VII. Dynamics of nitrogen mineralisation potentials and microbial biomass. *Aust. J. Soil Res.* 25:461-472.

Dalal, R.C., W.M. Strong, E.J. Weston, and J. Gaffney. 1991. Sustaining multiple production systems. 2. Soil fertility decline and restoration of cropping lands in subtropical Queensland. *Tropical Grasslands* 25:173-180.

Edwards, A.P. and J.M. Bremner. 1967. Microaggregates in soils. *J. Soil Sci.* 34:105-112.

Elliott, E.T., C.A. Palm, D.E. Reuss, and C.A. Monz. 1991. Organic matter contained in soil aggregates from a tropical chronosequence: correction for sand and light fraction. *Agric. Ecosyst. Environ.* 34:443-451.

Elliott, E.T. 1986. Aggregate structure and carbon, nitrogen and phosphorus in native and cultivated soils. *Soil Sci. Soc. Am. J.* 50:627-633.

Emerson, W.W., R.C. Foster, and J.M. Oades. 1986. Organo-mineral complexes in relation to soil aggregation and structure. In: P.M. Huang and M. Schnitzer (eds.), *Interactions of soil minerals with natural organics and microbes*. pp. 521-548. SSSA Spec. Pub. No.17 Soil Science Society of America, Madison.

Ford, G.W. and D.J. Greenland. 1968. The dynamics of partly humified organic matter in some arable soils. *Trans. 9th Int. Congr. Soil Sci.* 2:403-410.

Foster, R.C. 1988. Microenvironments of soil microorganisms. *Biol. Fertil. Soils*. 6:189-203.

Greenland, D.J. 1965. Interaction between clays and organic compounds in soils. *Soils Fert.* 28:415-425.

Greenland, D.J. 1981. Soil management and soil degradation. *J. Soil Sci.* 32:301-322.

Greenland, D.J. and G.W. Ford. 1964. Separation of partially humified organic materials from soils by ultrasonic dispersion. *Trans. 8th Int. Congr. Soil Sci* 3:137-148.

Haas, H.J., C.E. Evans, and E.F. Miles. 1957. Nitrogen and carbon changes in Great Plains soils as influenced by cropping and soil treatments. *U.S. Dep. Agric. Tech. Bull.* No.1164.

Hamblin, A.P. 1980. Changes in aggregate stability and associated organic matter properties after direct drilling and ploughing on some Australian soils. *Aust. J. Soil Res.* 18:27-36.

Hamilton, W.E. and Dindal, D.L. 1989. Influence of earthworms and leaf litter on edaphic variables in sewage-sludge-treated soil microcosms. *Biol. Fertil. Soils.* 7:129-133.

Harte, A.J. 1984. Effect of tillage on the stability of three red soils of the northern wheat belt. *J. Soil Conserv. Serv. NSW* 40:94-101.

Ike, I.F. 1986. Soil and crop responses to different tillage practices in a ferruginous soil in the Nigerian savanna. *Soil Tillage Res.* 6:261-272.

Insam, H., D. Parkinson, and K.H. Domsch. 1989. Influence of macro-climate on soil microbial biomass. *Soil Biol. Biochem.* 21:211-221.

Jastrow, J.D. 1987. Changes in soil aggregation associated with tallgrass prairie restoration. *Am. J. Bot.* 74:1656-1664.

Jenkinson, D.S. 1988. Soil organic matter and its dynamics. In: A. Wild (ed.), *Russell's soil conditions and plant growth.* pp. 564-607. Longman, Harlow.

Jenny, H. 1980. *Soil resources: origin and behaviour.* Springer-Verlag, New York.

Jenny, H. and S.P. Raychaudhuri. 1960. *Effect of climate and cultivation on nitrogen and organic matter reserves in Indian soils,* pp.1-146. Indian Council of Agric. Res.. New Delhi.

Kilbertus, G. 1980. Etude des microhabitats contenus dans les agregats du sol. Leur relation avec la biomasse bacterienne et la taille des pro caryotes present. *Reveu d' Ecologie et de Biologie du Sol.* 17:543-557.

Ladd, J.N., M. Amato, L. Jocteur Monrozier, and M. Van Gestel. 1990. Soil microhabitats and carbon and nitrogen metabolism. *Proc. 14th Intern. Cong. Soil Sci* 3:82-87.

Lal, R. 1989. Conservation tillage for sustainable agriculture: tropics versus temperate environments. *Adv. Agron.* 42:85-191.

Lee, K.E. and R.C. Foster. 1991. Soil fauna and soil structure. *Aust J. Soil Res.* 29:745-775.

Loch, R.J. and K.J. Couglan. 1984. Effect of zero tillage and stubble retention on some properties of a cracking clay. *Aust. J. Soil Res.* 22:91-98.

Loch, R.J., K.J. Coughlan, and J.C. Mulder. 1987. Effects of frequency of peanut *(Arachis hypogae* L.) cropping on some properties of krasnozem and euchrozem soils in the south Burnett area, Queensland. *Aust. J. Exp. Agric.* 27:585-589.

Lugo, A.E., M.J. Sanchez, and S. Brown. 1986. Land use and organic carbon content of some subtropical soils. *Plant Soil.* 96:185-196.

Lynch, J.M. and E. Bragg. 1985. Microorganisms and soil aggregate stability. *Adv. Soil Sci.* 2:133-171.

Martin, J.P., W.P. Martin, J.B. Page, W.A. Raney, and J.G. De Ment. 1955. Soil aggregation. *Adv. Agron.* 7:1-37.

Martin, A., A. Mariotti, J. Balesdent, P. Lavelle, and R. Vuattoux. 1990. Estimate of organic matter turnover rate in a savanna soil by ^{13}C natural abundance measurements. *Soil Biol. Biochem.* 22:517-523.

Mbakaya, D.S., W.H. Blackburn, J.M. Skovlin, and R.D. Child. 1988. Infiltration and sediment production of a bushed grassland as influenced by livestock grazing systems, Buchuma, Kenya. *Trop. Agric.* (Trinidad). 65:99-105.

McGill, W.B. and R.J.K. Myers. 1987. Controls on dynamics of soil and fertiliser nitrogen. In: R.F. Follett, J.W.B. Stewart, and C.V. Cole (eds.), *Soil fertility and organic matter as critical components of production systems.* pp.73-99. Soil Science Society of America, Madison.

McKenzie, D.C., T.S. Abbott, and F.R. Higginson. 1991. The effect of irrigated crop production on the properties of a sodic Vertisol. *Aust. J. Soil Res.* 29:443-453.

Mele, P.M. and M.R. Carter. 1993. Effect of climatic factors on the use of microbial biomass as an indicator of changes in soil organic matter. In: K. Mulongoy and R. Merckx (eds.), *Soil organic matter dynamics and sustainability of tropical agriculture.* pp. 57-63. John Wiley and Sons, New York.

Miller, R.M. and J.D. Jastrow. 1990. Hierarchy of root and mycorrhizal fungal interactions with soil aggregation. *Soil. Biol. Biochem.* 22:579-584.

Monnier, G. 1965. Action des matieres organiques sur la stabilite structurale des sols. *Ann. Agron.* 16:471-534.

Mortland, M.M. 1970. Clay organic complexes and interactions. *Adv. Agron.* 22:75-117.

Nye, P.H. and D.J. Greenland. 1960. The soil under shifting cultivation. *Tech. Comm.* 51, Commonwealth Bureau of Soils, Slough, England.

Oades, J.M. 1987. Aggregation in soils. In: P. Rengasamy (ed.), *Soil Structure and aggregate stability.* pp. 74-101. Vic. Dep. Agric. Rural Affairs, Melbourne.

Oades, J.M. 1988. The retention of organic matter in soils. *Biogeochemistry* 5:35-70.

Oades, J.M. and A.G. Waters. 1991. Aggregate hierarchy in soils. *Aust. J. Soil Res.* 29:815-828.

Parton, W.J., D.S. Schimel, C.V. Cole, and D.S. Ojima. 1987. Analysis of factors controlling soil organic matter levels in Great Plains grasslands. *Soil Sci Soc. Am. J.* 51:1173-1179.

Payne, D. 1988. Soil structure, tilth and mechanical behaviour. In: A. Wild (ed.), *Russell's soil conditions and plant growth.* pp. 378-411. Longman, Harlow.

Perfect, E., B.D. Kay, W.K.P. van Loon, R.W. Sheard, and T. Pojasok. 1990. Rates of change in soil structural stability under forages compared to corn. *Soil Sci. Soc. Am. J.* 54:179-186.

Post, W.M., J. Pastor, P.J. Zinke, and A.G. Stangenberger. 1985. Global patterns of soil nitrogen storage. *Nature* 317:613-616.

Powlson, D.S. 1980. The effects of grinding on microbial and non-microbial organic matter in soil. *J. Soil Sci.* 31:77-85.

Prebble, R.E. 1987. Effect of cultivation on aggregate stability, microaggregation and organic carbon of Vertisols. *Division of Soils, Divisional Report no.91.* pp. 108. CSIRO, Melbourne.

Prove, B.G., R.J. Loch, J.L. Foley, V.J. Anderson, and D.R. Younger. 1990. Improvements in aggregation and infiltration characteristics of a Krasnozem under maize with direct drill and stubble retention. *Aust. J. Soil Res.* 28:577-590.

Rovira, A.D. and E.L. Greacen. 1957. The effect of aggregate disruption on the activity of microorganisms in soil. *Aust. J. Agric. Res.* 8:659-673.

Rueda, E.B. and D.F. Viqueira. 1986. Estudio de algunos factores que influyen en la formacion de microagregados en los suelos del N.W. de Espana. *An. Edafol. Agrobiol.* 22:289-300.

Russell, J.S. 1981. Models of long term soil organic nitrogen change. In: M.J. Frissel and J.A. van Veen (eds.), *Simulation of nitrogen behaviour of soil-plant systems*, p. 222-232. Centre for Agricultural Publishing and Documentation, Wageningen.

Skidmore, E.L., J.B. Layton, D.V. Armbrust, and M.L. Hooker. 1986. Soil physical properties as influenced by cropping and residue management. *Soil Sci. Soc. Am. J.* 50:415-419.

Skjemsted, J.O. and R.C. Dalal. 1987. Spectroscopic and chemical differences in organic matter of two vertisols subjected to long periods of cultivation. *Aust. J. Soil Res.* 25:323-335.

Skjemsted, J.O., R.C. Dalal, and P.F. Barron. 1986. Spectroscopic investigations of cultivations effects on organic matter of Vertisols. *Soil Sci. Soc. Am.. J.* 50:354-359.

Skjemstad, J.O., R.P. Le Feuvre, and R.E. Prebble. 1990. Turnover of soil organic matter under pasture as deterrnined by ^{13}C natural abundance. *Aust. J. Soil Res.* 28:267-277.

Smettem, K.R.J., A.D. Rovira, S.A. Wace, B.R. Wilson, and A. Simon. 1992. Effect of tillage and crop rotation on the surface stability and chemical properties of a red-brown earth (Alfisol) under wheat. *Soil Tillage Res.* 22:27-40.

Smith, G.D., K.J. Coughlan, and W.E. Fox. 1978. The role of texture in soil structure. In: W.W. Emerson, R.D. Bond and A.R. Dexter (eds.), *Modification of Soil Structure.* pp. 79-86. John Wiley and Sons, New York.

Sorensen, L.H. 1981. Carbon: nitrogen relationships during the humification of cellulose in soils containing different amounts of clay. *Soil Biol. Biochem.* 13:313-321.

Sparling, G.P. 1992. Ratio of microbial biomass carbon to soil organic carbon as a sensitive indicator of changes in soil organic matter. *Aust. J. Soil Res.* 30:195-207.

Sparling, G.P., T.G. Shepherd, and H.A. Kettles. 1992. Changes in soil organic C, microbial C and aggregate stability under continuous maize and cereal cropping, and after restoration to pasture in soils from the Manawatu region, New Zealand. *Soil Tillage Res.* 24:225-241.

Stoneman, T.C. 1973. Soil structure changes under wheatbelt farming systems. *J. Agric. West. Aust.* 14:209-214.

Theng, B.K.G. 1987. Clay-humic interactions and soil aggregate stability. In: P. Rengasamy (ed.), *Soil structure and aggregate stability.* pp. 32-73. Vic. Dep. Agric. Rural Affairs, Melbourne.

Thomas, R.S., R.L. Franson, and G.J. Bethlenfalvay. 1993. Separation of vesicular arbuscular mycorrhizal fungus and root effects on soil aggregation. *Soil Sci. Soc. Am. J.* 57:77-81.

Tiessen, H. and J.W.B. Stewart. 1988. Light and electron microscopy of stained microaggregates: the role of organic matter and microbes in soil aggregation. *Biogeochemistry* 5:312-322.

Tisdall, J.M. 1991. Fungal hyphae and structural stability of soil. *Aust. J. Soil Res.* 29:729-743.

Tisdall, J.M. and J.M. Oades. 1980. The effect of crop rotation on aggregation in a red brown earth. *Aust. J. Soil. Res.* 18:423-433.

Tisdall, J.M. and J.M. Oades. 1982. Organic matter and water-stable aggregates in soils. *J. Soil Sci.* 33:141-163.

Tyagi, S.C., D.L. Sharma, and G.P. Nathani. 1982. Effect of different cropping patterns on the physical properties of medium black soils of Rajasthan. *Curr. Agric.* 6:172-176.

Van Veen, J.A. and P.J. Kuikman. 1990. Soil structural aspects of decomposition of organic matter by microorganisms. *Biogeochemistry* 11:213-233.

Wada, A. and T. Higashi. 1976. The categories of aluminium and iron-humus complexes in Ando soils determined by selective dissolution. *J. Soil Sci.* 27:357-368.

Yerima, B.P.K., L.P. Wilding, C.T. Hallmark, and F.G. Calhoun. 1989. Statistical relationships among selected properties of northern Cameroon Vertisols and associated Alfisols. *Soil Sci. Soc. Am. J.* 53:1758-1763.

Zantua, M.I. and J.M. Bremner. 1977. Stability of urease in soils. *Soil Biol. Biochem.* 9:135-140.

Zunino, H., F. Borie, S. Aguilera, J.P. Martin, and K Haider. 1982. Decomposition of ^{14}C- labelled glucose, plant and microbial products, and phenols in volcanic ash-derived soils of Chile. *Soil Biol Biochem.* 14:37-43.

Aggregation and Organic Matter Storage in Kaolinitic and Smectitic Tropical Soils

C. Feller, A. Albrecht, and D. Tessier

1-56670-033-7/96/$0.00+$.50

I. Introduction

In most of the tropical regions, the clearing of native vegetation and the development of the land for food crop production, generally leads to drastic changes in soil properties. Decreases in plant nutrient reserves, organic matter (OM) and organic carbon (OC) contents, as well as a structure degradation are principally observed in low activity clay (LAC) soils, even those managed by low-intensified cultivation practices (Nye and Greenland, 1960; Fauck et al., 1969; Morel and Quantin, 1972; Siband, 1974; Sanchez, 1976; Feller and Milleville, 1977; Boyer, 1982; Moreau, 1983; Albrecht and Rangon, 1988; Pieri, 1989). The agronomic and ecological consequences are a decline in crop production and an increase in soil erodibility or soil erosion (Lal and Greenland, 1979; Roose, 1981).

Plant nutrient reserves can be relatively well-controlled by inputs of fertilizers and manures (Sanchez, 1976), but it is more difficult to manage soil physical properties and especially their structure in a long-term perspective. The stability of soil structure is a consequence of numerous interactive factors including crop rotations, cultural practices, soil OM content, and soil biological activities. Thus, the study of the relationships between soil structure and organic carbon storage in tropical agricultural soils represents one of the major aspects of soil fertility and conservation.

In this chapter we shall consider the above relationships through the following four points: (i) The general relationships between structure, cultivation and OC storage with some emphasis on macromorphological observations and statistical correlations; (ii) The distribution of OM and aggregates in the surface soil with some emphasis on particle size approaches; (iii) The role of OM in the 'stabilization' of structure, and (iv) The correlative role of soil structure in the 'stabilization' of soil OM throughout the physical protection effect against mineralization processes or against detachability, erodibility and transportability.

Because soils exhibit very different chemical and physical properties, especially in relation to their mineralogy, three main groups of tropical soils can be distinguished:

(i) The 'low activity clay (LAC) soils,' the clay minerals of which are dominated by 1:1 type phyllites, as kaolinites and halloysites, associated with more or less crystallised Fe, Al, Mn oxides and/or hydroxides. In this paper they will be termed 'LAC' soils or 'Kaolinitic' soils. In the US Soil Taxonomy they are mainly represented by alfisols, ultisols and oxisols and secondarily by some inceptisols and entisols. LAC soils cover about 60% of the tropical areas and more than 70%, if aridic and mountainous lands are excluded (after Sanchez, 1976);

(ii) The 'high activity clay (HAC) soils,' the clay minerals of which are dominated by 2:1 clay minerals as smectites. They are mainly represented by vertisols (about 3% of tropical areas);

(iii) The 'allophanic soils,' the mineral constituents of which are dominated by amorphous or crypto-crystallized minerals, as allophanes and imogolites. They are well represented by andisols which cover about 1% of tropical lands.

LAC soils principally and, more specifically, well-drained LAC soils will be considered in this paper. Some comparisons will be made with vertisols, but andisols will not be discussed.

II. General Considerations on Soil Structure and Aggregate Stability

Different definitions of soil structure have been proposed, either in a static and descriptive sense (five definitions reported by Jastrow and Miller, 1991), or in a more dynamic and functional way with references to soil management (Lal, 1979; Cassel and Lal, 1992). In the first case, soil structure is 'the arrangement of primary particles into aggregates' with various detailed descriptions of soil aggregates as they naturally occur in natural conditions. The nature and magnitude of forces acting between soils particles is generally invoked. For examples, 'In aggregates, forces holding the particles together are much stronger than those occurring between adjacent aggregates' (Martin et al., 1955, quoted by Robert and Chenu, 1992). In the second case, soil structure corresponds to 'those properties of soil that regulate and reflect a continuous array of various sizes of interconnected pores, their stability and durability, capacity to retain and transmit fluids and ability to supply water and nutrients for supporting active root growth and development' (Cassel and Lal, 1992). These conceptions were integrated by Dexter (1988) who defined soil structure as 'the spatial heterogeneity of different components or properties of soil.'

In relation to the functionality definition, Lal (1979) proposed various methods to evaluate soil structure and its stability: rainfall acceptance, aggregate stability, soil erodibility, detachability and transportability, porosity and pore-size distribution. Some of these different aspects will also be considered in this paper.

Physical fractionation techniques are a powerful approach to characterize the relationships between soil OM and aggregation on the macro- and microaggregate scales. Historically it is clear that size fractionation of macro- and microaggregates was an early and common method to evaluate both distribution and stability of soil aggregates (Kemper and Chepil, 1965; Henin et al., 1958). From 1970, both size and density fractionation methods, using wet sieving, were extensively used to separate soil OM forms. Detailed reviews concerning historical concepts, methodological aspects and applications to the study of bio-organomineral interactions in the soil were recently published (Elliott and Cambardella, 1991; Christensen, 1992; Feller, 1995). However, the aggregate size fractionation (AGSF) and the organic matter size fractionation (OMSF) are based on completely different approaches. For AGSF only low energy is generally involved to limit both disaggregation and dispersion of organomineral colloids. For OMSF, the main methodological objective is to reach a maximum soil dispersion, close to that obtained by mechanical analysis (Bremner and Genrich, 1990; Feller et al., 1991a; Christensen, 1992). The simultaneous use of these methods provide the means to separate quantitatively and with limited OM solubilisation, different compartments of OM, such as plant debris associated to sands, organo-silt complexes composed of particulate organic debris and very stable organomineral microaggregates, and organo-clay fractions enriched in amorphous and humified OM, partly having a microbial origin.

For organomineral interactions studies, the above two approaches appear to be very complementary: naturally occuring water stable aggregates may be separated by AGSF and their composition studied by OMSF. In view of this, Christensen (1992) termed the organo-clay fraction obtained by OMSF as a 'primary organomineral complex,' while the aggregates obtained by AGSF as the 'secondary organomineral complexes.'

At present, this double approach of AGSF and OMSF to characterize organomineral interactions is beginning to be widely applied to soils of cold or temperate regions (see previous chapters). By contrast, few published data are available for tropical situations. Therefore, in this paper some unpublished data obtained recently by the authors on a collection of soil surface samples originated from West Africa, Antilles and Brazil, will be presented. The samples were selected in order to test the effects of OM content, texture, mineralogy and cultivation on different soil properties.

III. Sites, Soils and Methods

A. Sites and Soils

The results reported here were obtained from the pedological situations summarized in Tables 1 and 2. For each situation comparisons were made between plots corresponding to different types of soil management. Most of the analytical data were summarized in Feller (1994) and are not detailed here.

Changes in soil properties will be studied as a function of: (i) Clearing of the native vegetation and of the duration of continuous cropping on sites 3, 5, 11 and 12; (ii) Spontaneous grass- or tree-fallows, and eventually their duration, on sites 1, 2, 3, 5, 6 and 10; (iii) Artificial meadows after continuous annual cropping on sites 7, 8 and 9; (iv) Organic amendments (crop residues and/or animal manures) on sites 1, 2, 3, 8, 9, 10. With exception of some plots at site 9, all sites chosen were apparently not eroded. Each soil sample (0-10, 10-20, 20-40 cm) was constituted from 6 to 12 replicates. The coefficients of variation of OC contents ranged from 3 to 18% with a mean value of 11%. The mean value of confidence interval was calculated to be 9%.

These soils belong to the following orders of the US Soil Taxonomy: oxisols (7, 10, 11, 12), ultisols (5, 6), alfisols (3, 4), inceptisols (8), entisols (1, 2),

Table 1. Some soil, climatic, and land use characteristics of the study sites

Locality and reference[a]	Symbol and number[b]		----Mean annual climatic data---- P (mm)	T (°C)	Soil order	Type of vegetation and/or crops
Senegal (a)	Psl	1	700	29	Entisol	Bush fallow, peanut, millet, sorghum
Senegal (a)	Ftl	2	700	29	Entisol	Grassfallow, peanut, millet
Senegal (b)	Fll	3	800	29	Alfisol	Tree savanna, peanut, millet
Ivory Coast (c)	Fl2	4	1360	26	Alfisol	Tree, bush and grass savanna
Ivory Coast (c)	Fr2	5	1360	26	Ultisol	Tree savanna, rice, corn, manioc
Togo (d)	Fr3	6	1040	27	Ultisol	Forest, corn-bush fallow
Guadeloupe (e)	Fr4	7	3000	25	Oxisol	Artificial meadow[c], market gardening
Martinique (e)	Fi6	8	1820	26	Inceptisol	Tree savanna, sugarcane, artificial meadow[c], market gardening
Martinique (e)	Ve6	9	1200	27	Vertisol	Tree savanna, artificial meadow[c], market gardening
Ste Lucia (e)	Fr7	10	2700	25	Oxisol	Grass fallow, corn, yam, market gardening
Brazil (f)	Fo8	11	1200	21	Oxisol	Forest, sugarcane
Brazil (g)	Fo9	12	1530	18	Oxisol	Prairie-rice, wheat, corn, soybean

[a]References: (a) Feller, 1994; (b) Feller and Milleville, 1977; (c) Fritsch et al., 1989; (d) Poss, 1991; (e) Albrecht et al., 1988; (f) Cerri et al., 1991; (g) Feller, 1994.

[b]The symbols refer to nomenclature used by Feller, 1994. Underlined site numbers correspond to smectitic soils.

[c]Planted with *Digitaria decumbens*.

Table 2. Some soil (0-20 cm) mineralogical and chemical characteristics of the study sites

Number[a]	Soil horizon (0-20 cm)						
	Mineralogy	F_2O_3t	0-20 μm	C	TRB[b]	CEC	pH-H_2O
	(0-2 μm fraction[a])	(g/100 g soil)				(cmol/kg soil)	
1	S-K-Q	1.2	13.6	0.58	22	5.7	5.9
2	K-I-Q	0.7	3.5	0.42	9	1.9	6.1
3	K-Q	1.0	18.7	0.75	10	nd[c]	6.4
4	K-Go	2.7	nd	1.16	18	nd	7.0
5	K-Hm	5.6	nd	1.58	16	nd	5.7
6	K-Q-(IS)	1.6	19.1	1.67	7	9.6	7.2
7	K-H-(Cri)	12.8	92.3	3.86	nd	15.5	4.8
8	K-H-Go-(IS)	14.0	63.1	4.41	135	24.0	6.4
9	S-(K)	nd	71.0	3.32	118	56.7	6.2
10	K-H-Go-(Cri)	nd	76.3	2.09	nd	14.8	5.3
11	K-Go-Hm-(Gi)	12.8	64.6	2.60	12	6.8	5.3
12	Gi-Hm-K	17.9	17.0	3.99	13	3.0	4.9

[a]Q = quartz, Cri = cristobalite, k = kaolinite, H = halloysite, S = smectite, I = illite, IS = interstratified clay, Go = goethite, Hm = hematite, Gi = gibbsite.
[b]TRB = Total reserve in bases according to Herbillon, 1989.
[c]not determined.

vertisols (9). The sandy soil, 1, and the clayey soil, 9, exhibit a clay fraction rich in smectite. The vertisol (9) is qualified as 'magneso-sodic' because exchangeable Mg and Na represent 30 to 40% and 5 to 10% of the total exchangeable cations, respectively. The selected LAC soils cover a wide range of texture, from sandy (2) to clayey (8, 10, 11, 12), with OC contents ranging from 0.4% (2) to 4.4% (8). The clayey LAC soils differ by their high (8) or very low (11, 12) Total Reserve in Bases (TRB) (Herbillon, 1989) in relation to the amount of primary weatherable minerals. For simplification, we will often distinguish four groups of samples. For LAC soils: (i) The coarse textured samples of kaolinitic soils from West Africa (sites 2 to 6); (ii) The clayey and TRB rich samples of kaolinitic/halloysitic soils from Antilles (sites 7, 8, 10); (iii) The clayey and TRB poor samples of kaolinitic/oxidic soils from Brazil (sites 11, 12). For HAC soils, the sandy smectitic (site 1) and the magneso-sodic vertisol (site 9) were selected.

B. Soil Organic Matter and Soil Aggregate Fractionation

Various methods adapted to aggregate size fractionation and soil organic matter distribution in particle size fractions were recently developed or improved and extensively used in the study.

1. Organic Matter Size Fractionation (OMSF)

The main OMSF method used in this study was described in Feller et al. (1991a). Briefly, it consists in shaking for 2 to 16 hours the 0 to 2 mm soil sample (40 g) in water (300 ml). The presence of a cationic resin (R) saturated with Na^+ improves the soil dispersion. Under these conditions, the pH of the soil suspension during fractionation was close to neutrality, and the OC solubilization was lower than 4%. This was followed by wet sievings at 200 and 50 μm to separate the coarse (200 to 2000 μm) and fine (50 to 200 μm) sand fractions. An ultrasonic treatment (US) of the 0 to 50 μm suspension (100J/ml) improved the clay dispersion. The coarse silt fraction (20 to 50 μm) was obtained by sieving. The fine silt (2 to 20 μm) was separated from clay (0 to 2 μm) by repeated sedimentations and the fine clay (0 to 0.2 μm) was obtained by centrifuging the clay (< 2 μm). C and N analysis were performed by dry combustion with a CHN Analyser (Carlo Erba, Mod. 1106).

The above method provided high dispersion of the soil constituents even with no application of ultrasonic treatment of the whole 0 to 2 mm soil sample. Balesdent et al. (1991) showed that an ultrasonic treatment of the whole soil may lead to an artificial transfer (about 50%) of OM associated with sands (plant debris) into the fine fractions (< 50 μm). The fractionation procedure was applied to 43 surface samples of the selected situations. The OM characteristics associated with the different size fractions may be succinctly described by the following (Feller et al., 1991 a, c):

(i) *fractions > 20 μm.* Predominance of 'plant debris' at different stages of decomposition are dominant with carbon to nitrogen (C/N) ratios ranging between 12 to 33 (mean value, mv = 20);

(ii) *fractions 2-20 μm.* 'Organo-silt complex' consisting of very humified plant and fungi debris associated with stable organomineral microaggregates which have not been destroyed during the fractionation. C/N ratios vary from 10 to 21 (mv = 15);

(iii) *fractions < 2 μm.* 'Organo-clay fractions' with predominance of amorphous OM acting as a cement for the clay matrix. Sometimes, under forest or savanna, presence of plant cell walls occurs in the coarse clay fraction but usually not in the fine clay. Very often, bacterial cells or colonies at different stages of decomposition can be observed in both fractions. C/N ratios vary from 7 to 12 (mv = 10). The microbial origin of amorphous OM in the clay fractions of these tropical samples agree with their very low xylose/mannose ratios (Feller et al., 1991 c).

Depending on the type of soil and soil management, the 20 to 2000 μm fraction, described as the 'plant debris fraction,' represented from 8 to 51% (mv = 26%) of the total OC content. In the 2 to 20 μm fraction the total OC content averaged from 11 to 40% (mv = 26%) and in the 0 to 2 μm fraction OC ranged from 20 to 70% (mv = 44%).

2. Aggregate Size Fractionation (AGSF)

The method applied to the selected samples was described in Albrecht et al. (1992a,b) and derived from methods of Yoder (1936), Williams et al. (1966) and Kemper and Rosenau (1986). It is based on kinetics of soil disaggregation in water under different shaking times. Soil samples were taken with a cylinder to preserve their structure, wetted to field capacity before treatment and shaken in water (35 g/250 ml) end over end (50 t/min) during variable times: 0, 0.5, 1, 2, 6, 12 and 18 hours. A supplementary 'time,' between time 0 and 0.5 hour was applied corresponding to the test of Henin et al. (1960) and consisting of 30 manual turnings (during about 1 min) and was named '30 t.' After shaking, the samples were sieved under water at 1000, 500, 200, 50 and 20 μm, then fractionated at 5 μm by sedimentation. The 'mean weight diameter, MWD' was graphically obtained from the median of the cumulative frequencies curve. Thus, 50% of the soil (by weight) will be in aggregate sizes under the value of 'MWD.'

C. Other Determinations

1. Henin et al. (1960) described an Instability Test (Is) given by the formula:

$$Is = \frac{(A+LF)\ max\ \%}{1/3\ Ag\ \%\ -\ 0.9\ SG\ \%} \quad (1)$$

where '(A+LF) max %' represents the maximum amount of dispersed 0-20 μm fraction obtained after three treatments of the initial soil sample: without pretreatment (air) and with immersion in alcohol or in benzene. 'Ag %' refers to the > 200 μm aggregates (air, alcohol, benzene) obtained after shaking, i.e. 30 manual turnings and sieving under water of the 3 pre-treated samples. 'SG %' represents the content of coarse mineral sand (> 200 μm). The denominator of the formula (1/3 Ag % - 0.9 SG %) is an estimation of the 'mean percent stable aggregates.' It is often expressed by the decimal logarithm form of 10 x Is. This index is well related to the permeability estimated in the laboratory (Henin et al., 1960), and De Vleeschauwer et al. (1979) have shown that Is is also a very good index (among 14 others) of soil detachability in tropical soils.

2. Field rainfall simulation was applied on 1 m^2 surface at sites 8 and 9 with a mini-simulator (Asseline and Valentin, 1978). For the situations studied, the soil surface was prepared for a seed-bed using manual tillage at a 5 cm soil depth. Rainfall simulation was conducted on soils with high water contents to favor runoff (Le Bissonnais et al., 1990).

IV. Organic Matter, Structure and Cultivation in Kaolinitic and Smectitic Tropical Soils

Generally, clearing of the native vegetation followed by cultivation involves dramatic alterations in the morphology of the soil surface, together with a decrease in OC contents and aggregate stability.

A. Modifications of Soil Morphology in Soils under Cultivation

The type of structures encountered in different selected sites under native or long fallow vegetations and annual or sugarcane (*Saccharum officinarum* L.) crops are summarized in Table 3. Four types of situations can be distinguished:

(i) West African soils (sites 1 to 6), with coarse-textured surface horizons. The initial structure more or less massive/crumbly or polyhedral under native vegetation, appeared after 1 to 3 year-cultivation as being single-grain at the 5 to 10 cm soil depth accompanied by the formation of surface crust (Casenave and Valentin, 1989) and compaction of the horizon just below;

Table 3. Description of topsoil structure for the cultivated and non-cultivated study sites

Site	Non-cultivated		Cultivated	
number	Vegetation	Structure[a]	Vegetation	Structure[a]
1	Bush-fallow	Massive to cubic	Sorgho	Single grain and massive + crusts
2	Grass-fallow	Single grain to crumbly (fd)	Millet	Single grain and massive + crusts
3	Tree-savanna	Massive to crumbly (fd)	Peanut and millet	Single grain and massive + crusts
4	Tree-savanna	Massive to crumbly (fd)	nd[b]	
5	Forest	Crumbly (fd)	nd	
6	Forest	Crumbly to polyhedral (md)	Corn	Single grain and massive
7	Artificial meadow	Crumbly to polyhedral (wd)	Market gardening	Crumbly to polyhedral (wd)
8	Forest	Polyhedral (wd)	Food crops	Crumbly to polyhedral (wd)
9	Artificial meadow	Crumbly with prismatic over-structure (wd)	Market gardening	Cubic (large) and prismatic (large) (wd)
10	Grass-fallow	Crumbly to polyhedral (wd)	Food crops	Polyhedral (wd)
11	Forest	Polyhedral (wd)	Sugarcane	Pseudo-single grain and massive to lamelar
12	Natural meadow (campos)	Polyhedral (wd)	Corn, soybean	Pseudo-single grain and cubic to prismatic overstructure

[a]The development of structure is symbolized by : few (fd), moderately (md), or well (wd) developed.
[b]nd = not determined.

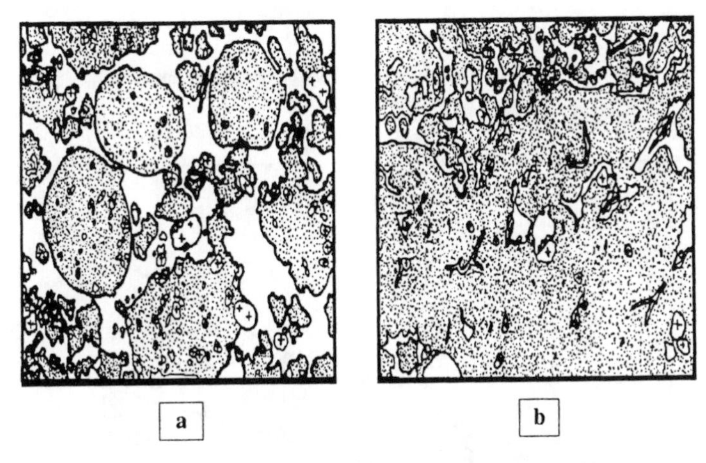

Figure 1. Optic microscopy (x 10) of the upper horizon (A1) of a clayey oxisol (Brazil, site 11) under (a) forest and after clearing and (b) 12 years of cultivation. Note the drastic diminution of macroporosity after cultivation. (Adapted from Cerri et al., 1991.)

(ii) In the clayey kaolinitic/halloysitic soils from volcanic origin (sites 7, 8, 10), a fragmental structure remained well developed with cultivation, more or less polyhedral, with some tendency to form larger overstructure;

(iii) In the highly weathered clayey oxisols of Brazil, the well-developed crumbly and polyhedral structure under native vegetation was rapidly changed (in 3 to 10 years), as in sandy soils, to a single grain-like structure over the 0-10 cm depth consisting of pseudo-sands. Below this depth the structure tended to be massive and/or prismatic or cubic;

(iv) In the magneso-sodic vertisol (site 9), the structure was dramatically modified with cultivation from crumbly in the 0-10 cm, under forest or meadows, to largely prismatic under annual crops.

These different morphological observations for kaolinitic soils are in agreement with those made in coarse-textured soils of Ivory Coast (De Blic, 1976; De Blic and Moreau, 1979), Brazil (Carvalho, 1990), in clayey oxisols of Congo (Mapangui, 1992) or Brazil (Cerri et al., 1991; Chauvel et al., 1991). These three last studies point out that the morphology and properties of the cultivated topsoil (Ap1 horizon) became similar to those of the initial savanna or forest subsoil (A/B horizons), and that after cultivation there was a dramatic reduction in meso- and macropores ($> 5\ \mu$m) (Figure 1). If oxisols and ultisols are considered to have water stable aggregates (Greenland, 1979), as for alfisols, this structural aggregation did not survive the cultivation effects. In contrast, a well-developed structure seemed to be maintained in the younger kaolinitic/halloysitic volcanic soils of the Antilles (sites 7, 8, 10). The structure of the magneso-sodic vertisol was very sensitive to cultivation, while a stable structure

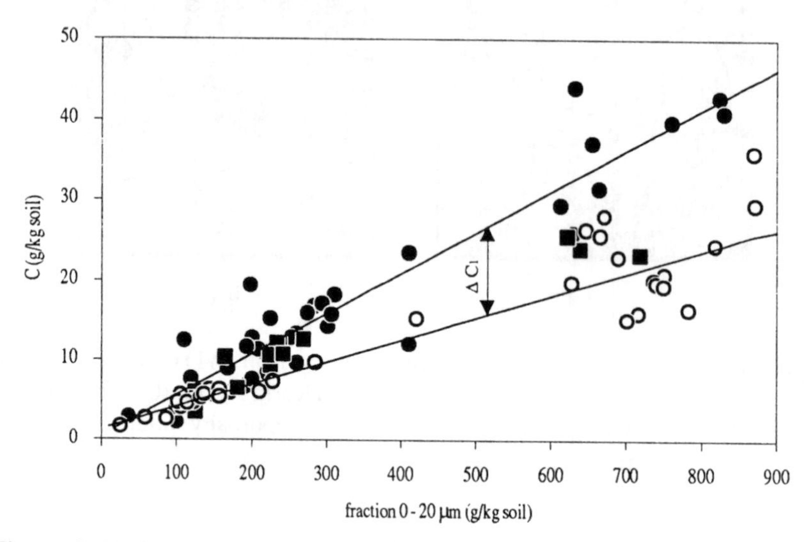

Figure 2. Variations in soil organic carbon (C) content of the 0-10 cm layer in relation with clay and fine silt fraction (0 to 20 μm) content of selected kaolinitic soils of West Africa, Antilles and Brazil. ΔC_1 represents the mean differences in C between the non-cultivated (forest, savanna, pasture) and the continuous cultivated situations.

was generally described (Greenland, 1979) for cultivated calcic or calcareous vertisols.

B. Modifications in OM Contents and Structural Stability with Cultivation

The carbon contents of the 0 to 10 cm soil layers of 59 plots in the selected non-eroded sites are presented in Figure 2 (Feller, 1991d). The relative decrease in OC with cultivation (ΔC_1) was about 40% that of the initial OC whatever the soil texture. This result agrees with numerous studies already quoted in Introduction.

For the cultivated plots the linear regression between OC and 0 to 20 μm fraction % was:

$$C\ (\text{g/kg soil}) = 0.294\ (0\text{-}20\ \mu\text{m}\ \%) + 0.31 \quad (n = 25,\ r = 0.95,\ p < 0.01) \qquad (2)$$

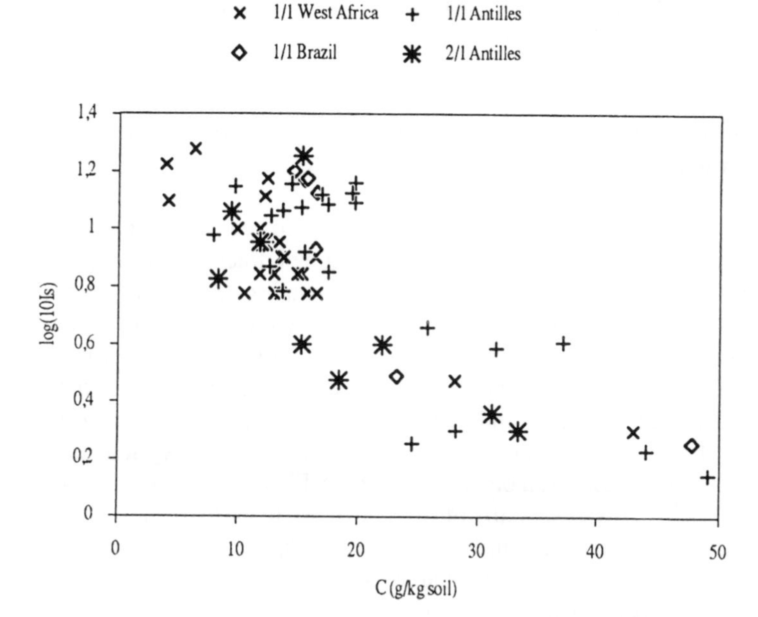

Figure 3. Relationships between the Henin's structural instability index (Is) and the organic carbon content (C) of selected kaolinitic (1/1) and smectitic (2/1) soils of West Africa, Antilles and Brazil. Surfaces horizons 5 to 15 cm. Is is expressed in form of $\log_{10}$ (10xIs). For the Is formula see text.

It was very close to that of Lepsch et al. (1982) for cultivated soils in Brazil:

$$C(\text{g/kg soil}) = 0.325\ (0\text{-}20\ \mu\text{m}\ \%) + 0.77 \quad (3)$$
$$(n = 87,\ r = 0.81,\ p < 0.01)$$

Henin's test (Is) was applied to 60 soil samples from the reported sites. Figure 3 illustrates the high correlation ($r = -0.79$) between log (10 x Is) and C % contents of the surface horizons, while the correlation between log (10 x Is) and 0-20 μm % was non-significant ($r = -0.08$). Significant correlations between Is and OC contents were also given by Combeau (1960), Combeau et al. (1961), Thomann (1963) and Martin (1963) for African oxisols. With the water stable aggregates (WSA) tests, Goldberg et al. (1988) showed significant correlations between WSA and OC for diverse Californian soils. Similar conclusions were provided by Alegre and Cassel (1986) for a fine loamy ultisol of Peruvian Amazon, by Dutartre et al. (1993) for sandy alfisols of West Africa, and by Arias and De Battista (1984) for Uruguyan vertisols.

The relationships between OM and aggregation for each type of soil may be summarized by the detailed analysis of the distributions of both OM and aggregates within the bulk soil.

C. Modifications in OM Distributions in Particle Size Fractions with Cultivation

As seen in Section III.B.1., size-fractionation allowed the separation, in a first approximation, of three types of organic compartments: (i) the plant debris fraction (> 20 μm); (ii) the organo-silt complex (2-20 μm) composed of humified plant and fungi debris and of very stable microaggregates; and (iii) the organo-clay fraction dominated by amorphous OM of partly microbial origin.

The OMSF procedure using Na-resin was applied to samples from the 0 to 10 cm depth on cultivated and non-cultivated plots corresponding to situations 1 to 11. The detailed results on the carbon contained (g C/kg soil) in the three size fractions of each sample were given in Feller et al. (1991d). Here, the differences, ΔC_2 (g C/kg soil), of OC contained in each fraction that appeared with changes in soil management are reported (Figure 4). Different types of soil management were studied (comparisons of Figure 4, a to d) for soils with different textures (Figure 4, a1 to a3, b1 to b3 etc...) or different mineralogies (Figure 4, c3 and c3 bis). The examples show: (i) the decrease in OC contents following clearing and continuous cultivation during 10 years (Figure 4a) or, (ii) the increase in OC contents following: spontaneous fallowing (6 to 10 years) (Figure 4b), 10 years of artificial meadow (Figure 4c) and 4 years of composted straw inputs (Figure 4d).

After continuous cultivation, the decrease in soil OC contents (negative ΔC_2) was mainly due to the decrease of the plant debris fraction in sandy soils (Figure 4a1) and of the OC-clay fraction in clayey soils (Figure 4a3). The 0 to 20 μm fraction and the sandy clayey soil (Figure 4a2) gave intermediate variations. After fallow or meadow, the increase in soil OC contents (positive ΔC_2) was mainly due to the amount of the plant debris fraction in sandy soils (Figure 4b1 and 5c1) and to both plant debris and organo-clay fractions in clayey soil (Figure 4b3 and 4c3). Increases in the 2 to 20 μm fraction were limited. The vertisol (Figure 4c3 bis) showed similar variations to ferrallitic soils (Figure 4c3). Following compost addition, the increase in soil OC contents in sandy soils (Figure 4d1) was largely due to the increase of the plant debris fraction.

All these results suggest that over a medium to a long term scale (> 5 years), the effects of soil management on soil OC variations will differ quantitatively and qualitatively in relation to soil texture. In coarse-textured soils a large part of the OC variations is mainly due to the variations in the plant debris fractions. Similar results were observed by Djegui et al. (1992) and Bacye (1993) for sandy to sandy clayey LAC soils of Benin and Burkina-Faso, respectively. This apparent high turnover rate of plant debris is confirmed for tropical situations, by the study of OC dynamics using the ^{13}C approach (Cerri et al., 1985;

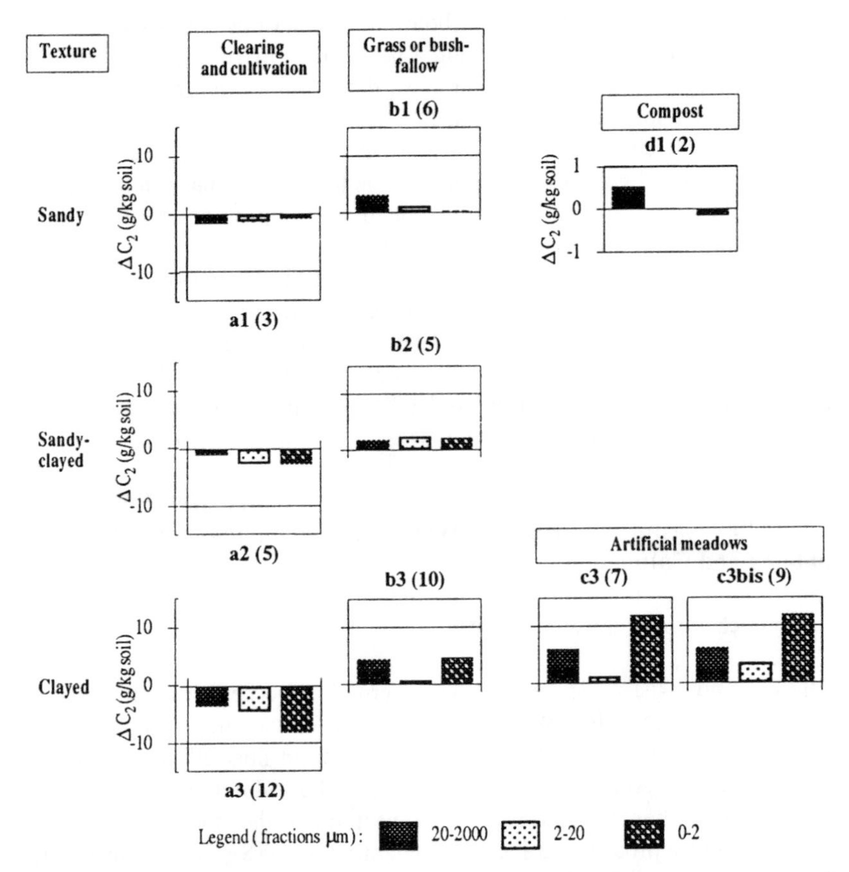

Figure 4. Effect of texture and different types of soil management on the variation (ΔC_2) of organic carbon contained in the 20 to 2000, 2 to 20 and 0 to 2 μm fractions (gC/kg soil) of the 0-10 cm horizons. The corresponding site numbers are indicated in parentheses and refer to Tables 1 and 2. $\Delta C_2 < 0$ = decrease of carbon, $\Delta C_2 > 0$ = increase of carbon. Compost refers to plots fertilized with millet straw compost during 4 years (about 10 t DM/ha/year).

Balesdent et al., 1987 and 1988), for coarse-textured LAC soils (Martin et al., 1990; Trouve et al., 1991; Desjardins et al., 1994) as well as for the clayey oxisol of site 11 (Cerri et al., 1985; Feller et al., 1991b; Bonde et al., 1992). In contrast, in fine-textured soils and especially in clayey soils, the high absolute OC variations are mainly due to the 0 to 2 μm fraction even if fallow or meadow effects explain the decrease or increase in plant debris fractions due to differences in root growth. This implies that a relatively large portion of OC associated with clay fractions is apparently labile. From the results of Martin et al. (1990), Feller et al. (1991b) and Desjardins et al. (1994) obtained using the

^{13}C approach, it was calculated that about 20 to 40% of the OC associated with the clay fraction were renewed during a period of 9 to 16 years.

Combining the data from different sites, the values of OC contained in each fraction (g C/kg soil) are summarized in Figure 5 for non-cultivated (NCULT) and cultivated (CULT) situations, in relation to soil texture (0-2 μm fraction weight). In terms of statistical relationships the following points can be stressed:

(i) In both CULT and NCULT situations there is a positive effect of texture on OC contained in the organo-clay fraction, but;

(ii) The CULT and NCULT situations differ for the organo-silt complex and plant debris fraction (i.e. no significant correlation for CULT, and tendency to significant correlation for NCULT). This might be interpreted in terms of a protective effect of aggregation on the plant debris fraction: significant effect for soils with high structural stability (NCULT situations), no effect for soils with low structural stability (CULT situations).

D. Modifications in Water Stable Aggregate Distributions with OM Contents and Cultivation

1. Approaches to Characterize Water Stable Aggregates

There does not exist a standardized and international accepted approach to the 'Water Stable Aggregate' (WSA) concept, both for the energy applied in the disaggregation technique as well as for the choice of normalized particle size classes. In regard to the energy applied, the shaking duration time may vary from minutes to hours according to the objectives of the studies.

Three main approaches are used to characterize WSA, based on specific aggregate size or whole soil analysis. The most currently used methods for the first approach are based on single- or multiple-sieve techniques, but the considered aggregate sizes may vary considerably. As examples for tropical situations, the smallest diameter is 100 μm for Oliveira et al. (1983) and Alegre and Cassel (1986), 250 μm for Arias and De Battista (1984), Goldberg et al. (1988) and 500 μm for Ike (1986) and Ekwue (1990). From the works of Edwards and Bremner (1967) the size of c.a. 250 μm is often considered as a boundary between macro- ($> 250\mu$m) and microaggregates (< 250 μm) and numerous recent studies of aggregate stability accord a great importance to the WSA larger than 200 or 250 μm (Tisdall and Oades, 1982; Utomo and Dexter, 1982; Arias and De Battista, 1984; Elliott, 1986; Goldberg et al., 1988; Pojasok and Kay, 1990; Miller and Jastrow, 1990; Haynes et al., 1991; Angers, 1992; Carter, 1992; Beare and Bruce, 1993). However, with such methods, information is generally lacking about the process of disaggregation: slaking or dispersion. The second approach to characterize WSA is used in some cases where soils are rich in swelling clays and/or exchangeable sodium and is based on the measurement of the 'dispersed' fractions (0 to 2 or 0 to 20 μm) (Oliveira et al., 1983; Goldberg et al., 1988; Dalal, 1989). Surprisingly, the use of the

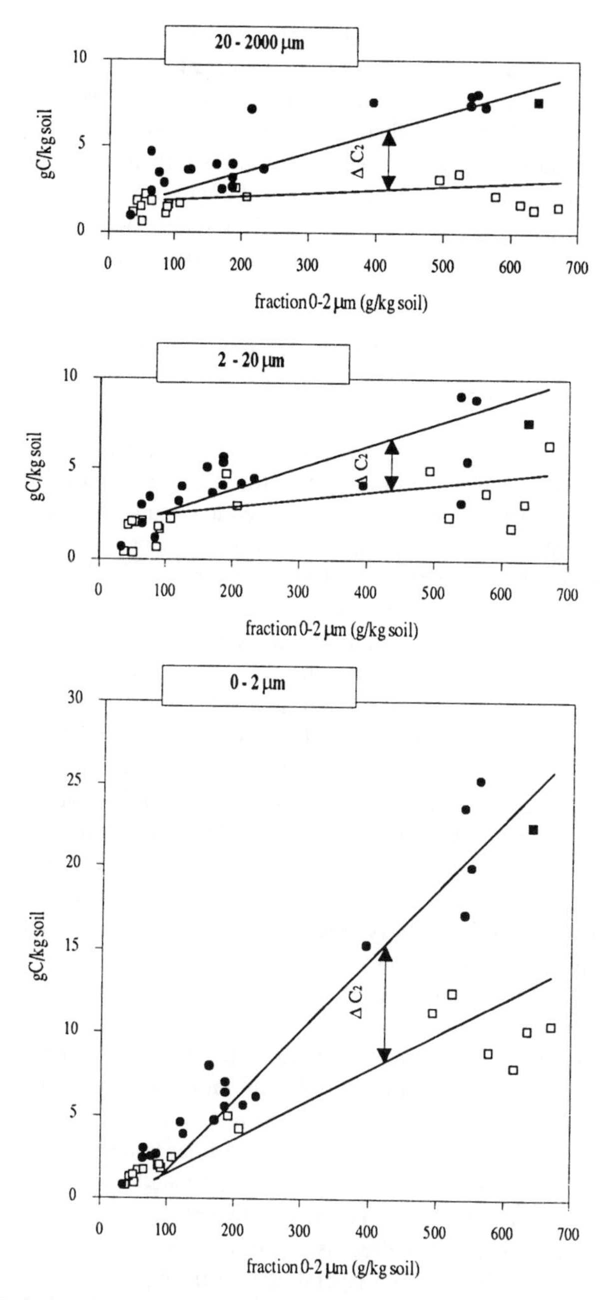

Figure 5. Relationships between the organic carbon (C) contained in the 20 to 2000, 2 to 20 and 0 to 2 μm fractions (gC/kg soil) from cultivated (□) or non-cultivated (●) situations. Horizons 0-10 cm of selected kaolinitic and smectitic soils of West Africa, Antilles and Brazil. ΔC_2 represents the mean differences in C between the non cultivated and the cultivated situations.

third approach to characterize WSA, which consists in whole aggregate size analysis from the macroaggregates to the dispersed 0 to 2 μm fraction on the same sample, and takes into account the energy input level, is relatively scarce. Oades and Waters (1991) published some examples of such a complete approach for WSA distribution, that allowed these authors to discuss the concept of 'aggregate hierarchy' according to soil type. We will present In the next section data (Albrecht et al., 1992b and unpublished) obtained with such an approach for cultivated and non-cultivated soils of West Africa, Antilles and Brazil.

2. Water Stability of Macro- and Microaggregates of Selected Kaolinitic and Smectitic Tropical Soils

The methodological aspects of this section are described in section III.B.2. Simplified results are presented in Figure 6 for three main aggregate classes: the macroaggregates larger than 200 μm, the secondary microaggregates 5 to 200 μm, and the primary microaggregates 0 to 5 μm. For each situation a non-cultivated (NCULT) soil (forest, savanna, grass or tree-fallow, meadow) rich in OC is compared with a cultivated (CULT) and poor in OC soil.

The kinetics of the disaggregation of macroaggregates differ according to the soil type and the soil OC content. For the sandy clayey West Africa soil and the vertisol, macroaggregates are destroyed in about 0.5 to 1.0 hour, while 6 hours at least are necessary for the clayey LAC soils of Antilles and Brazil. The effect of OC contents varies also with the soil type: it had a relatively low effect on soils with low macroaggregate stability (West Africa LAC soils and Vertisol), but a more important effect for the other soil types. The stability of secondary microaggregates (5 to 200 μm) differs with soil type also. For the sandy clayey soil of Africa, few microaggregates remain after 2 hours (values close to that of the mechanical analysis), whatever the OC content. For the vertisol, the stability of microaggregates depends on the OC content of the sample, whereas the clayey LAC soils of Antilles and Brazil display high microaggregates stability, whatever the OC content.

The examination of the curves for the primary microaggregates (0-5 μm) provide precise information on the relative importance of the slaking effect (disruption of macroaggregates into secondary microaggregates) and the dispersion effect (disruption with dispersion of finer colloidal materials) during the disaggregation process. For example, the cultivated vertisol is characterized by a large and immediate (t_0 or 30 turnings) dispersion effect compared with the other cultivated soils, but this effect is strongly reduced with a high OC content. In the other soils, significant dispersion effects appear with larger shaking durations.

Finally, to characterize the water stability of aggregates, it seems important to take into consideration: (i) the different classes of macro- (> 200 μm), secondary micro- (5 to 200 μm), and primary microaggregates (0 to 5 μm), and (ii) the level of energy input utilized in a kinetic approach. The latter may be

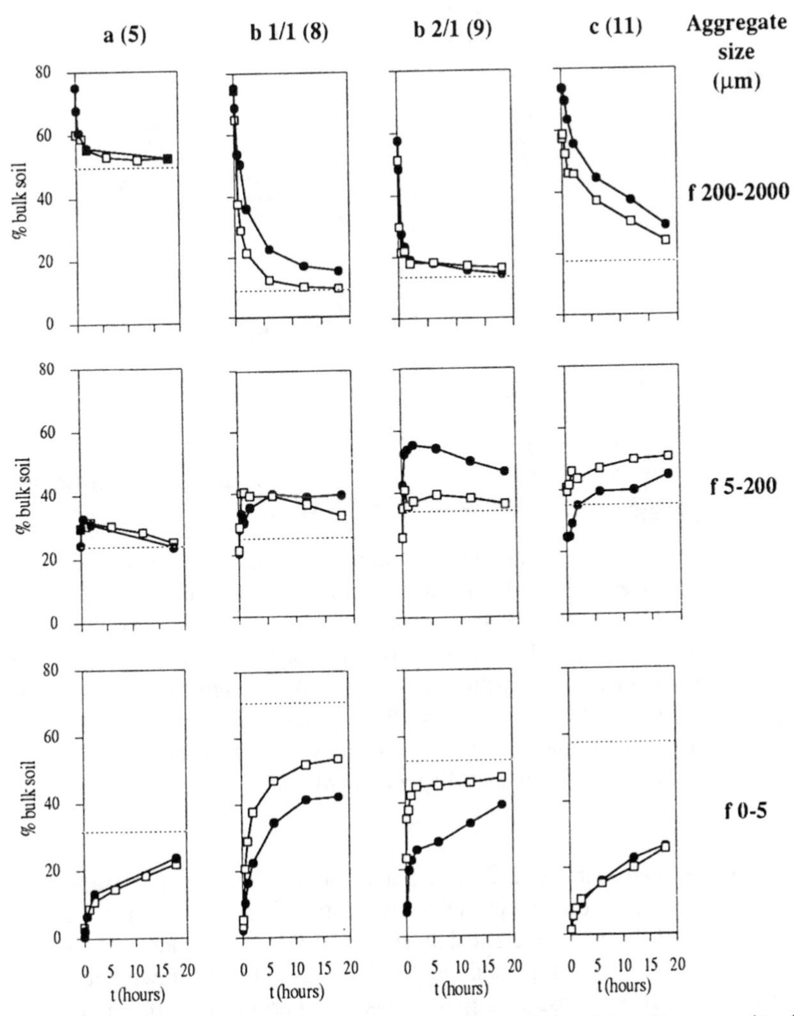

Figure 6. Variations with the shaking time (t) of the weight of macro- (> 200 μm) and micro- (5-200 and 0-5 μm) aggregates of the 0-10 cm layer of selected kaolinitic (1/1) and smectitic (2/1) soils of West Africa (a, site 5), Antilles (b, sites 8 and 9) and Brazil (c, site 11). Black signs (●) are rich OM samples: a = savanna, b = pasture, c = forest. White signs (□) are poor OM samples: a = 10-year food crops, b = 10-year market gardening, c = 12-year sugarcane. The dotted line (...) represents the weight of the corresponding size fraction obtained by mechanical analysis (H_2O_2 treatment and dispersion).

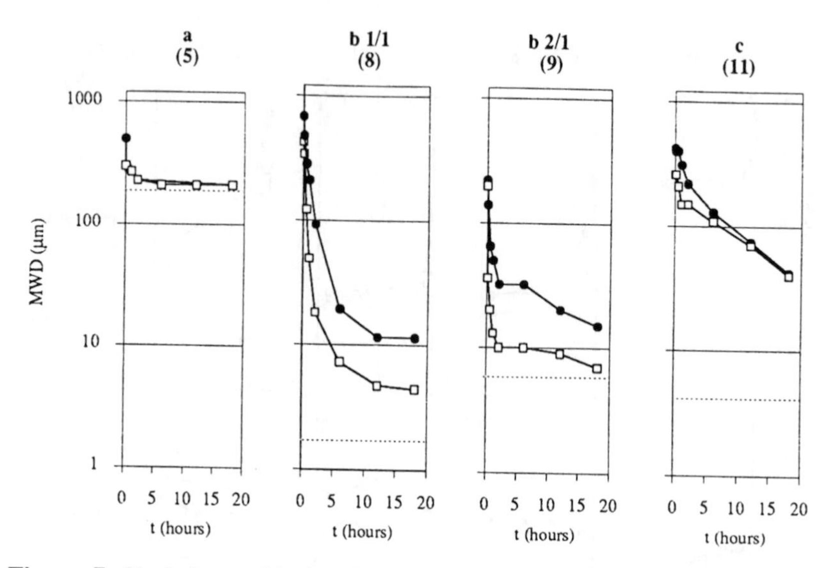

Figure 7. Variations with the shaking time (t) of the mean weight diameter MWD (μm) of selected samples from kaolinitic (1/1) and smectitic (2/1) soils of West Africa (a), Antilles (b) and Brazil (c). The corresponding site's number refers to Tables 1 and 2. Horizons 0-10 cm. Black signs (●) are rich OM samples: a = savanna, b = pasture, c = forest. White signs (□) are poor OM samples: a = 10-year food crops, b = 10-year market gardening, c = 12-year sugarcane. The dotted line (...) represents the MWD obtained by mechanical analysis after H_2O_2 treatment and dispersion (MWDm).

very informative in characterization of water stable aggregates. For example, an integrated representation of the WSA distribution may be obtained from the kinetic curve of the mean weight diameter (MWD) of the sample (Figure 7). The differences with soil type and OC level are also clearly shown with such a representation. If we consider a given shaking duration (i.e. 0.5 hour) there exists, for the LAC samples, a highly significant positive correlation ($p < 0.01$) between $\Delta MWD_{0.5}$ and OC content. ΔMWD is the mean weight diameter of the 'aggregated material' and is calculated by the difference between $MWD_{0.5}$ and MWD_m, if MWD_m represents MWD obtained by mechanical analysis (after H_2O_2 treatment and dispersion). Significant correlations ($p < 0.05$ or < 0.01) between ΔMWD and OC were also reported by Alegre and Cassel (1986) for an ultisol of Peruvian Amazon after clearing and a subsequent 2 year-cultivation, even with relatively low variations (20%) of OC contents; and by Arias and De Battista (1984) for vertisol of Argentina; and by Haynes and Swift (1990), Angers (1992), Carter (1992) for non-tropical soils.

3. 'Aggregate Hierarchy' in the Studied Tropical LAC Soils

Oades and Waters (1991) have proposed an evaluation of 'hierarchy' for aggregate stability based on the particle size distributions of aggregates destroyed between two treatments of different energy. According to these authors, the concept of 'aggregate hierarchy' is based on the 'principle that disaggregation will occur due to planes of weakness between stable aggregates'.... 'Therefore, it should be possible to determine whether an hierarchical order of aggregates exists or not by systematic studies of disaggregation as disruptive energy is increased. If aggregates breakdown in a stepwise fashion then hierarchy exists. If aggregates breakdown to release silt and clay size materials directly then there is not an hierarchical order '....' The concept may not apply to clods formed from strongly sodic clays in which planes of weakness are rare.' The Oades and Water's examples are reported in Figure 8d for selected oxisol, mollisol and alfisol. Two types of curve can be described, one with a 'maximum' (mollisol and alfisol), the other with a plateau (oxisol). The first type indicates that as the disruptive energy increases during the destruction of the macroaggregates a dominant class of stable microaggregates appears. Their mean diameter corresponds to the maximum diameter. For lower diameters, the regular decrease of the curve denotes the presence of different classes of smaller stable microaggregates but without significant release of clay particles or clay-size microaggregates. This hierarchical order of stable aggregates with limited dispersion processes corresponds to the concept of 'Aggregate Hierarchy.' In contrast, the presence of a plateau illustrates the absence of an hierarchical order for the stable microaggregates and the existence of important dispersion processes together with the apparition of clay particles or clay-size microaggregates.

Such an approach was applied to our situations for non-cultivated or cultivated situations under approximately the same experimental conditions as those of Oades and Waters (1991) in using data of the t_0 and t_{16} hour treatments (Figure 8a to 8c), after 16-hour shaking treatments. A tendency to an aggregate hierarchy appeared for the LAC soils (sites 5, 8, 11), cultivated or non-cultivated, but not at all for the vertisol (9). The hierarchy for LAC soils is not so clear as that in the mollisol and alfisol studied by Oades and Waters (1991). This contrasted behavior is due to the existence of an high dispersion effect at 16 hours for all the LAC soils as shown in Figure 6: the % weight differences for the 0-5 μm fraction t_0 and t_{16} hours vary from between 20 to 50 %. However an OC effect corresponding to the maximum of the curves (Figure 8) is visible on the aggregate size: for non-cultivated LAC soils (rich in OC) the stable aggregate diameter is close or larger than 100 μm, while for cultivated LAC soils (poor in OC) it is comprised between 20 to 50 μm. Using a 'fractal' approach, Bartoli et al. (1992) have demonstrated that the size of the elementary structural pattern of oxisols decreases as a function of OM content. This result is in agreement with the hypothesis that OM leads to the development of an

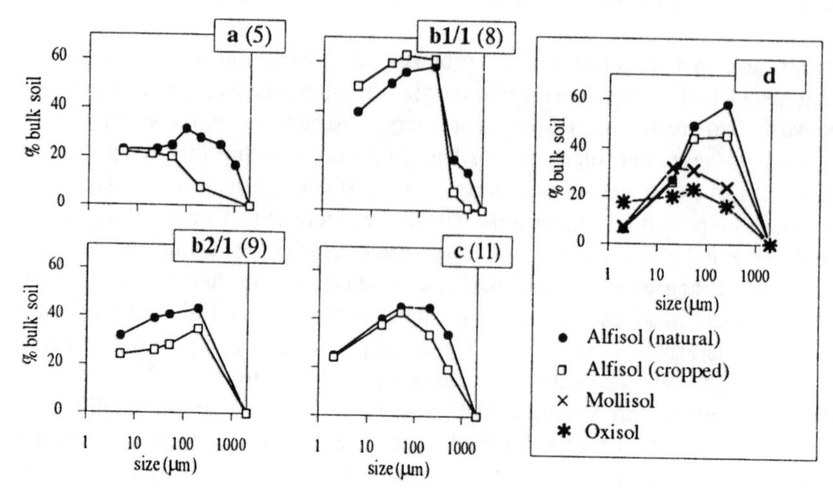

Figure 8. Application of the 'aggregate hierarchy' approach (Oades and Waters, 1991) to selected kaolinitic and smectitic soils of West Africa (a, site 5), Antilles (b, sites 8 and 9) and Brazil (c, site 11). Figures a, b and c may be compared with the figure d adapted from the Oades and Waters' results obtained with uncropped and cropped alfisol (South Australia), mollisol (Canada) and oxisol (Queensland, Australia). The alfisol and mollisol curves are characteristics of samples showing an "aggregate hierarchy"; it is not the case for the oxisol curve. The y scale (8d) represents the weight difference (%) between t_0 (with 0 is the shaking time in hours) and t_{18} (or t_{16} for the figure d) calculated from the corresponding cumulative frequencies curves. Black signs (●) are rich OM samples: a = savanna, b = pasture, c = forest. White signs (□) are poor OM samples: a = 10-year food crops, b = 10-year market gardening, c = 12-year sugarcane.

'aggregate hierarchy.' More studies are yet necessary to confirm the validity of such a concept for tropical soils.

4. Conclusions to the Effects of Cultivation on the Distributions of OM and Water Stable Aggregates

The different observations and results presented here for the tropical situations may be summarized as follows:

(i) With changes in soil management, the resulting changes in soil OC distribution are closely related to soil texture whatever the soil mineralogy: for coarse-textured soils, variations concern mainly the plant debris fraction, for fine-textured soils variations concerned both the plant debris and organo-clay fractions;

(ii) In terms of water stable aggregate distribution the consequences are more complex and may also take into consideration the characteristics of the mineralogical and ionic environment. For the LAC, coarse-textured West Africa kaolinitic soils (Figure 6a), the macroaggregates are unstable. The OC effect (plant debris effect) is only significant at a low energy input (up to 30 turnings) and acts on the slaking process (no differences in the dispersed 0 to 5 μm fraction). For the kaolinitic / halloysitic clayey soils of Antilles (Figure 6b 1/1), the macroaggregates are relatively stable. The OC effect acts first (plant debris OC effect) on the slaking process at a low energy input, then on the dispersion process (organo-clay OC effect) at high energy inputs. For the Brazilian oxisols the macroaggregates appear very stable. The OC effect (both plant debris and clay OC effects) acts only on slaking, the stabilization of clay fraction (0 to 5 μm) in secondary microaggregates (5 to 200 μm) being more controlled by oxides than by humified and amorphous OM. For all types of the studied LAC soils, OC content affects the 'aggregate hierarchy.' In regard to the magneso-sodic vertisol of Antilles (Figure 8b 2/1), the macroaggregates are very unstable. The OC effect acts mainly by a dramatic limitation of the dispersion process (clay OC effect).

E. OM Distribution in Water Stable Aggregates in Relation to Soil Type and Cultivation

WSA distributions vary according to the applied disruptive energy, the soil type and the cropping system (or soil OC content). Therefore the OC-aggregate distribution (OC in % of total soil OC) will depend on these three factors. Figure 9 gives examples for a magneso-sodic vertisol (site 9) and a clayey LAC soil (site 8). For each soil we compared plots corresponding to market-gardening (m) or intensified pastures (p) and soil surface samples submitted to low (AGSF at time 0, t_0) or high (AGSF at time 6 hours, t_6) disruptive energy, the maximum disaggregation effect being obtained with the organic matter size fractionation (OMSF).

In the case of the vertisol, at time 0 with market-gardening, the OC-aggregate distribution is already close to that obtained by OMSF, with the highest OC-aggregate percentage (55%) found in the primary microaggregates (> 5 μm). This result is completely different to that of the clayey LAC soil in which the highest OC-aggregate percentage (75%) is found among the macroaggregates (> 200 μm). Under pastures (high OC content sample) the OC-aggregate distributions at time 0 was rather similar for the two soils with a dominant location of OC in the macroaggregates (> 200 μm). Even after a 6-hour shaking the OC-aggregate distribution in the vertisol under pasture remains very different from the distribution obtained by OMSF, the highest OC-aggregate percentage (47%) being found in the microaggregates (5 to 200 μm).

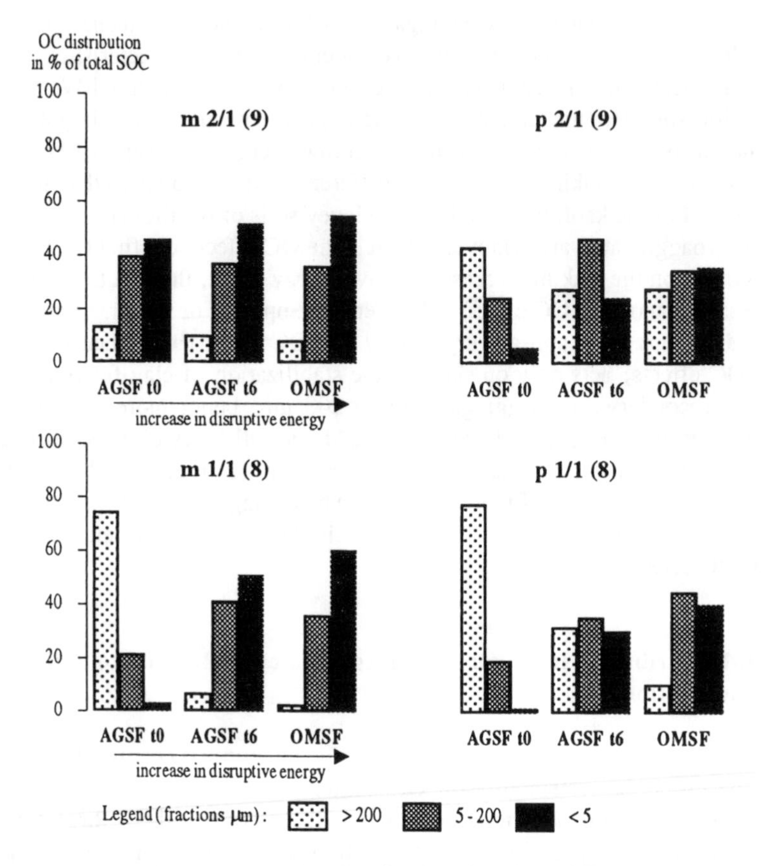

Figure 9. Organic carbon (C) distribution in the different water stable aggregates obtained after 0 (AGSF, t_0) and 6 hours (AGSF, t_6) shaking of a vertisol (site 9) and a clayey LAC soil (site 8). Plots corresponding to market gardening (m) or pasture (p). Comparison with the OC distribution obtained by organic matter size fractionation (OMSF). Results expressed in % of the total soil C.

V. Soil Biota and Organic Constituents and the Stabilization of Structure

Modifications in structural stability are apparently strongly dependent on OM contents. However, in several instances, changes in OM due to soil management are very often accompanied by changes in soil biological activity such as root development, microbial (algual, bacterial and fungal) or faunal activities. For example, in a vertisol (site 9) and a kaolinitic/halloysitic soil (site 7), the positive effects of meadow on aggregate stability (Is) may be due not only to an increase in total OM or in specific organic constituents but also to an increase in root biomass, bacterial activity (as reflected by C and N mineralization) and/or macrofauna activity such as earthworms (Table 4). The following illustrates some effects of soil biota and specific organic constituents on the structural stability of tropical soils.

A. Role of Soil Biota in the Stabilization of Soil Structure

1. Roots and Microbial Effects

Australian soil scientists have developed models of different organic and biological binding agents as well as basic mechanisms involved in the stability of soil structure (Tisdall and Oades, 1980 and 1982; Oades, 1984 and 1993; Oades and Waters, 1991). They also examined the implications as far as soil management is concerned. Two levels of soil aggregation are generally considered:

(i) The soil macroaggregates ($> 250\ \mu m$) which are mainly stabilized by a 'packing' effect due to the network of living roots and hyphae, particularly the vesicular-arbuscular mycorhizal (VAM) hyphae. These agents, classified as temporary binding agents, are more or less associated with transient polysaccharides. This approach emphasizes the role of soil management, i.e. crop rotation and agricultural practices, on soil structure;

(ii) The microaggregates ($< 250\ \mu m$) are mainly stabilized by more persistent organic or organometallic binding agents including small plant and fungal debris, aromatic humic materials and amorphous OM polysaccharides, originated from root, fungi and bacteria. The microaggregate stability could be an intrinsic characteristic of the soil, not greatly dependent on soil management.

This *in situ*-approach was confirmed by Dorioz et al. (1993) with *in vitro* laboratory experiments conducted to determine the respective effects of fungi, bacteria and roots on the arrangement of kaolinitic and smectitic clay particles. Dorioz et al. (1993) distinguished the following effects:

(i) 'Packing' effects due to the network of fungal hyphae and fine roots and their consequences on the 50 to 200 μm aggregate genesis;

(ii) 'Mechanical' effects which contribute to modified particle arrangements either by root and hyphae penetration and compaction associated with dehydration and reorientation of clay particles. It is stressed that bacterial and fungal cells and roots are located in the 5 to 20 μm and 50 to 200 μm sites, respectively;

(iii) 'Polysaccharidic' effects with the formation of a clay-polysaccharide complex (CL-PS) on a 1 μm scale for bacteria and 50 μm for roots.

Important changes in particle arrangement of kaolinite were observed while with montmorillonite an adhesion of cells and/or polysaccharides to the quasi-crystals was observed. The effect of polysaccharides from different origin (vegetal, fungal, microbial) was also reported in details by Chenu (1993).

Table 4. Comparisons of some soil properties (0-10 cm) of a vertisol (site 9) and a clayey LAC soil (site 7) under pasture (p) or market-gardening (m)

				Biomass		Mineralization tests	
			SOC[a]	Root	Macrofauna	C	N
Site number	Vegetation	Henin's test	(g C kg^{-1} soil)	(g C kg^{-1} soil)	(g m^{-2})	(mg kg^{-1})	(mg kg^{-1})
9	m	0.70	11.6	0.12	3.1	650	21
	p	0.27	32.9	10.1	366.4	1760	179
7	m	1.24	22.7	nd[b]	3.3	650	96
	p	0.14	40.1	nd	52.2	1430	255

[a]Soil organic carbon; [b]not determined.

These *in situ* or *in vitro* observations are in good agreement with our own descriptions of size OM fractions, their locations in soils or aggregates of tropical situations and the relation between OM and WSA distributions (this chapter, and Chotte et al., 1993). In contrast, the idea that macroaggregation would be mainly dependent on soil management and microaggregation on intrinsic soil characteristics does not fit well with results reported in section IV.C for the vertisol and oxisol (Figure 6b 2/1 and 6c). For these two situations the stability of secondary microaggregates is also largely dependent on OC content and soil management.

2. Faunal Effects

Earthworms and termites represent by their biomass and their activities the most important soil macrofauna communities in relation to soil behavior and properties as well as SOM dynamics and physical properties (Garnier-Sillam, 1987; Lavelle, 1987; Lal, 1988; Lavelle et al., 1992; Brussaard et al., 1993). With land management practices, soil macrofauna diversity and/or activity are assumed to change as mentioned above (Table 4) and reported recently by Lavelle and Pashanasi (1989) and Fragoso et al. (1992) for numerous tropical situations.

For both earthworms and termites, it is necessary to distinguish three main functional groups depending on feeding (surface litter or SOM including roots) and location of animals in the profile (Lavelle et al., 1992). *Epigeics* live and feed in the litter. *Endogeics* live in the soil and feed on SOM and mainly dead roots (living roots seem to be a poor resource). *Anecics* live in the soil (or in epigeic nests for some termites) but feed in the litter. Endogeics and anecics exert the most important influence on SOM dynamics and physical properties. Anecics can enrich the upper soil layer in OC by mixing soil with litter debris, while endogeics do not. Therefore, this last group, especially earthworms, can be used in controlled experiments to show specific faunal effects on aggregation with a limited interference with exogenic-soil OM inputs.

3. Earthworms and Soil Structure

In the field, earthworm casts are generally richer in OC and have a higher aggregate stability than the non-ingested soil due either to the incorporation of litter OM or to a selection of finer soil particles (Roose, 1976 and Fritsch, 1982 results quoted in Feller et al., 1993; Mulongoy and Bodoret, 1989). As structural stability is strongly correlated to OC, it is difficult to attribute a specific effect of earthworms on *in situ* structural stability. Hence, controlled experiments were generally conducted with endogeic earthworms which dominate in the humid tropics. Blanchart (1990) estimated that the weight of new endogeic earthworm aggregates represent about 50 to 60% of an alfisol surface horizon under savanna in Ivory Coast.

The specific effect of an endogeic pantropical earthworm (*Pontoscolex corethrurus*) on the structure of a vertisol was clearly demonstrated from electron microscopic strudies by Barois et al. (1993). They showed that the soil structure was first completely destroyed during the transit in the anterior gut by the swelling and dispersion of the clay particles in a semi-liquid medium rich in free polysaccharides. At the end part of the gut and in casts, the soil is restructured with the formation of new aggregates due to interactions between clay particles, polysaccharides and microbial colonies. Because the material was dispersed during the gut transit, a certain instability of the fresh casts compared to that of the non-ingested soil was observed. In contrast, after drying and ageing, the aggregate stability of the casts was higher in relation to their OC contents and colonization by fungal hyphae (Shipitalo and Protz, 1988; Marinissen and Dexter, 1990). This positive and specific effect of endogeic earthworm on aggregation and stability was also demonstrated for poorly structured alfisol in Ivory Coast by Blanchart (1992) and Blanchart et al. (1993). After 28 months, the water stability of aggregates (> 2 mm) increased in the different treatments according to the following order:

Milsonia anomala > *Eudrilidae* > Control (Natural Savanna) > No Earthworms.

This increase in water aggregate stability may be accompanied by an increase in bulk density and decrease in mean pore sizes (De Vleeschauwer and Lal, 1981; Blanchart et al. 1993), but Martin and Marinissen (1993) quoted other contradictory results in the literature (Joschko et al., 1989; Elliott et al., 1990). Brussaard et al. (1993) also observed in a Nigerian alfisol an important decrease in voids > 30 μm which was attributed to low earthworm activity especially when mulch practices were absent.

4. Termites and Soil Structure

The study of morphological characteristics of the edaphic material in termite mounds is well documented (Sleeman and Brewer, 1972 ; Kooyman and Onck, 1987), but there are few studies on termite soil aggregates. Garnier-Sillam et al. (1985) described the microaggregates from the feces of four species of termites, and Eschenbrenner (1986) emphasized the similarity of aggregates between termite-inhabited soils and mounds. According to Lavelle et al. (1992) much progress remains to be made in this area.

The effects of termites on structure may be variable. In a Congo ultisol, Garnier-Sillam et al. (1988) reported a positive effect of *Thoracotermes macrothorax* on aggregate stability in relation to an enrichment in OM. With *Macrotermes mülleri*, mound constructions were poor in OM and aggregate stability was low in comparison to those of the soil A horizons. According to Roose (1976) and Janeau and Valentin (1987) termites may also cause soil crusting and therefore increase runoff.

B. Role of Specific Soil Organic Constituents in the Stabilization of Soil Structure

This subject can be studied on various scales of soil organisation: (i) bulk soil sample, and (ii) aggregates or size fractions. Experimental models were also used to better understand involved mechanisms, especially organo-clay interactions. As few data are available on tropical soils, examples of both tropical and temperate situations will be mentioned.

1. Studies on Bulk Soil Samples

Diverse families of soil organic compounds were studied in respect to their aggregative properties. Using classical chemical studies of SOM, Combeau (1960), Martin (1963) and Thomann (1963) found that the structural stability of tropical LAC soils was more related to the humin fraction (total carbon minus acid and alkali extractable fractions) than to fulvic or humic acids. This result is in agreement with the data of Dutartre et al. (1993) for sandy tropical alfisols who showed that stable aggregation may be related to high contents of humin, uronic acids, osamines and polyphenols. In contrast, for non-tropical soils, Chaney and Swift (1984) and Piccolo and Mbagwu (1990) attributed a positive effect to humic and fulvic acids in the stabilization of structure.

Giovanni and Sequi (1976 a,b) and Wierzchos et al. (1992), used organic solvents (acetylacetone/benzene) to isolate organic constituents involved in aggregation. They emphasized the role of the associated mineral cations, especially iron and aluminium. Hamblin and Greenland (1977) showed that organic materials removed by acetylacetone or pyrophosphate had a more important effect on aggregate stability than polysaccharides. On the other hand, other organic constituents such as phenolic substances (Griffiths and Burns, 1972) or SOM 'alphatic fraction' extracted by super-critical hexane and characterized by hydrophobic properties (Capriel et al., 1990) were also considered efficient in the stabilization of structure. Hayes (1986) also estimated that the most efficient molecules in maintaining soil stability would have linear or linear helix conformations in solution, for such conformations would span the longest distances.

However, most of the studies on soil stucture stability were concerned with polysaccharides (PS), and it was shown that these compounds act as glues inside soil aggregates (review in Cheshire, 1979 and Tisdall and Oades, 1982). Nevertheless, their relative importance in comparison with others compounds (see above) may be questioned (Hamblin and Greenland, 1977). PS mainly act as transient compounds on a short-term scale (Tisdall and Oades, 1982), the correlations between total carbohydrate contents and aggregate stability were not necessarily significant (Baldolk et al., 1987). Based on the sugar composition of the PS fraction, Cheshire et al. (1983 and 1984) concluded that PS of microbial origin were more efficient than plant-derived PS. This agrees with the results of Sparling and Cheshire (1985) who observed that the polysaccharidic effect on aggregate stability was less important in rhizosphere than it was in the non-rhizospheric soil because a large part of PS might be in the form of plant remains and debris. However Benzing-Purdie and Nikuforuk (1989) stressed the importance of the PS of plant root origin for soil aggregation. For a New Zealand cambisol and inceptisol, Haynes and Swift (1990) and Haynes et al. (1991) found that aggregate stability was more closely correlated with hot water-extractable carbohydrate content than with total OC or hydrolysable (HCl) or NaOH-extractable carbohydrate contents. They concluded that hot water-extractable carbohydrate fraction may represent an important fraction for the formation of stable aggregates. For a clayey LAC soil of Antilles (site 8), Feller et al. (1991c) showed, using ultramicroscopic studies and metal-stained techniques (Thiery, 1967; Foster, 1981), that hot water extracted a large amount of the stained amorphous OM associated with clay matrix. This extract was very rich in HCl-hydrolysable carbon and its sugar composition was characterized by a very low (< 0.1) xylose to mannose ratio consistent with a microbial exudate origin for this fraction. A same tendency (low xylose / mannose ratio) was also observed (Feller, unpublished results) for a vertisol (site 9). This may emphasize the role of microbial PS origin in the stabilization of the structure of clayey tropical soils.

2. Studies on Water Stable Aggregates

The majority of studies on the organic composition of WSA generally deals with measurements of their OC and carbohydrate contents, and sometimes with their neutral sugar composition. Of course, other organic components have been studied, e.g. humic substances (Piccolo and Mbagwu, 1990).

The literature is rich in conflicting results for OC and carbohydrate contents as well as for sugar composition. In some studies, variations in OC and/or carbohydrate contents do not appear to follow a clear trend (Dormaar, 1987; Piccolo and Mbagwu, 1990). Tisdall and Oades (1980) reported for different plots on a red brown earth the following trend : high OC contents (1 to 2%) for > 250 μm, low OC contents (0.3 to 0.7%) for 20 to 250 μm and medium OC contents (0.7 to 1.7%) for < 20 μm aggregates. A decrease in OC and/or carbohydrate contents with a decrease in aggregate size was reported by Haynes and Swift (1990) for inceptisols (comparison of > 2.0 and < 0.25 mm), by Oades and Waters (1991) for a mollisol and an alfisol, and by Dormaar (1984) for a chernozem (size aggregates from 2.0 to 0.1mm). For a silt loam brunisol, Baldock et al. (1987) showed that the carbohydrate content remained constant for 8.0 to 1.0 mm aggregates but increased for lower aggregate sizes (1.0 to 0.1 mm). The same problems in variation appear in the comparison of the relative sugar composition (content of a single sugar in % of total sugar in each aggregate class) among different aggregate size classes. Dormaar (1984) did not find differences in the relative sugar composition between various aggregates. By contrast, Baldock et al. (1987) observed that in continuous bromegrass (*Bromus inermis* Leyss) the contribution of plant carbohydrates increased as aggregate size decreased, while in a continuous grain-corn (*Zea mays* L.) plot the reverse was noted.

For all these aspects, one problem encountered is that the studies generally do not take into consideration the mineral and organic heterogeneities in the composition of WSA, i.e. the mineral sand and the plant debris (or light) fractions. In order to interpret aggregate OM data, Elliott et al. (1991) proposed to correct the sand and organic light fractions of every aggregate size class. This was applied by these authors on a tropical cultivation chronosequence of Peruvian ultisols, and they demonstrated that the OC concentration of the 'heavy' fraction was, with few exceptions, not different among size classes or treatments. In the same way, Albrecht et al. (1992b) reported for a vertisol (site 9) and a clayey LAC soil (site 8) that the WSA obtained after 6 hours of end over end shaking presented about the same concentrations (20 mg C/g aggregate) for aggregate size ranging from 5 to 2000 μm when the light fraction was excluded. This might be also applied to the relative sugar composition of different aggregates for it was often shown, for temperate (Whitehead et al., 1975; Murayama et al., 1979; Turchenek and Oades, 1978; Cheshire and Mundie, 1981; Angers and Mehuys, 1990; Angers and N'dayegamiye, 1991) as well as for tropical soils (Feller et al., 1991c and unpublished results), that the *in situ* sugar composition of the size OM fraction varies systematically: a decrease in the xylose/mannose (or xylose + arabinose/galactose + mannose) ratio from the plant debris (or light) fraction to the organo-clay fraction. That decrease is consistent with a decrease in the participation of plant-PS (or an increase in the participating PS from microbial origin) in aggregate stabilization.

In conclusion, these apparently contradictory results have two main implications for soil structure research: (i) it is important to standardize the methodological and conceptual approaches to characterize the so-called 'water-stable aggregates;' and (ii) as water-stable aggregates are mixtures of mineral particles, particulate OM (plant and fungi debris) and organo-clay and silt complexes, research must be focused on aggregate composition in terms of size OM fractions. As the OMSF includes OC and mineral particle size distributions and the characterization of organic fraction of different qualities, it allows the direct studies of OM location in aggregates.

VI. Role of Soil Structure in the Stabilization of Soil OM

In the preceding section, we have seen that for a given pedological situation, the higher the OC content, the higher the structural stability. As a feedback effect, could a high aggregation stability stabilize OM? Two types of processes are usually invoked for the *in situ* stabilization of soil OM: (i) limitation of mineralization processes; and (ii) limitation of erosion processes. These two aspects will be now discussed for tropical situations.

A. Role in the Mineralization Processes

As outlined by Ladd et al. (1993), 'electron microscopy studies (SEM, TEM) have provided the visual evidence to reinforce conclusions drawn from other studies that physical protection mechanisms are important determinants of the stability of organic matter in soil.' Ultramicroscopic observations (TEM) of tropical vertisol under pasture (Figure 10) agree with this 'visual evidence:' plant cell wall debris, bacteria colonies or amorphous OM can be protected from decomposer organisms in microaggregates by a surrounding dense clay fabric. Diverse experimental approaches relevant to a physical protection of OM were recently reviewed by Ladd et al. (1993) for non-tropical situations. Their general conclusions are that in either increasing clay content and/or increasing structure stability, there are 'limitations in the accessibility of substrates to decomposer microflora and of microorganisms to microfaunal predators, by virtue of differences in their pore size location within aggregates.'

As both texture and structure affect the pore space and the pore size distribution in the soil, it is often difficult to distinguish between a 'texture effect' (clay or < 2 μm primary microaggregate contents) and an 'aggregation effect' (> 2 μm secondary stable aggregate contents) on the physical OM protection. Although enough significant data are lacking on these aspects for tropical situations we shall present some partial, indirect and perhaps contradictory results to illustrate the need of future research. We shall try arbitrarily to distinguish the 'texture effect' from the 'structure effect.'

1. The 'Texture Effect'

From Figure 2 it appears that clay content plays an important role in the storage of OC. For non-cultivated as well as for cultivated situation, the OC storage depends on the soil management and differs according to the soil OM size fraction (Figure 4). In *cultivated situations*, with a low aggregate stability, organic carbon storage is dependent on texture mainly for the organo-clay fraction, while OC contents of plant debris (20 to 2000 μm) and silt complex (2 to 20 μm) fractions are relatively constant in the range of the texture studied. There are no protective effects from the clay content on the > 2 μm fractions.

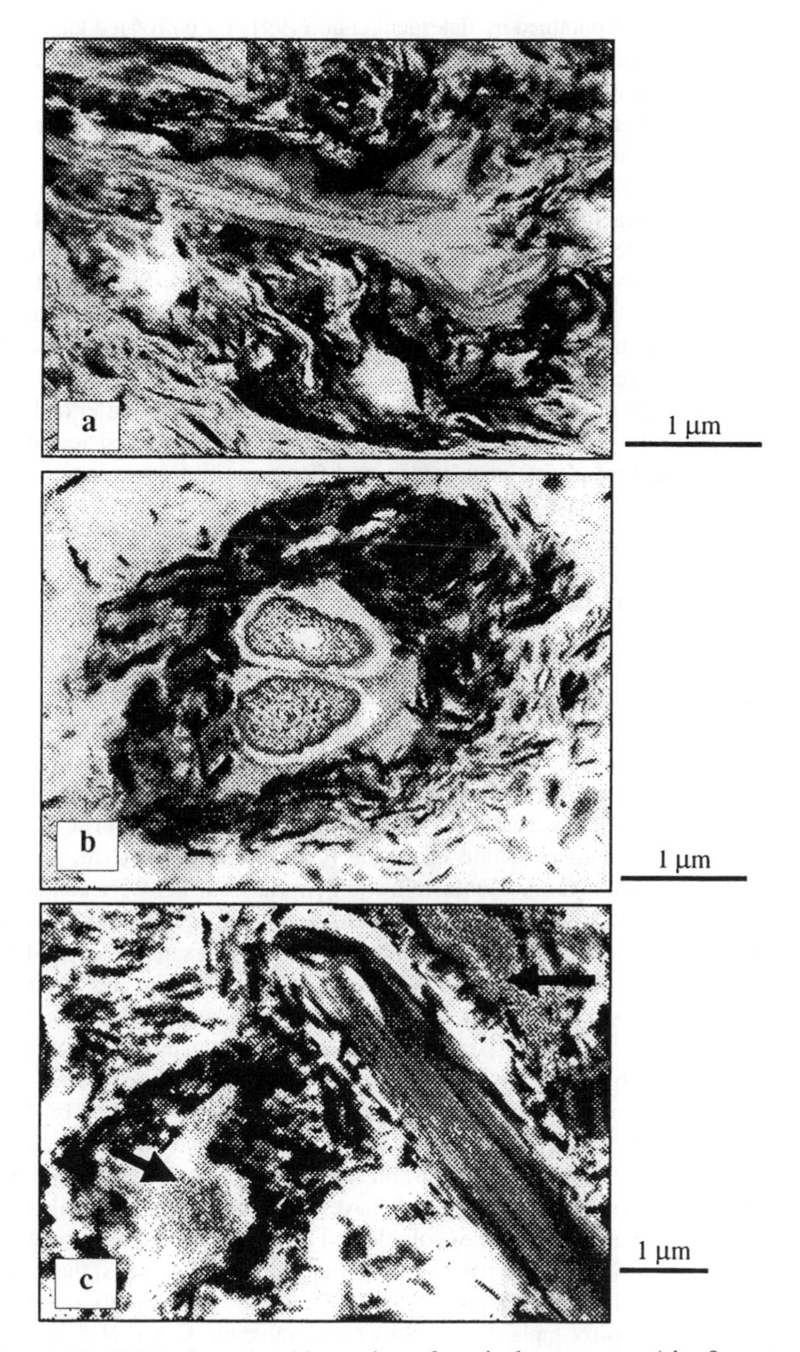

Figure 10. TEM of an ultrathin section of vertisol aggregates (site 9, pasture) showing (a) plant cell wall debris, (b) bacteria, (c) amorphous OM 'physically protected' by clays.

Similar results were obtained by Balesdent et al. (1991) for cultivated temperate soils. For *non-cultivated situations* (native vegetation, long-term fallows, artificial meadows), with a generally high aggregate stability, a trend is observed for an increase in OC content with clay content for all the size fractions, in particular for the plant debris fraction. Different hypotheses can be invoked to explain the increase in OC content of the plant debris fraction with clay content: (i) plant productivity and therefore OC inputs are higher in clayey soils than in coarse textured soils; and (ii) as the structural stability of rich in OC clayey soils is higher than that of coarse textured soils, the plant debris fraction can be protected in stable aggregates against mineralization. Moreover, a large fauna (earthworms) activity in non-cultivated clayey soils enhanced the stability of the structure.

Unfortunately, for the situations investigated in this chapter, the litter and root inputs to the soil were not quantified. Therefore, it is difficult to determine if the observed 'texture effect' may be due to differences in organic input levels or in a protection against mineralization (aggregation effect).

2. The 'Aggregation Effect'

The effect of aggregation on the accessibility of an agent (biotic or abiotic) to a substrate can be studied by the comparison of results obtained before and after disaggregation of a given soil sample. A different disaggregation status can be obtained by grinding the air-dried sample or dispersing it under water. We report results concerning the second approach, specifically the effect of aggregation on: (i) the accessibility to specific organomineral surface areas (SSA); and (ii) the mineralization of carbon (Cm) and nitrogen (Nm).

In both cases (SSA, C and N mineralization), measurements were conducted on the 'aggregated' soil 0 to 2 mm (bulk soil) and on its size OM fractions obtained after dispersion. Values for the 'dispersed' soil sample (sum of the fractions) were then calculated from the weight and the SSA, Cm and Nm of the separated fractions and compared to those of the 'aggregated' soil. An aggregation effect on the accessibility to SSA and mineralizable C and N might be expressed by the inequality: Bulk sample < Sum of fractions. SSA were measured from nitrogen gas adsorption isotherms (N_2-BET method) on LAC soils of sites 2, 8, 11, 12 (Feller et al., 1992) and Cm and Nm after 28 day aerobic incubations (Nicolardot, 1988) on soils of sites 1 and 8 (Feller, 1993). Data from site 1 (very poor sandy soil) gave a low value for Nm which can probably be attributed to losses of soluble and easily mineralizable organic N (Cortez, 1989) during the size fractionation. For the remaining soils, there does not appear from Figure 11a and 11b to be a clear positive effect of the dispersion of the soil sample on its SSA or mineralizable C and N :

$$\text{SSA bulk} = \text{SSA-Sum} \quad \text{and} \quad \text{Cm, Nm - bulk} \geq \text{Cm, Nm - Sum}$$

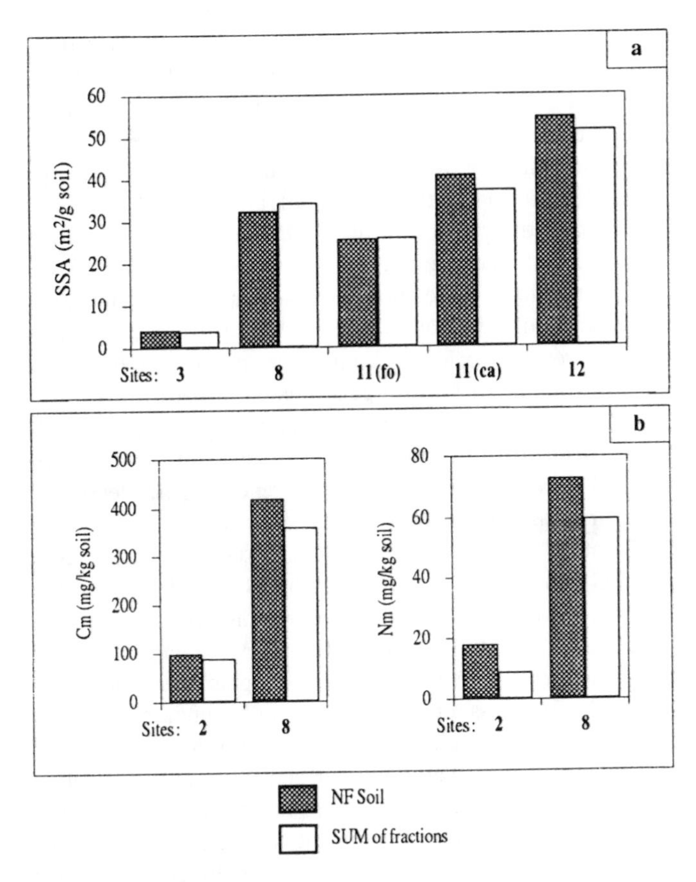

Figure 11. Effect of the dispersion of the soil on: (a) the accessibility of N_2 (BET-method) to organomineral specific surfaces (SSA) and (b) the quantities of carbon (Cm) and nitrogen (Nm) mineralized after 28 days incubation. The effect of dispersion is estimated by the comparison between the non-fractionated 0 to 2 mm soil sample (NF Soil) and the sum of size fractions obtained after dispersion (SUM). The site's number refers to Tables 1 and 2. For site 11, fo = forest and ca = 12-year sugarcane cultivation.

For these LAC soils, the destruction (without OM removal) of the secondary macro- and microaggregates (> 2 μm) does not reveal new organomineral surfaces. This is completely different from high activity clay soils where the N_2-BET external SSA is strongly dependent upon the mode of preparation of the sample (Feller, unpublished data). Similar results concerning the mode of preparation of the sample were observed on clay models (kaolinite, smectite) by Van Damme and Ben Ohoud (1990), and were already reported by Quirk (1978) from data of Fitzsimmons et al. (1970). This can be explained by the differences in the type of microporosity according to the clay mineralogy: always 'open' microporosity for kaolinitic soils or materials whatever the mode of preparation of the sample, both 'open' and 'closed' microporosity for smectitic samples. The ratio 'open'/'closed' is highly dependent on the hydric or ionic history of the sample (Tessier, 1991). Although there is a negligible effect of *secondary* aggregation on accessibility to organomineral surfaces, the comparison of cultivated (CULT) and non-cultivated (NCULT) sites of the clayey oxisol (8 Figure 10.a) indicate the possibility of an OM effect on SSA. However, the higher SSA for the cultivated sample is mainly due to a higher SSA of its fine and coarse clay fractions in relation to a lower carbon contents for these fractions (detailed results in Feller et al., 1992). Overall, it is only on the *primary* microaggregates scale ($< 2\ \mu m$) that an OM/aggregation effect can occur on the accessibility of N_2 to organomineral surfaces.

The results obtained here through mineralization experiments (Cm, Nm) seem to confirm the absence of a significant effect of aggregation on depressing OM mineralization. However, there are conflicting reports in the literature about this effect. For *tropical situations*, little or no effect was observed by Robinson (1967) and Bernhard-Reversat (1981). In contrast, an important effect of aggregation on C mineralization is reported by Martin (1992) and Lavelle and Martin (1992) for the casts of endogeic earthworms (*Millsonia anomala*) in an Ivory Coast alfisol. These authors show that after 400 days of incubation, the OC content of the earthworm casts was 20% higher than that of the control (non-ingested soil). This was attributed to the high compaction and stability of these casts (Blanchart, 1992; Blanchart et al., 1993). Moreover, the OMSF of samples show that differences between casts and control were mainly due to the plant debris fraction ($> 50\ \mu m$) which was protected from decomposition in the casts. This important result may explain the stabilization of the plant debris fraction ($> 20\ \mu m$) we also observed in the non-cultivated fine-textured soils (section VI.A.1. and Figure 5a). For a *subtropical oxisol*, Beare et al. (1994a, b) have clearly shown the protective effect of aggregation on soil C and N mineralization, with about 20% of SOM physically protected in water stable aggregates of the 0 to 5 cm soil layer, when soil was cultivated with no tillage management practices. In contrast, no effect was observed in soils under conventional tillage with lower OC contents and structural stability. For *temperate situations*, little or no effect of aggregation on C and N mineralization can be deduced from the results of Catroux and Schnitzer (1987) whereas Powlson (1980), Elliot (1986), Gupta and Germida (1988), Gregorich et al. (1989), Borchers and Perry (1992) and Hassink (1992) reported significant positive effects of disaggregation on C and/or N mineralization. This SOM protective effect of aggregation is generally relatively more important during the first days of incubation (Gupta and Germida, 1988; Gregorich et al., 1989) in fine textured soils than in the coarse textured (Borchers and Perry, 1992; Hassink, 1992) and sometimes (Gupta and Germida, 1988) for cultivated than for native soils.

In conclusion, on a *long-term scale* clay content appears as the main factor in the OC stabilization of non-eroded LAC soils, for cultivated as well as for non-cultivated situations. Whatever the soil management, OC stabilization occurs in the clay fraction, but a supplementary protective effect on plant debris fraction can appear in non-cultivated situations, probably due partly to the effect of the fauna on aggregation. This effect will be rapidly and largely reduced in many cultivated situations by the decrease of aggregate stability due to decreases in OM content and fauna activities. On a monthly or *seasonal scale*, our first results show that in LAC soils, secondary macro- and microaggregation (> 2 μm) by itself does not play an important role in the dissimulation of organo-mineral surfaces and in the C and N mineralization levels. On the other hand, on the primary microaggregate scale (< 2 μm), a significant part of the organo-clay surfaces can be protected from decomposition in the OC rich soil samples of the non-cultivated clay soils. It is clear that more systematic studies on tropical soils are necessary to ascertain which scale (size and time) of the protective effect of aggregation may be considered as important in regard to the mineralization processes.

B. Role in the Erosion Processes

The OC storage in tropical areas may be favored by diminishing the OC losses due to erosion and particularly sheet erosion (Roose, 1980/81; El Swaify and Dangler, 1982 ; Lal, 1982). For West Africa LAC soils, Roose (1977) measured the 'selective effect' of sheet erosion on fine element and nutrient losses (ratio of the composition of the eroded particles vs. composition of the 0 to 5 or 0 to 10 cm soil layer). For the OC content the selective ratio ranged between 1.5 to 12.8; while for the weight of the 0-2 μm fraction, it averaged from 1.1 to 6.0.

The higher the aggregate stability, the lower might be the soil erodibility. For instance, De Vleeschauwer et al. (1979) indicated that soil erosion (measured by rainfall simulator) and detachability (according to De Leenheer and De Boodt, 1959) were both determined by OC content (and CEC). Therefore, there exists a complex interaction between OC storage and aggregate stability: OC plays a major role in the stabilization of aggregation but aggregation can reinforce the OC storage by diminishing OC losses by sheet erosion. These aspects are illustrated with some recent results obtained on a vertisol (site 9) and a clayey LAC soil from Antilles (Albrecht et al., 1992b and unpublished data). The soil OC contents of the different plots in this study differed in relation to their historical land use and soil management practices: pasture (rich in soil OC) or market-gardening (poor in soil OC), the duration of the cropping system, and the degree of management intensification (irrigation, fertilization and duration of intercropping spontaneous fallows).

As OC content varies from 15 to 50 g C/kg soil in the 0 to 5 cm layers (Figure 12), these plots represent suitable field models to study the effect of OC on the physical behavior of surface soil. Field rainfall simulation was studied on

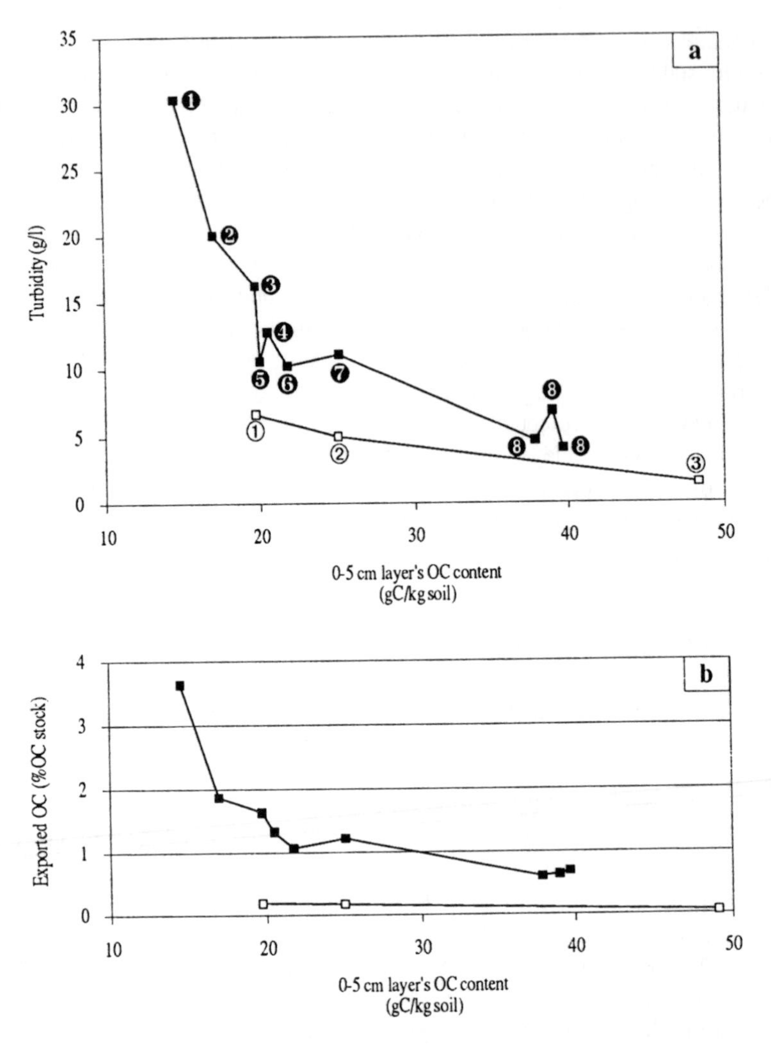

Figure 12. Relations between the OC content of the upper horizon and (a) the turbidity (obtained under rainfall simulation on 1 m^2 plots; rainfall and runoff intensities = 150 mm/h; the soil surface -hand ploughed- is the same in all the situations) and (b) the exported OC. Black signs are situations on vertisols (site 9): ❶ long-term market gardening, ❷ one-year spontaneous fallow after long-term market gardening, ❸ 3-year spontaneous fallow after long-term market gardening, ❹ 2-year cropping after long-term artificial meadows, ❺ 2-year artificial meadow after long-term market gardening, ❻ 4-year artificial meadow after long-term market gardening, ❼ 6-year artificial meadow after long-term market gardening, ❽ long-term artificial meadow. White signs are situations on ferrallitic soils (site 8): ① long-term market gardening, ② one-year spontaneous fallow after long-term market gardening, ③ long-term artificial meadow.

each plot as described in section III.C.2. Figure 12a clearly shows that for vertisol the runoff water turbidimetry was highly dependent on the OC content of the 0 to 5 cm layer. In all cases a selective exportation of fine elements was observed. When expressed in terms of percentage of exported OC (Figure 12b), the higher the OC content of the 0 to 5 cm layer, the lower the percentage of exported OC. For the clayey LAC soil (ferrallitic), the exported OC is much lower. On a medium to long-term scale (c.a. 10 years of intensified market-gardening), such sheet erosion on vertisol leads to a very low OC content (about 12 g C/kg soil) for a clayey soil, whereas the clayey LAC soil with the same cropping system displays a much higher OC content (18 g C/kg soil).

VII. Concluding Remarks

In tropical areas, the clearing of the native vegetation followed by cultivation is accompanied in LAC and HAC soils by rapid declines in SOM content and structural stability. Organic fractions involved in the observed variations in soil aggregation and aggregate stability were strongly related to soil texture: the plant debris fraction for coarse-textured soils, the plant debris and organo-clay fractions for fine-textured soils. The consequences in terms of aggregate stability were discussed in relation to water stable macro- and secondary microaggregates. This review has also demonstrated the use of multi-sieving and kinetic approaches for the WSA characterization of the studied tropical soils. In comparison with mollisols or alfisols in temperate areas, tropical soils exhibit only a low tendency towards an 'aggregate hierarchy' because the dispersion processes (rapid dispersion of clay particles or clay size microaggregates) are generally more important than the disaggregation processes (division of macroaggregates in smaller aggregates according to the increasing of disruptive energy).

Although the role of SOM in the stabilization of the soil structure in tropical soils is well documented, it lacks data to evaluate the effect of macro- and/or secondary microaggregation on OM storage by the protection of OM against mineralization. Some results such as electron microscopic observations or increase in the plant debris fractions with increase in the stability of the structure, or those of Beare et al. (1994b) on a subtropical oxisol, might agree with such an effect, while other results (for LAC soils) based on tests of disaggregation to study the accessibility to organo-mineral surfaces (SSA) or to mineralizable carbon and nitrogen (Cm, Nm), do not support the concept. Generally, the main important and evident effect of aggregation on OM storage was due to its positive effect on reducing soil erodibility, especially the limitation in OM losses by sheet erosion.

To address these different aspects, much research is needed to characterize the diverse relationships between soil structure and OC storage in LAC and HAC tropical soils. The following questions need to be answered:

(i) Does there exist a significant effect of aggregation on the protection of OM against mineralization? On what scale does this protection occur (macro-, secondary micro- or primary microaggregates) and what type of OM is involved (plant debris, amorphous fraction, etc) and for what type of soil (LAC, HAC)?

(ii) Do aggregates which differ in their stability, also differ in their OM composition? Which are the size OM fractions and/or the specific binding organic constituents involved?

(iii) Does there exist a threshold value for the SOM content or SOM fraction that might be significant in terms of erodibility of soils?

To answer these different questions, it is necessary to develop systematic approaches in terms of:

(i) Definition of the so-called water stable aggregates (WSA),

(ii) Location of OM within the WSA. For this the use of optic and electron microscopic techniques is necessary and the characterization of SOM by size fractionation besides chemical analysis might be powerful.

Acknowledgements

The authors are grateful to Mrs M.C. Larré-Larrouy, to Dr. M.R. Carter and to anonymous reviewers for reviewing the English manuscript and to Mrs A.L. Viala and G. Arnac for typing it.

References

Albrecht, A. and L. Rangon. 1988. Matière organique et propriétés physiques de quelques types de sols. p. 55-68. In: C. Feller (ed.), *Fertilité des sols dans les agricultures paysannes caribéennes. Effets des restitutions organiques. Chap. 4.* ORSTOM, Fort de France, Martinique.

Albrecht, A., M. Brossard, J.L. Chotte, C. Feller, A. Plenecassagne, J.P. Brizard, and L. Rangon. 1988. Systèmes de culture et propriétés générales de quelques types de sols. p. 20-46. In: C. Feller (ed.), *Fertilité des sols dans les agricultures paysannes caribéennes. Effets des restitutions organiques. Chap. 2.* ORSTOM, Fort de France, Martinique.

Albrecht, A., M. Brossard, J.L. Chotte, and C. Feller. 1992a. Les stocks organiques des principaux sols cultivés de la Martinique (Petites Antilles). *Cah. ORSTOM, sér. Pédol.* 27:23-36.

Albrecht, A., L. Rangon, and P. Barret. 1992b. Effet de la matière organique sur la stabilité structurale et la détachabilité d'un vertisol et d'un ferrisol (Martinique). *Cah. ORSTOM, sér. Pédol.* 27:121-133.

Alegre, J.C. and D.K. Cassel. 1986. Effect of land clearing methods and post clearing management on aggregate stability and organic carbon content of a soil in the humid tropics. *Soil Sci.* 142:289-295.

Angers, D.A. 1992. Changes in soil aggregation and organic carbon under corn and alfalfa. *Soil Sci. Soc. Am. J.* 56:1244-1249.

Angers, D.A. and G.R. Mehuys. 1990. Barley and alfalfa cropping effects on carbohydrate contents of clay soil and its size fractions. *Soil Biol. Biochem.* 22:285-288.

Angers, D.A. and S. N'dayegamiye. 1991. Effects of manure application on carbon, nitrogen and carbohydrate contents of silt loam and its particle-size fractions. *Biol. Fert. Soils* 11:79-82.

Arias, N.M. and J.J. De Battista. 1984. Evaluacion de metodos para la determinacion de estabilidad estructural en vertisoles de Entre Rios. *Ciencia Suelo* 2:87-92.

Asseline, J. and C. Valentin. 1978. Construction et mise au point d'un infiltromètre à aspersion. *Cah. ORSTOM, sér. Hydrol.* 15:321-350.

Bacye, B. 1993. *Influence des systèmes de culture sur le statut organique des sols et la dynamique de l'azote en zone soudano-sahélienne.* Thèse Doct. Univ., Aix-Marseille III.

Baldock, J.A., B.D. Kay, and M. Schnitzer. 1987. Influence of cropping treatments on the monosaccharide content of the hydrolysates of soil and its aggregate fractions. *Can. J. Soil Sci.* 67:489-499.

Balesdent, J., A. Mariotti, and B. Guillet. 1987. Natural ^{13}C abundance as a tracer for studies of soil organic matter dynamics. *Soil Biol. Biochem.* 19:25-30.

Balesdent, J., G.H. Wagner, and A. Mariotti. 1988. Soil organic matter turnover in long-term field experiments as revealed by the carbon-13 natural abundance. *Soil Sci. Soc. Am. J.* 52:118-124.

Balesdent, J., J.P. Petraud, and C. Feller. 1991. Effets des ultrasons sur la distribution granulométrique des matières organiques des sols. *Science Sol* 29:95-106.

Barois, I., G. Villemin, P. Lavelle, and F. Toutain. 1993. Transformation of the soil structure through *Pontoscolex carethrurus* (Oligochaeta) intestinal tract. *Geoderma* 56:57-66.

Bartoli, F., R. Philippy, and G. Burtin. 1992. Influences of organic matter on aggregation in oxisols rich in gibbsite or in goethite. 1 - Structures : the fractal approach. *Geoderma* 54:231-257.

Beare, M.H. and R.R. Bruce. 1993. A comparison of methods for measuring water-stable aggregates: implications for determining environmental effects on soil structure. *Geoderma* 56:87-104.

Beare, M.H., P.F. Hendrix, and D.C. Coleman. 1994a. Water-stable aggregates and organic matter fractions in conventional and no-tillage soils. *Soil Sci. Soc. Am. J.* 58:777-786.

Beare, M.H., M.L. Cabrera, P.F. Hendrix, and D.C. Coleman. 1994b. Aggregate-protected and unprotected pools of organic matter in conventional and no-tillage soils. *Soil Sci. Soc. Am. J.* 58:787-795.

Benzing-Purdie, L.M. and J.H. Nikiforuk. 1989. Carbohydrate composition of hay and maize soils and their possible importance in soil structure. *J. Soil Sci.* 40:125-130.

Bernhard-Reversat, F. 1981. Participation of light and organomineral fractions of soil organic matter in nitrogen mineralization in sahelian savanna soil. *Zbl. Bakt. II Abt.* 136:281-290.

Blanchart, E. 1990. *Rôle des vers de terre dans la formation et la conservation de la structure des sols de la savane de Lamto (Côte d'Ivoire)*. Thèse Doct., Univ. Rennes-1.

Blanchart, E. 1992. Restoration by earthworms (megascolecidae) of the macroaggregate structure of a destructured savanna soil under field conditions. *Soil Biol. Biochem.* 24:1587-1594.

Blanchart, E., A. Bruand, and P. Lavelle. 1993. The physical structure of casts of *Millsonia anomala* (Oligochaeta: Megascolecidae) in shrub savanna soil (Côte d'Ivoire). *Geoderma* 56:119-132.

Bonde, T.A., B.T. Christensen, and C.C. Cerri. 1992. Dynamics of soil organic matter as reflected by natural ^{13}C abundance in particle size fractions of forested and cultivated oxisols. *Soil Biol. Biochem.* 24:275-277.

Borchers, J.G. and D.A. Perry. 1992. The influence of soil texture and aggregation on carbon and nitrogen dynamics in southwest Oregon forests and clearcuts. *Can. J. Forest Res.* 22:298-305.

Boyer, J. 1982. *Les sols ferrallitiques. Tome X Facteurs de fertilité et utilisation*. IDT 52. ORSTOM, Paris.

Bremner, J.M. and D.A. Genrich. 1990. Characterization of the sand, silt and clay fractions of some mollisols. p. 598. In: M.F. De Boodt, M.H.B. Hayes and A. Herbillons (eds.), *Soils colloids and their association in aggregates.* Vol. 215. NATO ASI Series.

Brussaard, L., S. Hauser, and G. Tian. 1993. Soil faunal activity in relation to the sustainability of agricultural systems in the humid tropics. p. 241-256. In: K. Mulongoy and R. Merckx (eds.), *Soil organic matter dynamics and substainability of tropical agriculture.* J. Wiley-Sayce, Chichester.

Capriel, P., T. Beck, H. Borcher, and P. Härter. 1990. Relationship between soil alliphatic fraction extracted with supercritical hexane, soil microbial biomass and soil aggregate stability. *Soil Sci. Soc. Am. J.* 54:415-420.

Carter, M.R. 1992. Influence of reduced tillage systems on organic matter, microbial biomass, macro-aggregate distribution and structural stability of the surface soil in a humid climate. *Soil Tillage Res.* 23:361-372.

Carvalho, S.R.D. 1990. *Tassement des sols ferrallitiques mis en culture. Apport d'une analyse compartimentale de l'espace poral de sols podzolicos (Etat de Rio de Janeiro, Brésil)*. Thèse Doct. Univ. P. et M. Curie, Paris.

Cassel, D.K. and R. Lal. 1992. Soil physical properties of the tropics : common beliefs and management restraints. p. 61-89. In: R. Lal and P.A. Sanchez (eds.), *Myths and science of soils in the tropics. Chap. 5.* SSSA Special Publ. No. 20 SSSA, Madison WI.

Catroux, G. and M. Schnitzer. 1987. Chemical, spectroscopic, and biological characterization of the organic matter in particle size fractions separated from an Aquoll. *Soil Sci. Soc. Am. J.* 51:1200-1207.

Casenave, A. and C. Valentin. 1989. *Les états de surface de la zone sahélienne. Influence sur l'infiltration.* ORSTOM Paris, sér. Didactiques, 230 p. + Annexes.

Cerri, C.C., C. Feller, and A. Chauvel. 1991. Evoluçao das principais propriedades de um latossolo vermelho escuro apos desmatamento e cultivo por 12 e 50 anos com cana-de-açûcar. *Cah. ORSTOM, sér. Pédol.* 26:37-50.

Cerri, C.C., C. Feller, J. Balesdent, R. Victoria, and A. Plenecassagne. 1985. Application du traçage isotopique naturel en ^{13}C à l'étude de la dynamique de la matière organique dans les sols. *C. R. Acad. Sc. Paris* 9, Sér. 2:423-428.

Chaney, K. and R.S. Swift. 1984. The influence of organic matter on aggregate stability in some British soils. *J. Soil Sci.* 8:3-8.

Chauvel, A., M. Grimaldi, and D. Tessier. 1991. Changes in soil pore-space distribution following deforestation and revegetation: an example from the Central Amazon Basin, Brazil. *Forest Ecol. Manage.* 38:259-271.

Chenu, C. 1993. Clay or sand-polysaccharide associations as models for the interface between microorganisms and soil: water related properties and microstructure. *Geoderma* 56:143-156.

Cheshire, M.V. 1979. *Nature and origin of carbohydrates in soils.* Academic Press, London.

Cheshire, M.V. and C.M. Mundie. 1981. The distribution of labelled sugars in soil particle size fractions as a mean of distinguishing plant and microbial carbohydrate residues. *J. Soil Sci.* 32:605-618.

Cheshire, M.V., G.P. Sparling, and C.M. Mundie. 1983. Effect of periodate treatment of soil on carbohydrate constituents and soil aggregation. *J. Soil Sci.* 34:105-112.

Cheshire, M.V., G.P. Sparling, and C.M. Mundie. 1984. Influence of soil type, crop and air drying on residual carbohydrate content and aggregate stability after treatment with periodate and tetraborate. *Plant Soil* 76:339-347.

Chotte, J.L., L. Jocteur-Monrozier, G. Villemin, and A. Albrecht. 1993. Soil microhabitats and the importance of the fractionation method. p. 39-45. In: K. Mulongoy and R. Merckx (eds.), *Soil organic matter dynamics and substainability of tropical agriculture.* J. Wiley-Sayce, Chichester.

Christensen, B.T. 1992. Physical fractionation of soil and organic matter in primary particle size and density separates. p. 1-90. In: *Advances in Soil Science.* Vol. 20. Springer-Verlag, New-York.

Combeau, A. 1960. Quelques facteurs de la variation de l'indice d'instabilité stucturale dans certains sols ferrallitiques. *C.R. Acad. Agric. Fr.* 109-115.

Combeau, A., C. Ollat, and P. Quantin. 1961. Observations sur certaines caractéristiques des sols ferrallitiques. Relations entre les rendements et les résultats d'analyse des sols. *Fertilité* 13:27-40.

Cortez, J. 1989. Effect of drying and rewetting on mineralization and distribution of bacterial constituents in soil fractions. *Biol. Fert. Soils* 7:142-151.

Dalal, R.C. 1989. Long term effects of no-tillage, crop residues and nitrogen application on properties of a vertisol. *Soil Sci. Soc. Am. J.* 53:1511-1515.

De Blic, P. 1976. Le comportement de sols ferrallitiques de Côte d'Ivoire après défrichement et mise en culture mécanisé : rôle des traits hérités du milieu naturel. *Cah. ORSTOM, sér. Pédol.* 14:113-130.

De Blic, P. and R. Moreau. 1979. Structural characteristics of ferrallitic soils under mechanical cultivation in the marginal forest areas of the Ivory Coast. p. 106-111. In: R. Lal and D.J. Greenland (eds.), *Soil physical properties and crop production in the tropics*. J. Wiley & Sons, Chichester.

De Leenheer, L. and M. De Boodt. 1959. Determination of aggregate stability by the change in mean weight diameter. *Meded. Landb. Gent.* 24:290-351.

De Vleeschauwer, D. and R. Lal. 1981. Properties of worm casts under secondary forest regrowth. *Soil Sci.* 132:175-181.

De Vleeschauwer, D., R. Lal, and M. De Boodt. 1979. Comparison of detachability indices in relation to soil erodibility for some important Nigerian soils. *Pedology* 28:5-20.

Desjardins, T., F. Andreux, B. Volkoff, and C.C. Cerri. 1994. Organic carbon and ^{13}C contents in soils and soil size-fractions, and their changes due to deforestation and pasture installation in eastern Amazonia. *Geoderma* 61:103-118.

Dexter, A.R. 1988. Advances in characterization of soil structure. *Soil Tillage Res.* 11:199-238.

Djegui, N., P. De Boissezon, and E. Gavinelli. 1992. Statut organique d'un sol ferrallitique du Sud Benin sous forêt et différents systèmes de cultures. *Cah. ORSTOM, sér. Pédol.* 27:5-22.

Dorioz, J.M., M. Robert, and C. Chenu. 1993. The role of roots, fungi and bacteria on clay particle organization. An experimental approach. *Geoderma* 56:179-194.

Dormaar, J.F. 1984. Monosaccharides in hydrolysates of water-stable aggregates after 67 years of cropping to spring wheat as determined by capillary gas chromatography. *Can. J. Soil Sci.* 64:647-656.

Dormaar, J.F. 1987. Quality and value of wind movable aggregates in chernozemic Ap horizons. *Can. J. Soil Sci.* 67:601-607.

Dutartre, P., F. Bartoli, F. Andreux, and J.M. Portal. 1993. Influence of content and nature of organic matter on the structure of some sandy soils from West Africa. *Geoderma* 56:459-478.

Edwards, A.P. and J.M. Bremner. 1967. Microaggregates in soils. *J. Soil Sci.* 18:64-73.

Ekwue, E.I. 1990. Organic matter effects on soil strength properties. *Soil Tillage Res.* 16:289-297.

El-Swaify, S.A. and E.W. Dangler. 1982. Rainfall erosion in the tropics : a state-of-the-art. p. 1-26. In: ASA (ed.), *Soil Erosion and Conservation in the Tropics*. Special Publ. No. 43 ASA-SSSA, Madison (WI).

Elliott, E.T. 1986. Aggregate structure and carbon, nitrogen and phosphorus in native and cultivated soils. *Soil Sci. Soc. Am. J.* 50:627-633.

Elliott, E.T. and C.A. Cambardella. 1991. Physical separation of soil organic matter. *Agric. Ecosyst. Environ.* 34:407-419.

Elliott, E.T., C.A. Palm, D.E. Reuss, and C.A. Monz. 1991. Organic matter contained in soil aggregates from a tropical chronosequence : correction for sand and light fraction. *Agric. Ecosyst. Environ.* 34:443-451.

Elliott, P.W., D. Knight, and J.M. Anderson. 1990. Denitrification in earthworm casts and soil from pasture under different fertilizer and drainage regimes. *Soil Biol. Biochem.* 22:601-605.

Eschenbrenner, V. 1986. Contribution des termites à la macro-agrégation des sols tropicaux. *Cah. ORSTOM, sér. Pédol.* 22:397-408.

Fauck, R., C. Moureaux, and C. Thomann. 1969. Bilan de l'évolution des sols de Séfa (Casamance, Sénégal) après quinze années de culture continue. *L'Agron. Trop.* 29:1228-1248.

Feller, C. 1993. Organic inputs, soil organic matter and functional soil organic compartments in low activity clay soils in tropical zones. p. 77-88. In: K. Mulongoy et R. Merckx (ed.), *Soil organic matter dynamics and substainability of tropical agriculture.* J. Wiley-Sayce, Chichester.

Feller, C. 1994. La matière organiqne dans le sols tropicaux à argiles 1:1. Recherche de compartiments organiques fonctionnels. Une approche granulométrique. Thèse Doct. ès Sciences, Univ. Strasbourg (ULP). 393 p. + Annex, 236 p.

Feller, C. 1995. The concept of soil humus in the past three centuries. *CATENA, Special Issue* (in press).

Feller, C. and P. Milleville. 1977. Evolution des sols de défriche récente dans la région des Terres Neuves (Sénégal Oriental). 1 - Présentation de l'étude. *Cah. ORSTOM, sér. Biol.* 12:199-211.

Feller, C., G. Burtin, B. Gerard, and J. Balesdent. 1991a. Utilisation des résines sodiques et des ultrasons dans le fractionnement granulométrique de la matière organique des sols. Intérêt et limites. *Science Sol.* 29:77-94.

Feller, C., H. Casabianca, and C.C. Cerri. 1991b. Renouvellement du carbone associé aux différentes fractions granulométriques d'un sol ferrallitique forestier (Brésil) à la suite du défrichement et de cultures continues de canne à sucre. Etude avec ^{13}C en abondance naturelle. *Cah. ORSTOM, sér. Pédol.* 26:365-370.

Feller, C., C. François, G. Villemin, J.M. Portal, F. Toutain, and J.L. Morel. 1991c. Nature des matières organiques associées aux fractions argileuses d'un sol ferrallitique. *C.R. Acad. Sci. Paris* Sér. 2:1491-1497.

Feller, C., E. Fritsch, R. Poss, and C. Valentin. 1991d. Effets de la texture sur le stockage et la dynamique des matières organiques dans quelques sols ferrugineux et ferrallitiques (Afrique de l'Ouest, en particulier). *Cah. ORSTOM, sér. Pédol.* 26:25-36.

Feller, C., E. Schouller, F. Thomas, J. Rouiller, and A.J. Herbillon. 1992. N_2-BET specific surface areas of some low activity clay soils and their relationships with secondary constituents and organic matter contents. *Soil Sci.* 153:293-299.

Felier, C., P. Lavelle, A. Albrecht, and B. Nicolardot. 1993. La jachère et le fonctionnement des sols tropicaux. Rôle de l'activité biologique et des matières organiques. Quelques éléments de réflexion. p. 15-33. In: ORSTOM (ed.) *La Jachère en Afrique de l'Ouest*. Collection Colloques et Séminaires, ORSTOM. Paris.

Fitzsimmons, R.F., A.M. Posner, and J.P. Quirk. 1970. Electron microscopic and kinetic study of the flocculation of calcium montmorillonite. *Israel J. Chem.* 8:301-314.

Foster, R.C. 1981. Polysaccharides in soil fabrics. *Science* 214:665-667.

Fragoso, C., K.K.J. Kanyonyo, P. Lavelle, and A. Moreno. 1992. A survey of communities and selected species of earthworms for their potential use in low-input tropical agricultural systems. p. 7-34. In: P. Lavelle (coordinator) *Conservation of soil fertility in low-input agricultural systems of the humid tropics by manipulating earthworm communities (Macrofauna Project)*. CEE Project No. TSD2 0292 F(EDB), ORSTOM, Bondy.

Fritsch, E. 1982. Evolution des sols sous recru forestier après mise en culture traditionnelle dans le Sud-Ouest de la Côte d'Ivoire. ORSTOM, Abidjan.

Fritsch, E., A.J. Herbillon, E. Jeanroy, P. Pillon, and O. Barres. 1989. Variations minéralogiques et structurales accompagnant le passage "sols rouges-sols jaunes" dans un bassin versant caractéristique de la zone de contact forêt-savane de l'Afrique Occidentale (Booro-Borotou, Côte d'Ivoire). *Sci. Géol. Bull.* 42:65-89.

Garnier-Sillam, E. 1987. *Biologie et rôle des termites dans les processus d'humification des sols forestiers tropicaux du Congo*. Thèse d'Etat, Paris XIII.

Garnier-Sillam, E., F. Toutain, and J. Renoux. 1988. Comparaison de l'influence de deux termitières (humivore et champignonniste) sur la stabilité structurale de sols forestiers tropicaux. *Pedobiologia* 32:89-97.

Garnier-Sillam, E., F. Toutain, G. Villemin, and J. Renoux. 1985. Formation de micro-agrégats organo-minéraux dans les fécès de termites. *C.R. Acad. Sc. Paris*, Sér. 3:213-218.

Giovannini, G. and P. Sequi. 1976a. Iron and aluminium as cementing substances of soil aggregates. 1 - Acetylacetone in benzene as an extractant of fractions of soil iron and aluminium. *J. Soil Sci.* 27:140-147.

Giovannini, G. and P. Sequi. 1976b. Iron and aluminium as cementing substances of soil aggregates. 2 - Changes in stability of soil aggregates following extraction of iron and aluminium by acetylacetone in a non-polar solvent. *J. Soil Sci.* 27:148-153.

Goldberg, J., D.L. Suarez, and R.A. Glaubig. 1988. Factors affecting clay dispersion and aggregate stability of arid zone soils. *Soil Sci.* 146:317-325.

Greenland, D.J. 1979. Structural organization of soils and crop production. p. 106-111. In: R. Lal and D.J. Greenland (eds.), *Soil physical properties and crop production in the tropics*. J. Wiley & Sons, Chichester.

Gregorich, E.G., R.G. Kachanoski, and R.P. Voroney. 1989. Ultrasonic dispersion of organic matter in size fractions. *Can. J. Soil Sci.* 68:395-403.

Griffiths, E., and R.G. Burns. 1972. Interaction between phenolic substances and microbial polysaccharides in soil aggregation. *Plant Soil* 36:599-612.

Gupta, V.V.S.R. and J.J. Germida. 1988. Distribution of microbial biomass and its activity in different soil aggregate size classes as affected by cultivation. *Soil Biol. Biochem.* 20:777-786.

Hamblin, A.P. and D.J. Greenland. 1977. Effect of organic constituents and complexed metal ions on aggregate stability of some East Anglian soils. *J. Soil Sci.* 28:410-416.

Hassink, J. 1992. Effects of soil texture and structure on carbon and nitrogen mineralization in grassland soils. *Biol. Fert. Soils* 14:126-134.

Hayes, M.H.B. 1986. Soil organic matter extraction, fractionation, structure and effects on soil structure. p. 183-208. In: Y. Chen and Y. Avnimelech (eds.), *The role of organic matter in modern agriculture. Chap. 9.* Martinus Nighoff, Dordrecht.

Haynes, R.J., and R.S. Swift. 1990. Stability of soil aggregates in relation to organic constituents and soil water content. *J. Soil Sci.* 41:73-83.

Haynes, R.J., R.S. Swift, and R.C. Stephen. 1991. Influence of mixed cropping rotations (pasture - arable) on organic matter content, water stable aggregation and clod porosity in a group of soils. *Soil Tillage Res.* 19:77-87.

Henin, S., G. Monnier, and A. Combeau. 1958. Méthode pour l'étude de la stabilité stucturale des sols. *Ann. Agron.* 9:73-92.

Henin, S., J. Gras, and G. Monnier. 1960. Le profil cultural. *Sociétés d'éditions des Ingénieurs agricoles. Paris.*

Herbillon, A.J. 1989. Chemical estimation of weatherable minerals present in the diagnostic horizons of low activity clay soils. *Proc. 8th Int. Soil Classification Workshop, Rio de Janeiro*, 39-48.

Ike, I.F. 1986. Soil and crop responses to different tillage practices in a ferruginous soil in the Nigerian savanna. *Soil Tillage Res.* 6:261-272.

Janeau, J.L., and C. Valentin. 1987. Relations entre les termitières Trinervitermes s.p. et la surface du sol: réorganisations, ruissellement et érosion. *Rev. Ecol. Biol. Sol* 24:637-647.

Jastrow, J.D. and R.M. Miller. 1991. Methods for assessing the effects of biota on soil structure. *Agric. Ecosyst. Environ.* 34:279-303.

Joschko, M., H. Diestel, and O. Larink. 1989. Assessment of the burrowing efficiency in compacted soil with combination of morphological and soil physical measurements. *Biol. Fert. Soils* 8:158-164.

Kemper, W.D. and W.S. Chepil. 1965. Size distribution of aggregates. p. 499-519. In: C.A. Black, D.D. Evans, J.L. White, L.E. Ensminger and F.E. Clark (eds.), *Methods of soil analysis. Part.1. Physical and mineralogical properties.* Agronomy 9. ASA., Madison (WI).

Kemper, W.D. and R.C. Rosenau. 1986. Aggregate stability and size distribution. p. 425-442. In: A. Klute (ed.), *Methods of soil analysis, Part 1. Physical and mineralogical methods.* Agronomy 9 (2nd ed.), ASA-SSSA, Madison (WI).

Kooyman, C. and R.F.M. Onck. 1987. Distribution of termite (Isoptera) species in Southwestern Kenya in relation to land use and the morphology of their galleries. *Biol. Fert. Soils* 3:69-73.

Ladd, J.N., R.C. Foster, and J.O. Skjemstad. 1993. Soil structure : carbon and nitrogen metabolism. *Geoderma* 56:401-434.

Lal, R. 1979. Modification of soil fertility characteristics by management of soil physical properties. p. 397-406. In: R. Lal and D. Greenland (eds.), *Soil Physical Properties and Crop Production in the Tropics. Chap. 7.1.* J. Wiley & Sons, Chichester.

Lal, R. 1982. Effective conservation farming systems for the humid tropics. p. 57-76. In: ASA (ed.), *Soil Erosion and Conservation in the Tropics.* 149 p. Special Publ. No. 43 ASA-SSSA, Madison (WI).

Lal, R. 1988. Effects of macrofauna on soil properties in tropical ecosystems. *Agric. Ecosyst. Environ.* 24:101-116.

Lal, R. and D.J. Greenland. 1979. *Soil physical properties and crop production in the tropics.* J. Wiley & Sons, Chichester.

Lavelle, P. 1987. The importance of biological processes in productivity of soil in the humic tropics. p. 175-214. In: R.E. Dickinson et J. Lovelock (eds.), *Geophysiology of the Amazon.* J. Wiley & Sons, NY.

Lavelle, P. and A. Martin. 1992. Small-scale and large-scale effects of endogeic earthworms on soil organic matter dynamics in soils of the humid tropics. *Soil Biol. Biochem.* 24:1491-1498.

Lavelle, P. and B. Pashanasi. 1989. Soil macrofauna and land management in Peruvian Amazonia (Yurimaguas, Loreto). *Pedobiologia* 33:283-291.

Lavelle, P., E. Blanchart, A. Martin, A.V. Spain, and S. Martin. 1992. Impact of soil fauna on the properties of soils in the humid tropics. p. 157-185. In: R. Lal and P.A. Sanchez (eds.), *Myths and science of soils in the tropics. Chap. 9.* 185 p. SSSA Special Publ. No. 20 SSSA, Madison (WI).

Le Bissonnais, Y., A. Bruand, and M. Jamagne. 1990. Etude expérimentale sous pluie simulée de la formation des croûtes superficielles. Apport à la notion d'érodibilé des sols. *Cah. ORSTOM, sér. Pédol.* 25:31-40.

Lepsch, I.F., N.M. Da Silva, and A. Espironelo. 1982. Relacão entre matéria orgânica e textura de solos sob cultivo de algodão e cana-de-acucar, no estado de São Paulo. *Bragantia* 41:231-236.

Mapangui, A. 1992. *Etude de l'organisation et du comportement des sols ferrallitiques argileux de la vallée du Niari (Congo). Conséquence sur l'évolution physique sous culture de manioc en mécanisé depuis 15 ans.* Thèse Doct. Univ. P. et M. Curie, Paris.

Marinissen, J.C.Y. and A.R. Dexter. 1990. Mechanisms of stabilization of earthworm casts and artificial casts. *Biol. Fert. Soils* 9:163-167.

Martin, G. 1963. Dégradation de la structure des sols sous culture mécanisée dans la vallée du Niari. *Cah. ORSTOM, sé r. Pédol.* 2:8-14.

Martin, A. 1992. Short-term and long-term effect of the endogeic earthworm *Millsonia anomala* (Omodeo Megascolecidae, Oligochaeta) of a tropical savanna, on soil organic matter. *Biol. Fert. Soils* 11:234-238.

Martin, A., and J.C.Y. Marinissen. 1993. Biological and physico-chemical processes in excrements of soil animals. *Geoderma* 56:331-347.

Martin, A., A. Mariotti, J. Balesdent, P. Lavelle, and R. Vuattoux. 1990. Estimate of organic matter turnover rate in a savanna soil by ^{13}C natural abundance measurements. *Soil Biol. Biochem.* 22:517-523.

Martin, J.P. W.P. Martin, J.B. Page, W.A. Laney, and J.D.D. Ment. 1955. Soil aggregation. *Adv. Agron.* 7:1-37.

Miller, R.M. and J.D. Jastrow. 1990. Hierarchy of root and mycorhizal fungal interactions with soil aggregation. *Soil Biol. Biochem.* 22:579-584.

Moreau, R. 1983. Evolution des sols sous différents modes de mise en culture en Côte d'Ivoire forestière et préforestière. *Cah. ORSTOM, sér. Pédol.* 20:311-325.

Morel, R. and P. Quantin. 1972. Observations sur l'évolution à long terme de la fertilité des sols cultivés à Grimari (République Centrafricaine). *L'Agron. Trop.* 27:667-739.

Mulongoy, K. and A. Bodoret. 1989. Properties of worm casts and surface soils under various plant covers in the humid tropics. *Soil Biol. Biochem.* 21:197-203.

Murayama, S. M.V. Cheshire, C.M. Mundie, G.P. Sparling, and H. Shepherd. 1979. Comparison of the contribution to soil organic matter fractions particularly carbohydrates, made by plant residues and microbial products. *J. Sci. Food Agric.* 30:1025-1034.

Nicolardot, B. 1988. Evolution du niveau de biomasse microbienne du sol au cours d'une incubation de longue durée: relations avec la minéralisation du carbone et de l'azote organique. *Rev. Ecol. Biol. Sol* 25:287-304.

Nye, P.H. and D.J. Greenland. 1960. *The soil under shifting cultivation.* Vol. 51. Techn. Comm. Commonwealth. Bur. Soils (Harpenden). Bucks, England.

Oades, J.M. 1984. Soil organic matter and structural stability: Mechanisms and implications for management. *Plant Soil* 76:319-337.

Oades, J.M. 1993. The role of biology in the formation, stabilization and degradation of soil structure. *Geoderma* 56:337-400.

Oades, J.M., and A.G. Waters. 1991. Aggregate hierarchy in soils. *Aust. J. Soil Res.* 29:815-828.

Oliveira, M., N. Curi, and J.C. Freire. 1983. Influencia de cultivo na agregação de um podzolico vermelho amarelo textura media/argilosa de região de Lavras (MG). *R. bras. Ci. Solo* 7:317-322.

Piccolo, A., and J.S.C. Mbagwu. 1990. Effect of different organic waste amendments on soil microaggregates stability and molecular sizes of humic substances. *Plant Soil* 123:27-37.

Pieri, C. 1989. *Fertilité des terres de savanes.* Ministère de la Coopération-CIRAD (eds.), Paris.

Pojasok, T., and B.D. Kay. 1990. Assessment of a combination of wet sieving and turbidimetry to characterize the structural stability of moist aggregates. *Can. J. Soil Sci.* 70:32-42.

Poss, R. 1991. *Transferts de l'eau et des éléments minéraux dans les terres de barre du Togo. Conséquences agronomiques*. Thèse Doct., Univ. Paris-6.

Powlson, D.S. 1980. The effects of grinding on microbial and non-microbial organic matter in soil. *J. Soil Sci.* 31:77-85.

Quirk, J.P. 1978. Some physico-chemical aspects of soil structural stability. A review. p. 3-16. In: W.W. Emerson et al. (eds.), *Modification of soil structure. Chap. 17.* J. Wiley & Sons, Chichester.

Robert, M. and C. Chenu. 1992. Interactions between soil minerals and microorganisms. p. 307-404. In: G. Stotzki and J.M. Ballag (eds.), *Soil Biochemistry.* Vol. 7. Marcel Dekker, NY.

Robinson, J.B.D. 1967. Soil particle size fractions and nitrogen mineralization. *J. Soil Sci.* 18:109-117.

Roose, E. 1976. Contribution à l'étude de l'influence de la mésofaune sur la pédogenèse actuelle en milieu tropical. ORSTOM, Abidjan.

Roose, E. 1977. *Erosion et ruissellement en Afrique de l'Ouest. Vingt années de mesures en petites parcelles expérimentales*. Vol. 78. Travaux et Documents. ORSTOM, Paris.

Roose, E. 1980-81. Dynamique actuelle d'un sol ferrallitique très désaturé sur sédiments sablo-argileux sous culture et sous forêt dense humide subéquatoriale du sud de la Côte d'Ivoire. Adiopodoumé 1964-1976. 2ème Partie : Les transferts de matière. *Cah. ORSTOM, sér. Pédol.* 18:3-28.

Roose, E. 1981. *Dynamique actuelle des sols ferrallitiques et ferrugineux tropicaux d'Afrique Occidentale*. Vol. 130. Travaux et Documents. ORSTOM, Paris.

Sanchez, P.A. 1976. *Properties and management of soils in the tropics*. J. Wiley & Sons, NY.

Shipitalo, M.J., and R. Protz. 1988. Factors influencing the dispersibility of clay in worm casts. *Soil Sci. Soc. Am. J.* 52:764-769.

Siband, P. 1974. Evolution des caractères et de la fertilité d'un sol rouge de Casamance. *l'Agron. Trop.* 29:1228-1248.

Sleeman, J.R., and R. Brewer. 1972. Microstructures of some Australian termite nests. *Pedobiologia* 12:347-373.

Sparling, G.P. and M.V. Cheshire. 1985. Effect of periodate oxidation on the polysaccharide content and microaggregate stability of rhizosphere and non-rhizosphere soils. *Plant Soil* 88:115-122.

Tessier, D. 1991. Behaviour and microstructure of clay minerals. p. 387-415. In: M. De Boodt, M. Hayes and A. Herbillon (eds.), *Soil colloids and their associations in aggregates. Chap. 14.* Plenum Press, NY.

Thiery, J.P. 1967. Mise en évidence de polysaccharides sur coupes fines en microscopie électronique. *J Microscopie* 6:987-1017.

Thomann, C. 1963. Quelques observations sur l'extraction de l'humus dans les sols . Méthode au pyrophosphate de sodium. *Cah. ORSTOM, sér. Pédol.* 3:43-72.

Tisdall, J.M. and J.M. Oades. 1980. The effect of crop rotation on aggregation in a red brown earth. *Aust. J. Soil Res.* 18:423-433.

Tisdall, J.M. and J.M. Oades. 1982. Organic matter and water stable aggregates in soils. *J. Soil Sci.* 33:141-163.

Trouve, C., A. Mariotti, D. Schwartz, and B. Guillet. 1991. Etude par le traçage naturel en ^{13}C de la dynamique de renouvellement des matières organiques des sols de savane après plantation de peuplements de pins et d'Eucalyptus au Congo. *Cah. ORSTOM, sér. Pédol.* 26:357-364.

Turchenek, L.W. and J.M. Oades. 1978. Organo-mineral particles in soils. p. 137-144. In: W.W. Emerson et al. (eds.), *Modification of soil stucture. Chap. 16*. J. Wiley & Sons, Chichester.

Utomo, W.H. and A.R. Dexter. 1982. Changes in soil aggregate water stability induced by wetting and drying cycles in non-saturated soil. *J. Soil Sci.* 33:623-637.

Van Damme, H., and M. Ben Ohoud. 1990. From flow to fracture and fragmentation in colloidal media. Part II. Local order and fragmentation geometry. In: J.C. Charmet, S. Roux, and E. Guillou (eds.), *Disorder and fracture*. Vol. 235. NATO ASI series, B, Plenum, Press, NY.

Whitehead, D.C., H. Buchan, and R.D. Hartley. 1975. Component of soil organic matter under grass and arable cropping. *Soil Biol. Biochem.* 7:65-71.

Williams, B.G., D.J. Greenland, G.R. Lindstrom, and J.P. Quirk. 1966. Techniques for the determination of the stability of soil aggregates. *Soil Sci.* 101:157-163.

Wierzchos, J., C. Ascaso-Ciria, and M.T. Garcia-Gonzalez. 1992. Changes in microstructure of soils following extraction of organically bonded metals and organic matter. *J. Soil Sci.* 43:505-515.

Yoder, R.E. 1936. A direct method of aggregate analysis of soils and study of the physical nature of erosion looses. *J. Am. Soc. Agron.* 28:337-351.

Organic Carbon Storage in Tropical Hydromorphic Soils

H.W. Scharpenseel, E.-M. Pfeiffer, and P. Becker-Heidmann

I. Introduction

Wet and dry soils have about the same area of distribution and organic carbon load (C) per m^2. Wetland soils (i.e. soils with aquic soil moisture regime) (Soil Survey Staff, 1992; Moormann, 1980; Moormann and van de Wetering, 1985; Kyuma, 1985; Scharpenseel, 1993) and soils under a regime of 300-550 mm annual rainfall (with caliches/calcretes) both cover ca. 10^9 ha and contain about 50 to 100 kg C m^{-2}. The caliches, as $\delta^{13}C$ indicates, contain roughly 50% C of biogenic origin. Carbon sequestration in peat and soil organic matter (SOM) of other wetlands represents a C sink of ca. 650 Pg C, equal to the C pool of the living biomass, and slightly less than the C pool of the atmosphere (740 Pg C) or of the remaining economically accessable fossil fuel C (1000 Pg C). Thus, wetland-C is a major factor controlling the soil born Greenhouse (GH) forcing trace gas emissions of CO_2 and CH_4. This applies especially to the ca. 50% of wetland-C derived from the tropics.

This chapter tries to assess the different C compartments in wetlands/hydromorphic soils. A certain emphasis is placed on the agronomically utilised lowland rice soils; their SOM dynamic is especially well explored (Neue, 1989; Neue and Singh, 1984; Neue et al., 1995). Close to 90% of the total ricelands

1-56670-033-7/96/$0.00+$.50

are wetlands. Due to a continued conversion into cropland for both rice (*Oryza sativa* L.) and taro (*Colocasia esculenta*), as well as drainage for diverse economic projects, natural wetlands were in the more recent past under a trend of areal shrinkage. However, their value for C sequestration (i.e. control of GH forcing CO_2 - emission) and for sustaining genetic diversity has led to an almost ideological effort of defending the remaining wetlands, especially in the densely populated industrial countries.

II. Overview of Carbon Pools and Compartments

The pool size of the two major organic C compartments of the global C cycle, the economically accessible fossil fuel C and the soil organic matter (SOM) C pools are rather poorly quantified (Scharpenseel, 1993). The former, according to the International Energy Agency, comprises about 1000 Pg C (P = Peta = 10^{15}) (Graßl and Klingholz, 1990), while in Batjes and Bridges (1992) it is estimated it to be about 10,000 Pg C. Estimates of SOM C pools range between 1000 and 3000 Pg C (Batjes, 1992). Recently, Eswaran et al. (1993) estimated that SOM-C pools, by soil order, amounted to a total of 1,555 Pg C (Table 1 and 2). Carbon in the compartments of the pedosphere was tentatively quantified by Scharpenseel (1993) to range from 1,300 to 1,500 Pg, with about 40 Pg outside the epipedons.

The global inventory of wetland's by Aselmann and Crutzen (1990) suggests a total of 570 million ha natural wetlands plus 130 million ha of riceland. The natural wetlands amount to 187 million ha of bog, 148 million ha of fen, 113 million ha of swamp, 0.27 million ha of marsh, 0.82 million ha of floodplain, and 0.11 million ha of shallow lake soils. These natural wetlands generate 4 to 8 Pg of dry matter production per year.

Aselmann and Crutzen (1990) indicated that the 130 million ha of riceland produce a net plant production (NPP) of 1400 million t dry matter. While Neue et al. (1990, 1991) report a production of 477 to 510 million t rice grain on an area of 136 million ha. They also differentiate between agroecosystems of varying productivity for rice such as 73.76 million ha irrigated, 38.25 million ha rainfed, 11.45 million ha of deep water, and 19.7 million of upland rice fields (current estimation of the International Rice Research Institute in the Philippines). More recent studies on wetland ecosystems arrive at higher total wetland areas and C pool sizes than those given above (Gorham, 1991a,b; Twilley et al., 1992; 1993) (Table 3).

Tans et al. (1990) drew attention to the 'missing C fraction' of 1.5 to 2.0 Pg C, remaining from the annual anthropogenic C inputs of 7.0 Pg C (5.4 Pg fossil fuel C, 1.6 Pg forest clearing / slash and burn released C), after compensation for 3.4 Pg of C emission to the atmosphere and about 2.0 Pg of C input to the oceans. However, the hypothesised promotion of higher biomass production in the terrestrial northern higher latitudes or the tropics (Enting and Mansbridge,

Table 1. Organic carbon in different soil orders of the world in 0 to 100 cm depth

Soil order	Organic carbon Pg (Gt)
Ultisols	101
Andosols	69
Aridisols	110
Oxisols	150
Inceptisols	267
Alfisols	136
Mollisols	72
Vertisols	38
Spodosols	98
Entisols	106
Histosols	390

(According to Eswaran et al., 1993.)

Table 2. Carbon reservoirs in pedosphere including kg C m^{-2} and some C–inputs

C in soil epipedon	Million ha	C (Pg)	Kg C m^{-2}	Pg C yr^{-1}
Ice free continental surface	13,400	ca. 2,000		
Terrestrial soils	12,800	ca. 1,500	11.7	
Wetlands ecosystems (peat, swamp, marsh, rice)	1,024	ca. 646	63.1	
Fluvial C, input to oceans				0.8
Cropland (ca. 12% of terr. SOM)	1,500	ca. 150	10.0	
Rice land (a)	140	ca. 12	8.6	
Grassland (b)	>3,000			
Halophytes	130			0.7
Woodlands (incl. C sequestration) (3,4)	>4,000	300- 400	8-10	1-2.5
Pedolith, calcrete, and caliche	1,000	1,000	100	<0.1
Charcoal from combustion of forest and savannah (4)				0.2-0.3

(a) Small SOM rise until 28°C; (b) some sequestration of C in SOM, high C residence time.

(From (1) Barnwell et al., 1992; (2) Glenn et al., 1992; (3) Winjum et al., 1992; (4) Brown et al., 1992.)

Table 3. Estimate of wetlands by Working Group at USEPA—German Environmental Office jointly sponsored workshop, March 1993, Bad Harzburg, Germany

Wetland ecosystems	Area (10^6 ha)	Pg C	kg C m^{-2}
Temperate peat	346	455	131
Tropical wetlands (w/o ricelands, marshes)	500	150	30
Riceland	140	12	8.6
Marshes	38	29	76.3
Total wetlands	1024	646	

(From Twilley et al., 1993; Gorham, 1991b.)

Table 4. Inventory and fluxes of C in coastal ocean sediments

Pool size (Pg C yr^{-1})	Major ecosystem for storage of C
0.16	Stored in mangrove wood
0.02	Accumulates in mangrove peat
0.05	Exported to coastal oceans
0.30 and 0.08	Total from rivers and wetlands
0.025	C sinks in coastal wetlands
0.41	C sinks in coastal reefs, shelf plus estuarine sediments
0.6	Total C storage potential

(From Twilley et al., 1992; Lugo and Wisniewski, 1992; Wollast and MacKenzie, 1989.)

1991) is drawn into question by a suggested back-transport to lower latitudes via north-Atlantic conveyer circulation (Keeling et al., 1989; Degens, 1989; Broecker and Peng, 1992). Barnwell et al. (1992) suggest alternatively that partial storage of C can occur in mangrove wood and peat, and in coastal ecosystem sand sediments. Natural sinks of CO_2 (i.e. annual C accumulation) were calculated by Lugo and Wisniewski (1992) to be 0.3 to 0.6 Pg yr^{-1} for coastal wetland systems, and 1.2 to 2.8 Pg yr^{-1} for oceans (Twilley et al., 1992; Orr and Sarmiento, 1992) (Table 4).

Ritchard (1992) suggested enhancing C sequestration through strengthening growth of phytoplankton by Fe fertilization. As algae rank second to forests in C sequestration, a CO_2-C removal of 0.7 to 3.0 Pg C yr^{-1} from the atmosphere would appear conceivable. Similarly, Twilley et al. (1992) emphasized that the tropical estuarine and shelf ecosystems are the most biogeochemically active and important C sinks. As 50% of terrigenous material is being delivered to the oceans through 21 major river systems, C storage in mangroves becomes important. Globally it comprises about 4.03 Pg C; 70% of it in the latitudinal belt of 0° to 10°. The authors estimate wood production as being of the order of 12.08 Mg ha^{-1} yr^{-1}, giving a global C storage of 0.16 Pg yr^{-1} in mangrove biomass. This, plus about 0.02 Pg C yr^{-1} in mangrove sediments, adds up to a net ecosystem production in mangroves of about 0.18 Pg C yr^{-1}.

Carbon dynamics in coastlines near wetlands are considerable: C accumulation is about 0.3 to 0.6 Pg per year. The biogeochemical activity in tropical estuarine and shelf ecosystems leads to a storage of the riverine C transport with a pool of ca. 4 Pg C and to annual gains by wood production and sediment enrichment of 0.18 Pg C. According to Twilley et al. (1992) the export from coastal wetlands amounts to 0.08 Pg C yr^{-1}, compared with an input of 0.36 Pg C yr^{-1} from the rivers to the coastal ecosystems. The total allochthonous input of 0.44 Pg C yr^{-1} is lower than the *in situ* production of 6.65 Pg C yr^{-1}. Thus, the total supply amounts to 7.09 Pg C yr^{-1}. Only about 6% of the total input (about 0.41 Pg C yr^{-1}) is respired or accumulated in coastal sediments. The C accumulation in wood and sediments of the coastal wetlands is according to Twilley et al. (1992) 0.205 Pg C yr^{-1}, about half of the assumed amount of C sequestered in coastal sediments. The assessment of the fluvial C by Sunquist (1985) indicates an input to the oceans of 0.37 to 0.45 Pg yr^{-1} as DIC (dissolved inorganic matter), 0.08 to 0.32 Pg yr^{-1} as DOC (dissolved organic matter), 0.04 to 0.3 Pg C yr^{-1} as POC (particular organic matter), and 0.02 to 0.2 Pg C yr^{-1} as PIC (particular inorganic matter). Twilley's analog estimates are 0.35 Pg C yr^{-1} DOC plus POC. From the total fluvial C input to the oceans, amounting to 0.8 Pg C yr^{-1}, about 0.6 Pg C yr^{-1} are brought about by human activities, mainly by enhanced erosion (Wollast and MacKenzie, 1989).

III. The Carbon Cycle in Tropical Hydromorphic Soils and Sediments

The carbon cycle in the tropics is shown in Figure 1. Carbon pools, sources and sinks for tropical soils are given in Table 5, while the areas and total C pools are shown in Tables 6, 7, and 8. Differences in the hydrological system between rain fed agriculture and irrigation are given in Figure 2 (see also Bishop, 1975). The lateral water influx and the water compartments associated with excess delivery, as well as the consequences on gas fluxes and soil redox reactions, express the major differences between land use systems.

The Soil Fertility Capability Classification considers for tropical wetland soils only organic soils of > 30% SOM to a depth of 50 cm or more (Buol et al., 1975), and wetland soils with < 20% organic matter (Sanchez and Buol, 1985). For rice production, the above classification considers deep organic peat soils with little or no potential, or shallow organic soils with a minimal layer of less than 50 cm depth and a potential for rice production (Sanchez and Buol, 1987). They identify 4 soil types (in case of riceland 6 soil types). Soil organic matter is, however, not listed among the '13 Major Condition Modifiers' for tropical

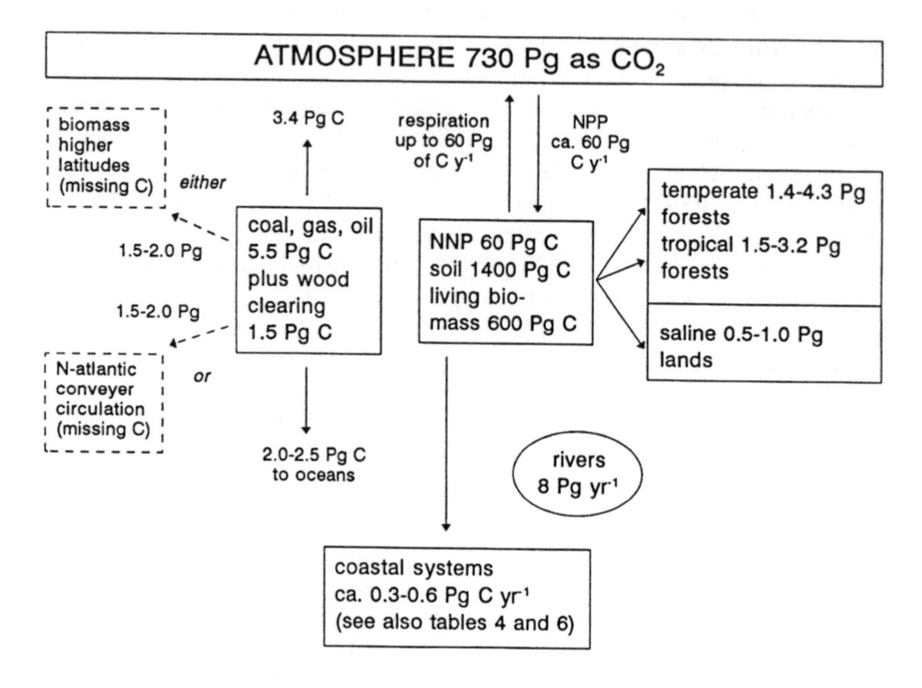

Figure 1. Major compartments of carbon cycle, including tropics.

wetlands. Similarly, SOM is not considered in the 14 'modifiers' for riceland, while low SOM is mentioned only in connection with degraded paddy soils. However, SOM is of great importance for cation exchange capacity (CEC) in low activity clay (LAC) soils (Uehara 1982, Scharpenseel and Miehlich, 1989), as a source of methanogenesis (Martin et al., 1983; Braatz and Hogan, 1991; Batjes, 1992; Denier van der Gon et al., 1992) and as a source of N_2O generation (due to its N content) in connection with nitrification, denitrification, and diazotrophic systems (Conrad et al., 1983; Umarov, 1990; Bouwman, 1990, 1992). SOM further acts as fuel to energise the reduction of Fe and SO_4 in soils with iron-toxicity, ferrolysis or acid sulfate soil development (Moormann and Van Breemen, 1978; Brinkman, 1979; Ottow et al., 1982).

Neue (1985), evaluating all available data for field experiments with SOM and N, concluded, that, dependent on the N-fertilizer rate, soil C contents of 2 to 2.5% C (slightly rising with N supply) are optimal for highest crop yields in rice fields.

Table 5. Carbon pools, size, and conditions for C source and sink in tropical soils

Ecosystems reservoir of C	Area (10^6 ha)	C-pool (Pg C)	C-source (Pg C yr^{-1})	C-sink (Pg C yr^{-1})
Pedosphere w/o wetlands	12,800	1,400		
Including wetlands	13,400	2,000		
Cropland	1,500	150	temporarily	(after rise of CO_2)
Andisols: total	107		$>$10,000	1,000 to
tropics	50	12	yr age	10,000 yr
Riceland	140	1,030	(paddy $>28^{\circ}C$)	(paddy $<28^{\circ}C$)
Wetlands: total	1,030	500		
-tropics, w/o riceland	500			
Tropical woodland plus savannah		9.3% of C pool		
Tropical forests (35%)[a]	1,700			
Tropical rainforest (38%)[b]	655			
Deciduous forest, wet (37%)[b]	626			
Deciduous forest, dry (14%)[b]	252			
Mountain forest (78%)[b]	178			
Deforestation annually (1981-1990)	16.9		1.5	
Grassland, temperate	1,500			
		ca. 300	0.3-0.5	(high SOM resilience)
Grassland, tropical	1,700			

[a] Percent of tropical land surface; [b] percent of tropical forests.
(From Greenland, (personal communication); Scharpenseel and Miehlich, 1989; Twilley et al., 1993; Wisniewski and Lugo, 1992; Sombroek et al., 1993.)

IV. Carbon Cycle in Riceland Systems

An initiation of intensified research on SOM in riceland ecosystems was generated by the International Conference on Organic Matter and Rice at the International Rice Research Center(IRRI) in Los Banos, Philippines in 1982. Riceland ecosystems were divided as follows: 52% irrigated riceland, 27% rainfed rice, 13% upland/dryland rice, and 8% deep water riceland (Neue et al., 1990). A follow-up of even greater intensity was geared by public support of the forthcoming International Global Change Program (IGBP) with emphasis on the greenhouse forcing trace gases, where wetlands are major contributors (IGAC and GCTE Core Projects of the IGBP).

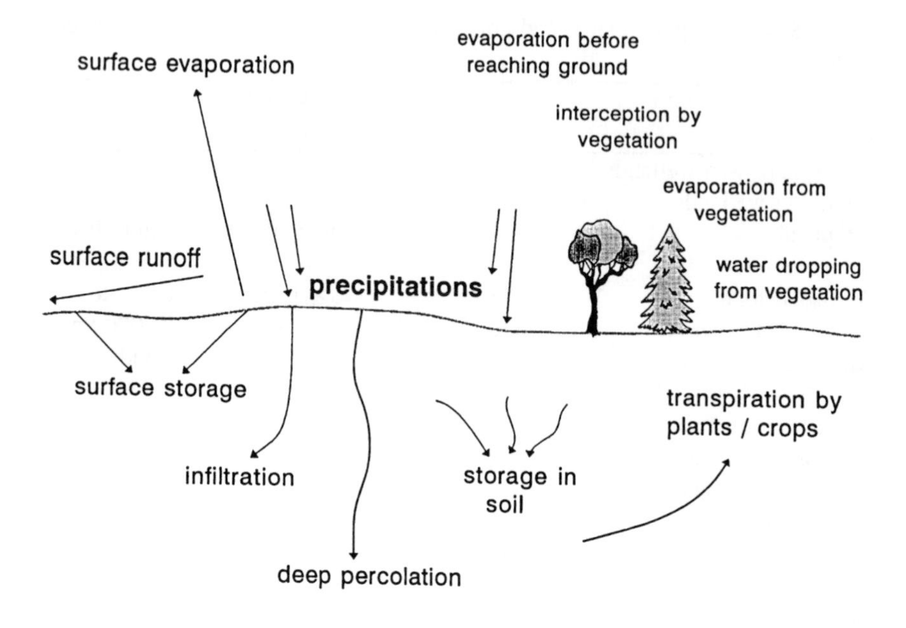

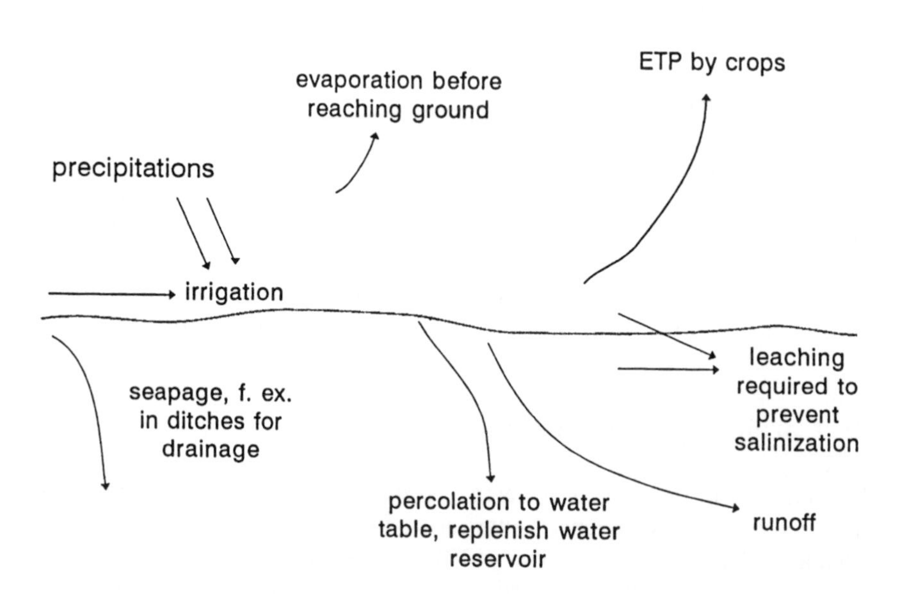

Figure 2. Hydrological system in rain-fed agriculture (top) and irrigated land use (bottom).

Table 6. Amounts and forms of fluxes to and from mainly terrestrial ecosystems

Carbon sources or fluxes	Area (10^6 ha)	Pg C total yr^{-1}
1) CO_2: concentration up 30% since 1980		
In 1980s anthropogenic emissions were:		
By fossil fuel		5.4 ±0.5
By land use change		1.6 ±1.0
Total		7.0 ±1.5
Remaining in atmosphere		3.4 ±1.5
Uptake in oceans		2.0 ±0.8
'Missing C fraction'		1.6 ±1.4
2) Agroecosystems (12% of terrestrial soil-C)[a]	1,400	ca. 200
3) C-sequestration potential in typical forests	4,000	2.5
Below ground biomass and debris		0.8
4) Fluvial C input to ocean; due to human activity, enhancing erosion		0.6
5) Charcoal from combustion (forest, savannah); addition of refracting C to the soils		0.2-0.3
6) Forestry and agroforestry practices		1-2
Anthropogenic emission in atmosphere[b]		3-4
7) Halophytes[c]	130	0.7
Sequestration potential for C		0.7

[a] Rise of CO_2 + effect of 140 million t mineral fertilizer yr^{-1} increase biomass production and % share of C in agroecosystems; [b] could sequester 50 to 100 Pg C over 50 years; compared to anthropogenic emission of 300 to 400 Pg C, from which 50% goes into the atmosphere; [c] growth potential (coastal desert, inland salt desert, secondary salinization by irrigation).
(From (1) Downing and Cataldo, 1992; (2) Barnwell et al., 1992; (3) Brown et al., 1992; (4) Wollast and MacKenzie, 1989; (5) Johnson, 1992 and Schlesinger, 1990; (6) Winjum et al., 1992; (7) Glenn et al., 1992.)

Table 7. Areas of different wet ecosystems in the tropics

Ecosystem	Area in 10^6 ha or %
Lakes (1)	
area > 10 ha, ca. 4.3 m deep	485
area 1 to 10 ha, < 3 m deep	157
Total upstream basin area[a]	2,350
Internal drainage	350
External drainage	2,000
Reservoirs	
Total area	40
Total upstream reservoir area[b]	1,200
Rivers	
Temperate climate	0.5
Wetlands	570 (2)-880 (3)
Ricelands	140-143 (4)
Ecosystems, % of total area	
Irrigated rice	
Rainfed rice	51.6%
Deep water rice	27.2%
Upland rice	8.0%
	13.8%
Continental shelves	
	2824 (5)
	7.8% of area of oceans

[a] Basin area is lake area multiplied by 15; [b] basin area is total area multiplied by 30.
(From (1) Downing and Cataldo, 1993; (2) Aselmann and Crutzen, 1990; (3) Twilley et al., 1993; (4) Neue et al., 1990; (5) Bartels and Angenheister, 1969.)

In the state of art reports of the 1982 IRRI conference anaerobic decomposition of organic matter (Watanabe, 1984), gaseous products of organic matter decomposition in submerged soil (Neue and Scharpenseel, 1984), volatile products and low molecular weight phenolic products of anaerobic decomposition (Tsutsuki, 1984), and research advances on the effects of straw application (Ponnamperuma, 1984a) were debated. Discussion on the latter remained inconclusive (Thenobadu et al., 1985; Voss, 1988). Existing results, as well as potential uses for both stable and radioisotopes, were analysed (Scharpenseel and Neue, 1984) along with the possibilities for their enhanced use at the IRRI

Table 8. Carbon pools of different wet ecosystems in the tropics

C-pools[a]	C-fluxes in g C m^{-2} yr^{-1} or in g C yr^{-1}
Lakes, reservoirs, rivers:	
POC export of rivers	1.8
Originate from soil	1.0
DIC in carbonate lakes	100
DIC in non-carbonate lakes	5
POC storage in lakes	36 x 10^{12}
DOC stored or added to TOC retention	15 x 10^{12}
TOC in lakes with $CaCO_3$	51 x 10^{12}
DIC total retention in lakes	26 x 10^{12}
Total organic and inorganic C	75 x 10^{12}
Storage in reservoirs	19 x 10^{12}
Wetlands (1,2)	- C-fluxes atmosphere to temperate/boreal wetlands 10^{14} g yr^{-1}; - atmosphere C-storage of temperate/boreal wetlands 0.15 x 10^{15} g yr^{-1};
Ricelands (3)	- 8000 g C m^{-2} - 11.2 x 10^{15} g C totally - SOM-C yield optimum 2-2.4% - CH_4-C emission ca. 60 x 10^{12} g yr^{-1}
Continental shelves (4)	- NPP g C m^{-2} yr^{-1}; average 274, Alasca 20; - Primary production minus respiration of organic matter 84 x 10^{12} g yr^{-1}; - Global C export from continental shelves 2.5 x 10^{15} g C yr^{-1}

[a]POC, particular organic matter; DIC, dissolved inorganic matter; DOC, dissolved organic matter; TOC, total organic carbon; NPP, net plant production.
(Source: (1) Twilley et al., 1993; (2) Downing and Cataldo, 1993; (3) Neue et al., 1990; (4) Bartels and Angenheister, 1969.)

especially for turnover studies with uniformly ^{14}C labelled rice straw (Neue and Scharpenseel, 1987). Results and further possibilities for use of thin layer wise ^{14}C-age and δ ^{13}C analysis of selected rice soil profiles were also discussed (Neue et al., 1990; Scharpenseel et al., 1989; Scharpenseel and Becker-Heidmann, 1992).

As Table 2 indicates, the 140 million ha rice land sequester about 12 Pg of C (i.e. an average of close to 9 kg C m^{-2}). Trends for organic matter turnover in upland and lowland/paddy ricefields (using uniformly ^{14}C-labelled rice straw) were found to be similar (Neue and Scharpenseel, 1987). Submergence retarded SOM turnover slightly in the commonly neutral or alkaline lowland soils. In upland sites, SOM turnover in an acid Humult was similar to the reduced level in the submerged soils. Apparently for the upland rice environment, the additional influence of pH must be added to the dominant effect of temperature and moisture as determining factors for biomass turnover in humid (in both temperate as well as tropical climates) soils (Jenkinson and Ayanaba, 1977), along with the strong prevalence of soil moisture and LAC or HAC-clay minerals as driving forces for organic matter turnover in semi-arid Alfisols and Vertisols (Singer, 1993; Scharpenseel et al., 1992). Since the submerged lowland/paddy soils will generally approach a pH near neutral (Ponnamperuma, 1972; 1984b; 1985) temperature becomes the main factor in SOM turnover. Greenland (1992, personal communication) estimates, that 28°C is the temperature boundary, where rice soils turn from being a weak carbon sink into a carbon source.

Neue (1989) as well as Neue and Snitwongse (1989), in listing the field conditions for high biomass decomposition rates in ricefields, arrived at the following 11 important factors:

- soil is intensively puddled each cropping season
- soil temperature of the puddled layer is 30 to 35°C
- neutral pH
- low soil bulk density and wide soil: water ratio
- shallow floodwater
- high and balanced nutrient supply
- no long lasting accumulation of organic acids
- permanent supply of energy rich photosynthetic aquatic and benthic biomass
- high diversity of micro- and macroorganisms that provide successive fermentation down to CO_2, CH_4, H_2, and NH_3 end products
- supply of O_2 into the reduced layer by rice root excretion and oligochaetes population
- diurnial oversaturation of the floodwater with O_2 due to photosynthetic aquatic biomass enhancing the aeration function of oligochaetes.

Because O_2 diffusion into submerged soil is not commensurate to the demand of the aerobic soil organisms, the redox potential drops and is adjusted and controlled by the dominant redox buffer system (i.e. $Fe^{3+} <-> Fe^{2+}$), plus the organic compounds between Eh +100 and -100 mV. The pH in the soil solution of a reduced rice soil stabilises at about pH 6.5 to 7.0 (Ponnamperuma, 1985; Patrick and Reddy, 1978; Neue and Bloom, 1989). Addition of organic matter accelerates Eh and pH changes and leads to high concentrations of water soluble Fe, Mn and CO_2 (Katyal, 1977). In the submerged water layer, O_2 equilibration is controlled by photosynthesis-related O_2 delivery, diffusion between air and water and the consumption by processes of respiration and oxidation. As demonstrated by Roger and Kulasooriya (1980), O_2 and pH are positively correlated (CO_2 inversely). Neue and Bloom (1989) showed that biomass decomposition and humification in flooded rice soils was delayed by low energy, which was related to fermentation processes synthesising fewer microbial cells per unit of degraded C. This resulted in incomplete decomposition and subsequently less humification. Decomposition curves showed during an extended initial phase that rice straw had about a 2-year half life period, while differences in initial decomposition were influenced by alternate drying and wetting.

The merit of rice straw application (or application of other organic amendments) needed to be critically evaluated, since a higher production intensity, with continuous rice cropping per year or multiple cropping with a follow-up crop after rice, leaves less fallow time for incorporation and decomposition of the harvested rice straw. Studies conducted by Ponnamperuma (1984a) gave initial results, especially regarding N- recycling. Nagarajah et al. (1989) found soil incorporation of *Sesbania rostrata* (legume) or *Azolla anabaena* (water fern plus blue-green algae, see also Scharpenseel and Knuth, 1987) enhanced the NH_4-N both in soil solution and exchangeable fraction, whereas rice straw depressed its level. Voss (1988), compared field experiment treatments of composted straw, no straw, straw mixed, straw burned (ash only), and found very slight benefits for straw application on soil bulk density (at the 0 to 10 cm depth) which improved available water capacity and total pore volume (Figure 3 and Table 9). Neue and Bloom (1989) concluded that the immediate agronomic benefits of organic amendments were related to the C:N ratio, the lignin content and the N-equivalent added. Generally, the results of high straw application are still uncertain and not recommended at present for general practice.

Azolla anabaena, because of its high protein content (25-30%) and usefulness in mixed farming (paddy rice plus *Tilapia nilotica* in rice-fish culture, or rice plus duck raising) initially caused much interest (Scharpenseel and Knuth, 1987). However, problems were encountered related to its high dependence on P and Ca availability, as well as due to its extremely high water content (approx. 95%), which caused further problems in regard to harvest and transportation. In comparison with *Sesbania rostrata* (with its stem nodules), Azolla was always lower in NH_4-N release during its decomposition (Nagarajah et al., 1989; Neue and Bloom, 1989). In considering the organic sources, Watanabe and Roger (1985) observed a higher potential productivity of aquatic weeds (about 500 kg dry weight ha^{-1}) than of blue-green algae (150 to 250 kg dry weight ha^{-1}).

A. The Importance of Methanogenesis in Hydromorphic Soils

Soil organic matter (SOM) decomposition and humification (formation of stable humic substances) proceeds in hydromorphic soils at a slower rate than in well-drained, aerated soils. However, these processes are of special interest because in wetland soils (e.g. flooded rice fields) the anaerobic fermentation of SOM leads to the release of methane which is an important greenhouse forcing trace gas. Depending on the main soil and site parameter, a basic emission rate of about 0.5 g CH_4 m^{-2} day^{-1} is reported for wetland rice areas in Asia (Bachelet

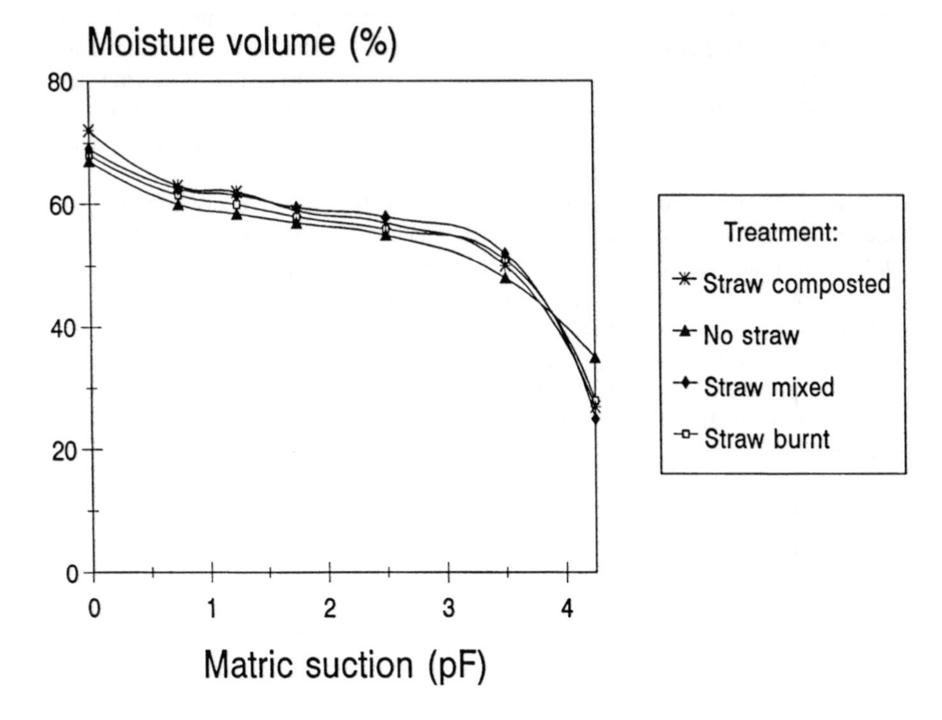

Figure 3. Moisture release pattern as affected by straw amendments of soils. (From Voss, 1988.)

Table 9. Impact of straw amendment on some soil properties in the topsoil (0-10 cm) of IRRI rice fields

Treatment	Bulk density ($g\ cm^{-3}$)	Available water capacity (vol. %)	Total pore volume (vol. %)
Straw composted	0.80	30.88	71.46
No straw	0.88	21.30	66.48
Straw mixed	0.84	33.03	67.99
Straw burned (ash)	0.87	29.38	66.93

(According to Voss, 1988.)

and Neue, 1993). Recent global estimations of methane emission rates from rice fields range from 60 to 140 Tg yr^{-1} (T = tera = 10^{12}) (Aselmann and Crutzen, 1990). Estimated methane emission rates from different natural wetlands are given in Table 10.

Hydromorphic soils are divided into an upper aerobic soil zone (zone of methane oxidation) and a deeper anaerobic zone (zone of methane formation).

Table 10. Estimated methane emission of different wetlands

Wetland type	Methane emission rate ($mg\ CH_4\ m^{-2}\ d^{-1}$)	Global methane emission ($Tg\ yr^{-1}$)
Bogs	1-50	0.4-1.8
Fens	28-216	7-52
Swamps	57-112	18-35
Marshes	137-399	12-30
Lakes	17-89	1-4
Ricefields (flooded)	300-1000	20-100

(According to Aselmann and Crutzen, 1990.)

The aerobic soil zone is characterized by a distinct structure, redox values higher than -100 mV, and high nitrate content. The main processes are aerobic decay of organic matter, biological nitrogen fixation and nitrification of ammonium and nitrite. In the aerobic layers strong reductive compounds, like CH_4 and NH_4, are oxidized as they move through the soil surface to the atmosphere. In deepwater rice soils there is also a thin aerobic layer between the deeper anaerobic layers and the upper water layer as a result of flooding. The soil temperature regime in tropical hydromorphic soil is characterized by a high temperature of about 20 to 25°C. The anaerobic condition in the lower parts of hydromorphic soils is characterized by a low redox potential less than -150 mV, pH values between 6 and 7 (Wang et al., 1993), increase in denitrification and accumulation of ammonium, and increase in reduced and soluble iron, manganese and sulfur content (see Figure 4). The main end products of organic matter decay in hydromorphic soils are CO_2 , CH_4 , NH_4, H_2, H_2S and partially humified residues.

A typical composition of a soil gas produced in the anaerobic soil zone contains 50 to 70% CH_4, 5 to 10% CO_2, 5 to 30% N_2, and <1% O_2. The main compound methane is the most reduced organic compound and its formation is the terminal step of the anaerobic food chain (see Figure 5). The methanogens (archaebacteria) play an important role in the degradation of the organic matter in hydromorphic soils. Methanogens cannot utilize complex organic compounds. The main substrates are C1 compounds like formate, methanol, methylamines and CO_2 (see equation 1), while acetate is the only C2 compound (see equation 2). The above can be described by the following simple fermentation equations:

$$CO_2 + 4H_2 \rightarrow CH_4 + 2H_2O \quad (-130\ kJ/mol\ CH_4) \qquad (1)$$

$$CH_3\text{-}COO^- + H^+ \rightarrow CH_4 + CO_2 \quad (-33\ kJ/mol\ CH_4) \qquad (2)$$

Seventy eight percent of the methane is produced by the transmethylation of acetic acid dependent methanogenesis, while 22% is produced by the CO_2/H_2 dependent methanogenesis.

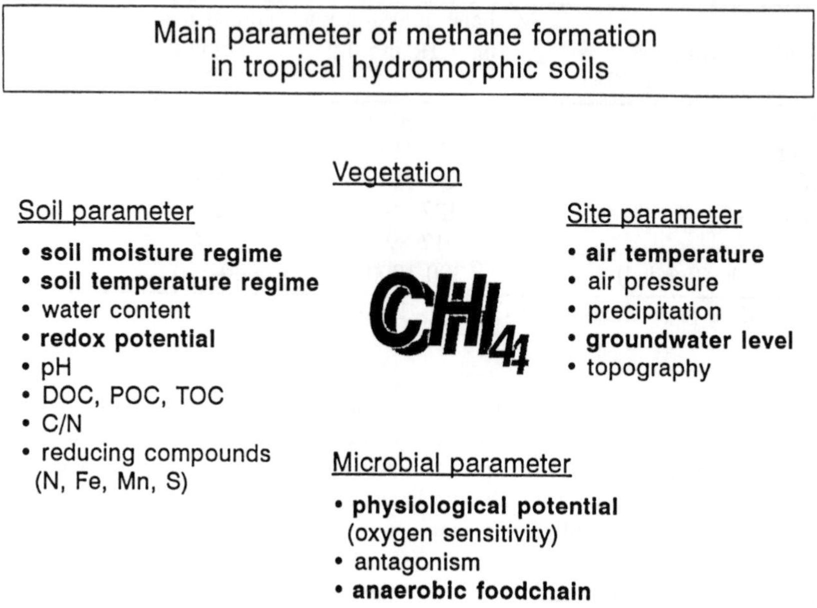

Main parameter of methane formation in tropical hydromorphic soils

Vegetation

Soil parameter

- **soil moisture regime**
- **soil temperature regime**
- water content
- **redox potential**
- pH
- DOC, POC, TOC
- C/N
- reducing compounds (N, Fe, Mn, S)

CH_4

Site parameter

- **air temperature**
- air pressure
- precipitation
- **groundwater level**
- topography

Microbial parameter

- **physiological potential** (oxygen sensitivity)
- antagonism
- **anaerobic foodchain**

Figure 4. Main parameter of methane formation in tropical hydromorphic soils.

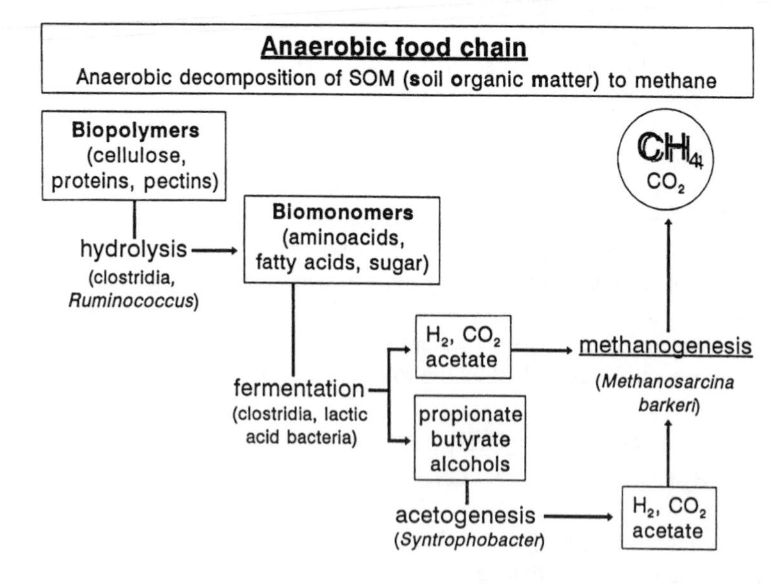

Figure 5. Anaerobic decomposition of SOM in tropical hydromorphic soils.

Methane formation is not only limited by temperature but also by the availability of the methanogenic substrates. Low temperature does not only limit the methanogens themselves, but also limits those bacteria which produce the methanogenic substrates (Conrad et al., 1987, 1989; Conrad and Babbel, 1989).

Methane formation as well as CH_4 oxidation varies with the involved species of microorganisms, the soil biochemical conditions, and the quality of the degradable soil organic matter. Nearly 95% of the produced methane is oxidized in the upper aerobic soil zone or in the aerobic water overlying the flooded rice soils. The fact that the methanogens survive the periods of desiccation when the rice soil is under aerobic conditions, indicates their resistence against O_2. Under *in vitro* conditions, about 10% of the initial methanogens survived aerobic desiccation in paddy soil samples (Fetzer et al., 1993).

The bacterial processes of methane formation and oxidation lead to significant carbon isotope fractionation by favoring the light ^{12}C over the heavier ^{13}C. The result is a strongly ^{13}C enriched residual soil organic matter (fresh rice straw, -27 ; decomposed SOM, -15 ; distinct ^{13}C depleted CH_4, -60 ; and CO_2, -18 (Takai and Wada, 1990). The $\delta^{13}C$-carbon distribution in hydromorphic soils is shown in Figure 6. The strongest enrichment of $\delta^{13}C$ in SOM is found in the anaerobic soil zones and in the main CH_4 formation zone which is characterized by high amounts of methanogenic bacteria (high MPN, most probable number) and a high potential CH_4 production. Using isotope mass balances it could be shown that up to 22% of the organic carbon supplied to rice paddies was transformed to methane (Uzaki et al., 1991).

B. Interaction with Soil Fauna in Rice Soils

While the average of organic C in rice soils should be close to 9 kg m^{-2} (Table 2), the total dry weight of organic biomass rarely exceeds 1 t ha^{-1}, showing a productivity similar to that of eutrophic lakes (Roger, 1986). In rice fields, biomass and in correlation nutrient recycling are a consequence of the coexistence of microorganisms, protozoa, zooplankton and benthos, which also include bottom-dwelling animals and invertebrate fauna, such as oligochaetes, chironomid larvae and tubificids (Roger et al., 1987; Roger and Kurihara, 1988; Neue and Bloom, 1989). Grant and Seegers (1985) observed in mineralization studies a doubling of NH_4-production in just 7 days, as a consequence of tubificial, mineralizing activities alone. However, zooplankton with population densities of 10 to 20,000 m^{-2}, along with invertebrate populations, such as chyronomid larvae, and similarly the tubeficids (Oligochaetes Tubeficidae) with population densities of about 20,000 m^{-2} produce only a few kg of dry weight ha^{-1} (Watanabe and Roger, 1985).

Snails however, especially efficient lignin decomposers, may contribute a biomass of about 1,000 kg fresh weight ha^{-1}. Together with Protozoa, Rotifera and the bacterial primary decomposers, the combined activity of rice soil fauna and flora produces faster decomposition than could be expected from the sum of subsequent species related activities.

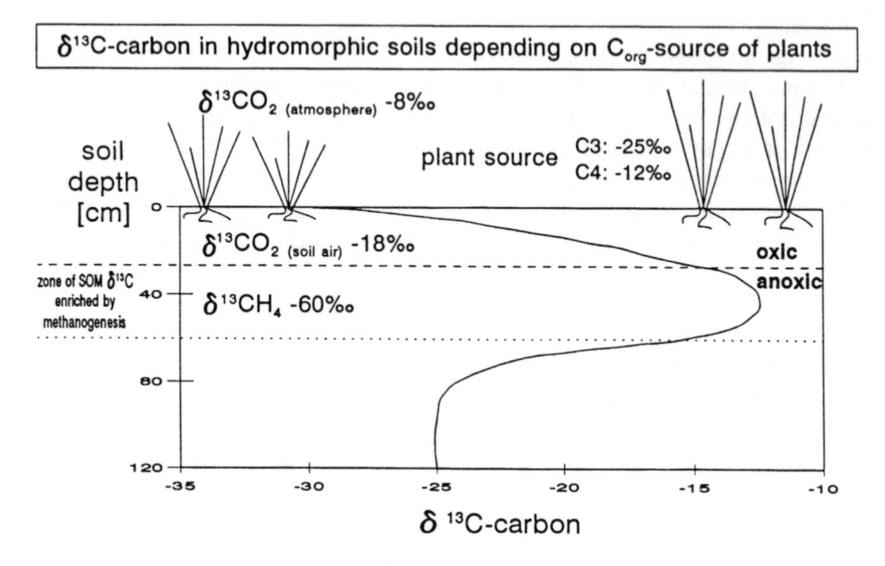

Figure 6. Scheme of the δ^{13}C-carbon pattern (‰ PDB) in hydromorphic soils depending on C_{org}-source of plants. (From Pfeiffer, 1995.)

V. Benefits of Isotope Studies

The use of isotopes in soil organic matter studies of wetlands was last reviewed by Scharpenseel and Neue (1984), including tracer studies on decomposition of ^{14}C labeled materials as well as the early investigations of natural abundances of ^{14}C and ^{13}C. While the tracer methods reveal the short-term dynamics, the natural abundance methods provide information about the long-term status of soil organic matter dynamics. An attempt to use both methods and create an integrated model of soil organic matter dynamics for a specific rice soil was described by Becker-Heidmann et al. (1985).

Due to the increase of its atmospheric concentration by atomic bomb tests between 1945 and 1962, ^{14}C has been acting as a 'natural' tracer entering the soil, allowing the calculation of net input of organic carbon. Since Becker-Heidmann and Scharpenseel (1986) introduced the thin layer sampling method, the percolation of fresh organic material into the deeper part of the soil can be detected by 'bomb carbon' found at horizon boundaries (Becker-Heidmann, 1990; Scharpenseel et al., 1989). Therefore, the loss of fresh carbon by percolation can be estimated. Other processes of organic matter dynamics, which were deduced from ^{14}C depth profiles using the thin layer sampling method on wetland soil profiles, are adsorption by clay, management practices, and climate impact (Becker-Heidmann and Scharpenseel, 1989; 1992a). The adsorption of organic matter by clay minerals is one of the major factors of carbon storage not only in terrestrial but also in wetland soils.

The $\delta^{13}C$ depth distributions, acquired simultaneously with ^{14}C, not only supports the interpretation of ^{14}C curves, but gives also independent information on soil organic matter, especially in tropical soils and soils with anaerobic conditions. The $\delta^{13}C$ value, defined as the relative deviation of the $^{13}C/^{12}C$ ratio from that of the marine carbonate standard PDB (= *Belemnitella americana* of the Cretaceous Peed De Formation) according to Craig (1953), reflects different C sources as well as processes of SOM.

$$\delta^{13}C = \frac{\frac{^{13}C}{^{12}C}(\text{sample}) - \frac{^{13}C}{^{12}C}(\text{PDB})}{\frac{^{13}C}{^{12}C}(\text{PDB})} \cdot 1000‰$$

A variety of tropical wetland soils (Inceptisols, Entisols and Ultisols) under different types of management were studied in the Philippines, Taiwan and Thailand (Becker-Heidmann, 1990; Becker-Heidmann and Scharpenseel, 1989; 1992a). The $\delta^{13}C$ values of SOM found in these soils are generally higher than in terrestrial soils, either as a result of a higher contribution of a carbon source with higher $\delta^{13}C$ (e.g. C_4 plant) or as a result of a more pronounced isotope discriminating decomposition process (methanogenesis).

Many tropical grasses assimilate carbon via the C_4 photosynthesis mechanism, which is more effective than the usual C_3 type, and in the assimilation process discriminates less against ^{13}C. Thus the plant material of these species and their decomposition products within the soil have a significantly higher $\delta^{13}C$ value. For example, $\delta^{13}C$ values of soil organic matter higher than -25 PDB in a rice paddy can result from a carbon input by C_4 weeds which may compete significantly with rice (C_3 plant). For example, in Taiwan, *E. crus-galli* caused 71 to 92% yield loss of transplanted rice in a paddy, while grasses, sedges and broadleaf weeds have been found to reduce yield of paddy rice in the Philippines by 67 to 86% (Smith, 1983). Under anaerobic conditions prevailing in wetlands, more methane than carbon dioxide is produced by decomposition of plant residues. The discrimination of ^{13}C by methane producing microorganisms is the strongest observed in nature (Rosenfeld and Silverman, 1959; Games and Hayes, 1976; Heyer et al., 1976; Whiticar et al., 1986) leading to methane with $\delta^{13}C$ values as low as -90 to -45 PDB, depending on oxygen availability among others, and correspondingly ^{13}C enriched residues remaining in the soil (Neue et al., 1990). The wetland soil prone to methane production is characterized by

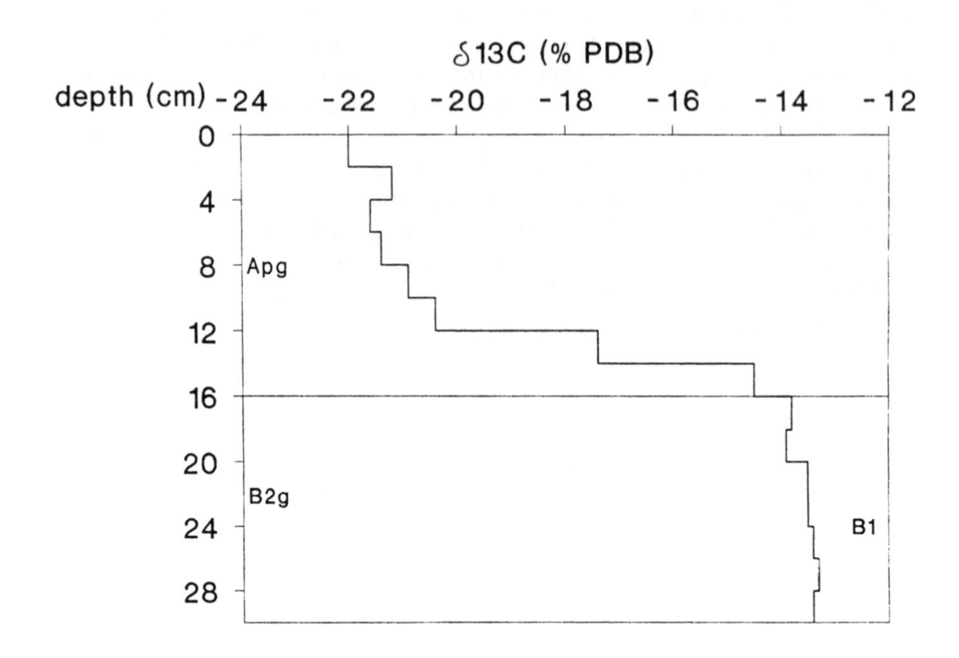

Figure 7. Depth distribution of $\delta^{13}C$ in a Haplic Hydraquent at Bugallon, Philippines, under paddy rice. (From Becker-Heidmann, 1990.)

either a $\delta^{13}C$ value above -25 PDB within the topsoil or a steep increase of δ13C with depth or both (Figure 7). To estimate carbon storage in the soil simple equations might be applied to the ^{14}C and ^{13}C data (Becker-Heidmann and Scharpenseel, 1992b), however, for more comprising assessment the development of a model approach seems preferable.

VI. Carbon Storage in Hydromorphic Soils

Wetlands, ranging from aquic conditions of terrestrial soils via bogs, fens, swamps, marshes, temporarily submerged rice fields and aquatic systems, store about 650 Pg C in about 1 billion ha area. Depending on their state and length of water saturation, the ca. 300 million ha of organic deposits in the world (Brady, 1984), which are dominantly Histosols with about 390 Pg C (Eswaran et al., 1993) existing mainly as peat and muck or organic litter horizons, are the major terrestrial wetlands. Their capacity for water storage is mostly 4 to 6 times, but can be up to 12 times the weight of the dry biomass. In the course

Table 11. Distribution of principal wetland soils of the world

Soil suborder	Area (ha 10^6)	% of land surface
Aquent	77.7	0.6
Histosol	129.5	0.9
Aquept	984.2	7.5
Alboll	51.8	0.3
Aquod	7.8	0.1
Aquult	51.8	0.4
Total	1302.8	9.7

(From Guthrie, 1985.)

of continued decomposition of organic deposits and their change from peat to muck, fragments of plants as parent material become undistinguishable. The classification criteria for Histosols according the US Soil Taxomony (Soil Survey Staff, 1992) is differentiated into fibrist, saprist, hemist and folist suborders based on their degree of decomposition.

The next best investigated wetland systems are the flooded rice soils with ca. 13 to 14 Pg C stored in 140 million ha of riceland. Just as conversion of woodland and grassland into cropland has initially released some 30 to 40 Pg C, wetland conversion into riceland seems first to deplete SOM, then under sustained use for rice production it seems to become a mild C sink at temperatures < 28°C; beyond 28°C it turns into a mild C source (Greenland, personal communication, 1992). Approximate amounts of C storage, annual C inputs/outputs and methane emission rates are given in Tables 1-8 and Table 10.

The distribution of principal wetland soils in the world and their diagnostic and associated properties are given in Tables 11 and 12. In the World Soil Resources (1991) of the FAO, wetlands and hydromorphic soils of the tropics and subtropics cover an area of together 723.36 million ha. They belong mainly to the four soil groups given in Table 13. Undoubtedly not all areas of the Histosols, Fluvisols, Gleysols and Planosols would qualify for all wetland criteria. Thus, the 723 million ha for (sub)tropical wetlands would be a high estimate.

Wet forests seem to excel in C dynamics and sequestration (also N_2O emission), although Whittaker and Likens (1973) emphasize the importance also of C sequestration in temperate grassland. Soil epipedons of the latter show similar levels of C sequestration as the living biomass of tropical rain forests. Use of 2 cm-wise thin layer sampling of different wetland soils indicates, especially for puddled riceland soils, that although considered seemingly mild C-sinks, most have (Figure 8) bomb-^{14}C to considerable depths and show low residence time for SOM-C. This must be the consequence of rapid C cycling, which may be due to the dispersion by puddling and reduced formation of stable clay/oxide-humus-complexes. The latter, in terrestrial soils, form the major part of the so called 'passive C compartment' of several thousand years C residence

Table 12. Diagnostic and associated properties of some wetland soils

Classification	Morphology	Physical/chemical properties
Tropaqualfs	Long periods of wetness each year	High base saturation
Tropaquents	Long periods of wetness each year	Variable texture, moderate fertility
Fluvaquents	Annual flooding, variable wetness	High fertility, stratified textures, moderate OM
Hydraquents	Permanent wetness	Low strength, irreversible hardening upon drying
Sulfaquents	Long periods of wetness each year, tidal influence	Extremely low pH after drainage
Tropaquepts	Long periods of wetness each year	Moderate fertility, variable textures
Halaquepts	Long periods of wetness each year	High sodium, salt spray
Argialbolls	Seasonal wetness, eluvial/illuvial horizon sequence	High fertility, high organic matter
Ochraquox	Seasonal wetness, minimal profile differentiation	Low CEC, low fertility
Tropaquods	Long periods of wetness each year, podzolic profile	Variable charge clays, sandy, low fertility
Plinthaquults	Long periods of wetness each year, plinthite, eluvial/illuvial	Low fertility, low pH
Tropaquults	Seasonal wetness, eluvial/illuvial horizon sequence	Low fertility, low pH

(According to Guthrie, 1985.)

Table 13. Areas of different hydromorphic soil groups in the tropics and subtropics (ha x 10^3)

Soil group	Mediterranean	Arid	Seasonally dry tropics	Humid	Total area
Histosols	1,823	3,410	12,232	32,449	49,914
Fluvisols	14,430	90,074	84,360	66,207	255,071
Gleysols	4,925	34,492	111,543	167,704	318,664
Planosols	15,598	3,762	74,083	6,267	99,710
Total					723,359

(According to World Soil Resources, 1991.)

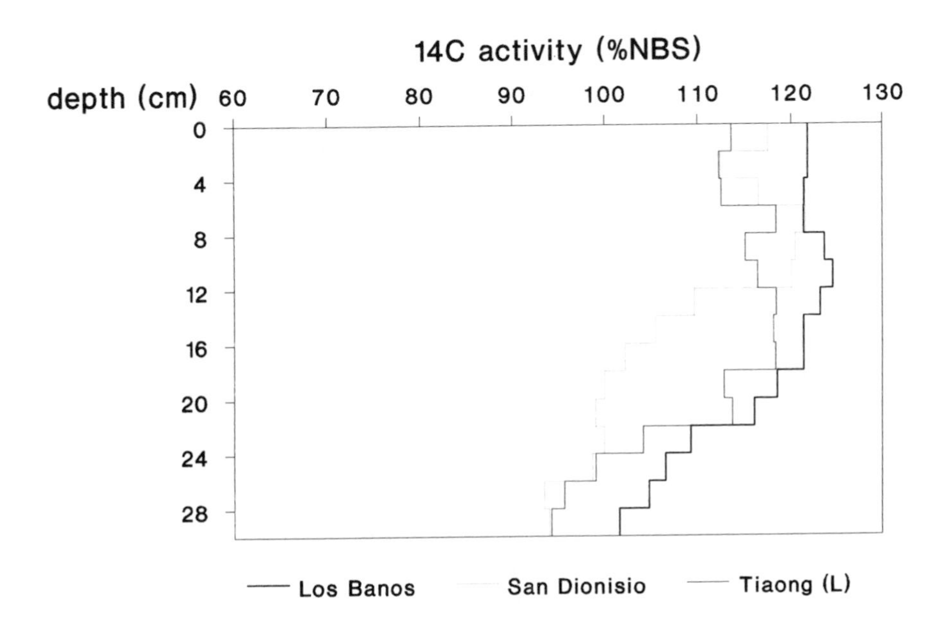

Figure 8. Depth distribution of ^{14}C-activity (% NBS) of three Philippine rice soils: Los Banos (Aeric Tropaquept), San Dionisio (Typic Paleudult), Tiaong L (Haplic Hydraquent).

time (Parton et al., 1987). In comparison, for terrestrial gley soils (with aquic conditions) the C residence time gradient with depth follows the usual trend of a steady rise of SOM residence time (^{14}C-age) with depth (Figure 9).

A tentative assessment of the carbon storage in hydromorphic soils is reflected in Tables 1, 2, 5, 7, and 8.

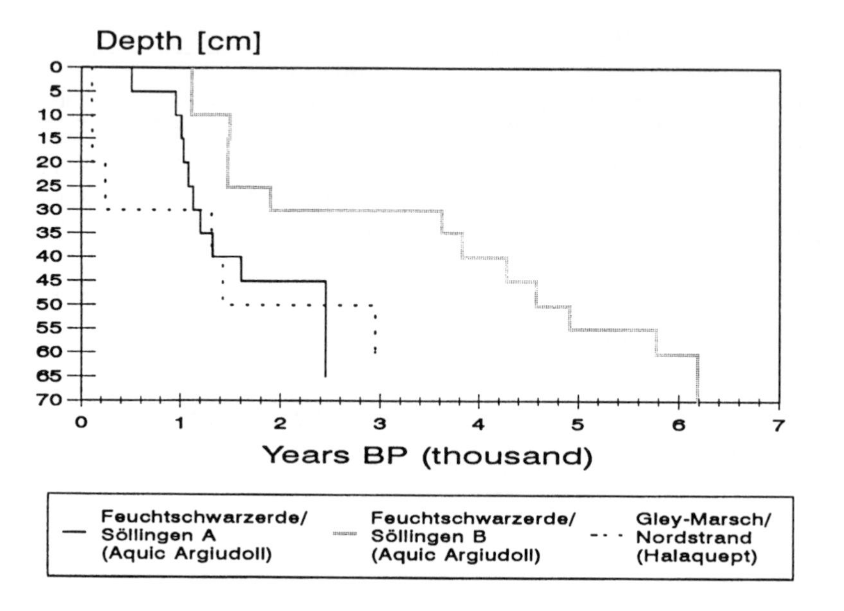

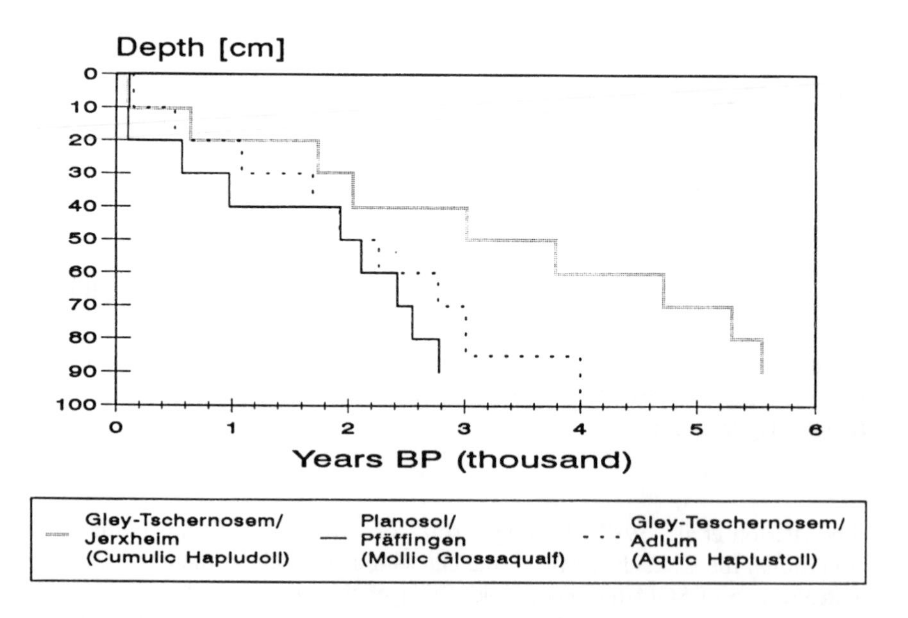

Figure 9. Depth function of the ^{14}C-age (x 1000 years BP) of the SOM of different soils with aquic moisture regime.

VII. Summary and Conclusions

Second to the largest compartment of the organic carbon cycle (i.e. the fossil fuel - C with more than 10,000 Pg C pool size and ca. 1000 Pg C economically accessable oil-, gas- and coal-C resources) is the C of the SOM. The SOM compartment has ca. 1550 Pg C plus near to 50% of the ca. 1000 Pg C in almost one billion ha of caliches/calcretes (carbonate). Wetlands are estimated worldwide to comprise about 1 billion ha and contain ca. 650 Pg C. Approximately, 700 million ha of these wetlands are found in the tropics, and store some 180 to 200 Pg C.

Most tropical soils, with the exception of those under woodland, are in steady state. Andisols, however, with initial C accumulation in allophane (young volcanic ash minerals)- humus-complexes act as C-sinks in the first 1000 to 10,000 years of their development. Subsequent respiratory C release (C- source) occurs upon change of allophane into clay minerals (often ca. 10,000 years of age). Riceland can also be a mild C sink below a paddy temperature of 28°C, but at higher temperatures can serve as a rather mild C source, with pronounced fluctuations. Generally, as for soils in temperate regions, the remaining humus-C after microbial and soil faunal attack is protected by formation of clay/oxide-humus- complexes, which are the passive C compart- ment with > 1000 years of ^{14}C-residence time.

In tropical LAC soils (low activity clay soil) SOM is especially important as a major matrix of the CEC (cation exchange capacity). In tropical wetlands, SOM is involved in N_2O-release in the processes of denitrification, nitrification and diazotropic N-collection. In riceland soils the zone of methanogenesis is characterised by δ^{13}C increase in the residual SOM (as demonstrated by soil profiles scan for δ^{13}C) due to emission of the strongly ^{13}C depleted CH_4 (δ^{13}C values about -60). Furthermore, natural ^{14}C measurements indicate a very low carbon residence time in the upper part of puddled riceland due to fast oxidative replacement of SOM, mainly due to the puddling process or rotavator technics.

Optimal SOM content in riceland for best yields appears to be 4 to 4.5%. Amendment studies indicated that straw input in the paddy seems to produce slight yield increases, mainly by improving soil bulk density and water holding capacity. Introduction of *Azolla* and *Sesbania* diazotrophs require adequate PO_4 and Ca levels for optimal amendment of soil fertility. However, the relatively small dry weight derived from zooplankton, invertebrates and tubeficids are important for nutrient recycling. Overall, the mixed population of fauna and flora in rice soil is especially efficient in organic matter decomposition.

References

Aselmann, J. and P.J. Crutzen. 1990. A global inventory of wetland distribution and seasonality, net primary productivity and estimated methane emissions. p.441-455. In: A.F.Bouwman (ed.), *Soils and the greenhouse effect*. J. Wiley & Sons, New York.

Bachelet, D. and H.U. Neue. 1993. Methane emission from wetland rice areas of Asia. *Chemosphere* 26:219-237.

Barnwell, T.O., R.B. Jackson, E.T. Elliott, I.C. Burke, C.V. Cole, K. Paustian, E.A. Paul, A.S. Donigian, A.S. Patwardhan, A. Rowell, and K. Weinrich. 1992. An approach to assessment of management impacts on agricultural soil carbon.In: J. Wisniewski and A.E. Lugo (eds.), Natural sinks of CO_2. *Water, Air, Soil Pollut.* 64:423-435.

Bartels, J. and G. Angenheister.1969. *Geophysik*. Fischer, Frankfurt.

Batjes, N.H. 1992. Methane. p. 37-66. In: N.H. Batjes and E.M. Bridges (eds.), *A review of soil factors and processes that control fluxes of heat, moisture and greenhouse gases*. ISRIC, Wageningen.

Batjes, N.H. and E.M. Bridges (eds.). 1992. *World inventory of soil emissions potentials*. ISRIC. Wageningen.

Becker-Heidmann, P. 1990. Carbon fluxes in important soil classes, with emphasis on Lessivé soils and on soils of the terrestrial, of the hydromorphic, and temporarily submerged environment. Final report to GTZ and DFG (unpl.).

Becker-Heidmann, P., U. Martin, and H.W. Scharpenseel.1985. Radiokohlenstoffdatierung und Abbau von 14C-markiertem Reisstroh zur Modellierung der Kohlenstoffdynamik eines Reisbodens. *Mitteilungen der Deutschen Bodenkundlichen Gesellschaft* 43/2:525-530.

Becker-Heidmann, P. and H.W. Scharpenseel. 1986. Thin layer $\delta^{13}C$ and $D^{14}C$ monitoring of "Lessivé" soil profiles. *Radiocarbon* 28:383-390.

Becker-Heidmann, P., and H.W. Scharpenseel. 1989. Carbon isotope dynamics in some tropical soils. *Radiocarbon* 31:672-679.

Becker-Heidmann, P., and H.W. Scharpenseel. 1992a. Studies of soil organic matter dynamics using natural carbon isotopes. *The Science of the Total Environ- ment* 117,118:305-312.

Becker-Heidmann, P., and H.W. Scharpenseel. 1992b. The use of natural ^{14}C and ^{13}C in soils for studies on global climatic change. *Radiocarbon* 34/3:535-540.

Bishop, A.A. 1975. On farm water management in soil management in tropical America. p. 186-194. In: E. Bornemisza and A. Alverado (eds.), North Carolina State Univ. Publ.

Bouwman, A.F. (ed.). 1990. *Soils and the greenhouse effect*. John Wiley & Sons, New York.

Bouwman, A.F. 1992. Methodology and data used to estimate natural N_2O emissions. p. 93-106. In: N.H. Batjes and E.M. Bridges (eds.), *World inventory of soil emissions potentials*. ISRIC. Wageningen.

Braatz, B.V. and K.B. Hogan. 1991. Sustainable rice productivity and methane reduction. p.1-7. In: Research Plan US EPA. Office Air and Radiation. Washington D.C., (Draft).

Brady, N.C. 1984. *The nature and properties of soils*. Macmillan Publ. NY. 478 pp.

Brinkman, R. 1979. *Ferrolysis, a soil forming process in hydromorphic conditions*. PUDOC, Wageningen No. 106.

Broecker, W.S. and T.H. Peng. 1992. Interhemispheric transport of carbon dioxide by ocean circulation. *Nature* 356:587-589.

Brown, R., A.E. Lugo, and L.R. Iverson. 1992. Processes and lands for sequestering carbon in the tropical forest landscape. In: J. Wisniewski and A.E. Lugo (eds.), Natural sinks of CO_2. *Water, Air, Soil Pollut.* 64:139-155.

Buol, St. W., P.A. Sanchez, R.B. Cate, and M.A. Granger. 1975. Soil fertility capability classification in soil management in tropical America. p. 126-141. In: E. Bornemisza and A. Alvaredo (eds.), North Carolina State University Publ. of Proc. CIAT. Cali-Seminar Febr. 1974.

Conrad, R., W. Seiler and G. Bunze. 1983. Factors influencing the loss of fertilizer nitrogen into the atmosphere as N_2O. *J. Geophysical Research* 88:6709-6718.

Conrad, R., H. Schütz and M. Babbel. 1987. Temperature limitation of hydrogen turnover and methanogenesis in anoxic paddy soil. *FEMS Microbiol. Ecol.* 45:281-289.

Conrad, R. and M. Babbel. 1989. Effect of dilution on methanogenesis, hydrogen turnover and interspecies hydrogen transfer in anoxic paddy soil. *FEMS Microbiol. Ecol.* 62:21-28.

Conrad, R., H.-P. Mayer, and M. Wüst. 1989. Temporal change of gas metabolism by hydrogen-syntrophic methanogenic bacterial associations in anoxic paddy soil. *FEMS Microbiol. Ecol.* 62:265-274.

Craig, H. 1953. The geochemistry of stable carbon isotopes. *Geochim. Cosmochim. Acta* 3:53-92.

Degens, E.T. 1989. *Perspectives on biogeochemistry*. Springer. Berlin. 423 pp.

Denier van der Gon, H.A.C., H.U. Neue, R.S. Lantin, R. Wassmann, M.C.R. Alberto, J.B. Aduna, and M.J.P. Tan. 1992. Controlling factors of methane emission from rice fields. p. 81-92. In: N.H. Batjes and E.M. Bridges (eds.), *World inventory of soil emission potentials.* ISRIC. Wageningen.

Downing, J.P. and D.A. Cataldo. 1992. Natural sinks of CO_2: Technical synthesis from the Palmas Del Mar Workshop. In: J. Wisniewski and A.E. Lugo (eds.), Natural sinks of CO_2. *Water, Air, Soil Pollut.* 64:439-454.

Enting J.G. and J.V. Mansbridge. 1991. Latitudinal distribution of sources and sinks of CO_2: Results of an inversion study. *Tellus* 43B:156-170.

Eswaran, H., E. Van Den Berg, and P.F. Reich. 1993. Organic carbon in soils of the world. *Soil Sci. Soc. Am. J. 57:192-194.*

Fetzer, S., F. Bak, and R. Conrag. 1993. Sensitivity of methanogenic bacteria from paddy soil to oxygen and desiccation. *FEMS Microbiol. Ecol.* 12:107-115.

Games, L.M. and J.L. Hayes. 1976. On the natural mechanism of CO_2 and CH_4 production in natural anaerobic environments. p. 51-73. In: J.O. Nriagu (ed.), *Environmental biogeochemistry. Vol. 1*, Ann Arbor Science, Ann Arbor, MI.

Glenn, E.P., L.F. Pitelka, and M.W. Olsen. 1992. The use of halophytes to sequester carbon. In: J. Wisniewski and A.E. Lugo (eds.), Natural sinks of CO_2. *Water, Air, Soil Pollut.* 64:251-263.

Gorham, E. 1991a. Northern peatlands: role in the carbon cycle and probable responses to climate warming. *Ecol. Appl.* 1:182-195.

Gorham, E. 1991b. Ecological Applications 1. p. 182. In: J.P. Downing et al. (eds.), 1993. *Land and water interface zones.* Workshop Proceedings Terrestrial biospheric carbon fluxes: quantification of sinks and sources of CO_2. Bad Harzburg- Workshop, Kluwer Academic Publishers.

Grant, J.F., and R. Seegers. 1985. Tubificial role in soil mineralization and recover of algal nitrogen by lowland rice. *Soil Biol. Biochem.* 17:559-563.

Graßl, H., and R. Klingholz. 1990. *Wir Klimamacher.* Fischer Publ., Frankfurt. 70 pp.

Guthrie, R.L. 1985. Characterizing and classifying wetland soils in relation to food production. p. 11-20. In: *Wetland soils: characterization, classification and utilisation.* IRRI Publ. Manila.

Heyer, J., H. Hübner, and I. Maaß. 1976.Isotopenfraktionierung des Kohlenstoffs bei der mikrobiellen Methanbildung. *Isotopenpraxis* 5:202-205.

Jenkinson, D.S. and A. Ayanaba. 1977. Decomposition of ^{14}C labelled plant material under tropical conditions. *Soil Sci. Soc. Am. J.* 41:912-915.

Johnson, D.W. 1992. Effects of forest management on soil carbon storage. In: J. Wisniewski and A.E. Lugo (eds.), Natural sinks of CO_2. *Water, Air, Soil Pollut.* 64:83-120.

Katyal, J.C. 1977. Influence of organic matter on chemical and electrochemical properties of some flooded soils. *Soil Biol. Biochem.* 9:259-266.

Keeling, C.D., S.C. Piper, and M. Heimann. 1989. p. 305-363. In: D.H. Peterson (ed.), *Aspects of climate variability in the Pacific and in the Western Americas.* Geophys. Monograph 55. American Geophysical Union, Washington D.C.

Kyuma, K. 1985. Fundamental characteristics of wetland soils. p. 191-206. In: *Wetland soils: characterization, classification and utilization.* IRRI Publ. Manila.

Lugo, A.E. and J. Wisniewski. 1992. Natural sinks of CO_2: conclusions, key findings and research recommendations from the Palmas Del Mar Workshop. In: J. Wisniewski and A. Lugo (eds.), Natural sinks of CO_2. *Water, Air, Soil Pollut.* 64:455-460.

Martin, U., H.U. Neue, H.W. Scharpenseel, and P. Becker. 1983. Anaerobe Zersetzung von Reisstroh in einem gefluteten Reisboden auf den Philippinen. *Mitteil. Deutsche Bodenkundlicher Gesellschaft* 38:245-250.

Moormann, F.R. 1980. Future activities of International Committee on Soil with Aquic Moisture Regime (ICOMAQ). p. 537-540. In: *Wetland soils: characterization, classification and utilization.* IRRI Publ. Manila.

Moormann, F.R. and N. van Breemen. 1978. *Rice: soil, water, land.* IRRI Publ. Manila.

Moormann, F.R. and H.T.J. van de Wetering. 1985. Problems in characterizing and classifying wetland soils. p. 53-70. In: *Wetland soils: characterization, classification and utilization.* IRRI Publ. Manila.

Nagarajah, S., H.U. Neue and M.C.R. Alberto. 1989. Effect of Sesbania, Azolla and Rice straw incorporation on the kinetics of NH_4, K, Fe, Mn, Zn, and P in some flooded rice soils. *Plant Soil* 116:37-48.

Neue, H.U. 1985. Organic matter dynamics in wetland soils. p. 109-122. In: *Wetland soils: characterization, classification and utilization.* IRRI Publ. Manila.

Neue, H.U. 1989. Holistic view of chemistry of flooded soils. In: Proceedings of the First International Symposium on Paddy Rice Fertility. International Board for Soil Research and Management (IBSRAM). Bangkok. 33pp.

Neue, H.U. and P.R. Bloom. 1989. Nutrient kinetics and availability in flooded rice soils. p. 173-190. In: *Progress in irrigated rice research.* IRRI Publ. Manila.

Neue, H.U. and H.W. Scharpenseel. 1984. Gaseous products of the decomposition of organic matter in submerged soils. Int. Conf. *SOM in Rice,* IRRI Publ. Manila.

Neue, H.U. and V.P. Singh. 1984. Management of wetland rice and fishponds on problem soils in the tropics. Problem Soils in Asia. *FFTC Book Series* 27:352-366.

Neue, H.U. and H.W. Scharpenseel. 1987. Decomposition pattern of ^{14}C-labelled rice straw in aerobic and submerged rice soils of Philippines. *The Science of the Total Environment* 62:431-434.

Neue, H.U. and P. Snitwongse. 1989. Organic matter and nutrient kinetics. In: Proceedings of the International Rice Research Conference of 1988. IRRI Publ. Manila. 22 pp.

Neue, H.U., P. Becker-Heidmann and H.W. Scharpenseel. 1990. Organic matter dynamics, soil properties and cultural practices in rice lands and their relatioship to methane production. p. 457-466. In: A.F. Bouwman (ed.), *Soils and greenhouse effect*. J. Wiley & Sons, New York.

Neue, H.U., C. Quijano and H.W. Scharpenseel. 1991. Conservation and enhancement of wetland soils. In: *Evaluation for Sustainable Land Management in the Developing World.* Technical Papers, IBSRAM Proceedings No. 12, Vol.II:279-303.

Neue, H.U., C.L. Gaunt, Z.V. Wang, P. Becker-Heidmann and C. Quijano. 1995. Soil carbon in tropical wetlands. Proc. Nairobi Workshop, E.T. Elliott (ed.), *Geoderma* (in press).

Orr, J.C. and J.L. Sarmiento. 1992. Potential of marine macroalgae as a sink for CO_2: Constraints from a 3-D general circulation model of the global ocean.In: J. Wisniewski and A.E. Lugo (eds.), Natural sinks of CO_2. *Water, Air, Soil Pollut.* 64:405-422.

Ottow, J.C.G., G. Benckiser and I. Watanabe. 1982. Iron toxicity as a multiple nutritional soil stress. *Trop. Agric. Research Series* 15:167-179.

Parton, W.J., D.S. Schimmel,C.V. Cole and D.S. Ojima. 1987. Analysis of factors controlling soil organic matter levels in Great Plains grassland. *Soil Sci. Soc. Am. J.* 51:1173-1179.

Patrick, W.H., Jr. and C.N. Reddy. 1978. Chemical changes in rice soils. In: *Soils and rice.* IRRI Publ. Manila.

Pfeiffer, E.-M. 1995. *δ13*C-Analysis of soil carbon and its relation to methane fluxes in freshwater marshes of the Elbe River in Northern Germany. International Symposium on Soil Processes and Management Systems, Greenhouse Gas Emissions and Carbon Sequestration. Columbus, Ohio. 5-9 April 1993. (in press)

Ponnamperuma, F.N. 1972: The chemistry of submerged soils. *Adv. Agron.* 24:29-96.

Ponnamperuma, F.N. 1984a. Straw as a source of nutrients for wetland rice. Intern. Conf. on Organic Matter and Rice (1982). IRRI Publ. Manila.

Ponnamperuma, F.N. 1984b. Effects of flooding on soils. p. 9-45. In: T.T. Koslowski (ed.), *Flooding and plant growth.* Academic Press, N.Y.

Ponnamperuma, F.N. 1985. Chemical kinetics of wetland rice soils relative to soil fertility. p.71-89. In: *Wetland soils: characterization, classification and utilization.* IRRI Publ. Manila.

Ritchard, R.L. 1992. Marine algae as a CO_2 sink. In: J. Wisniewski and A.E. Lugo (eds.). Natural Sinks of CO_2. *Water, Air, Soil Pollut.* 64:289-303.

Roger, P.A., and Kulasooriya. 1980. *Blue-green algae and rice.* IRRI Publ. Manila. 112 pp.

Roger, P.A. 1986. Effect of algae and aquatic macrophytes on nitrogen dynamics in wetland rice fields. p. 21-31. Proc. Congr. Intern. Soil Science Soc. Hamburg.

Roger, P.A., J.F. Grant, P.M. Reddy and I. Watanabe 1987. The photosynthetic aquatic biomass in wetland rice fields and its effects on nitrogen dynamics. p. 43-68. In: *Efficiency of nitrogen fertilizers for rice.* IRRI Publ. Manila.

Roger, P.A. and Y. Kurihara. 1988. Floodwater biology of tropical wetland rice-fields. Proc. Symposium on Paddy Soil Fertility. Chiang Mai, Thailand.

Rosenfeld, W.D. and S.R. Silverman. 1959. Carbon isotope fractionation in bacterial production of methane. *Science* 130:1658-1659.

Sanchez, P.A. and S.W. Buol. 1985. Agronomic taxonomy for wetland soils. p.207-227. In: *Wetland soils: characterization, classification and utilization.* IRRI Publ., Manila.

Sanchez, P.A. and S.W. Buol. 1987. FCC Adaption to wetland soils. p. 74-76. In: N. Candle and C.B. McCants (eds.), Tropsoils techn. reports 1985-86. North Carolina State University Publ.

Scharpenseel, H.W. 1993. Major carbon reservoirs of the pedosphere; source - sink relations, potential of $\delta^{14}C$ and $\delta^{13}C$ as supporting methodologies. In: J. Wisniewski (ed.), Proc. of Bad Harzburg Workshops. *Water, Air, Soil Pollut.* 70:431-442.

Scharpenseel, H.W. and H.U. Neue. 1984. Use of isotopes in studying the dynamics of organic matter in soils. p. 273-310. In: *Organic matter and rice.* International Rice Research Institute. IRRI Publ. Manila.

Scharpenseel, H.W. and K. Knuth. 1987. Use and importance of *Azolla anabaena* in industrial countries. p. 153-167. Proc. Workshop on Azolla Use, Fuzhou, Fujian, China. IRRI Publ. Manila.

Scharpenseel, H.W. and G. Miehlich. 1989. Soil fertility and organic fractions. p. 111-119. Proc. IBSRAM Pacificland Workshop. vol. 8. IBSRAM Publ. Bangkok.

Scharpenseel, H.W., P. Becker-Heidmann, H.U. Neue and K. Tsutsuki. 1989. Bomb-carbon, 14C-dating and 13C-measurements as tracers of organic matter dynamics as well as of morphogenetic and turbation processes. *The Science of the Total Environment* 81/82:99-110.

Scharpenseel, H.W. and P. Becker-Heidmann. 1992. Twenty-five years of radiocarbon dating soils: paradigm of erring and learning. *Radiocarbon* 34:541-549.

Scharpenseel, H.W., H.U. Neue, and St. Singer. 1992. Biotransformations in different climate belts, source-sink relationships. p. 91-105. In: J. Kubat (ed.), *Humus, its structure and role in agriculture and environment*. Elsevier, Amsterdam.

Schlesinger, W. 1990. Evidence from chronosequence studies for a low carbon storage potential of soils. *Nature* 348:232-239.

Singer, St. 1993. The turnover of ^{14}C-labelled groundnut straw, soil organic matter dynamics and CO_2 evolution in an Alfisol and a Vertisol of semiarid tropical India. (Dissertation) Hamburger Bodenkundliche Arbeiten. Vol.19.

Soil Survey Staff. 1992. *Keys to Soil Taxonomy*. SMSS technical monograph No. 19. Blacksburg, Virginia.

Smith, R.J., Jr. 1983. Weeds of major economic importance in rice and yield losses due to weed competition. p. 19-36. In: International Rice Research Institute, Proc. Conf. on Weed Control in Rice, 31 Aug - 4 Sep.

Sombroek, W.G., F.O. Nachtergaele, and A. Hebel. 1993. Amounts, dynamics and sequestering of carbon in tropical and subtropical soils. *Ambio* 22:417-425.

Sunquist, E.T. 1985. Geological perspectives on carbon dioxide and the carbon cycle. In: E.T. Sundquist and W.S. Broecker (eds.), *The carbon cycle and atmospheric CO_2; natural variations archean to present*. Geophysical Monograph 32. American Geophysical Union, Washington D.C.

Takai, Y. and E. Wada. 1990. Methane formation in waterlogged paddy soils and its controlling factors. p. 101-107. In: Scharpenseel et al. (eds.), *Soils on a warmer earth*. Developments in Soil Science 20. Elsevier, Amsterdam.

Tans, P., I.Y. Fung, and T. Takahashi. 1990. Observational constraints on the global CO_2 budget. *Science* 247:1431-1438.

Thenobadu, M.W., M.C. Rubielos, and H.U. Neue. 1985. Effect of straw incorporation on growth and yield of rice in three soils. IRRI, Special Seminar, Soil Chemistry Department. IRRI Publ. Manila.

Tsutsuki, K. 1984. Volatile products and low molecular weight phenolic products of the aerobic decomposition of organic matter. In: *Organic matter and rice*. IRRI Publ. Manila.

Twilley, R.R., R.H. Chen and T. Hargis. 1992. Carbon sinks in mangroves and their implications to carbon budget of tropical coastel ecosystems. In: J. Wisniewski and A.E. Lugo (eds.). Natural sinks of CO_2. *Water, Air Soil Pollut.* 64:265-288.

Twilley, R.R., J. Downing, M. Maybeck, J. Orr and H.W. Scharpenseel. 1993. Report on land and water interface zones. In: Terrestrial biospheric carbon fluxes: Quantification of sinks and sources of CO_2. Wetland disscusion group document. *Water, Air, Soil Pollut.* 70:1-4.

Uehara, G. 1982. Soil clay minerals in tropical soils and fertility. Transaction 12th Intern. Congr. Soil Science, New Delhi, Comm. VII. 404 pp.

Umarov, M.M. 1990. Biotic sources of nitrous oxide (N_2O) in the context of the global budget of nitrous oxide. p. 263-268. In: A.F. Bouwman (ed.), *Soils and the greenhouse effect*. J. Wiley & Sons, New York.

Uzaki, M., H. Mizutani and E. Wada. 1991. Carbon isotope composition of CH_4 from rice paddies in Japan. *Biogeochemistry* 13:159-175.

Voss, G. 1988. Effect of organic amendments on soil properties and rice production. Terminal Report to IRRI. IRRI Publ., Manila. 12 pp.

Wang, Z. P., R.D. DeLaune, P.H. Masscheloyn, and W.H. Patrick, jr. 1993. Soil redox and pH effects on methane production in a flooded rice soil. *Soil Sci. Soc. Am. J.* 57:382-385.

Watanabe, I. 1984. Anaerobic decomposition of organic matter in flooded rice soils. p. 237-258. In: *Organic matter and rice*. IRRI Publ. Manila.

Watanabe, I. and P.A. Roger. 1985. Ecology of flooded ricefields. p. 229-243. In: *Wetland soils: characterization, classification and utilization*. Proc. IRRI Workshop. IRRI Publ. Manila.

Whiticar, M.J., E. Faber, M. Schoell. 1986. Biogenic methane formation in marine and freshwater environments; CO_2 reduction vs. acetate fermentation - Isotope evidence. Geochem. *Cosmochem Acta* 50:693-709.

Whittaker, R.H. and G.E. Likens. 1973. The primary production of the biosphere. *Human Ecol.* 1:299-369.

Winjum, J.K., R.K. Dixon, and P.E. Schroeder. 1992. Estimating the global potential of forest and agroforest management practices to sequester carbon.In: J. Wisniewski and A.E. Lugo (eds.), Natural sinks of CO_2. *Water, Air and Soil Pollution* 64:213-227.

Wisniewski, J. and A.E. Lugo (eds.), 1992. *Natural Sinks of* CO_2. Kluwer Academic Publishers, Dordrecht.

Wollast, R. and F.T. MacKenzie. 1989. Global biogeochemical cycles and climate. p. 453-473. In: A. Berger (ed.), *Climate and geosciences*. Kluwer Academic Publishers, Dordrecht.

World Soil Resources. 1991. p. 31-32. In: *World soil resources*. Report 66. FAO. Rome.

Assessment of Soil Organic Matter Storage

Conservation Strategies for Improving Soil Quality and Organic Matter Storage

D.L. Karlen and C.A. Cambardella

I. Introduction

Agricultural soils can function as both a sink and source for atmospheric C. Their role is dynamic, often fluctuating between the two processes as a result of how they are being managed. Changes in soil organic matter (SOM) provide the primary measure for determining which direction current management practices are headed. SOM concentrations affect plant nutrition (Stevenson, 1982), soil structure and compactibility (Soane, 1990), and water retention (DeJong et al., 1983). Changes in the latter parameters often have a large influence on the long-term sustainability of soil (Follett et al., 1987; Childs et al., 1989). Soil C concentrations are also considered a primary indicator of soil quality (Karlen et al., 1992; Karlen et al., 1994), which has been defined as the capacity of soil to function, within ecosystem boundaries, to sustain biological productivity, maintain environmental quality, and promote plant, animal, and human health (Doran and Parkin, 1994).

The objectives for this chapter are to review some of the characteristics of the soil C pool and to suggest how changes in this pool may affect global warming. Potential effects of four management strategies: (1) Conservation tillage; (2) Cover cropping; (3) Crop rotations; and (4) Manure and fertilizer applications

1-56670-033-7/96/$0.00+$.50

are then discussed with regard to how they might affect the soil C pools and soil quality, and the processes that determine whether a soil is functioning as a source or sink for atmospheric C.

II. Soil C Pool

A. Characteristics

The soil C pool is approximately two or three times larger than the amount of C stored in living vegetation (Schlesinger, 1990). Estimated at 1500 Pg (P = peta = 10^{15}), this pool plays a dynamic part in the geochemical C cycle. The C cycle involves the interaction of many complicated processes, each operating at different rates (Hobbie et al., 1984). Soil factors affecting biotransformation rates among various C pools include temperature, texture, water content, soil pH, redox potential, and the nature of the inorganic complexing matrix (Scharpenseel et al., 1992).

Changes in land use over the past two centuries have caused a large release of CO_2 to the atmosphere from the terrestrial biota and soils. Houghton et al. (1983) concluded that worldwide, there has been an annual net release of C from terrestrial ecosystems since at least 1860. Prior to approximately 1960, the annual release associated with changes in land use exceeded that from fossil fuels. Houghton et al. (1983) estimated the total release of C from the world's terrestrial ecosystems since 1860 has been between 135 and 228 Pg. Release of C since 1958 was estimated to range from 38 to 76 Pg. The large range in these numbers is primarily due to uncertainty regarding the effects of clearing tropical forests.

Source and sink relationships vary within different climatic zones and hemispheres. Houghton (1987) concluded that the northern hemisphere was more important because it has almost twice as much land area, and presumably, twice as much terrestrial metabolism as the southern hemisphere. He also reported that the seasonal amplitude in CO_2 concentration is larger in the northern hemisphere, and that geophysical models calculate the net seasonal flux to be three to four times greater than in the southern hemisphere.

Surface soils are conceptualized to contain a pool of very stable or recalcitrant organic C that has a mean age of several thousand years (van Veen and Paul, 1981). The Morrow Plots at the University of Illinois and Sanborn Field at the University of Missouri were found to have stable organic matter pools in the surface 0- to 200-mm depth that were approximately 11.0 and 7.3 g kg^{-1} in size. The mean ages for the stable organic matter pools at each location were approximately 2973 and 853 years, respectively. Surface soils are also conceptualized to contain a slow or intermediate organic matter pool that turns over every 20 to 50 years, and a very labile, active pool with a turnover time of 1 to 5 years (Parton et al., 1987; Hsieh, 1992).

Active, slow and stable organic C pools have very different functional roles in SOM dynamics and nutrient cycling. Each pool, therefore, may respond differently to various soil and crop management practices. Chemical or physical isolation of these pools has proved to be very difficult (Hsieh, 1992). Chemical extraction methods produce organic matter fractions that are neither clearly nor consistently related to the dynamics of organic matter turnover (Jenkinson, 1971; Oades and Ladd, 1977; Duxbury et al., 1989). Physical disruption of SOM reduces the chemical alteration of organic material that may accompany chemical extractions, and these methods have been used with some success to elucidate SOM dynamics (Turchenek and Oades, 1979; Tiessen and Stewart, 1983; Cambardella and Elliott, 1993; Cambardella and Elliott, 1994). For example, Cambardella and Elliott (1992) reported that a physically isolated particulate organic matter (POM) fraction, which may be analogous to light fraction (LF) organic matter (Greenland and Ford, 1964), represented 39% of the total soil C in a western Nebraska grassland soil that had never been plowed. Twenty years of cropping this soil to winter wheat using bare, stubble-mulch, or no-till fallow management reduced the C content of the POM fraction to 18, 19, and 25%, respectively, of the total organic C pool.

Crop selection and sequence can have a definite effect on characteristics of the SOM pools. Haynes et al. (1991) found that although total SOM content may be relatively stable, microbial biomass C, hot water-extractable carbohydrate, and aggregate stability increased markedly during the pasture phase and declined during the arable period of a mixed cropping rotation. Cambardella and Elliott (1992) found that POM derived from wheat (*Triticum aestivum* L.) turned over faster than POM derived from native grasses. They estimated the half-life of grass-derived POM to be about 13 years. Using delta ^{13}C as a tracer, Balesdent et al. (1988) identified a similar, easily mineralized SOM fraction with a half-life of 10 to 15 years that was exhausted after 30 to 40 years of cultivation. Greenland and Ford (1964) suggested that exponential decline of SOM during the first years of cultivation is probably related to the decomposition of LF organic matter. Developing a better understanding of the POM and other identifiable SOM fractions may provide tools for quantifying the effects of alternate soil and crop management practices on both C storage and soil quality.

B. Global Warming Implications

Atmospheric CO_2 concentration is a dominant factor controlling autotrophic biomass productivity and C sequestration in the soil humus and biomass pools. The CO_2 concentrations have changed considerably over the millennia, but are generally positively correlated with temperature of the earth (Woodward et al., 1991). Geologic records suggest that until the beginning of the industrial age, there was a gradual decline in atmospheric CO_2 (Scharpenseel et al., 1992), but, during the past 100 years, the concentration has increased by 15 to 25%

(Houghton et al., 1983). The primary source for the increasing CO_2 concentrations is uncertain. Geochemists argue that combustion of fossil fuel, plus interactions of CO_2 with the sea explain the increase. Ecologists argue that deforestation and land use change over the past century have been an important additional source (Hobbie et al., 1984).

With regard to global warming, some have suggested that terrestrial vegetation and soils would act as a large sink for atmospheric CO_2, even if its concentration increased to twice the present level. Schlesinger (1990) used data from chronosequence studies to show that production of refractory humus substances in soils could account for only 0.7% of terrestrial net primary productivity. He also stated that agricultural practices generally result in a release of soil C to the atmosphere.

Jenkinson et al. (1991) stated that one effect of global warming would be to accelerate the decomposition of SOM, thereby releasing CO_2 to the atmosphere and further enhancing the warming trend. Using the Rothamsted model, they suggest that if global temperatures increase as predicted and the annual return of plant residue to the soil remains constant, decomposition of SOM over the next 60 years will total 61 Pg C. This would be equivalent to approximately 19% of the CO_2 that would be released by combustion of fossil fuel during the same period if current use remains constant. Schlesinger (1990) concluded that if the terrestrial biosphere was to act as a C sink, it would more likely occur through changes in the distribution and amount of terrestrial vegetation than through changes in the accumulation of SOM.

The soil humus pool associated with legume and grass covered Mollisols and the living biomass C pool found in tropical rain forests are two terrestrial ecosystems with the highest potential for C sequestration (Scharpenseel et al., 1992). To understand the impact of various soil and plant management practices on the global C cycle, they suggest several potential processes and mechanisms for sequestering C into SOM that must be evaluated. These include evaluating the use of genetically engineered legume and grass vegetation to channel excessive CO_2 into humus through photosynthesis. Scharpenseel et al. (1992) are also using ecosystem analysis, field experiments, and computer simulations to trace SOM dynamics for these evaluations. Continued study of atmospheric CO_2 content, better estimates of pre-industrial-age atmospheric CO_2 concentrations, more measurements of C isotopes in oceans and tree rings, better estimates of forest clearing rates, and additional experiments with CO_2 fertilization of natural ecosystems are all needed to validate and confirm these simulation models.

Conservation tillage practices and selection of crop sequences provide management alternatives that can alter long-term C accumulation, release, and storage. Understanding these relationships provides strong justification for examining effects of soil and crop management practices on the various soil C pools.

III. Management Strategies for Changing Soil C Concentrations

No single soil and crop management strategy can be developed and implemented to change biotransformation processes and soil C storage. Practices must be tailored with respect to inherent soil, climatic, management, and social constraints. By understanding the basic processes occurring in the soil it may be possible to select practices that will maintain or improve soil quality. This would help ensure an economically viable, environmentally safe, and socially acceptable form of sustainable agriculture (Karlen et al., 1992; Karlen and Sharpley, 1994).

SOM declines rapidly when virgin prairie sod is broken for cultivation (Melsted, 1954; Haas et al., 1957; Bauer and Black, 1981), and continues to decline when cropping systems involving fallow are used rather than when crops are grown continuously (Haas et al., 1957; Unger, 1982). Fortunately, SOM can also be increased by selecting appropriate soil and crop management practices for the area of concern. When selecting practices, it is critical to prevent further decline in SOM and the concomitant negative consequences on soil quality indicators such as aggregate stability (Tisdall and Oades, 1982; Karlen et al., 1990; Kay, 1990; Karlen et al., 1992).

The global release of soil organic C from agriculture has been estimated at 800 Tg C yr^{-1} (T = tera = 10^{12}) (Schlesinger, 1990). Changes in several soil and crop management practices could presumably reduce this loss and perhaps change the entire soil system to a C sink. They include adoption of conservation or reduced tillage practices, increased use of cover crops and crop rotations, utilization of animal waste as an organic fertilizer, and, perhaps on a limited scale, application of non-traditional C sources such as municipal sewage sludge, newsprint and other cardboard materials, or lawn wastes such as grass clippings and leaves. These practices can not only influence C storage, but also soil tilth and soil quality factors such as aggregation, aeration, water retention, and nutrient cycling.

A. Conservation Tillage

Conservation or reduced tillage, where crop residues are left on the soil surface, can generally reduce soil loss by erosion (Moldenhauer and Wischmeier, 1960; Taylor et al., 1964), improve water use efficiency (Smika and Wicks, 1968; Unger and Phillips, 1973; Smika and Unger, 1986), and increase surface soil C concentrations (Agenbag and Maree, 1989; Karlen et al., 1989; Wood et al., 1991; Bauer and Black, 1992) compared to conventional tillage where crop residues are incorporated. Surface residues reduce raindrop impact and provide a surface mulch, but simply reducing tillage intensity can help maintain soil C if there are other demands for the straw or stover. Campbell et al. (1991a) reported that baling and removing straw did not affect soil organic C on a thin Black Chernozem soil near Saskatchewan, Canada. In most environments, however, removal of crop residues will decrease soil C levels, especially in dryland (non-irrigated) cropping systems (Unger et al., 1990). In these climatic regions, there may be inadequate amounts of crop residues to protect the soil against erosion, improve water conservation, and provide organic materials for microbiological activity and accumulation of soil C.

Tillage influences C storage, soil tilth, and soil quality because it inverts and mixes the soil. Soil disturbance by tillage introduces large amounts of oxygen into the soil which stimulates the consumption of organic matter as a food source by aerobic microorganisms (Doran and Smith, 1987). The impact of tillage on the global C balance is currently unknown. Mann (1986) compiled information from 50 studies concerning changes in soil C storage due to clearing of forests and agricultural conversion of native vegetation. These studies showed changes in soil organic C ranging from a 70% loss to a 200% gain, depending on soil type, depth of sampling, manure, lime and fertilizer additions, irrigation, cultivation, crop history, and erosion. Using regression analyses to evaluate C storage in relation to years of cultivation, Mann (1986) confirmed that the greatest rates of change occurred during the first 20 years. He also showed that C losses from cultivated soils were significantly influenced by the initial C content, and suggested that changes in microclimate may result in more rapid mineralization of C in soils with a high initial C content.

Rasmussen et al. (1989) reported that when virgin eastern Oregon soils were cultivated, many lost more than 25% of their organic matter in the first 20 years, with 35 to 40% being lost in 60 years. This was consistent with the general conclusion by Mann (1986) that C losses from agricultural soils having high C content will be at least 20% when cultivation is begun. In Oregon (Rasmussen et al., 1989), tillage for weed control during the fallow period was hypothesized as the primary cause for this C loss. Similar results published by Ridley and Hedlin (1968) showed that after 37 years, soils which had initial organic matter concentrations of nearly 10%, had 7.2% organic matter if cropped every year, and 3.7% if fallowed every other year. Soils fallowed after every two or three crops had intermediate SOM concentrations.

Use of no-till or conservation tillage has been suggested as a method to change the entire soil system from a source of atmospheric C to a net C sink (Kern and Johnson, 1993). Using geographic databases of conservation tillage usage and published relationships showing agricultural soil organic C response to conservation tillage, they predicted changes in soil organic C storage between 1990 and the year 2020 (Table 1). Their analyses suggested that by increasing conservation tillage in the U.S. from current levels to 76% of the planted cropland by 2010, agricultural systems will be changed from sources of atmospheric C (188 to 209 Tg C) to C sinks (131 to 306 Tg C). Soils projected to have the greatest potential for C sequestration through conversion from conventional tillage to no-till were those with high amounts of C in the surface

Table 1. Projected increase in soil organic C for no-till management by the year 2020 assuming three levels of conservation tillage (CT) adoption

	Soil organic C conctent								
	-------Current 27% use of CT------			-----57% adoption of CT-----			------76% adoption of CT------		
Depth	Mean	Minimum	Maximum	Mean	Minimum	Maximum	Mean	Minimum	Maximum
cm	----------Tg----------								
0 to 8	143	107	178	212	160	263	372	281	462
8 to 15	112	84	139	147	111	183	260	196	323
0 to 30	255	191	317	359	271	446	632	477	785

(Adapted from Kern and Johnson, 1993.)

Table 2. Projected changes in soil C after 100 years of four conservation tillage (CT) management scenarios

	C from original soil surface (kg m^{-2})	
Scenario	Top 15 cm	Top 1 m
Current tillage mixture	-0.04	-1.23
Current trend of conversion to CT	+0.13	-0.86
Increased adoption of no-till plus a	+0.37	-0.39
winter cover crop	+0.73	+0.03

(Adapted from Lee et al., 1993.)

15 cm. Those soils were generally found in cool and moist regions such as the Pacific Northwest, the Upper Midwest, and Northern New England.

In similar studies, Lee et al. (1993) used the Erosion Productivity Impact Calculator (EPIC) model to simulate soil erosion and soil C content under four management scenarios. They included: (1) Maintaining the current mix of tillage practices; (2) Continuing the current trend of conversion to mulch-till and no-till management; (3) Increasing adoption of no-till; and (4) Increasing adoption of no-till with the addition of a winter wheat (*Triticum aestivum* L. emend. Thell.) cover crop. Lee et al. (1993) concluded that adoption of reduced tillage would decrease soil erosion compared to the current mix of tillage practices, and increase soil C content in the upper 15 cm. Their simulations projected that adoption of no-till practices would have significant erosion control benefits, but that planting a winter cover crop would have little additional effect on erosion control. With respect to C sequestration, the current trend towards conversion to conservation tillage was projected to increase the C content of the top 15 cm of soil by 0.2 kg m^{-2} during the next 100 years, compared to the current mix of tillage practices. Adoption of no-till or no-till plus a winter cover crop was projected to cause increases of 0.4 or 0.8 kg m^{-2}, respectively (Table 2).

Gaston et al. (1993) used pre-cultivation estimates of soil C for the seven major soil classes found in agricultural areas of the former Soviet Union (FSU) and equations from Kern and Johnson (1993) to project the effects of adopting no-till management on soil C content. Their rationale was that since the onset of cultivation the 211.5 million ha of agricultural soils in the FSU have lost 10.2 Gt of C (G = giga = 10^9). A substantial portion of this loss was attributed to soil erosion. Although accurate data are not available, the rich Chernozems which produce 80% of the grain crops in the FSU are known to have been especially vulnerable to soil erosion and to have had a substantial loss of SOM (Priputina, 1989).

Gaston et al. (1993) concluded that almost 86% of the agricultural land in the FSU (181 million ha) was climatically suitable for no-till. They projected that a 10% increase in soil C content, achieved by complete conversion of all climatically suitable land to no-till management, would sequester 3.3 Gt of C. These changes would also improve soil quality by increasing SOM, leaving crop residues on the soil surface to reduce wind speed and raindrop energy, improving the rate of water infiltration, increasing the water stability of soil aggregates, and thus decreasing soil erosion. Unfortunately, they also concluded that it was unlikely that individual farmers or individual collective farms would be able to finance the cost of the new equipment required for no-till production within the next decade.

The effects of conservation tillage on C balance within the entire soil profile, as compared to the surface 15 cm, have been questioned with arguments being raised that adoption of conservation tillage would simply result in a stratification of the SOM. Total soil C data, collected incrementally to a depth of 600 mm following 12-year of continuous corn (*Zea mays* L.) using moldboard plowing, chisel plowing, or no-till practices to prepare the seedbed (Karlen et al., 1994), showed significantly higher levels in the 0- to 25-mm layer for the no-till treatment and significantly lower levels at 25- to 75-, and 75- to 100-mm increments for the plow treatment. There were no other significant differences, and no indication that long-term conservation tillage resulted in an accumulation of C near the surface at the expense of C deeper in the profile. Simulations by Lee et al. (1993) suggested that when examined to a depth of 1 m, all scenarios except the no-till plus winter cover crop would result in a net loss of C. Soil erosion accounted for 21, 24, and 28% of the total C loss from the top 1 m for the current mix, base trend, and no-till scenarios, respectively (Table 2). They also projected that 79% of the soil C gained by using no-till plus a winter cover crop would be lost through soil erosion. Lee et al. (1993) attributed the remaining C losses to dynamics within the soil profile, an area in which considerably more research is certainly needed.

The use of no-till or conservation tillage systems can generally reduce rates of SOM loss, but may not completely stop it. Collins et al. (1992) reported that after 58 years, total soil and microbial biomass C and N were significantly greater in annual-cropping treatments than for wheat-fallow rotations. They concluded that residue management (i.e. reduced tillage) significantly affected the level of microbial biomass C and that annual cropping significantly reduced declines in both SOM and soil microbial biomass. Lamb et al. (1985) described soil organic C and N changes for a long-term tillage experiment located in Western Nebraska where a native grassland site was cultivated for winter wheat in a crop-fallow rotation under three tillage systems. They reported that total soil C and N in the top 30 cm was significantly greater for no-till compared with stubble-mulch and conventional bare fallow tillage, with most of the differences occurring in the 0 to 10 cm depth increment. Similarly, Havlin et al. (1990) found that compared to native grassland, a 12-year wheat and fallow rotation resulted in total SOM levels that were 4, 14, and 16% lower with no-till, stubble mulch, and conventional tillage, respectively.

B. Cover Cropping

Growing cover crops between cash or grain crops is a management strategy that may improve soil tilth and quality while increasing C storage. The cover crops protect the soil from the impact of raindrops and slow runoff. When dead, they add organic matter to the soil. This generally increases both permeability and infiltration, slows runoff, and decreases erosion (Pieters and McKee, 1938). Leaching of N, K, and possibly other nutrients, especially on sandy textured soils, can also be decreased by growing cover crops. On fine-textured soils, they have been shown to improve aggregation, porosity, bulk density, and permeability (Rogers and Giddens, 1957). Cover crops can increase microbial activity and often increase soil aggregation by providing an additional C source and recycling nutrients (Ram et al., 1960).

Cover crops such as crimson clover (*Trifolium incarnatum* L.) are being successfully used in many locations in the southern U.S. (Frye et al., 1988), but limited germplasm currently prevents wide usage in the northern Corn Belt (Power, 1987). In Missouri, common chickweed (*Stellaria media* L.), Canada bluegrass (*Poa compressa* L.), and downy brome (*Bromus tectorum* L.) were used as winter cover crops by Zhu et al. (1989) prior to planting a no-till soybean [*Glycine max* (L.) Merr.] crop. They found that the winter cover crops increased soil cover by 30 to 50% during the critical erosion period of late spring to early summer. As a result, the chickweed, downy brome, and Canada bluegrass decreased mean annual soil loss by 87, 95, and 96% compared to a check treatment. Similarly, runoff was reduced 44, 53, and 45%, respectively. This shows that cover crops can also improve both water quality and soil quality by reducing soil erosion while accumulating C and retaining nutrients.

Studies with winter rye (*Secale cereale* L.) and Italian ryegrass (*Lolium perenne* L.) have shown that growing these plants as cover crops can reduce leaching losses of residual N (Brinsfield et al., 1988; Staver and Brinsfield, 1990; Martinez and Guiraud, 1990). They also provide substantial C inputs into the soil because of their high biomass production. A disadvantage of growing winter rye as a cover crop is its rapid growth during early spring. This can deplete available soil water supplies and stress the primary crop (Campbell et al., 1984; Karlen and Doran, 1991). Another disadvantage is that cover crops may contribute to lower mineralization of organic matter and/or greater denitrification losses before planting. This presumably reduces the availability of soil N and can create apparent N deficiencies in subsequent corn crops (Doran et al., 1989; Karlen and Doran, 1991). An alternative to rye in the northern Corn Belt is the use of oats (*Avena sativa* L.). This cover crop grows rapidly during autumn but freezes during winter and does not compete for soil water reserves in spring.

When legumes are used as cover crops, they provide additional N through biological fixation for subsequent grain crops (Smith et al., 1987). These crops frequently increase N uptake and yield of subsequent crops (McCracken et al.,

1989) and can add considerable amounts of C to the system (Langdale et al., 1990).

C. Crop Rotations

Increasing temporal and spatial diversity by using crop rotations may improve soil quality and increase C storage by mimicking natural ecosystems more closely than current agroecosystems. Kay (1990) suggested that a major goal for agricultural research should be to identify and promote cropping systems that sustain soil productivity and minimize deterioration of the environment. Factors that can be influenced by crop selection and rotation include soil structure, aggregation, bulk density, water infiltration, water retention, soil erodibility, and organic matter content.

The effectiveness of different crops for improving soil structure is related to the amount of water extracted and photosynthate deposited, as well as the persistence of photosynthate C at different soil depths (Kay, 1990). For example, Russell (1973) stated that, in perennial grasslands, over 2.5 Mg ha^{-1} dry matter can be added to the soil each year as roots, and total grass root systems may contain more than 12 Mg ha^{-1} of dry matter. This compares to only 2 to 5 Mg ha^{-1} of above-ground material. Elkins (1985) demonstrated how different crops can affect soil quality by creating biopores in compacted soils at depths that could not be economically tilled. He found that bahiagrass (*Papsalum notatum* Flugge 'Pensacola') roots could penetrate soil layers that impeded cotton (*Gossypium hirsutum* L.) roots and that including bahiagrass in a rotation increased the number of soil pores that were 1.0 mm or larger in diameter. The biopores enabled the cotton to obtain water and nutrients from a depth of at least 60 cm and were effective for 3 years after the grass had been killed by plowing.

The total amount of photosynthate deposited below ground, the composition of this photosynthate (Blum et al., 1991), and the resistance of this material to complete mineralization varies considerably between plant species (Kay, 1990). Studies with bromegrass (*Bromus inermis* Leyss.) showed that the total amount of rhizodeposition can be twice as high as the amount deposited by corn (Davenport and Thomas, 1988). Furthermore, C originating from bromegrass was more persistent in the soil than that from corn (Davenport et al., 1988).

Frequently, there is a synergistic effect between reduced tillage practices and increased cropping intensity on SOM accumulation. Wood et al. (1991) imposed no-till management on western Great Plain's soils with a long-term history of tillage or frequent fallow management. The cropping systems were: (1) Wheat and fallow; (2) Wheat and corn or sorghum [*Sorghum bicolor* (L.) Moench] and millet (*Panicum miliaceum* L.) and fallow; and (3) Perennial grass. Results indicated that no-till management had rapidly (within 4 years) altered the depth distribution and content of organic C and N for all cropping systems. Organic C and N accumulated in the 0- to 25-mm soil layer was maintained in the 25- to 50-mm soil layer, and declined in the 50- to 100-mm soil layer. Increasing cropping frequency from the more traditional wheat and fallow rotation significantly increased total organic C content throughout the 0- to 100-mm surface layer. In an earlier paper, they also report that potential C and N mineralization were 61 and 39% greater under System 2 than System 1 (Wood et al., 1990). Differences in potential C and N activity between cropping systems were attributed to greater surface organic C concentrations in System 2, which in turn were related to higher cumulative plant residue additions. Four years of these alternate cropping sequences also decreased total profile NO_3-N content in the upper 180 cm from 120 kg ha^{-1} for the wheat and fallow rotation to 60 kg ha^{-1} for System 2 and 10 kg ha^{-1} for the perennial grass. Wood et al. (1990) concluded that initiation of no-till crop production practices on semi-arid soils previously managed using tillage and frequent-fallow practices would not only promote higher equilibrium levels of organic C and N, but also decrease profile NO_3-N content.

Abandonment of multi-year crop rotations in favor of short rotations has generally resulted in a degradation of soil structure and presumably decreased C storage, as measured by soil aggregate stability, bulk density, water infiltration rate, and soil erosion (Bullock, 1992). Much of the blame for this degradation is attributed to decreases in SOM content. However, Bruce et al. (1990) stated that relationships between soil structure and crop rotation were complex and easily modified, frequently depending upon the amount and intensity of tillage used for the various crops.

Crop rotations that include legumes and/or grasses are generally beneficial to aggregate stability and formation of favorable soil structure (Kay, 1990; Robinson et al., 1994). Using measurements of mean weight diameter, van Bavel and Schaller (1950) and Wilson and Browning (1945) showed that soil aggregation with continuous corn was half of that found with a corn, oat, meadow rotation. Bullock (1992) argued that it is convenient to suggest that crop rotations beneficially affect soil aggregate formation and stabilization, but quantifying the relationships is not that simple. The addition of a red clover (*Trifolium pratense* L.) rotation into continuous barley (*Hordeum vulgare* L.) had no effect on soil aggregation or on soil organic C for the top 7.5 cm of soil, but no-till increased the mean weight diameter of water-stable aggregates by 40% and organic C by 20% when compared to plowed treatments (Angers et al., 1993a,b). When rotations involve numerous years of sod, pasture, or hay crops, improvements in soil structure do occur (Olmstead, 1947; Strickling, 1950; Adams and Dawson, 1964; Tisdall and Oades, 1982; Power, 1990). However, short 2-year corn and soybean rotations often reduce SOM and aggregation compared to well-fertilized continuous corn, presumably because soybean returns less crop residue than corn (Dick et al., 1986a,b; Power, 1990).

Havlin et al. (1990) demonstrated that including grain sorghum in a rotation, rather than growing continuous soybean, increased organic C and N in the soil. They concluded that increasing the quantity of residue returned to the soil through higher yields or through greater use of high residue crops in the rotation, combined with reduced tillage, could improve soil productivity. Juma et al. (1993) concluded that after 50 years of research on Gray Luvisolic soils at the Breton Plots in Alberta, Canada, SOM content is about 20% higher where a 5-year rotation has been used than where a 2-year wheat and fallow rotation was followed. Similarly, with constant tillage treatments, continuous cropping resulted in significant SOM increases compared to a crop-fallow system (Unger, 1968; Janzen, 1987; Campbell et al., 1991a; Wood et al., 1990). The effect of multi-year rotations on estimates of labile or mineralizable organic C has been shown to be more pronounced than the effect on soil organic C (Campbell et al., 1991b,c; Janzen, 1987). Increases in mineralizable C have been attributed to increased cropping frequency even when no changes in soil organic C were observed (Campbell et al., 1992). Those findings agree with those reported by Haynes et al. (1991). They concluded that in many short-term arable and pasture rotations total SOM may remain relatively constant, but microbial biomass C, hot water-extractable carbohydrate, and aggregate stability increase significantly during the pasture phase and decline during the arable phase of the rotation.

Angers (1992) documented changes in soil structure and organic matter under continuous silage corn, alfalfa (*Medicago sativa* L.), and fallow conditions. During a 5-year period, the mean weight diameter of water-stable aggregates in soils planted to alfalfa increased from 1.5- to 2.3-mm. The C content increased from 26 g kg^{-1} the first year to 30 g kg^{-1} in the last year, even though all of the aboveground alfalfa biomass was harvested. This implies large inputs to the soil organic C pool exclusively from below-ground root biomass as reported for wheat by Campbell et al. (1991a). With silage corn or fallow, changes in mean weight diameter and C content were minimal during the same period. About one-half of the temporal variation in mean weight diameter under silage corn or fallow could be explained by the variation in soil water content at the time of sampling. However, water content and mean weight diameter of aggregates collected from alfalfa treatments showed no significant relationships (Angers, 1992).

Other studies have shown decreases in soil C under conventional corn cropping, but when minimum tillage (Angers, 1992) or no-tillage are used, this trend can be reversed (Costamagna et al., 1982). Angers (1992) also concluded that soil aggregates from the alfalfa treatment were not as vulnerable to slaking as those from the other treatments.

Growing bromegrass (*Bromus inermis* Leyss.) for 3 years, following 20 years of continuous corn, was found to improve the stability of soil aggregates that were greater than 250 μm in diameter as water content decreased (Caron et al., 1992). This short-term beneficial effect of bromegrass on structural stability was attributed to organic materials from the grass, supporting conclusions reached by Haynes et al. (1991).

Raimbault and Vyn (1991) observed improved aggregate stability due to crop rotation in Ontario. They also found that first-year corn grown in rotation yielded 3.9% more than continuous corn grown using conventional tillage, and 7.9% more than continuous corn grown using minimum tillage practices. Hussain et al. (1988) concluded that crop rotation increased soil aggregation over time based on geometric mean diameter values. Their calculations for continuous corn, corn following soybean, soybean following corn, and an oat/red and yellow sweetclover (*Melilotus alba* L.) following corn showed average geometric mean diameters of 150, 211, 225, and 311 mm, respectively.

Evaluations of farming systems that include crop rotation have also shown significant differences in C storage and soil aggregation. In central Iowa, Jordahl and Karlen (1993) found that combined effects of alternative practices (i.e. a 5-year crop rotation including oats and meadow, manure/municipal sludge application, and ridge-tillage) resulted in greater water stability of soil aggregates than the conventional practices (i.e. a 2-year corn-soybean rotation with reduced tillage). They attributed the increased soil C content and more stable aggregation to the longer crop rotation, which included an oat and meadow crop and to the application of 45 Mg ha^{-1} of a mixture of animal manure and municipal sewage sludge during the first 3 years of each 5-year rotation.

Soil structure and temporal variation in aggregate stability were measured at three slope positions on soils from adjacent eastern Washington farms that had been managed for many years using conventional short-term rotations or alternative long-term rotations (Reganold, 1988; Mulla et al., 1992). The crop rotation on the conventional farm, which was first cultivated in 1908, currently includes winter wheat and spring pea (*Pisum sativum* L.) with summer fallow every sixth year. The alternative farm, which was first cultivated in 1909, is managed without using commercial fertilizers and with only limited amounts of pesticide by following a longer winter wheat, spring pea, Austrian winter pea [*Pisum sativa* subsp. *sativum* var. *arvense* (L.) Poiret], spring wheat, spring pea, and summer fallow rotation. Also, an alfalfa and wheatgrass [*Elymus trachycaulus* (Link) Gould ex Shinn.] mixture was planted every 19th and 20th year. Soil from the alternative farms had a more granular structure and friable consistence than soil from the conventional farm, presumably because the longer rotation period returned more C and also resulted in less frequent periods of summer fallow.

Temporally, aggregate stability in soils from both farms decreased significantly from October to March. This was in response to precipitation and cycles of freezing and thawing (Mulla et al., 1992). Differences in aggregate stability for soils from the conventional and alternative farms were not significant, although organic C content in samples from the alternative farm were significantly higher than those in samples from the conventional farm.

An important advantage of using crop rotations to increase C storage and thus improve soil tilth and quality is that it increases the temporal and spatial diversity of the landscape. This can help reduce both wind and water erosion by having shorter slope lengths and more diverse ground cover across large fields. Spatial and temporal diversity can also be increased on a very small scale by using narrow strip intercropping systems. Several studies have focused on two crop strips, particularly corn and soybean (Francis et al., 1986). By including a third (small grain) and possibly a fourth (forage) crop, unique micrometeorological and soil conditions are created. These generally result in favorable crop yield responses, especially in the border positions (Garcia et al., 1990) and may thus increase C accumulation by the cropping system. Cruse (1990) reported that yields for both corn and oat tend to increase in narrow strips compared to large, single-crop production fields. Soybean response in his studies was variable depending on seasonal weather patterns.

Alley cropping, narrow strips of complementary crops including woody or herbaceous perennials, is a management strategy that may be useful for increasing C storage and improving soil quality, especially in tropical areas. Species such as *Cassia siamea*, *Gliricidia sepium*, and *Leucaena leucocephala* can provide nutrients to corn and other grain crops and protect the soil from erosion by increasing residue cover. Chen et al. (1989) reported that growing *Leucaena* for 14 months at the International Institute for Tropical Agriculture (IITA) in Nigeria increased soil pH, K, organic C, and corn yield, while decreasing soil erosion. Kang et al. (1985), also working at IITA, reported that *Leucaena* stands established 8 years earlier remained viable and produced approximately 200 kg N ha^{-1} yr^{-1} from 1981 to 1983.

Alley crops can be used as a mulch to increase soil water retention, germination rates, and to decrease soil temperature (Wilson et al., 1986). When incorporated as green manure, the N they contribute is often used more efficiently. In areas of low rainfall, alley cropping can compete with corn for soil water, but Kang et al. (1985) reported that *Leucaena* develops a deep rooting system and extracts most of its water from depths greater than 50 cm. They concluded that competition between *Leucaena* and corn for water is minimal in deep soils without root restricting layers.

The impact of alley cropping on crop productivity, C storage, and soil quality is dependent upon the alley crop species. Yamoah et al. (1986) reported that *Cassia* maintained a high soil microbial biomass C, presumably due to its slow rate of decomposition. Soil bulk density was lower under all alley crop species evaluated. Soil N increased with *Cassia* and *Flemingia* but decreased with *Gliridia*. *Cassia* and *Gliridia* appeared to be promising species for use in Nigeria. Hulugalle and Kang (1990) also concluded that alley crops which produce large quantities of prunings that decompose slowly will have a beneficial effect on soil properties.

Effects of long-term organic residue additions on the active pool of SOM, microbial biomass C and N, water soluble C, and total C and N were evaluated by Mazzarino et al. (1993) in the humid tropics of Costa Rica. Their evaluations were made within annual corn and bean rotations that were conducted within four, 9-year old cropping systems. Two of the systems involved alley cropping with leguminous trees (*Erythrina* and *Gliricidia*) and two had no trees. They found higher total C and N, microbial C and N, water-soluble C and soil water content in the alley cropping treatment areas than in areas without trees. Seasonal changes in microbial biomass appeared to depend more on crop phenology, and management practices (pesticide and fertilizer applications) than on pruning additions. They concluded that an interaction between crop stovers and tree prunings results in a system that oscillates between N immobilization and mineralization.

D. Manure and Fertilizer Applications

To better understand the biological and chemical interactions causing yield declines for monoculture soybean and grain sorghum, Roder et al. (1988) measured microbial biomass, crop root dry weight, SOM, and total N content in soils managed for 5 years with various crop rotations and fertilizer management treatments. The crop rotations included continuous monoculture plantings of both crops and both years of the 2-year rotation. Fertilizer treatments included a non-fertilized control, a N-fertilized treatment for both crops, and a treatment that received 15.8 Mg ha^{-1} of manure each year. They found that soil microbial biomass C was 11 to 14% higher and that SOM contents were 6 to 16% higher for manured treatments than for the unfertilized controls. Roder et al. (1988) also reported that previous crop, present crop, and fertilizer treatment affected microbial biomass and dry matter partitioning between above-ground and below-ground plant parts. With sorghum as the previous crop, a higher proportion of the total production occurred below ground as roots and microbial biomass.

Campbell and Zentner (1993) reported that SOM increased in the top 15 cm of soil under well-fertilized annually cropped rotations of cereal grains, and remained constant in rotations that had periods of fallow or if the continuous wheat received inadequate N fertilizer. Their study showed that the amount of SOM can vary depending on the amount and the N content of the crop residues returned to the soil. Similarly, Paustian et al. (1992) found C inputs to be the single most important factor determining SOM levels. They also report significant differences in SOM levels at about the same C input rate that were closely associated with the type of organic matter added. High-lignin organic amendments like farmyard manure, which were more recalcitrant to decomposition, resulted in higher soil C accumulations per unit C input than low-lignin amendments like wheat straw. They attribute the residue-quality effects largely to a direct stabilization of lignin degradation products in the slow organic matter pool as opposed to the more complete metabolism of nonlignin residue fractions by the soil microbial biomass (active pool).

Juma et al. (1993) concluded that application of N, P, K, and S fertilizer and animal manure to Gray Luvisolic soils increased SOM by increasing crop yields. They also reported that application of manure increased SOM even more than fertilizer. This presumably occurred because in addition to its nutrient value, the 9 Mg ha^{-1} of manure that was added each year represented an additional source of organic C. The report by Juma et al. (1993) supports conclusions by Boyle et al. (1989) who suggested that returning C to the soil is 'a necessary expense that insures a sustainable harvest.' Both reports support suggestions by Karlen et al. (1992) that crop rotation, cover crops, and conservation tillage are the practices that most likely will lead to greater C storage and improved soil tilth and quality.

IV. Summary and Conclusions

Agricultural soils can function as both a source and a sink for atmospheric C. The direction of this dynamic equilibrium is generally determined by how the soil resource is being managed. The basis for this literature review is two-fold: (1) Addressing the potential for C sequestration as it is affected by soil and crop management practices; and (2) Evaluating the effects of those management practices on soil quality. Research studies evaluating the effects of four soil and crop management strategies: (1) Conservation tillage; (2) Cover cropping; (3) Crop rotations; and (4) Manure and fertilizer applications are discussed with regard to how they might affect soil C pools and processes related to these two issues.

The use of conservation tillage can increase surface soil C concentrations compared to more conventional tillage practices, where crop residues are incorporated. If there are other demands for the straw or stover, conservation tillage can help maintain soil C, although in most environments, and especially in dryland (non-irrigated) cropping systems, removal of crop residues will decrease soil C levels. With respect to C sequestration, continuing the current trend of conversion from conventional to conservation tillage was projected to increase soil C in the upper 15 cm by 0.2 kg m^{-2} during the next 100 years.

Questions concerning the effects of conservation tillage on the profile C balance were also addressed. Total soil C data, collected incrementally to a depth of 600 mm showed significantly higher levels in the 0- to 25-mm layer following 12-years of no-till and significantly lower levels at 25- to 75-, and 75- to 100-mm increments for the long-term plow treatment. Below 100 mm, there were no significant differences and no indication that long-term conservation tillage resulted in an accumulation of C near the surface at the expense of deeper in the profile.

Growing cover crops may improve soil quality while increasing C sequestration by increasing microbially driven nutrient cycling processes and enhancing soil aggregation. Cover crops such as crimson clover are being successfully used in many locations in the southern U.S., but limited germplasm currently prevents wide usage in the northern Corn Belt. Studies with winter rye and Italian ryegrass have shown that growing these plants can provide substantial C inputs into the soil because of their high biomass production.

Crop rotations may improve soil quality and increase C storage by mimicking natural ecosystems more closely than current agroecosystems. Factors that can be influenced by crop selection and rotation include soil structure, aggregation, bulk density, water infiltration, water retention, erodibility, and SOM levels. Effectiveness of different crops for C sequestration and improving soil structure is related to the amount of water extracted and photosynthate deposited, as well as the persistence of photosynthate C at different soil depths.

Evaluations of alternate farming systems that include crop rotation have also shown significant differences in C storage and soil aggregation. Increased soil C content and more stable aggregation were generally attributed to longer crop rotation periods and the application of animal manure. High-lignin organic amendments like farmyard manure, were reported to be more recalcitrant to decomposition and resulted in higher soil C accumulations per unit C input than low-lignin amendments like wheat straw. These effects were attributed to a direct stabilization of lignin degradation products in the slow organic matter pool as opposed to the more complete metabolism of nonlignin residue fractions by the soil microbial biomass (active pool).

This review documents that implementation of conservation tillage practices and crop rotations, application of organic C sources such as animal manure, and planting of cover crops where feasible, could result in greater C storage, reduced CO_2 loss to the atmosphere, and improved soil quality. These results highlight the importance of developing governmental policy that encourages the adoption of these practices, not only for soil conservation, but also for the implications that increasing the soil's capacity to sequester C have on the global C budget, especially to ameliorate the effects of global warming.

References

Adams, W.E. and R.N. Dawson. 1964. Cropping system studies on Cecil soils, Watkinsville, GA. 1943-62. USDA-ARS-41-83, South Piedmont Conserv. Res. Ctr., Watkinsville, GA.

Agenbag, G.A. and P.C.J. Maree. 1989. The effect of tillage on soil carbon, nitrogen and soil strength of simulated surface crusts in two cropping systems for wheat (*Triticum aestivum*). *Soil Tillage Res.* 14:53-65.

Angers, D.A. 1992. Changes in soil aggregation and organic carbon under corn and alfalfa. *Soil Sci. Soc. Am. J.* 56:1244-1249.

Angers, D.A., N. Bissonnette, A. Légère, and N. Samson. 1993a. Microbial and biochemical changes induced by rotation and tillage in a soil under barley production. *Can. J. Soil Sci.* 73:39-50.

Angers, D.A., N. Samson, and A. Légère. 1993b. Early changes in water-stable aggregation induced by rotation and tillage in a soil under barley production. *Can. J. Soil Sci.* 73:51-59.

Balesdent, J.A., G.H. Wagner, and A. Mariotti. 1988. Soil organic matter turnover in long-term field experiments as revealed by carbon-13 abundance. *Soil Sci. Soc. Am. J.* 52:118-124.

Bauer, A. and A.L. Black. 1981. Soil carbon, nitrogen, and bulk density comparisons in two cropland tillage systems after 25 years and in virgin grassland. *Soil Sci. Soc. Am. J.* 45:1166-1170.

Bauer, A. and A.L. Black. 1992. Organic carbon concentration effects on available water capacity of three soil textural groups. *Soil Sci. Soc. Am. J.* 56:248-254.

Blum, U., T.R. Wentworth, K. Klein, A.D. Worsham, L.D. King, T.M. Gerig, and S.W. Lyu. 1991. Phenolic acid content of soils from wheat-no till, wheat-conventional till, and fallow-conventional till soybean cropping systems. *J. Chem. Ecol.* 17:1045-1068.

Boyle, M., W.T. Frankenberger, Jr., and L.H. Stolzy. 1989. The influence of organic matter on soil aggregation and water infiltration. *J. Prod. Agric.* 2:290-299.

Brinsfield, R., K. Staver, and W. Magette. 1988. The role of cover crops in reducing nitrate leaching to groundwater. p. 127-146. In: *Agricultural Impacts on Groundwater.* Nat'l. Well Water Assoc., Dublin, OH.

Bruce, R.R., G.W. Langdale, and L.T. West. 1990. Modification of soil characteristics of degraded soil surfaces by biomass input and tillage affecting soil water regime. p. 4-9. In: *Transactions of the 14th International Congress of Soil Science.* Vol. VI. Int'l. Soc. Soil Sci., Kyoto, Japan.

Bullock, D.G. 1992. Crop rotation. *Critical Rev. Plt. Sci.* 11:309-326.

Cambardella, C.A. and E.T. Elliott. 1992. Particulate soil organic-matter changes across a grassland cultivation sequence. *Soil Sci. Soc. Am. J.* 56:777-783.

Cambardella, C.A. and E.T. Elliott. 1993. Carbon and nitrogen distribution in aggregates from cultivated and native grassland soils. *Soil Sci. Soc. Am. J.* 57:1071-1076.

Cambardella, C.A. and E.T. Elliott. 1994. Carbon and nitrogen dynamics in soil organic matter fractions from cultivated grassland soils. *Soil Sci. Soc. Am. J.* 58:123-130.

Campbell, C.A. and R.P. Zentner. 1993. Soil organic matter as influenced by crop rotations and fertilizations. *Soil Sci. Soc. Am. J.* 57:1034-1040.

Campbell, C.A., S.A. Brandt, V.O. Biederbeck, R.P. Zentner, and M. Schnitzer. 1992. Effect of crop rotations and rotation phase on characteristics of soil organic matter in a Dark Brown Chernozemic soil. *Can. J. Soil Sci.* 72:403-416.

Campbell, C.A., G.P. Lafond, R.P. Zentner, and V.O. Biederbeck. 1991a. Influence of fertilizer and straw baling on soil organic matter in a thin Black Chernozem in Western Canada. *Soil Biol. Biochem.* 23:443-446.

Campbell, C.A., K.E. Bowen, M. Schnitzer, R.P. Zentner, and L. Townley-Smith. 1991b. Effect of crop rotations and fertilization on soil organic matter and some biochemical properties of a thick Black Chernozem. *Can. J. Soil Sci.* 71:377-387.

Campbell, C.A., V.O. Biederbeck, R.P. Zentner, and G.P. Lafond. 1991c. Effect of crop rotations and cultural practices on soil organic matter, microbial biomass and respiration in a thin Black Chernozem. *Can. J. Soil Sci.* 71:363-376.

Campbell, R.B., D.L. Karlen, R.E. Sojka. 1984. Conservation tillage for maize production in the U.S. southeastern Coastal Plain. *Soil Tillage Res.* 4:511-529.

Caron, J., B.D. Kay, and J.A. Stone. 1992. Improvement of structural stability of a clay loam with drying. *Soil Sci. Soc. Am. J.* 56:1583-1590.

Chen, Y.S., B.T. Kang, and F.E. Caveness. 1989. Alley cropping vegetable crops with Leucaena in Southern Nigeria. *HortScience* 24:839-840.

Childs, S.W., S.P. Shade, D.W.R. Miles, E. Shepard, and H.A. Froehlich. 1989. Soil physical properties: importance to long-term forest productivity. p. 53-66. In: D.A. Perry, R. Meurisse, B. Thomas, R. Miller, J. Boyle, J. Means, C.A. Perry, and R.F. Powers (eds.), *Maintaining the Long-Term Productivity of Pacific Northwest Forest Ecosystems*. Timber Press, Portland, OR, U.S.A.

Collins, H.P., P.E. Rasmussen, and C.L. Douglas, Jr. 1992. Crop rotation and residue management effects on soil carbon and microbial dynamics. *Soil Sci. Soc. Am. J.* 56:783-788.

Costamagna, O.A., R.K. Stivers, H.M. Galloway, and S.A. Barber. 1982. Three tillage systems affect selected properties of a tiled, naturally poorly-drained soil. *Agron. J.* 74:442-444.

Cruse, R.M. 1990. Strip intercropping. Proc. Leopold Center for Sustainable Ag. Conf., Iowa State University, Ames, IA, p. 39-41.

Davenport, J.R. and R.L. Thomas. 1988. Carbon partitioning and rhizodeposition in corn and bromegrass. *Can. J. Soil Sci.* 68:693-701.

Davenport, J.R., R.L. Thomas, and S.C. Mott. 1988. Carbon mineralization of corn (*Zea mays* L.) and bromegrass (*Bromus inermis* Leyss.) components with an emphasis on the below-ground carbon. *Soil Biol. Biochem.* 20: 471-476.

DeJong, R.C., C.A. Campbell, and N. Nicholaichuk. 1983. Water retention equations and their relationship to soil organic matter and particle size distribution for disturbed samples. *Can. J. Soil Sci.* 63:291-302.

Dick, W.A., D.M. van Doren, Jr., G.B. Triplett, Jr., and J.E. Henry. 1986a. Influence of long-term tillage and rotation combinations on crop yields and selected soil parameters. I. Results obtained for a Mollic Ochraqualf soil. Res. Bull. No. 1180. Ohio Agric. Res. and Dev. Ctr., Ohio State Univ., Wooster, OH.

Dick, W.A., D.M. van Doren, Jr., G.B. Triplett, Jr., and J.E. Henry. 1986b. Influence of long-term tillage and rotation combinations on crop yields and selected soil parameters. II. Results obtained for a Typic Fragiudalf soil. Res. Bull. No. 1181. Ohio Agric. Res. and Dev. Ctr., Ohio State Univ., Wooster, OH.

Doran, J.W. and M.S. Smith. 1987. Organic matter management and utilization of soil and fertilizer nutrients. p. 53-72. In: R.F. Follett, J.W.B. Stewart, and C.V. Cole (eds.), *Soil Fertility and Organic Matter as Critical Components of Production Systems*, Spec. Publ. 19. ASA-CSSA-SSSA, Madison, WI.

Doran, J.W., D.L. Karlen, and R.L. Thompson. 1989. Seasonal variations in soil microbial biomass and available N with varying tillage and cover crop management. p. 213. Agron. Abstr., ASA-CSSA-SSSA, Madison, WI.

Doran, J.W. and T.B. Parkin. 1994. Defining and assessing soil quality. In: J.W. Doran, D.C. Coleman, D.F. Bezdicek, and B.A. Stewart. (eds.), *Defining Soil Quality for a Sustainable Environment*. Special Publ. 35, SSSA, Madison, WI.

Duxbury, J.M., M.S. Smith, and J.W. Doran. 1989. Soil organic matter as a source and a sink of plant nutrients. p. 33-67. In: D.C. Coleman and G. Uehara (eds.), *Tropical soil organic matter.* Univ. of Hawaii Press, Honolulu, HI.

Elkins, C.B. 1985. Plant roots as tillage tools. Proc. Int. Conf. on Soil Dynamics, Auburn Univ., Auburn, AL, p. 519-523.

Follet, R.F., S.C. Gupta, and P.G. Hunt. 1987. Conservation practices: relation to the management of plant nutrients for fertility and organic matter as critical components of production systems. p. 19-51. In: R.F. Follett. J.W.B. Stewart, and C.V. Cole (eds.), *Soil Fertility and Organic Matter as Critical Components of Production Systems*. Soil Sci Soc. Am., Spec. Publ. 19, American Soc. Agron., Madison, WI, U.S.A.

Francis, C.A., A. Jones, K. Crookston, K. Wittler, and S. Goodman. 1986. Strip cropping corn and grain legumes: A review. *Am. J. Alt. Agric.* 1:159-164.

Frye, W.W., R.L. Blevins, M.S. Smith, S.J. Corak, and J.J. Varco. 1988. Role of annual legume cover crops in efficient use of water and nitrogen. p. 129-154. In: W.L. Hargrove (ed.), *Cropping strategies for efficient use of water and nitrogen.* ASA Spec. Publ. 51. ASA-CSSA-SSSA, Madison, WI.

Garcia, R., R.M. Cruse, M. Owen, and D.C. Erbach. 1990. Competition for soil water between components of a strip intercropping rotation, as affected by tillage and position in the strip. p. 103. In: *Farming Systems for Iowa: Seeking Alternatives.* Leopold Center for Sustainable Agriculture 1990 Conf. Proc. Leopold Center, Iowa State Univ., Ames, IA.

Gaston, G.G., T. Kolchugina, and T.S. Vinson. 1993. Potential effect of no-till management on carbon in the agricultural soils of the former Soviet Union. *Agric., Ecosyst., Environ.* 45:295-309.

Greenland, D.J. and G.W. Ford. 1964. Separation of partially humified organic materials from soils by ultrasonic dispersion. p. 137-148. In: Trans. Int. Congr. Soil Sci. 8th, Budapest. 31 Aug.-9 Sept. 1964. Vol. 3. Rompresfilatelia, Bucharest.

Haas, H.J., C.E. Evans, and E.F. Miles. 1957. Nitrogen and carbon changes in Great Plains soils as influenced by cropping and soil treatments. Tech. Bull. No. 1164. U.S. Dept. Agric., Washington, DC. 111 p.

Havlin, J.L., D.E. Kissel, L.E. Maddus, M.M. Claassen, and J.H. Long. 1990. Crop rotation and tillage effects on soil organic carbon and nitrogen. *Soil Sci. Soc. Am. J.* 54:448-452.

Haynes, R.J., R.S. Swift, and R.C. Stephen. 1991. Influence of mixed cropping rotations (pasture-arable) on organic matter content, water stable aggregation and clod porosity in a group of soils. *Soil Tillage Res.* 19:77-87.

Hobbie, J., J. Cole, J. Dungan, R.A. Houghton, and B. Peterson. 1984. Role of biota in global CO_2 balance: The controversy. *BioScience* 34:492-498.

Houghton, R.A., J.E. Hobbie, J.M. Melillo, B. Moore, B.J. Peterson, G.R. Shaver, and G.M. Woodwell. 1983. Changes in the carbon content of terrestrial biota and soils between 1860 and 1980: A net release of CO_2 to the atmosphere. *Ecol. Monogr.* 53:235-262.

Houghton, R.A. 1987. Terrestrial metabolism and atmospheric CO_2 concentrations. *BioScience* 37:672-678.

Hsieh, Y.P. 1992. Pool size and mean age of stable soil organic carbon in cropland. *Soil Sci. Soc. Am. J.* 56:460-464.

Hulugalle, N.R. and B.T. Kang. 1990. Effect of hedgerow species in alley cropping systems on surface soil properties of an Oxic Paleustalf in Southwest Nigeria. *J. Agric. Sci.* 114:301-307.

Hussain, S.K., L.N. Mielke, and J. Skopp. 1988. Detachment of soil as affected by fertility management and crop rotations. *Soil Sci. Soc. Am. J.* 52:1463-1468.

Janzen, H.H. 1987. Soil organic matter characteristics after long-term cropping to various spring wheat rotations. *Can. J. Soil Sci.* 67:845-856.

Jenkinson, D.S. 1971. Studies on the decomposition of 14C-labeled organic matter in soil. *Soil Sci.* 111:64-70.

Jenkinson, D.S., D.E. Adams, and A. Wild. 1991. Model estimates of CO_2 emissions from soil in response to global warming. *Nature* 351:304-306.

Jordahl, J.L. and D.L. Karlen. 1993. Comparison of alternative farming systems. III. Soil aggregate stability. *Am. J. Alt. Agric.* 8:27-33.

Juma, N.G., R.C. Izaurralde, J.A. Robertson, and W.B. McGill. 1993. Crop yield and soil organic matter trends over 60 years in a Typic Cryoboralf at Breton, Alberta. In: The Breton Plots, Dept. Soil Sci., Univ. of Alberta, Edmonton, Alberta, Canada.

Kang, B.T., H. Grimme, and T.L. Lawson. 1985. Alley cropping sequentially cropped maize and cowpea with *Leucaena* on a sandy soil in Southern Nigeria. *Plant Soil* 85:267-277.

Karlen, D.L., D.C. Erbach, T.C. Kaspar, T.S. Colvin, E.C. Berry, and D.R. Timmons. 1990. Soil tilth: A review of past perceptions and future needs. *Soil Sci. Soc. Am. J.* 54:153-161.

Karlen, D.L. and J.W. Doran. 1991. Cover crop management effects on soybean and corn growth and nitrogen dynamics in an on-farm study. *Am. J. Alt. Agric.* 6(2):71-82.

Karlen, D.L., W.R. Berti, P.G. Hunt, and T.A. Matheny. 1989. Soil-test values after eight years of tillage research on a Norfolk loamy sand. *Commun. Soil Sci. Plant Anal.* 20:1413-1426.

Karlen, D.L., N.S. Eash, and P.W. Unger. 1992. Soil and crop management effects on soil quality indicators. *Am. J. Alt. Agric.* 7:48-55.

Karlen, D.L. and A.N. Sharpley. 1994. Management strategies for sustainable soil fertility. p. 47-108. In: J.L. Hatfield and D.L. Karlen (eds.), *Sustainable agricultural systems*. Lewis Publishers. CRC Press, Inc., Boca Raton, FL.

Karlen, D.L., N.C. Wollenhaupt, D.C. Erbach, E.C. Berry, J.B. Swan, N.S. Eash, and J.L. Jordahl. 1994. Crop residue effects on soil quality following 10-years of no-till corn. *Soil Tillage Res.* 31:149-167.

Kay, B.D. 1990. Rates of change of soil structure under different cropping systems. *Adv. Soil Sci.* 12:1-52.

Kern, J.S. and M.G. Johnson. 1993. Conservation tillage impacts on national soil and atmospheric carbon levels. *Soil Sci. Soc. Am. J.* 57:200-210.

Lamb, J.A., G.A. Peterson, and C.R. Fenster. 1985. Wheat fallow tillage systems' effect on a newly cultivated grassland soils' nitrogen budget. *Soil Sci. Soc. Am. J.* 49:352-356.

Langdale, G.W., R.L. Wilson, Jr., and R.R. Bruce. 1990. Cropping frequencies to sustain long-term conservation tillage systems. *Soil Sci. Soc. Am. J.* 54:193-198.

Lee, J.J., D.L. Phillips, and R. Liu. 1993. The effect of trends in tillage practices on erosion and carbon content of soils in the U.S. Corn Belt. *Water, Air, Soil Pollut.* 70:389-401.

Mann, L.K. 1986. Changes in soil carbon storage after cultivation. *Soil Sci.* 142:279-288.

Martinez, J. and G. Guiraud. 1990. A lysimeter study of the effects of a ryegrass catch crop, during a winter wheat/maize rotation, on nitrate leaching and on the following crop. *J. Soil Sci.* 41:5-16.

Mazzarino, M.J., L. Szott, and M. Jimenez. 1993. Dynamics of soil total C and N, microbial biomass, and water soluble C in tropical agroecosystems. *Soil Biol. Biochem.* 25:205-214.

McCracken, D.V., S.J. Corak, M.S. Smith, W.W. Frye, and R.L. Blevins. 1989. Residual effects of nitrogen fertilization and winter cover cropping on nitrogen availability. *Soil Sci. Soc. Am. J.* 53:1459-1464.

Melsted, S.W. 1954. New concepts of management of Corn Belt soils. *Adv. Agron.* 6:121-142.

Moldenhauer, W.C. and W.H. Wischmeier. 1960. Soil and water losses and infiltration rates on Ida silt loam as influenced by cropping systems, tillage practices, and rainfall characteristics. *Soil Sci. Soc. Am. Proc.* 24:409-413.

Mulla, D.J., L.M. Huyck, and J.P. Reganold. 1992. Temporal variation in aggregate stability on conventional and alternative farms. *Soil Sci. Soc. Am. J.* 56:1620-1624.

Oades, J.M. and J.N. Ladd. 1977. Biochemical properties: Carbon and nitrogen metabolism. p.127-162. In: J.S. Russell and E.L. Greacen (eds.), *Soil factors in crop production in a semiarid environment*. Univ. of Queenland Press, St. Lucia, Queensland.

Olmstead, L.B. 1947. The effect of long-time cropping systems and tillage practices upon soil aggregation at Hays, Kansas. *Soil Soc. Am. Proc.* 11:89-92.

Parton, W.J., D.S. Schimel, C.V. Cole, and D.S. Ojima. 1987. Analysis of factors controlling soil organic matter levels in Great Plains grasslands. *Soil Sci. Soc. Am. J.* 51:1173-1179.

Paustian, K., W.J. Parton, and J. Persson. 1992. Modeling soil organic matter in organic-amended and nitrogen-fertilized long-term plots. *Soil Sci. Soc. Am. J.* 56:476-488.

Pieters, A.J. and R. McKee. 1938. The use of cover and green-manure crops. p. 431-444. In: Soils and Men, The Yearbook of Agriculture. USDA. U.S. Gov't. Print. Off., Washington, D.C.

Power, J.F. 1987. The role of legumes in conservation tillage systems. *Soil Cons. Soc. Am.* Ankeny, IA.

Power, J.F. 1990. Legumes and crop rotations. p. 178-204. In: C.A. Francis, C.B. Flora, and L.D. King (eds.), *Sustainable Agriculture in Temperate Zones*. J. Wiley & Sons, NY.

Priputina, I.V. 1989. Lowering of the humus content of chernozem soils of the russian plain as a result of human action. *Soviet Geography* 30:759-762.

Raimbault, B.A. and T.J. Vyn. 1991. Crop rotation and tillage effects on corn growth and soil structural stability. *Agron. J.* 83:979-985.

Ram, D.N., M.T. Vittum, and P.J. Zwerman. 1960. An evaluation of certain winter cover crops for the control of splash erosion. *Agron. J.* 52:479-482.

Rasmussen, P.E., H.P. Collins, and R.W. Smiley. 1989. Long-term management effects on soil productivity and crop yield in semi-arid regions of eastern Oregon. Station Bull. 675. USDA-ARS and Oregon State Univ. Agric. Expt. Stn., Pendleton, OR.

Reganold, J.P. 1988. Comparison of soil properties as influenced by organic and conventional farming systems. *Am. J. Alt. Agric.* 3:144-145.

Ridley, A.O. and R.A. Hedlin. 1968. Soil organic matter and crop yields as influenced by frequency of summer fallowing. *Can. J. Soil Sci.* 48:315-322.

Robinson, C.A., R.M. Cruse, and K.A. Kohler. 1994. Soil management. p. 109-134. In: J.L. Hatfield and D. L. Karlen (eds.), *Sustainable agriculture Systems*. Lewis Publishers, CRC Press, Inc., Boca Raton, FL.

Roder, W., S.C. Mason, M.D. Clegg, J.W. Doran, and K.R. Kniep. 1988. Plant and microbial responses to sorghum-soybean cropping systems and fertility management. *Soil Sci. Soc. Am. J.* 52:1337-1342.

Rogers, T.H. and J.E. Giddens. 1957. Green manure and cover crops. p. 252-257. In: Soil, the 1957 Yearbook of Agriculture. USDA. U.S. Gov't. Print. Off., Washington, D.C.

Russell, E.W. 1973. *Soil conditions and plant growth.* 10th ed. Longman Ltd., NY.

Scharpenseel, H.W., H.U. Neue, and S. Singer. 1992. Biotransformations in different climate belts; source--sink relationships. p. 91-105. In: J. Kubat (ed.), *Developments in agricultural and managed-forest ecology*. Vol. 25. Elsevier Sci. Pub. Co., Amsterdam.

Schlesinger, W.H. 1990. Evidence from chronosequence studies for a low carbon-storage potential of soils. *Nature* 348:232-234.

Smika, D.E. and G.A. Wicks. 1968. Soil water storage during fallow in the Central Great Plains as influenced by tillage and herbicide treatments. *Soil Sci. Soc. Am. Proc.* 32:591-595.

Smika, D.E. and P.W. Unger. 1986. Effect of surface residues on soil water storage. *Adv. Soil Sci.* 5:111-138.

Staver, K.W. and R.B. Brinsfield. 1990. Patterns of soil nitrate availability in corn production systems: Implications for reducing groundwater contamination. *J. Soil Water Conserv.* 45:318-323.

Smith, M.S., W.W. Frye, and J.J. Varco. 1987. Legume winter cover crops. *Adv. Soil Sci.* 7:95-139.

Soane, B.D. 1990. The role of organic matter in soil compactibility: A review of some practical aspects. *Soil Tillage Res.* 16:179-201.

Stevenson, F.J. 1982. *Humus chemistry.* John Wiley & Sons, New York.

Strickling, E. 1950. The effect of soybeans on volume weight, and water stability of aggregates, soil organic matter content and crop yield. *Soil Sci. Soc. Am. Proc.* 15:30-34.

Taylor, R.E., O.E. Hays, C.E. Bay, and R.M. Dixon. 1964. Corn stover mulch for control of runoff and erosion on land planted to corn after corn. *Soil Sci. Soc. Am. Proc.* 28:123-125.

Tiessen, H. and J.W.B. Stewart. 1983. Particle-size fractions and their use in studies of soil organic matter. II. Cultivation effects on organic matter composition in size fractions. *Soil Sci. Soc. Am. J.* 47:509-514.

Tisdall, J.M. and J.M. Oades. 1982. Organic matter and water-stable aggregates in soils. *J. Soil Sci.* 33:141-163.

Turchenek, L.W. and J.M. Oades. 1979. Fractionation of organo-mineral complexes by sedimentation and density techniques. *Geoderma* 21:311-343.

Unger, P.W. 1968. Soil organic matter and nitrogen changes during 24 years of dryland wheat tillage and cropping practices. *Soil Sci. Soc. Am. Proc.* 32:426-429.

Unger, P.W. 1982. Surface soil physical properties after 36 years of cropping to winter wheat. *Soil Sci. Soc. Am. J.* 46:796-801.

Unger, P.W., D.W. Anderson, C.A. Campbell, and C.W. Lindwall. 1990. Mechanized farming systems for sustaining crop production and maintaining soil quality in semiarid regions. p. 56-88. In: J.W.B. Stewart (ed.), *Proc. Int. Conf. on Soil Quality in Semiarid Agriculture,* Vol. 1. Saskatchewan Inst. Pedology, Univ. Saskatchewan, Saskatoon, Sask., Canada.

Unger, P.W. and R.E. Phillips. 1973. Soil water evaporation and storage. p. 42-54. In: Conservation Tillage. *Proc. Nat'l. Conf.* Soil Conserv. Soc. Am., Ankeny, IA.

van Bavel, C.H.M. and F.W. Schaller. 1950. Soil aggregation, organic matter, and yields in a long-time experiment as affected by crop management. *Soil Sci. Soc. Am. Proc.* 15:399-408.

van Veen, J.A. and E. A. Paul. 1981. Organic carbon dynamics in grassland soils. I. Background information and computer simulation. *Can. J. Soil Sci.* 61:185-201.

Wilson, H.A. and G.M. Browning. 1945. Soil aggregation, yields, runoff, and erosion as affected by cropping systems. *Soil Sci. Soc. Am. Proc.* 10:51-57.

Wilson, G.F., B.T. Kang, and K. Mulongoy. 1986. Alley cropping: Trees as sources of green-manure and mulch in the tropics. *Biol Agr. Hort.* 3:251-267.

Wood, C.W., D.G. Westfall, G.A. Peterson, and I.C. Burke. 1990. Impacts of cropping intensity on carbon and nitrogen mineralization under no-till dryland agroecosystems. *Agron. J.* 82:1115-1120.

Wood, C.W., D.G. Westfall, and G.A. Peterson. 1991. Soil carbon and nitrogen changes on initiation of no-till cropping systems. *Soil Sci. Soc. Am. J.* 55:470-476.

Woodward, F.I., G.B. Thompson, and I.F. McKee. 1991. The effects of elevated concentrations of carbon dioxide on individual plants, populations, communities and ecosystems. *Ann. Botany* 67:23-38.

Yamoah, C.F., A.A. Agboola, G.F. Wilson, and K. Mulongoy. 1986. Soil properties as affected by the use of leguminous shrubs for alley cropping with maize. *Agric. Ecosyst. Environ.* 18:167-177.

Zhu, J.C., C.J. Gantzer, S.H. Anderson, E.E. Alberts, and P.R. Beuselinck P. 1989. Runoff, soil, and dissolved nutrient losses from no-till soybean with winter cover crops. *Soil Sci. Soc. Am. J.* 53:1210-1214.

Models to Evaluate Soil Organic Matter Storage and Dynamics

W.J. Parton, D.S. Ojima, and D.S. Schimel

I. Introduction

Although the factors which control soil organic matter (SOM) dynamics have been identified, it has only been during the last 30 years that significant progress has been made in quantifying the relative importance of these factors. Critical experiments have included adding ^{14}C labeled and unlabeled plant material to different soils (Pinck et al., 1950; Sørensen, 1981; Ladd et al., 1981; Stott et al., 1983), residue applications of different lignin and nitrogen content, soil biota and microbial studies, ^{14}C dating of different soil fractions (Martel and Paul, 1974a,b; Paul and van Veen, 1978), separation of SOM into different fractions (Tisdall and Oades, 1982; Turchenek and Oades, 1979; Cambardella and Elliott, 1993), tracer studies of atmospheric ^{14}C carbon inputs (atmospheric bomb test ^{14}C inputs) (Trumbore et al., 1989, 1990; Trumbore, 1993; Jenkinson et al., 1991) and long-term (5-150 year) field experiments (Jenkinson and Rayner, 1977; Rasmussen and Collins, 1991; Paustian et al., 1992) where the fate of added plant material was followed. Much of the recent progress in our understanding about SOM dynamics has come from combining information about long-term SOM field experiments with short-term laboratory experiments.

1-56670-033-7/96/$0.00+$.50

Computer models have allowed us to quantify the impact of the different factors which control SOM dynamics and help interpret results for both short- and long-term field and laboratory experiments.

One of the difficulties encountered in studies of SOM dynamics has been the lack of a physical or chemical methods to isolate directly the various organic matter pools suggested by various decomposition studies. The chemically determined pools of organic matter, such as fulvic or humic acids, humins, polyphenols, etc., do not neatly define SOM compounds with a uniform kinetic behavior (Anderson and Paul, 1984). Physical fractionation of SOM determined by density fractionation identify 'light' and heavy components of SOM, but linking these fractions to different kinetic pools have also met with mixed success (Stevenson, 1982; Anderson and Paul, 1984; Cambardella and Elliott, 1993; Scharpenseel et al., 1984; Trumbore et al., 1990). A method has been developed that measures bomb ^{14}C in the different density fractions (Anderson and Paul, 1984; Jenkinson et al., 1991; Trumbore, 1993). In the study by Trumbore (1993) it was determined that 'dense' fraction (specific gravity > 2.0 g m^{-3}) contained the oldest soil C in the soils analyzed. This technique was able to also determine that C turnover rates in tropical soils were much higher relative to temperate soils (Trumbore, 1993). However, the study by Trumbore could not isolate material directly relative to the more recalcitrant pool with acid hydrolysis of the dense fraction. So work still needs to be done to develop a technique to identify more definitively the parts of SOM.

In this chapter we will review the historical development of SOM models and will discuss how these models have been used to help interpret field and laboratory data, identify gaps in our knowledge about SOM dynamics, to quantify the relative importance of different soil forming factors, and to predict the long-term consequence of different management practices and global change. In the first half of the paper we will present a historical review of the development of SOM models and in the second part we will discuss how the different SOM models have been used for scientific investigation and as management tools.

II. Mathematical Model for SOM

Bouwman (1990) has described the development of models for plant residue decomposition and SOM dynamics. The earliest models were single pool first-order differential equations. This type of model was useful in describing the initial rapid decomposition of leaf litter and for readily metabolizable substrates such as carbohydrates (Swift et al., 1979). The basic equation applied to the formation of SOM is:

$$dC/dt = hA - kC$$

where C is soil carbon level, k is the decomposition rate of soil C (y^{-1}), h is the C storage constant (fraction) and A is the organic addition of C to the soil (g C

m^{-2} yr^{-1}). Bouwman (1990) cites many authors who used this approach including Jenny (1941), Jenny et al. (1949) and Greenland and Nye (1959). The major difficulty with this approach is the assumption that organic matter is composed of a single type of material and doesn't allow for the observed formation of more recalcitrant fractions as material decomposes (Sørensen, 1981). This approach can be refined by altering the decomposition rate (k) with time and the apparent age of the material (Janssen, 1984). A problem with this approach is that you must empirically evaluate k for each substrate and every environment.

Jenkinson and Rayner (1977) developed one of the first widely used SOM models which divided soil C into active, slow and passive pools with different turnover times (2, 50 and 1980 years). Campbell (1978) divided SOM into labile and stabilized fractions with turnover times (1/k) of 53 and 1429 years, respectively. Paul and van Veen (1978) and van Veen and Paul (1981) improved on these SOM models by dividing plant material into recalcitrant and decomposable fractions and including the concept of physical protection of SOM (see Figure 1a). They assumed that physically protected SOM has much lower decomposition rate than nonphysically protected SOM and that cultivation of the soil will increase the fraction of SOM that is nonphysically protected. Van Veen and Paul (1981) also included the impact of soil erosion on SOM dynamics.

More recently Parton et al. (1987, 1993) used similar SOM pools as Paul and van Veen (1978) and Jenkinson and Rayner (1977) and added the impact of soil texture on SOM dynamics; developed generalized nutrient cycling submodels which simulate soil C and inorganic and organic N, P, and S dynamics (Parton et al., 1988); added a simplified plant growth model; and developed a general approach for splitting plant residue into structural and metabolic material as a function of the initial lignin:nitrogen ratio of the material. Parton et al. (1987) suggest that the soil silt plus clay content influence the turnover rate of active SOM (higher for sandy soils) and the stabilization of active SOM into slow SOM (more stabilization for fine textured soils). More recently Parton et al. (1993) improved the CENTURY model by adding the impact of clay on formation of passive SOM (higher amount with high clay content), and including a surface litter decomposition submodel, and the leaching of soluble SOM (Figure 2). Jenkinson (1990) recently changed his SOM model (Figure 1b) by assuming that SOM stabilization during decomposition is a function of the clay content or cation exchange capacity, and assuming that there is a small amount of inert SOM with a long turnover time (> 10000 years). These changes were implemented in order to represent the incorporation of ^{14}C bomb carbon at Rothamsted (Jenkinson et al., 1991). Both Jenkinson (Jenkinson et al., 1991) and Parton (Parton et al., 1994) suggest that our knowledge is limited about the factors which control the formation of passive SOM.

Another set of models are based on the assumption that organic matter inputs and by-products form a continuum of substrate (Ågren and Bosatta, 1987; Bosatta and Ågren, 1991). So litter material enters the soil system and as it passes through various stages of decomposition the pattern of its decomposition is set by its inherent quality. Each cohort of litter needs to be characterized

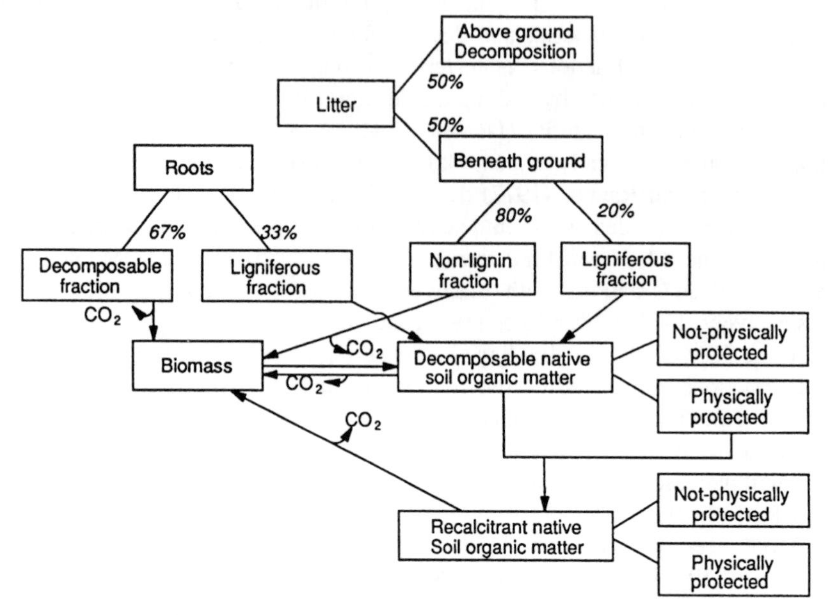

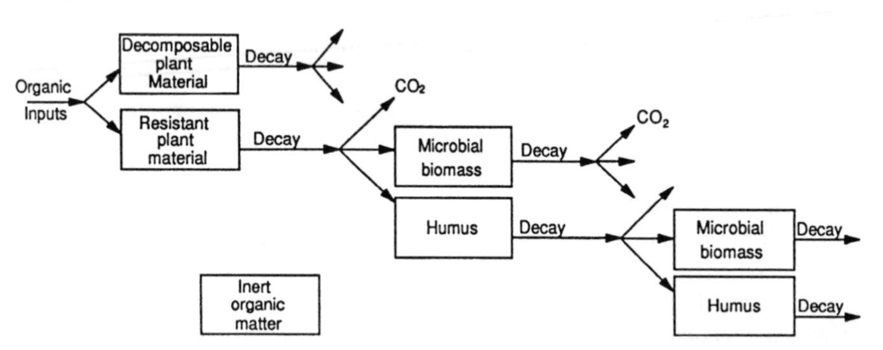

Figure 1. Flow diagram for the van Veen and Paul (van Veen and Paul, 1981) soil organic matter model (a) and the Jenkinson (1990) soil organic matter model (b).

relative to its quality in terms of its decomposability and this characterizes the pattern of decomposition it will follow (Bosatta and Ågren, 1991). Pastor and Post (1988) use a similar approach where decomposition of each cohort of added plant material is represented in the model.

During the last 20 years a large number of detailed process oriented soil nutrient cycling models have been developed (Hunt, 1977; McGill et al., 1981;

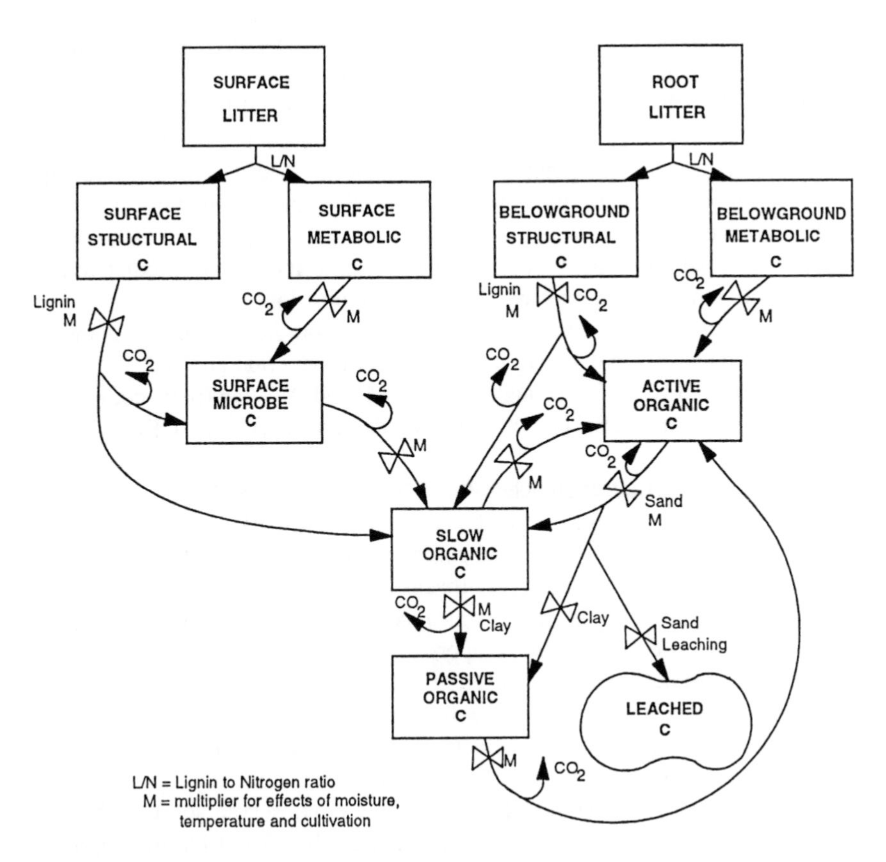

Figure 2. Flow diagram of the CENTURY soil organic matter model. (Modified from Parton et al., 1993.)

Molina et al., 1983; Li et al., 1992; and Sharpley and Williams, 1990a,b). These models have incorporated many of the concepts of the SOM models, however, they have focused on simulating detailed nutrient cycling processes using second to daily time steps and very detailed structural description of the soil system. For example, the McGill et al. (1981) model includes a detailed description of C and N dynamics for soil bacteria, fungi and actinomicytes and one resistant SOM pool. This contrasts with the SOM models where the detailed microbial dynamics are simplified using an active SOM pool (Parton et al., 1987), while the resistant SOM pools are divided into slow and passive pools that are further divided into physically protected and unprotected pools by van Veen and Paul (1981) (see Figure 1a). The detailed nutrient cycling models have typically been used to simulate the dynamics of C and N for a growing season, while the SOM models are used to simulate dynamics for longer time periods

(10 to 1000 years). The exception to this is the EPIC model (Sharpley and Williams, 1990a,b) which was designed to simulate the impact of erosion on plant production in 10 to 100 years time period. Detailed comparison of observed and simulated dynamics of SOM for long-term SOM plots in Sweden (Paustian et al., 1992), Australia (Probert et al., 1995) and Pendleton, Oregon (Parton and Rasmussen, 1994) suggest the need to combine detailed nutrient cycling SOM models and plant production models in order to improve our understanding about SOM dynamics. Specifically they show that errors in simulating water storage, surface runoff, leaching of NO_3^-, and allocation of C and N to plant parts caused the CENTURY model to incorrectly predict soil C and N inputs into the soil and the abiotic decomposition rate for some of the treatments. Parton et al. (1994) has recently incorporated a daily version of CENTURY model into a detailed grassland plant production and soil water and temperature model (GRASS - Coughenour, 1984) and Paustian (pers. commun.) has incorporated the CENTURY SOM model concepts into CERES crop models.

The various modeling approaches for investigations of SOM formation and dynamics have been derived from the experimental work that attempted to identify the chemical nature of SOM components, used physical fractionation characteristics, or used the kinetic approach for determining the decomposition rates without regard to the specific chemical components of the SOM. The major weakness of the present modeling approaches is that there is no generally acceptable way to determine the different SOM fractions, either chemically or physically, and thus it is impossible to directly measure SOM pools included in the models. Thus, most model validation work has compared observed and simulated total soil SOM levels for different treatments. More recently Jenkinson et al. (1991) tested his SOM model by simulating the observed soil C stabilization of the atmospheric bomb test ^{14}C added during the 1950-1970 time period.

III. Scientific Use of SOM Model

This section of the chapter will discuss how SOM models have been used as tools to improve our understanding about nutrient cycling and SOM dynamics, and how these models have been used as policy and management tools to evaluate the long-term implication of different management practices. We will primarily focus on how the CENTURY model has been used for evaluating management questions and scientific understanding, however, other SOM models (Jenkinson, 1990; van Veen and Paul, 1981) have also been used in similar ways.

The CENTURY model has been tested in great detail using data from several long-term SOM experiments and used as a scientific tool for integrating and interpreting data. The 55 year winter wheat plots from Pendleton, Oregon (Parton and Rasmussen, 1994; Rasmussen and Parton, 1994) were used to look at the impact of different types and amounts of soil organic matter additions (pea vine, manure), straw burning, and fertilizer levels on plant production, and soil C and N levels for a wheat fallow system. Data from a 30 year experiment in Sweden (Paustian et al., 1992) were used to evaluate the impact of adding different types of organic matter (sawdust, straw, green manure, and manure) and fertilizer levels on plant production and nutrient uptake, N budgets, and soil C and N levels. Observations (1970-present) from the Sidney, Nebraska site (Fenster and Peterson, 1979) were used to test the simulated impact of different tillage practices (no-till, stubble mulch, and plow) and fertilizer levels on soil water budgets, soil temperature, soil C and N levels, and plant production for a wheat-fallow system (Metherell et al., 1994). Data from continuous wheat system (1970-present) in Australia have been recently (Probert et al., 1995) used to test CENTURY's ability to simulate tillage management (no-till and conventional tillage), fertilizer levels (0, 40, and 80 kg N ha^{-1} y^{-1}), and straw burning on soil C and N levels, plant production, soil N mineralization rates and soil water storage. Carter et al. (1993) also recently used CENTURY to simulate changes in soil C and N levels under wheat and pasture rotations at five sites in southern Australia. Results from these papers will be summarized and will focus on how well the model simulated the data from the sites, identifying the factors that have a large impact on SOM dynamics, and suggesting future research needed to improve our understanding about SOM dynamics.

A. Soil C and N Levels

Figure 3 presents the observed vs simulated comparison of soil C and N (0-20 cm) for Pendleton, Oregon site; the Swedish site; Sidney, Nebraska site, and five long-term plots from southern Australia and shows that the comparison of observed and simulated soil C and N levels is quite good for a diverse set of sites and management practices. The model results and data from the sites suggested that soil C increased linearly with increasing C input levels and that the slope of the line was lowest for wheat-fallow systems and highest for the continuous wheat system in Sweden (compare Figures 4b with Figures 4a and 4c). The higher C stabilization rate for Swedish site is caused by low decomposition rate resulting from lower soil temperatures. The lower C stabilization rates for the wheat-fallow systems is caused by high decomposition rates during the fallow year (higher soil water and temperature). A detailed comparison of the observed and simulated C and N stabilization rates for the Pendleton, Oregon site show the model tended to overestimate stabilization efficiencies of C and N. In general, this overestimation was not found for the other sites.

B. Litter Quality and N Fertilizer

Litter quality and N fertilizer influence the stabilization of soil C and N. Model results and field data from Paustian et al. (1992) suggest that increasing the

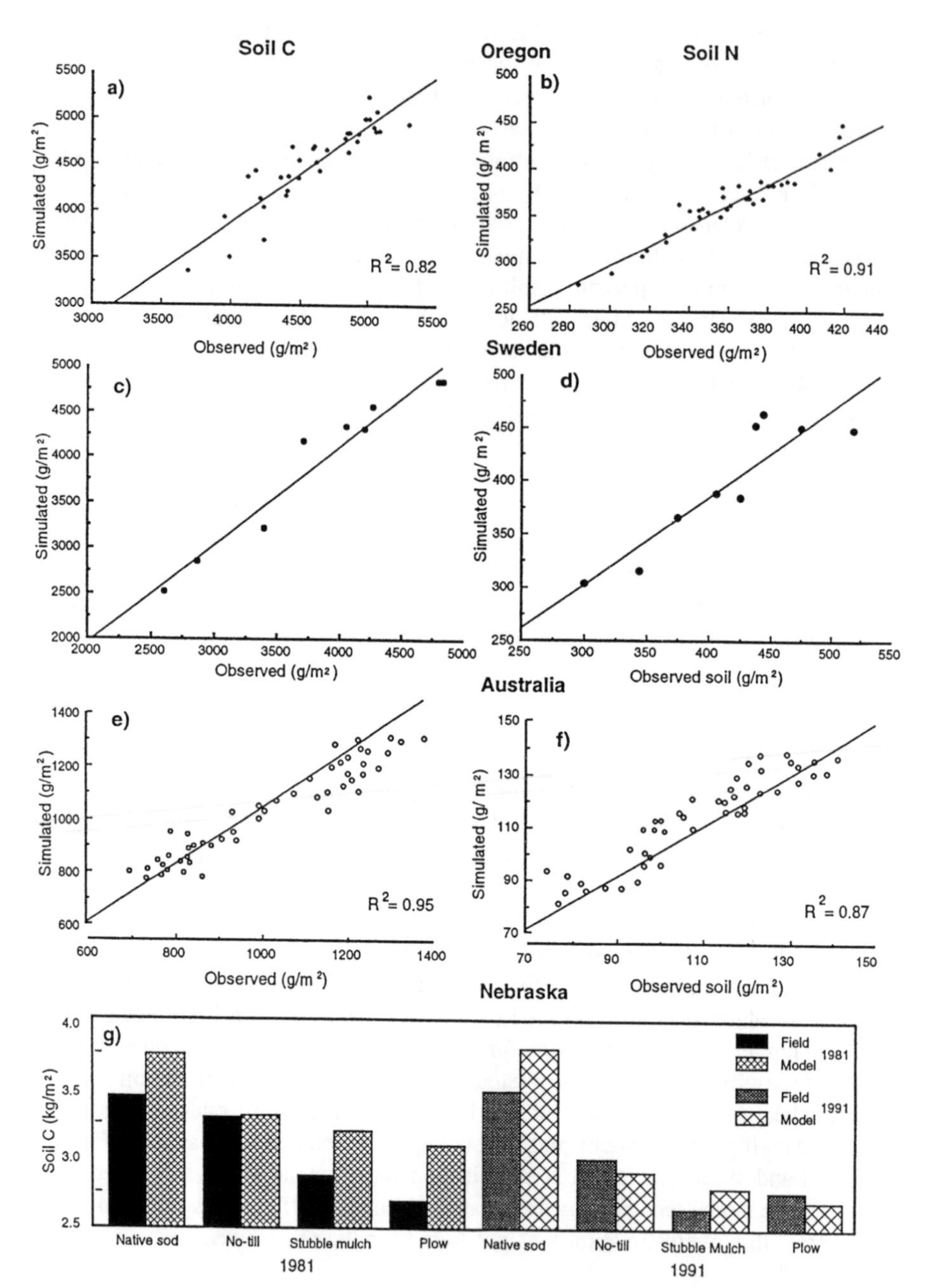

Figure 3. Comparison of observed vs simulated (using the CENTURY model) soil C (a) and N (b) levels at Pendleton, Oregon site (Parton and Rasmussen, 1994), soil C (c) and N (d) levels at the Swedish site (Paustian et al., 1992), soil C (e) and N (f) for long-term sites in Australia (Carter et al., 1993) and soil C levels at the Sidney, Nebraska site (Metherell, 1992) (g).

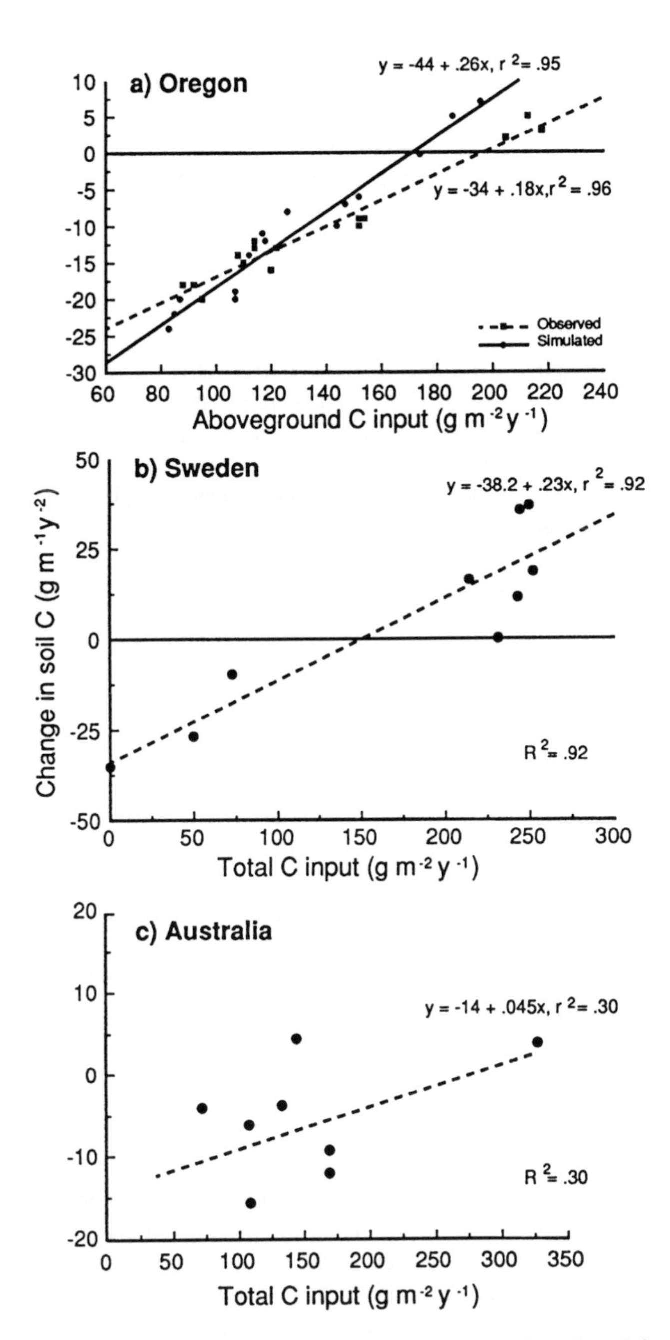

Figure 4. Comparison of changes in soil C levels as a function C input levels for sites in Pendleton, Oregon (a) (Parton and Rasmussen, 1994), Sweden (b) (Paustian et al., 1992) and Australia (c) (Carter et al., 1993).

lignin content of added plant material increases both soil C and N stabilized with soil C levels increasing by more than 500 g m^{-2} after 30 years where the lignin content of plant material increased from 15% to 30%. The highest stabilization of soil C and N was with the manure treatment where both lignin and nitrogen content of the added organic matter was high. Adding inorganic fertilizer to the system generally increases plant production and thus increases soil C inputs to the system from either root biomass or aboveground plant material. This was observed for all of the sites where the CENTURY model was tested. Adding inorganic N fertilizer or organic matter with high N contents also increases the stabilization of soil C. At the Pendleton, Oregon site the observations and model results showed that highest soil C stabilization efficiencies (26%) were observed for the high inorganic N fertilizer treatment and the lowest value for the pea vine addition treatment (12%). These results are consistent with data from Berg and Ekbohm (1991), however, it is not clear what process is causing higher soil C stabilization for the high N treatments. The data from Sweden and Pendleton, Oregon also show that soil N stabilization efficiency is higher for organic N additions as compared to inorganic N additions (40 to 60% vs 10 to 30%). The major discrepancies between the model results and observations were that the model tended to underestimate soil N levels when high lignin content plant material is added to the soil and that the model underestimated the increase in soil C storage when inorganic fertilizer is added to the soil (Parton et al., 1994), thus suggesting the need for model development and further research.

C. Tillage

Field and model results suggest that the major impact of tillage on SOM dynamics is (1) increased decay rates due to stirring of the soil and (2) alterations of soil microclimate (soil water and temperature) which impacts the decomposition rate of the SOM pools. The CENTURY model assumes that for each cultivation event the decomposition rate of the SOM pools will be increased for the month that it occurred with the decomposition rate increased 10% for minimal soil stirring and 30% for conventional plowing. The results from the Sidney, Nebraska site (Metherell, 1992; and Metherell et al., 1995) tested these assumptions and suggested that the increase in decomposition with plowing needs to be greater. The conceptual basis for increasing decomposition with tillage is sound, however, it is difficult to quantify using existing field or laboratory data. Tillage influences soil microclimate by changing the amount of standing dead and surface litter. Increasing the amounts of surface and standing dead litter causes the soil temperature to decrease and soil water storage to increase. The current model results (Metherell, 1992; and Probert et al., 1995) suggest that we need to do a better job representing these effects in CENTURY and that we will need to combine detailed field observations of soil microclimate (soil water and temperature) with daily heat and water flow models in order to better quantify the effect of surface plant material on the microclimate. Recent results from using the model in Australia (Probert et al., 1995) suggest that we need to include surface runoff in the model and decrease the shading effect of standing dead on surface soil temperatures.

D. Plant Production

Simulation of aboveground plant production (straw and grain) and N uptake by the plant were evaluated for most of the sites where CENTURY has been utilized. Figure 5 shows a comparison of observed and simulated plant production for Pendleton, Oregon site (Figure 5a), the Swedish site (Figure 5b), and the Sidney, Nebraska site (Figure 5c). This comparison includes different organic and inorganic N and C additions treatments, and treatments where different soil tillage practices were used. The results show that the model did an adequate job of simulating long-term patterns in plant production for the Pendleton, Oregon site (see Figure 5a), however the model didn't simulate year to year changes in plant production very well (see Figures 5b and 5c). Long-term average nutrient uptake by the plants was fairly well represented, however, year to year changes were not as well simulated (data presented for the Swedish study by Paustian et al., 1992). These results suggest that daily time step plant physiologically based plant growth models are needed to simulate year to year changes in plant production, however, the long-term trends in plant production and response to different management practices can be reasonably well simulated by simplified plant growth model in CENTURY. The results from the Pendleton, Oregon site (Parton and Rasmussen, 1994) also suggest that dynamic C and N allocation functions should be included in plant growth models in order to improve on results from some of the treatments. There was also an observed model bias for underestimating plant production for low soil fertility sites. The most likely explanation for this discrepancy is that the model does not consider nutrient mineralization below 20 cm in the soil. Data from the Pendleton, Oregon site suggest that a substantial amount of N is mineralized from 30 to 60 cm soil layer (0.5-1.5 g N m^{-2} y^{-1}). Model results show that adding this N as fertilizer can correct the underestimate of plant production for the low fertility treatments.

E. Nitrogen Budgets

Detailed N budgets were calculated for the Pendleton, Oregon site and the Swedish site based on the data and compared to simulated N budgets for the different treatments. Both the model results and observed N budgets show that highest N uptake in the plants occurs when inorganic N is added to the system (40 to 54% of the added N) and that adding N as an organic material results in lower N uptake by the plants (25 to 40% of the added N). Addition of very low N content plant material (straw and sawdust) at the Swedish site reduced plant

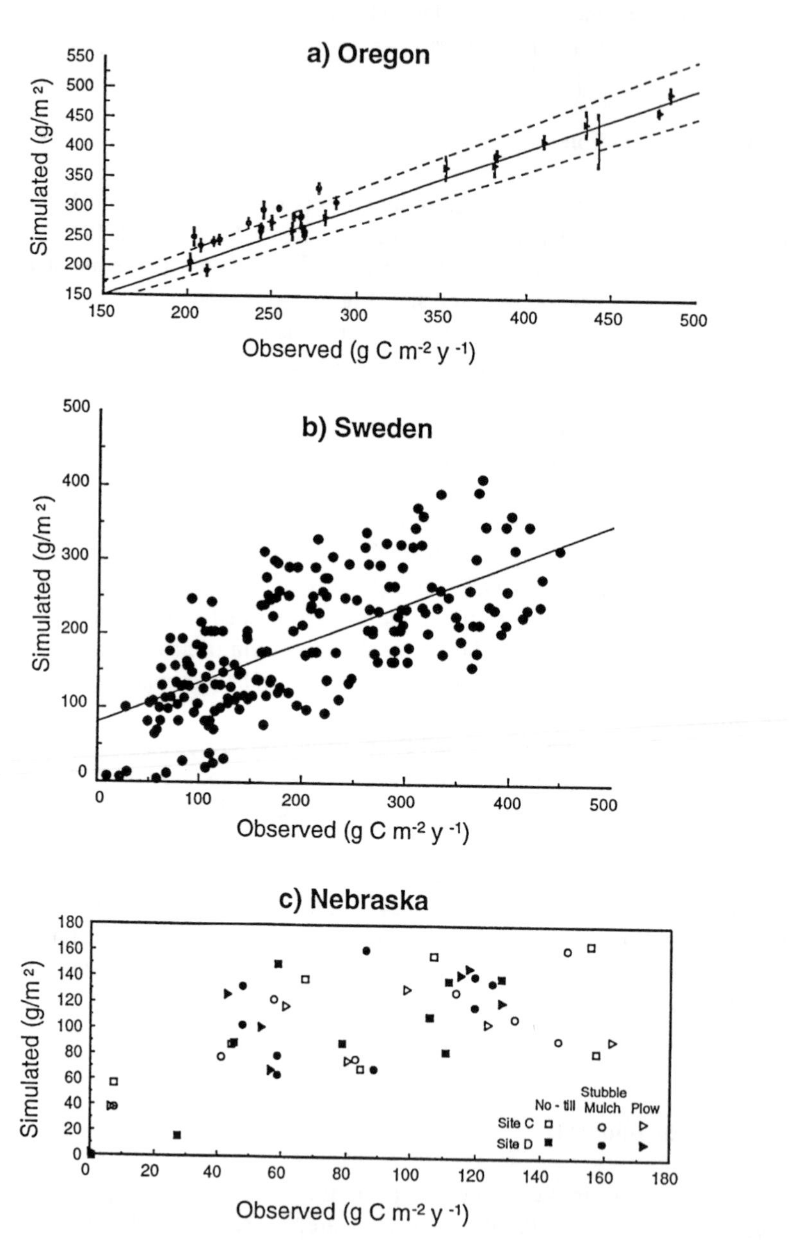

Figure 5. Simulated (using the CENTURY model) vs observed grain yield for the Pendleton, Oregon site (a) (Parton and Rasmussen, 1994), simulated vs observed plant production for Swedish site (b) (Paustian et al., 1992), and simulated vs observed grain yield for the Sidney, Nebraska site (c) (Metherell, 1992).

N uptake even after 30 years of adding plant material. Inorganic N additions generally have lower N stabilized in the soil (12 to 26%), while organic N additions have high soil N stabilization (30 to 80%) with higher amounts of N stabilized in the soil when the N content of the plant material is low. Leaching and gaseous N losses are higher for inorganic N additions as compared to organic N additions, and N losses resulting from inorganic N addition can be reduced by up to 50% if low N content organic matter (straw and sawdust) is combined with inorganic N fertilizer. A comparison of observed and simulated N budgets for the treatments at the two sites showed that the model correctly predicted the major differences between the different treatments however, the model tended to overestimate N stabilization in the soil at the Pendleton, Oregon site. This discrepancy was not apparent for the Swedish site. The N budget results at Pendleton, Oregon; Sidney, Nebraska, and the Hermitage plots in Australia all showed that substantial amounts of N were mineralized in soil below the 20 cm soil depth and that the N mineralized in the deep soil layers needed to be considered.

IV. Climate Change Assessment

During the last 15 years, SOM models have been used as tools for evaluating the impact of different management practices and the potential impact of global climatic change on natural and managed ecosystems. The SOM models are most useful for simulating the long-term impacts of climatic change on plant production, soil fertility, and soil C, N, and P storage. In this section we will primarily discuss how the CENTURY model has been used to evaluate long-term impact of climatic change scenarios on natural and managed agroecosystems. We will also discuss how Jenkinson's SOM model and the TEM ecosystem model (Raich et al., 1991; Melillo et al., 1993) have been used to look at the potential impact of climatic change scenarios on plant production and soil C levels at the global scale.

Schimel et al. (1990) used the CENTURY model (Version 1.0-Parton et al., 1987) to simulate plant production, nutrient cycling and SOM dynamics for the grasslands in the US Great Plains under current climatic conditions and simulated the change in ecosystem properties under different General Circulation Model (GCM) climatic change predictions. The results showed that the model did a good job of simulating regional patterns in plant production and SOM levels for the Great Plains. For most of the grassland sites, the climatic change scenarios caused plant production and N cycling to increase and SOM levels to decrease. The predicted increase in air temperature by the GCM's (2-4°C) had the largest effect on the system and was responsible for increasing N availability for plants and decreases in SOM levels. Higher plant production resulted from the higher N availability. Field experiments indicate that adding N fertilizer to most grasslands results in increased plant production (Dodd and Lauenroth, 1978; Owensby et al., 1970).

This analysis was improved on in a later paper (Schimel et al., 1991a) where the impact of climatic change scenarios was evaluated for the whole Great Plains using the Goddard Institute of Space Science (GISS) general circulation model of double CO_2 climate (GCM 2xCO_2) climatic change scenarios. Soil texture was held constant for the region (sandy loam soil) and it was assumed that natural grasslands occupied the whole region. The GISS GCM model predicted increased air temperature for the whole region (2 to 4°C), decreased precipitation for the central Texas and Oklahoma region and increased precipitation for the northern part of the Great Plains. The model results showed (see Figure 6) that plant production increased and soil C levels decreased for most of the region, except the Texas and Oklahoma region where plant production decreased by > 10%. These new results confirm the previous results showing an increased plant production in response to increases in soil N mineralization rates and that the increased N mineralization rates would compensate for up to a 10% decrease in precipitation.

Burke et al. (1991) improved on the previous study by using observed patterns in soil texture, climate and the GISS 2 x CO_2 climatic changes for the southern part of the Great Plains. The results showed that including the observed patterns in soil texture substantially altered regional patterns of SOM for the current climate, and had relatively little impact on regional patterns of plant production and response to climatic change scenarios. They also showed that the model could simulate year to year changes in the regional patterns of plant production by comparing simulated patterns with plant production estimated using satellite derived estimates. In the future we hope to improve on climate change assessment for the Great Plains and the U.S. by representing the actual land use, observed soil texture data, and ramped changes in the altered climate during the next 50 years. The most important improvement is the use of actual land use data, instead of potential vegetation since much of the land in the U.S. is used for agriculture or timber production.

The CENTURY grassland model (Version 4.0) has recently been tested using plant production and soil C and N data from tropical and temperate grassland sites around the world (Parton et al., 1993). A version of the model set up for all of the major grassland types in the world was used to assess the potential effect of different GCM climatic change scenarios on plant production and soil C and N levels (Ojima et al., 1993). In this assessment we assumed the projected GCM 2 x CO_2 climatic change would occur as a 50-year linear increase and also included the direct impact of increasing atmospheric CO_2 levels on the system. The two major ecosystem effects of increasing atmospheric CO_2 levels are increasing the plant water use and nitrogen use efficiencies (Ojima et al., 1993). The results (Table 1) from this analysis (Ojima et al., 1993) show that the direct effect of increasing atmospheric CO_2 under current climate conditions is to increase the plant production and soil C levels for all of the major grassland regions of the world. The largest increases in soil C are in the savanna grassland, while the largest increases in plant production are in the continental steppe grasslands.

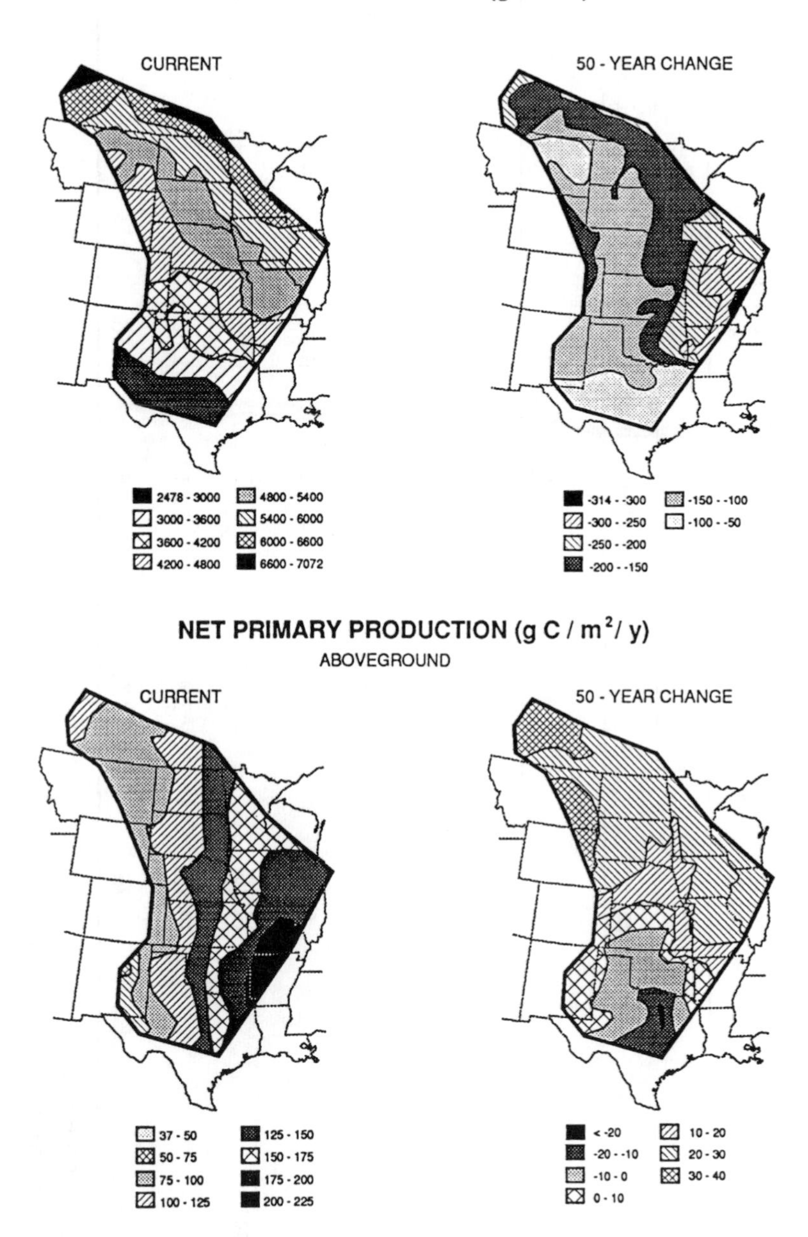

Figure 6. CENTURY model results of equilibrium and 50 year change in grassland soil C levels (a) and net primary production for native grassland in the Great Plains (Schimel et al., 1991b).

Table 1. Simulated change in soil C (average change during the 50-75 year period following initiation of climatic change) of altered climatic conditions and CO_2 levels (doubling in 50 years) using climate change scenarios derived from the CCC and GFDL GCM 2 x CO_2 runs

	Current soil C	% Change				
			Canadian Climate Center		GFDL Model	
	GT[a] to 20 cm	2 x CO_2	CC[b] only	CC[b] + 2 x CO_2	CC[b]	CC[b] + 2 x CO_2
Extreme continental steppe	12.65	1.42	-11.64	-12.21	-9.08	-9.05
Dry continental steppe	7.72	1.03	-13.16	-12.92	-14.11	-13.10
Humid temperate	28.89	2.45	-4.39	-1.95	-4.19	-2.12
Mediterranean	0.25	2.94	-9.23	-6.44	-9.14	-6.23
Dry savanna	12.79	-0.09	-0.37	0.04	0.49	1.03
Savanna	39.46	3.44	-1.61	1.62	-1.13	2.31
Humid savanna	3.76	8.55	-4.36	3.56	-3.44	5.59

[a] 1×10^{15} g C; [b] Climate change.
(Based on data presented by Ojima et al., 1993.)

The direct effect of climatic change on plant production is to increase production for most regions with the exception of the extreme continental steppe regions where production decreased dramatically. The decrease in these regions resulted from increased drought stress as a result of higher air temperatures. Soil C generally decreased in response to climatic change, with the largest decreases in the continental steppe and mediterranean grasslands. A net global decrease in soil C of 4 x 10^{15} g C due to climatic change effect alone is simulated. The model results show that the combined impact of increasing atmospheric CO_2 levels and climatic change scenarios is approximately equal to the sum of the independent effects. Thus, including a direct CO_2 effect on the plant system results in increased plant production and SOM level for most regions. The combined effect of climatic change and CO_2 changes in global soil C was a net decline of 2 x 10^{15} g C, compared to 4 x 10^{15} g C with climatic change alone (Parton et al., 1995).

The TEM global ecosystem models and Jenkinson's (1990) recent SOM model have been used to assess the impact of different climatic change scenarios on plant production and soil C and N levels for potential natural vegetation of the world estimated from the current climate. Jenkinson's study (Jenkinson et al., 1991) shows that predicted changes in climate would result in a 32 x 10^{15} g C release from the soil with a 3°C increase in air temperature. Jenkinson's model assumed that plant production would not change during this time period and did not include any direct CO_2 impacts. The TEM model included the iterative feedbacks of climate and nutrient availability on plant production and the direct impact of atmospheric CO_2 levels on plant production. The results showed that climatic change alone resulted in decreased soil C levels and an increase in plant production at the global scale (Melillo et al., 1993). Adding the direct effect of atmospheric CO_2 levels further increased plant production and greatly reduced the predicted decrease in loss of soil C predicted with climate change alone.

In subsequent analyses with CENTURY, a sequence of grassland and forest sites spanning global climate and textural gradients were simulated, using sites where at least minimal validation data were available (Schimel et al., 1994). The results were then extrapolated globally using soil and climate data bases. A sensitivity to soil texture was done at each site, using 8 different textural combinations. Soil C increased linearly with clay content, almost independently of temperature, largely because of the effect of clay on the stabilization of passive SOM. This results in higher C storage than expected in the tropics (i.e., Oxisols, Spodosols, Ultisols, and Verisols), as most of the clay soils in the world are in tropical regions (Buol et al., 1980). The distribution of soil organic matter between detrital, microbial, slow and passive pools (approximately 3, 8, 55 and 35% respectively) was roughly independent of climate but was influenced by ecosystem type. Specifically, forests have more slow (60 vs 50%) SOM than grasslands, and less passive (~30 vs 40%). Heterotrophic respiration is partitioned very differently, with 50% originating from recent detritus, and essentially none from passive. The effect of temperature at steady state was estimated from the derivative of the SOM vs mean annual temperate (MAT) curve, estimated from the ensemble of model simulations. The predicted relationship is dSOM/dMAT = $183e^{-0.034}$ MAT in g m^{-2}. When this is integrated globally, using global temperature data, it suggests a global sensitivity to warming of -11.1 Gt SOM/°C. This is comparable to the estimate obtained similarly from observations and less than that estimated by uncoupled models. Again, the CENTURY-predicted sensitivity is lower than other models because of the N cycle feedback that tends to increase net primary production (NPP) as SOM is lost, increasing detrital inputs and reducing SOM losses.

Increasing carbon dioxide increases nutrient and water use efficiencies in CENTURY. While CENTURY shows increased NPP and C storage when CO_2 is increased, the N cycle interactions are again critical to understanding the model's response. While NPP can increase by 40% at doubled CO_2 in CENTURY, in practice it increases about 20%. As NPP increases, C storage increases, resulting in the sequestration of N. Increasing nutrient use efficiency also results in wider tissue C:N ratios in detritus and a larger sink for N in soils. The steady state increase in NPP reflects largely the change in nutrient use efficiency. However, these effects are additive when climate change and CO_2 are considered together, because the N released by warming can be used at the higher CO_2-fertilized efficiencies. The synergism of these two effects suggests that CO_2 would reduce warming-caused losses of soil organic C, and in some systems could even cause C to be sequestered (Parton et al., 1995). Responsiveness to CO_2 is also influenced by N inputs, as systems with high atmospheric N inputs can also show sustained responses to CO_2. Similar results from the G'Day model (Comins and McMurtrie, 1993) support analysis of this mechanism and highlight the importance of understanding the N cycle and ecosystem N budgets in modeling the response of terrestrial ecosystems to climate and CO_2. Studies of the role of terrestrial ecosystems in global change suggest several important conclusions. First is that transient and steady state responses may be very different (Schimel et al., 1990; Smith and Shugart, 1993). Second, the transient responses are strongly influenced by the interactions of production and decomposition through nutrient availability. Third, the effect of CO_2 on ecosystem C storage is as sensitive to interactions with ecosystem processes and biogeochemistry as it is to variations in the physiological responses of organisms to elevated CO_2. Flexibility in C:N ratios influencing decomposition, the balance of sources and sinks of nutrients and interactions with warming are key to understanding and predicting ecosystem responses to CO_2.

V. Use of Models for Agricultural Management

The Environmental Protection Agency (EPA) recently decided to use the CENTURY model (Barnwell et al., 1992; Donigian et al., 1994) for assessing the impact of different management practices on soil C sequestration in agricultural soils in central United States (corn belt, Great Lakes and part of the

Great Plains). The model was selected because of its ability to simulate complex cropping rotation and the interactive impact of management practices on soil water dynamics, soil temperature, nutrient cycling, plant production and SOM dynamics. Figure 7 shows the simulated total soil C levels for the central U.S. from 1910 to the present and the projected levels for 1990 to 2030 under different management assumptions. A summary of the major conclusions from this study include:

(1) extrapolation of current agricultural practices and trends until year 2030 would result in 1 to 2 GT increase in soil C assuming crop yields increase from 0.5% to 1.5% per year during that time period.

(2) conservation tillage can increase soil C levels but the results are variable across the region.

(3) utilizing cover crops will increase soil C levels and the south and southeastern U.S. are most suitable regions for this management practice.

(4) expanding the current conservation reserve program (CRP) had mixed results depending on the rotation. The major factor controlling the results being the relative amount of C added to the soil.

Under the current CRP program farmers are not allowed to add fertilizer to the system. CENTURY model results (Johnson and Kern, 1991) showed that adding N fertilizer to CRP land will greatly increase plant production and soil C sequestration. This is an example of how model results could possibly be used to give guidance for modifying national programs like the CRP program. This EPA study is an excellent example of how ecosystem models can be used to assess the impact of different management practices and provide information for policy makers.

In recent CENTURY model analysis of cropping systems in eastern Colorado, Metherell (1992) found that projected climate changes (i.e., 3.5 °C increase in mean annual temperature and 13 and 25% change in annual rainfall) generally increased SOM levels across a variety of management practices. The critical factors he identified influencing SOM dynamics under different climate change scenarios included organic matter inputs and cropping system. The wheat-fallow systems tended to accumulate less or even lose SOM under climate change, compared to wheat-corn systems which tended to accumulate greater amounts of SOM under climate change. Metherell (1992) also simulated a resodding of cropped land to grasslands and found the greatest level of SOM accumulation, although in 100 years of simulated grasslands, SOM levels reached only 50 to 60% of the original sod levels of pre-cropping simulations.

Lee et al. (1993) recently used the EPIC model (Sharpley and Williams, 1990a,b) to simulate the impact of different tillage management practices on soil C level in the U.S. Corn Belt using the current corn-soybean rotations. The four management practices simulated include: (1) current tillage practices maintained, (2) extending current trends from mulch-till to no-till, (3) increase in the conversion to no-till, and (4) increased no-till along with winter cover crop. As expected, increasing the amount of no-till and adding a winter cover crop decreased the soil erosion rates and increased the soil C levels (see Figure 8a)

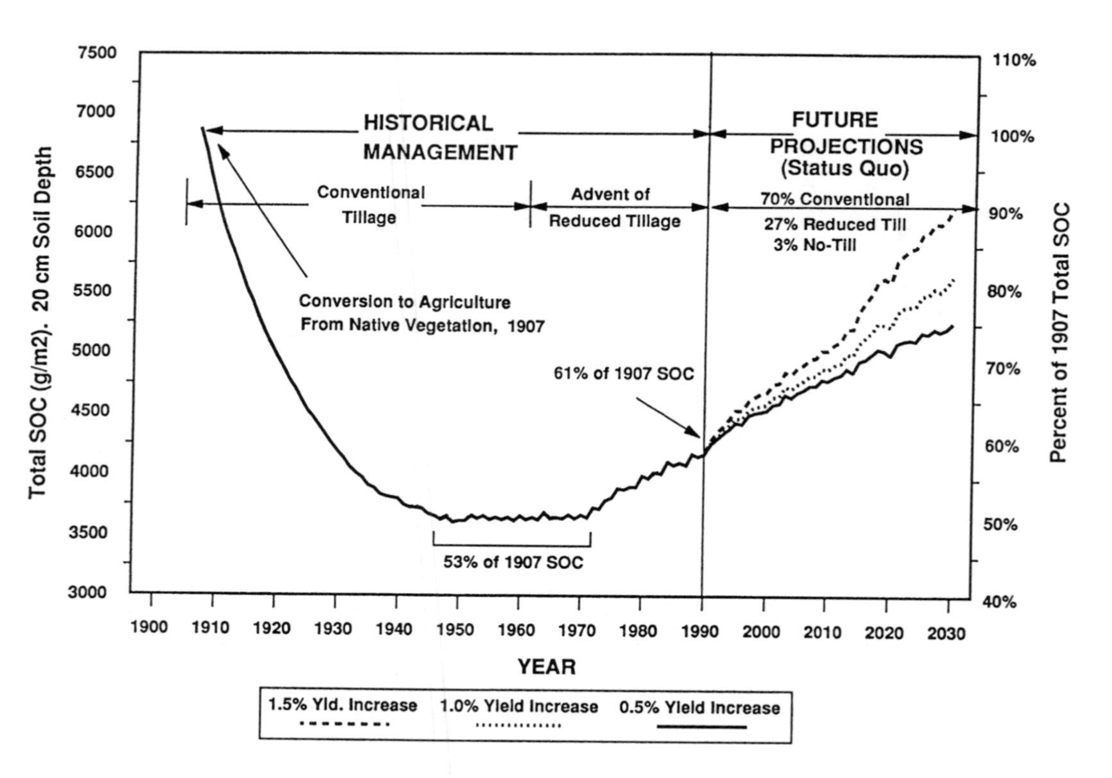

Figure 7. Simulated (using CENTURY model) total soil carbon levels for the central U.S. under the status quo scenario for three alternative levels of future crop yield increases. (Modified from Donigian et al., 1994.)

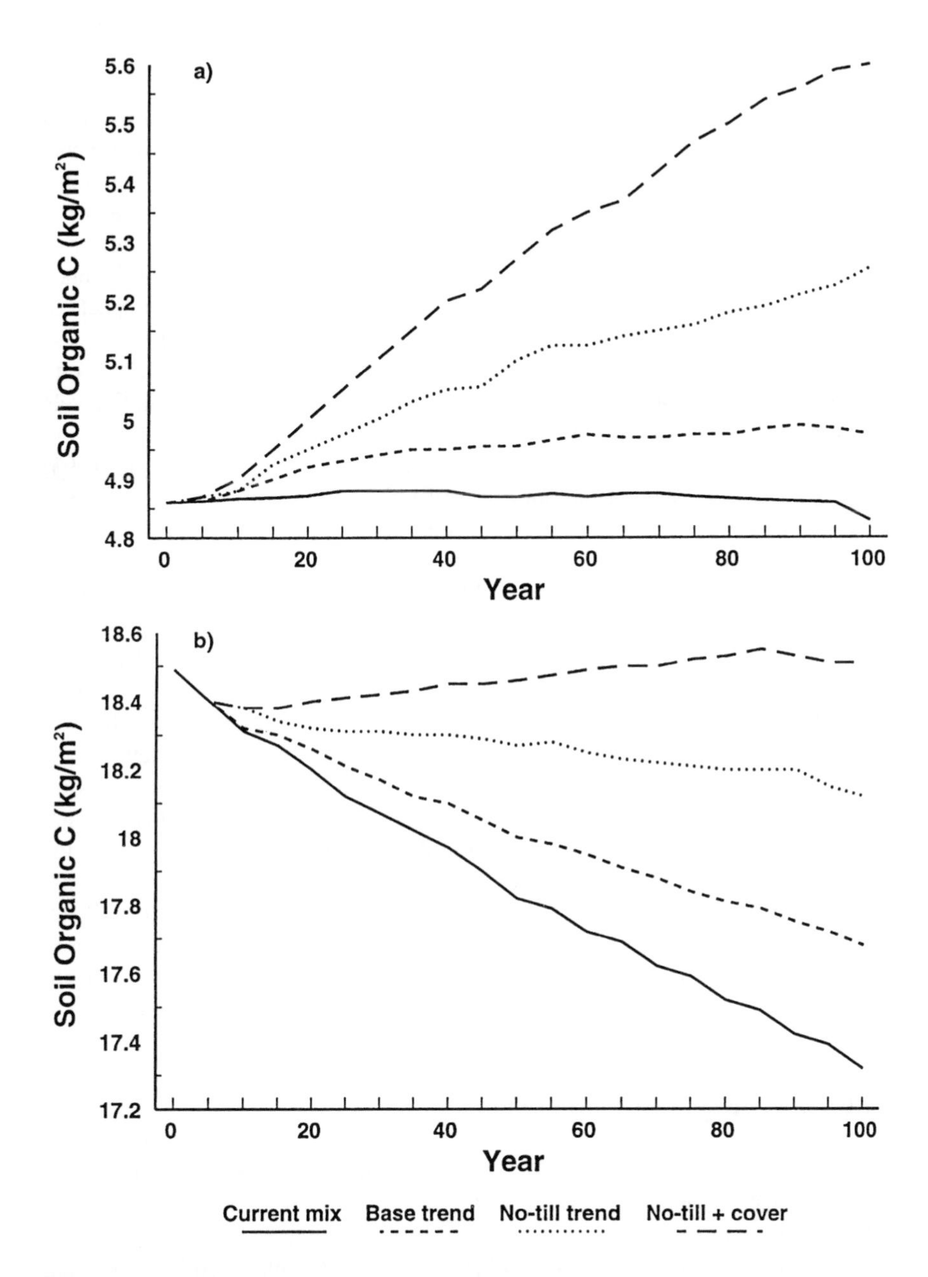

Figure 8. Simulated impact of different management practices on soil carbon for (a) 0-15 cm and (b) 0-100 cm in the U.S. Corn Belt (Modified from Lee et al., 1993).

relative to extending the current tillage practice for a 100-year period. For the100-year simulation soil C levels in the 0 to 15 cm layer remained constant for the current tillage run and increased for all of the other treatments with the highest level for the combined increased no-till and winter cover crop run. Results for the top 1 m of the soil (Figure 8b) show that soil C levels decrease for all of the treatments except the combined no-till and cover crop treatment. A detailed comparison of EPIC and CENTURY model results is difficult since the models were not set up to simulate the same management practices. However, in general terms the results seem comparable for the 0 to 15 cm layer and CENTURY does not simulate dynamics of soil C in the deep soil layers.

Extensive data bases exist to test model predictions of the impact of cultivation practices in the 0 to 30 cm soil layer, however, there is much less data for changes in soil C levels in the deep soil layers. Lee et al. (1993) cite a paper by Smith et al. (1990) to verify their predicted decrease in soil C levels in deep soil layers (30 to 100 cm depth). However, an analysis of carbon loss from paired cultivated and uncultivated soils (Davidson and Ackerman, 1993) shows there is substantial variation in soil C loss from deep soil layers (30 to 100 cm depth) with some sites showing substantial C loss (30-50%), while other sites have slight increases in soil C in the deep layers. We selected data for six paired sites presented by Davidson and Ackerman (1993) that were typical of the soils found in the Corn Belt and found that 78% of the total soil C loss came from the 0 to 30 cm soil layer and 22% from the 30 to 100 cm layer. The EPIC model results and the observed soils data show that there is considerable uncertainty about the impact of cultivation on deep soil C dynamics and that we need to improve our understanding about C dynamics in deep soil layers.

Burke et al. (1991) compared the impact of current agricultural practices on soil C losses with the potential impact of climatic change on soil C levels. They estimated soil C losses due to cultivation of rangeland soil in the Great Plains and indicated that 800-2000 g C m^{-2} losses of C since settlement of the plains. This compares with CENTURY model simulated loss of 100 to 300 g C m^{-2} over a 50 year period of climatic change from native grasslands. Other authors have also shown that land use has had a substantial impact on release of C from the soil (Houghton et al., 1987; Post et al., 1990; Schlesinger, 1990).

VI. Summary and Conclusions

SOM comprises about 75% of terrestrial C storage and plays an important role in the global C cycle, as a reservoir of C, and by regulating nutrient dynamics. The major factors which control soil organic matter dynamics have been identified for many years (Jenny, 1941) and include C input, soil texture, soil mineralogy, soil erosion, litter quality, soil biota, and soil microclimate (Allison, 1973; Swift et al., 1979). This understanding has been promoted by extensive field and laboratory experiments and the development of computer models. Both model results and observed data suggest that soil texture and C inputs are the primary factor controlling stabilization of soil C, while plant lignin content, soil microclimate and soil mineral N content also impact C stabilization. The N budget studies show that plant N uptake is higher for inorganic N addition, while soil N stabilization is higher for organic N addition.

The CENTURY model has been used extensively to illustrate the role of SOM in the response of grassland ecosystems to global change. These studies showed that while warming resulted in slow, asymptotic reductions in SOM, Net Primary Production (NPP) increased initially, but in the long term decreased to near or below the initial steady state. This occurred because N released during the oxidation of SOM initially fertilized the vegetation, but was then lost or incorporated into inactive fractions. Similar studies of warming by Jenkinson (1990) using a carbon-only model with fixed NPP showed greater sensitivity to warming than did CENTURY. These results showed clearly that the coupling of production and decomposition was an important factor in controlling the response of ecosystems to global change, and that initial transient responses to climate change could be quite different in magnitude or even direction from steady state responses because of biogeochemical interactions.

Studies of the role of terrestrial ecosystems in global change suggest several important conclusions. First is that transient and steady state responses may be very different (Schimel et al., 1990; Smith and Shugart, 1993). Second, the transient responses are strongly influenced by the interactions of production and decomposition through nutrient availability. Third, the effect of CO_2 on ecosystem C storage is as sensitive to interactions with ecosystem processes and biogeochemistry as it is to variations in the physiological responses of organisms to elevated CO_2 (Ojima et al., 1993). Flexibility in C:N ratios influencing decomposition, the balance of sources and sinks of nutrients and interactions with warming are key to understanding and predicting ecosystem responses to CO_2.

Different SOM models have been used to simulate the impact of management practices on soil C storage. The models suggest that increasing the amount of no-till agriculture and inputs of C into the soil increases soil C storage in agricultural soils. Methods of increasing C inputs to soil include increasing crop production and the amount of plant residue returned to the soil, and include winter cover crops. Most of these model predictions have not been well tested using observed data and need to be verified.

References

Ågren, G.I. and E. Bosatta. 1987. Theoretical analysis of the long-term dynamics of carbon and nitrogen in soils. *Ecology* 68:1181-1189.

Anderson, D.W. and E.A. Paul. 1984. Organo-mineral complexes and their study by radiocarbon dating. *Soil Sci. Soc. Am. J.* 48:298-301.

Allison, F.E. 1973. *Soil organic matter and its role in crop production.* Development in Soil Science 3. Elsevier, Amsterdam-London-New York.

Barnwell, T.O. Jr., R.B. Jackson, IV, E.T. Elliott, E.A. Paul, K. Paustian, A.S. Donigian, A.S. Patwardhan, A. Rowell, and K. Weinrich. 1992. An approach to assessment of management impacts on agricultural soil carbon. *Water, Air, Soil Pollut.* 64:423-435.

Berg, B., and G. Ekbohm. 1991. Litter mass-loss rates and decomposition patterns in some needle and leaf litter types. Long-term decomposition in a Scots pine forest. VII. *Can. J. Bot.* 69:1449-1456.

Bosatta E. and G.I. Ågren. 1991. Dynamics of carbon and nitrogen in the organic matter of the soil: a generic theory. *Am. Nat.* 138:227-245.

Bouwman, A.F. 1990. Exchange of greenhouse gases between terrestrial ecosystems and the atmosphere. p. 61-127. In: A. F. Bouwman (ed.), *Soils and the Greenhouse Effect*. John Wiley & Sons, New York.

Buol, S.W., F.D. Hole, and R.J. McCracken. 1980. *Soil Genesis and Classification*. Second Edition, The Iowa State University Press, Ames.

Burke, I. C., T. G. F. Kittel, W. K. Lauenroth, P. Snook, C. M. Yonker, and W. J. Parton. 1991. Regional analysis of the Central Great Plains: Sensitivity to climate variability. *BioScience* 41:685-692.

Cambardella, C.A. and E.T. Elliott. 1993. Carbon and nitrogen distribution in aggregates from cultivated and native grassland soils. *Soil Sci. Soc. Am. J.* 57:1017-1076.

Campbell, C.A. 1978. Soil organic carbon, nitrogen, and fertility. p. 173-272. In: M. Schnitzer and S.U. Khan (eds.), *Soil Organic Matter*. Elsevier Scientific Publ. Co. Amsterdam.

Carter, M.R., W.J. Parton, I.C. Rowland, J.E. Schultz, and G.R. Steed. 1993. Simulation of soil organic carbon and nitrogen changes in cereal and pasture systems of Southern Australia. *Aust. J. Soil Res.* 31:481-491.

Comins, H.N. and R.E. McMurtrie. 1993. Long-term response of nutrient-limited forest to CO_2 enrichment; equilibrium behaviour of plant-soil models. *Ecol. Appl.* 3:666-681.

Coughenour, M.B. 1984. A mechanistic simulation analysis of water use, leaf angles, and grazing in east African graminoids. *Ecol. Modelling* 26:203-230.

Davidson, E.A. and I.L. Ackerman. 1993. Changes in soil carbon inventories following cultivation of previously untilled soils. *Biogeochemistry* 20:161-193.

Dodd, J.D. and W.K. Lauenroth. 1978. Analyses of the response of grassland ecosystem to stress. p. 43-58. In: N.R. French (ed.), *Perspectives in Grassland Ecology*. Springer-Verlag. New York.

Donigian, A.S. Jr., T.O. Barnwell, Jr., R.B. Jackson, IV, A.S. Patwardhan, K.B. Weinrich, A.L. Rowell, R.V. Chinnaswamy, and C.V. Cole. 1994. *Assessment of Alternative Management Practices and Policies Affecting Soil Carbon in Agroecosystems of the Central United States*. U.S. Environmental Protection Agency Technical Report EPA/600/R-94/067, Athens, GA.

Donigian, A.S., Jr., A.S. Patwardhan, R.B. Jackson, IV., T.O. Barnwell, Jr., K.B. Weinrich, and A.L. Rowell. 1995. Modeling the impacts of agricultural management practices on soil carbon in the Central U.S. p. 121-145. In: R. Lal, J. Kimble, E. Levine, and B.A. Stewart (eds.), *Soil Management and Greenhouse Effect*. Advances in Soil Science. CRC Press. Boca Raton, FL.

Fenster, C.R. and G.A. Peterson. 1979. Effects of no-tillage fallow as compared to conventional tillage in a wheat-fallow system. University of Nebraska; Agricultural Experiment Station Research Bulletin 289. 28pp.

Greenland, D.J. and P.H. Nye. 1959. Increases in carbon and nitrogen contents of tropical soils under natural fallows. *J. Soil. Sci.* 10:284-299.

Houghton, R.A., R.D. Boone, J.R. Fruci, J.E. Hobbie, J.M. Melillo, C.A. Palm, B.J. Peterson, F.R. Shaver, and G.M. Woodwell. 1987. The flux of carbon from terrestrial ecosystems to the atmosphere in 1980 due to changes in land use: geographic distribution of the global flux. *Tellus* 39B:122-139.

Hunt, H.W. 1977. A simulation model for decomposition in grasslands. *Ecology* 58:469-484.

Janssen, B.H. 1984. A simple method for calculating decomposition and accumulation of 'young' soil organic carbon. *Plant Soil* 76:297-304.

Jenkinson, D.S. 1990. The turnover of organic carbon and nitrogen in soil. *Phil. Trans. R. Soc. Lond. B.* 329:361-368.

Jenkinson, D.S. and J.H. Rayner. 1977. The turnover of soil organic matter in some of the Rothamsted Classical Experiments. *Soil Sci.* 123:298-305.

Jenkinson, D.S., D.E. Adams, and A. Wild. 1991. Model estimates of CO_2 emissions from soil in response to global warming. *Nature* 351:304-306.

Jenny, H., 1941. *Factors of soil formation*. McGraw Hill, New York.

Jenny, H., S.P. Gessel, and F.T. Bingham. 1949. Comparative study of decomposition of organic matter in temperate and tropical regions. *Soil Sci.* 67:419-432.

Johnson, M.G. and J.S. Kern. 1991. Sequestering Carbon Soils: A Workshop to Explore the Potential for Mitigating Global Climate Change. Proceedings of U.S. Environmental Protection Agency Workshop, Corvallis Oregon, 26-28 February 1990.

Ladd, J.N., J.M. Oades, and M. Amato. 1981. Microbial biomass formed from ^{14}C ^{15}N-labelled plant material decomposing in the field. *Soil Biol. Biochem.* 13:119-126.

Lee, J.J., D.L. Phillips, and R. Liu. 1993. The effect of trends in tillage practices on erosion and carbon content of soils in the U.S. Corn Belt. *Water, Air, Soil Pollut.* 70:389-401.

Li, C., S. Frolking, and T.A. Frolking. 1992. A model of nitrous oxide evolution from soil driven by rainfall events: Model structure and sensitivity. *J. Geophys. Res. (Atmospheres)* 97:9759-9776.

Martel, Y.A. and E.A. Paul. 1974a. Effects of cultivation on the organic matter of grassland soils as determined by fractionation and radiocarbon dating. *Can. J. Soil Sci.* 54:419-426.

Martel, Y.A. and E.A. Paul. 1974b. The use of radiocarbon dating or organic matter in the study of soil genesis. *Soil Sci. Soc. Am. Proc.* 38:501-506.

McGill, W.H., H.W. Hunt, R.G. Woodmansee, and J.O. Reuss. 1981. PHOENIX, a model of the dynamics of carbon and nitrogen in grassland soils. *Ecol. Bull. NFR* 33:49-115.

Melillo, J.M., A.D. McGuire, D.W. Kicklighter, B. Moore, C.J. Vörösmarty, and A.L. Schloss. 1993. Global climate change and terrestrial net primary production. *Nature* 363:234-240.

Metherell, A.K. 1992. Simulation of soil organic matter dynamics and nutrient cycling in agroecosystems. Ph.D. Dissertation, Colorado State University, Fort Collins.

Metherell, A.K., C.A. Cambardella, W.J. Parton, G.A. Peterson, L.A. Harding, and C.V. Cole. 1995. Simulation of soil organic matter dynamics in dryland wheat-fallow cropping systems. p. 259-270. In: R. Lal, John Kimble, Elissa Levine, and B.A. Stewart (eds.), *Soil Management and Greenhouse Effect*. Advances in Soil Science, CRC Press, Inc., Boca Raton, FL.

Molina, J.A.E., C.E. Clapp, M.J. Shaffer, F.W. Chichester, and W.E. Larson. 1983. NCSOIL, a model of nitrogen and carbon transformations in soil: description, calibration, and behavior. *Soil Sci. Soc. Am. J.* 47:85-91.

Ojima, D.S., W.J. Parton, D.S. Schimel, J.M.O. Scurlock, and T.G.F. Kittel. 1993. Modelling the effects of climatic and CO_2 changes on grassland storage of soil C. *Water, Air, Soil Pollut.* 70:643-657.

Owensby, C.E., R.M. Hyde and K.L. Anderson. 1970. Effects of clipping and supplemental nitrogen and water on loamy upland range. *J. Range Manage.* 23:341-346.

Parton, W.J. and P.E. Rasmussen. 1994. Long-term effects of crop management in wheat-fallow: II. CENTURY model simulations. *Soil Sci. Soc. Am. J.* 58:530-536.

Parton, W.J., D.S. Schimel, C.V. Cole, and D.S. Ojima. 1987. Analysis of factors controlling soil organic matter levels in Great Plains Grasslands. *Soil Sci. Soc. Am. J.* 51:1173-1179.

Parton, W.J., J.W.B. Stewart, and C.V. Cole. 1988. Dynamics of C, N, P, and S in grassland soils: A model. *Biogeochemistry* 5:109-131.

Parton, W.J., D.S. Ojima, C.V. Cole, and D.S. Schimel. 1994. A general model for soil organic matter dynamics: sensitivity to litter chemistry, texture and management. p. 147-167. In: R.B. Bryant and R.W. Arnold (eds.), *Quantitative modeling of soil forming processes*. Soil Sci. Soc. Am. Special Pub. 39, SSSA, Madison, WI.

Parton, W.J., J.M.O. Scurlock, D.S. Ojima, D.S. Schimel, D.O. Hall, and SCOPEGRAM GROUP MEMBERS. 1995. Impact of climate change on grassland production and soil carbon worldwide. *Global Change Biology* (in press).

Parton, W.J., J.M.O. Scurlock, D.S. Ojima, T.G. Gilmanov, R.J. Scholes, D.S. Schimel, T. Kirchner, J.-C. Menaut, T. Seastedt, E. Garcia Moya, A. Kamnalrut, and J.L. Kinyamario. 1993. Observations and modeling of biomass and soil organic matter dynamics for the grasslands biome worldwide. *Global Biochem. Cycles* 7:785-809.

Pastor, J. and W.M. Post. 1988. Response of northern forests to induced climate change. *Nature* 334:55-58.

Paul, E.A. and J. van Veen. 1978. The use of tracers to determine the dynamic nature of organic matter. *Trans. Int. Congr. Soil Sci.* 11th 3:61-102.

Paustian, K., W.J. Parton, and J. Persson. 1992. Modeling soil organic matter in organic-amended and nitrogen-fertilized long-term plots. *Soil Sci. Soc. Am. J.* 56:476-488.

Pinck, L.A., F.E. Allison, and M.S. Sherman. 1950. Maintenance of soil organic matter: II. Losses of carbon and nitrogen from young and mature plant material during decomposition in soil. *Soil Sci.* 69:391-401.

Post, W.M., T. Peng, W.R. Emanuel, A.W. King, V.H. Dale, and D.L. DeAngelis. 1990. The global carbon cycle. *Am. Sci.* 78:310-326.

Probert, M., W.J. Parton, B. Keating. 1995. Soil organic matter dynamics for the Hermitage Plots. *Aust. J. Soil. Res.* (in press).

Raich, J.W., E.B. Rastetter, J.M. Melillo, D.W. Kicklighter, P.A. Steudler, B.J. Peterson, A.L. Grace, B. Moore III., and C. J. Vörösmarty. 1991. Potential net primary productivity in South America: Application of global model. *Ecol. Appl.* 1(4):399-429.

Rasmussen, P.E. and H.P. Collins. 1991. Long-term impacts of tillage, fertilizer and crop residue on soil organic matter in temperate semi-arid regions. *Adv. Agron.* 45:93-134.

Rasmussen, P.E. and W.J. Parton. 1994. Long-term effects of residue management in wheat/fallow system: I. Inputs, yield, and soil organic matter. *Soil Sci. Soc. Am. J.* 58:523-530.

Scharpenseel, H.W., H. Schiffmaxnn, and B. Hintze. 1984. Hamburg University radiocarbon dates III. *Radiocarbon* 26:196-205.

Schimel, D.S., W.J. Parton, T.G.K. Kittel, D.S. Ojima, and C.V. Cole. 1990. Grassland biogeochemistry: links to atmospheric processes. *Climate Change* 17:13-25.

Schimel, D.S., T.G.F. Kittel, and W.J. Parton. 1991a. Terrestrial biogeochemistry cycles: global interactions with the atmosphere and hydrology. *Tellus* 43ab:188-203.

Schimel, D.S., T.G.F. Kittel, A.K. Knapp, T.R. Seastedt, W.J. Parton, and V.B. Brown. 1991b. Physiological interactions along resource gradients in a tallgrass prairie. *Ecology* 72:672-684.

Schimel, D.S., B.H. Braswell, Jr., E.A. Holland, R. McKeown, D.S. Ojima, T.H. Painter, W.J. Parton, and A.R. Townsend. 1994. Climatic, edaphic and biotic controls over storage and turnover of carbon in soils. *Global Biogeochem. Cycles* 8:279-293.

Schlesinger, W.H. 1990. Evidence from chronosequence studies for a low carbon-storage potential of soils. *Nature* 348:232-234.

Sharpley, A.N., and J.R. Williams (eds.). 1990a. *EPIC--Erosion/Productivity Impact Calculator: 1. Model Documentation*. U.S. Department of Agriculture Technical Bulletin No. 1768, Washington, DC, 235 pp.

Sharpley, A.N., and J.R. Williams (eds.). 1990b. *EPIC--Erosion/Productivity Impact Calculator: 2. User Manual*. U.S. Department of Agriculture Technical Bulletin No. 1768, Washington, DC, 127 pp.

Smith, S.J., A.N. Sharpley, and A.D. Nicks. 1990. Evaluation of EPIC nutrient projections using soil profiles for virgin and cultivated lands of the same soil series. p. 217-219. In: A.N. Sharpley and J.R. Williams (eds.), *EPIC - Erosion/Productivity Impact Calculator: 2. Model Documentation*. U. S. Dep. Agric. Tech. Bull. No. 1768, Washington, D.C.

Smith, T.M. and H.H. Shugart. 1993. The transient response of terrestrial carbon storage to a perturbed climate. *Nature* 361:523-526.

Sørensen, L.H. 1981. Carbon-nitrogen relationships during the humification of cellulose in soils containing different amounts of clay. *Soil Biol. Biochem.* 13:313-321.

Stevenson, F.J. 1982. *Humus Chemistry*, John Wiley & Sons, New York.

Stott, E, G. Kassin, W.M. Jarrell, J.P. Martin, and K. Haider. 1983. Stabilization and incorporation into biomass of specific plant carbons during biodegradation in soil. *Plant Soil* 70:15-26.

Swift, M.J., O.W. Heal, and J.M. Anderson. 1979. *Decomposition in Terrestrial Ecosystems*. Studies in Ecology, Vol. 5. Blackwell Scientific Publications, Oxford.

Tisdall, J.M. and J.M. Oades. 1982. Organic matter and water stable aggregates in soils. *J. Soil Sci.* 33:141-163.

Trumbore, S.E. 1993. Comparison of carbon dynamics in tropical and temperate soils using radiocarbon measurements. *Global Biogeochem. Cycles* 7(2):275-290.

Trumbore, S.E., J.S. Vogel, and J.R. Southon. 1989. AMS ^{14}C measurements of fractionated soil organic matter: an approach to deciphering the soil carbon cycle. *Radiocarbon* 31:644-654.

Trumbore, S.E., G. Bonani, and W. Wölfli. 1990. The rates of carbon cycling in several soils from AMS ^{14}C measurements of fractionated soil organic matter, p. 405-414. In: A.F. Bouwman (ed.), *Soils and the Greenhouse Effect*. John Wiley & Sons, New York.

Turchenek, L.W. and J.M. Oades. 1979. Fractionation of organo-mineral complexes by sedimentation and density techniques. *Geoderma* 21:311-343.

van Veen, J.A. and E.A. Paul. 1981. Organic carbon dynamics in grassland soils. I. Background information and computer simulation. *Can. J. Soil Sci.* 61:185-201.

Methods to Characterize and Quantify Organic Matter Storage in Soil Fractions and Aggregates

M.R. Carter and E.G. Gregorich

I. Introduction

The various conceptual frameworks and models used to characterize the storage and sequestration of organic matter in soil require suitable methods to verify the underlying concepts. The developing methodology should consider the relationship between organic matter accumulation and the physical location of organic matter within the soil matrix. Techniques are also required to characterize soil organic matter fractions that have various turnover times in the soil system and thus can affect soil structure or aggregation in different ways. The objective of this chapter is to provide an overview of the approaches used to characterize and quantify soil organic matter storage, as reported in the recent literature, and to provide some examples of the most promising techniques.

Several recent reviews have covered a range of methods that apply to measurement of organic matter storage in soil in relation to soil structure. Stevenson and Elliott (1989) discussed methods used to determine both the chemical composition of organic matter and the physical location of organic matter in soil particles and aggregates. Methods to assess the latter, specifically

1-56670-033-7/96/$0.00+$.50

the distribution of organic matter within physical fractions of the soil, were further described by Elliott and Cambardella (1991). Methods for determining soil structure and aggregation that allow evaluation of the role of biota in soil structure formation and stabilization have been described by Jastrow and Miller (1991). Beare and Bruce (1993) presented methods for measuring water-stable aggregation that provide an approach to describing environmental influences (e.g., soil disturbance, organic matter additions) on soil structure. Imhof (1988) provided a useful overview of the various methods of wet-sieving for mainly macroaggregate assessment.

Terminology concerning the storage of organic matter in soil should take into consideration and reflect soil behavior. For example, the total organic matter stored in soil can be categorized simply into two pools: *inert* and *labile* (Hsieh, 1992). The total amount of organic matter in the soil can be considered as a measure of *stored organic matter*. In a sense, the latter is a *mean organic matter store* or *standing stock of organic matter* because it reflects the net product or balance between ongoing accumulation and decomposition processes and is thus greatly influenced by crop management and productivity. In contrast, a portion of the stored organic matter is somewhat physically or, to a lesser degree chemically, protected from decomposition processes. Organic matter stored in this fashion, in situations where decomposition is suppressed, can be considered *sequestered*. Generally, only organic matter in soil microaggregates would be considered as sequestered, because most organic matter found in inter-microaggregate locations (i.e., within macroaggregates) has a relatively short turnover and residence time (Buyanovsky et al., 1994). Organic matter associated with microaggregates is 'old' because of its residence time in soil. A small portion of this organic matter may be labile and readily decomposable, but most is stable, either because of its chemical structure (i.e., humified material) or because of its association (i.e., bonding) with clay minerals.

II. Total Soil Organic Matter Storage

Generally, estimations of overall total soil organic matter storage use straightforward techniques that are well established. Assessment of total organic matter storage employ either wet or dry oxidation techniques and take into consideration the thickness and bulk density of the soil layers to express stored organic matter on a mass per unit area or equivalent mass basis (Gregorich et al., 1994). Total organic matter storage has been asssessed recently in a range of soils throughout the world (Davidson and Ackerman, 1993; Eswaran et al., 1993; Stone et al., 1993; Kern, 1994; Moraes et al., 1995).

III. Organic Matter in Soil Aggregates and Fractions

Relationships between soil structure and organic matter are of current interest because soil structure, which itself is mainly formed by biological processes, exercises strong control over organic matter dynamics and storage in soil systems (Van Veen and Kuikman, 1990; Juma, 1993, 1994). Ladd et al. (1993) showed how soil structural factors can confer stability to organic matter by limiting reactions between substrate and enzyme, as well as other processes that constrain decomposition.

In contrast to estimations of total organic matter storage in soil, soil organic matter fractions or pools and organic matter stored in specific locations and associated with soil particles are measured using a wide range of methods. Certain organic matter fractions (e.g., microbial biomass and carbohydrates) can be quantified by chemical extraction procedures, whereas organic matter stored in various locations within the soil matrix is generally characterized and measured using physical separation techniques (Table 1).

Physical separation disrupts soil aggregates and particles that contain organic matter by supplying enough energy to overcome the strength of the various binding agents or mechanisms that underly the aggregation process. Stevenson and Elliott (1989) and Elliott and Cambardella (1991) indicate that most physical separation techniques follow a sequence of soil disruption (using sonication or shaking) followed by particle separation (using sieving, sedimentation, or densimetry). Sieving, especially wet-sieving, is often used to separate macro-organic matter and to estimate organic matter in macro- and microaggregates, whereas sedimentation is used to separate and characterize organo-mineral particles. Measurement of organic matter and organic matter fractions in sand, silt, and clay fractions may involve a sequence of methods, such as sonication (for disruption of soil particles), sieving, centrifugation, and filtration to isolate specific particle-size fractions (Stevenson and Elliott, 1989; Bremner and Genrich, 1990; Christensen, 1992). The separated fractions can be analyzed chemically, and in some cases isotopic techniques or incubation studies are used to trace or biologically characterize the organic matter or added substrate. Light fraction soil organic matter can be separated by both wet sieving and sedimentation, followed by densimetric techniques (Elliott and Cambardella, 1991).

In summary, various approaches and techniques may be used to assess soil organic matter storage, usually by characterizing soil aggregates, density, and size fractions. Table 2 provides an overview of some approaches used to measure structure and organic matter storage in agricultural soils.

IV. Chemical Structure of Stored Organic Matter

In some cases, characterization of the chemical structure of the organic matter stored in different fractions or pools is of interest. Past studies involving the

Table 1. Methodology commonly used to characterize and quantify organic matter fractions and evaluate carbon stored in agricultural soil

Type of organic matter	Examples of selected methodology
Litter	Litter harvested and expressed on oven-dry basis per area (Stevenson and Elliott, 1989).
Plant roots	Soil cores; washing and sieving (0.15 mm); expressed on oven-dry (80 °C) basis per area (Atkinson and Mackie-Dawson, 1991).
Water-soluble organic C	Extraction (H_2O)/centrifugation (McGill et al., 1986).
Microbial biomass	Fumigation extraction (Voroney et al., 1993).
Macroorganic matter	Dispersion/sieving (53 μm) (Cambardella and Elliott, 1992).
Light fraction	Densimetric fractionation (Gregorich and Ellert, 1993).
Carbohydrates	Extraction or acid hydrolysis/colorimetric estimation of saccharide monomers (Lowe, 1993).
Inter-microaggregate[a]	Wet-sieving macroaggregates (obtained by dry sieving whole soil); organic C (dry combustion: Tiessen and Moir, 1993) by difference between macro- and microaggregates.
Intra-microaggregate[b]	Dry or wet-sieving; organic C in microaggregate by dry combustion.
Organo-mineral particles	Dispersion/sedimentation (Tiessen and Stewart, 1983); dispersion/densimetry (Turchenek and Oades, 1979).

[a] Organic matter located within macroaggregates (> 250 μm diameter), but external to microaggregates (< 250 μm diameter)

[b] Organic matter located within microaggregates

Table 2. Some examples of different approaches to evaluate soil organic matter storage in relation to soil structure in agricultural soils

Site location and reference	Purpose of the study	Soil disruption and separation methods used	Some conclusions of the study in regard to soil organic matter storage
Nebraska, U.S. (1)	Evaluate aggregation characteristics and models.	Wet sieving to separate aggregates followed by incubation of aggregates.	Organic matter concentration was lower in microaggregates than macroaggregates, and less labile.
Amazon basin, Peru (2)	Compare organic matter in soil aggregates and particles.	Shaking followed by wet sieving. Aggregates subject to dispersion and densimetric techniques.	Total C was highest in the largest and smallest particle sizes. Need to correct aggregates for light fraction and sand to estimate mineral associated organic matter.
Nebraska, U.S. (3)	Isolate macroorganic matter.	Dispersion followed by wet sieving; ^{13}C analysis.	Macroorganic matter was a relatively large (39%) fraction of total stored C in grassland; reduced to about 20% by arable farming.
Southeastern U.S. (4)	Describe size of organic matter pools in various aggregates.	Wet sieving followed by laboratory incubations of intact and crushed aggregates.	Organic matter in aggregates placed in unprotected, protected and resistant fractions.
Eastern Canada (5)	Evaluate soil structure and organic matter in aggregates.	Wet sieving and turbidimetry.	Organic matter increase positively related to soil macroaggregation and structural stability. Turbidimetric analysis characterized micro-aggregate dynamics.
Missouri, U.S. (6)	Relate age of organic matter in soilfractions.	Wet sieving; sonification and sedimentation; ^{14}C analysis.	Organic matter degradation rate decreased in following order: coarse macroaggregates, macroaggregates, and primary particles.

References: (1) Elliott (1986); (2) Elliott et al. (1991); (3) Cambardella and Elliott (1992); (4) Beare et al. (1994); (5) Angers (1992), Carter (1992), Carter and Kunelius (1993); (6) Buyanovsky et al. (1994).

chemical fractionation of soil mainly concentrated on the extraction of humic materials, such as fulvic and humic acid. The main disadvantages of this approach are that quantitative extraction is impossible, artifacts may be produced during extraction and purification, and characterization of the extracted material is difficult. Also, the biological or agronomic relevance of the extracted organic matter has not been clearly established. Stevenson and Elliott (1989) discuss the limitations of the chemical fractionation approach in providing information on organic matter fractions and pool sizes. Anderson et al. (1974), Stevenson and Elliott (1989), and Theng et al. (1989) discuss the ideas and approaches underlying chemical fractionation of whole soil or soil particles, mainly in regard to characterization of relatively stable humic substances that are involved with soil aggregate stability.

New techniques eliminate the need to extract organic matter from the mineral fraction with various reactive reagents. With these techniques the whole organic fraction is analyzed (in whole soil or fractions) rather than a portion soluble in a specific extractant. Oades et al. (1987) and Baldock et al. (1992) used ^{13}C nuclear magnetic resonance (NMR) to determine the chemical structure of organic materials in soil particles (from soil disrupted by sonication) separated by sedimentation and densimetric techniques. This method can elucidate certain aspects of organic matter storage in soil. For example, Golchin et al. (1995) showed that the amount of O-alkyl C material within aggregates was inversely related to aggregate stability, whereas that of aromatic C material was directly related to aggregate stability. Using high energy ultraviolet photo-oxidation, Skjemstad et al. (1993) demonstrated that organic materials external to both clay and silt-sized particles were mainly proteinaecous in nature, whereas physically protected humic materials occupied the interior of these microaggregates. Baldock et al. (1992) observed that the extent of decomposition was greatest in fine fractions and noted the compositional differences in different particle and density fractions. Using lignin parameters and carbohydrate ratios, Guggenberger et al. (1994) characterized the composition of organic matter in particle-size separates (separated by sonication and sedimentation) in relation to land use changes.

Another recently developed technique, pyrolysis field ionization-mass spectrometry (Py-FIMS), has been used to characterize the molecular subunits of organic matter associated with soil particle-size fractions (Schulten and Leinweber, 1991). Monreal et al. (1995) found that Py-FIMS confirmed ^{13}C NMR results and observed that the majority of C was found in aliphatic structures in whole soil and particle-size fractions of a forest and cultivated soil.

The relationship between organic matter storage and the aggregation process, noted above, is dependent on specific binding agents that function at different scales. It is commonly accepted that fungal hyphae and fine roots bind microaggregates together to form large aggregates (i.e., macroaggregates) (Tisdall, 1994). In some studies, use of selective chemicals has provided information about the kinds of organic matter or other agents that function as binding agents in soil particles and microaggregates. For example, sodium periodate has been used to selectively sever polysaccharide molecules, while sodium pyrophosphate can displace organic molecules from complexes of Fe, Al, and Ca cations (Hamblin and Greenland, 1977; Baldock and Kay, 1987).

Micromorphological techniques combined with fluorescent staining have also proved useful to examine soil microstructure and the location of microbes and plant residues (Juma, 1994). Such techniques can provide information on the interaction between organic matter and the soil matrix, and on the location of organic materials within soil aggregates and particles. Ladd et al. (1993) showed that, in addition to spectroscopy (NMR), electron microscopy was a useful technique for *in situ* characterization of organic matter within soil aggregates and fractions.

V. Biological Nature of Stored Organic Matter

The biological nature of the organic matter in aggregate fractions and particle-size and density fractions can be evaluated in laboratory studies in which the soil is fractionated and subsequently incubated under controlled conditions. These laboratory incubations are essentially bioassays that let the microorganisms and fauna define what is decomposable (Ellert and Gregorich, 1995). The rate and extent of mineralization measured in these studies reflect a potential bioavailability of the soil organic matter in these fractions. The compounds released in a bioassay may function as plant nutrients or as precursors for more stable organic matter and therefore are also agronomically important and relevant to organic matter storage.

The biological nature of soil organic matter associated with size separates has been assessed during incubation under aerobic (Lowe and Hinds, 1983; Christensen, 1987) and anaerobic conditions (Cameron and Posner, 1979; Christensen and Christensen, 1991). In general these studies have shown that the potential bioavailability of soil organic matter increases as particle size decreases, and the rate of mineralization of organic matter associated with clay is twice that observed for silt (Christensen, 1992). Also, the results of these studies do not indicate the presence of a large pool of *labile* soil organic matter physically protected within aggregates, which is released and made available by dispersing the soil (Christensen, 1992).

This type of assay may also give information into the location and accessibility of organic matter within the soil matrix. For example, Christensen and Christensen (1991) measured the potential denitrification (N_2O evolution) from clay and silt during anaerobic incubation with excess nitrate. The results suggested that reduced availability of soil organic matter could result from the formation of microaggregates in which the organic matter is less accessible to denitrifiers.

Incubation of separated aggregates can assist in determining the role of protected soil organic matter and substrate C availability in soil. For example, Beare et al. (1994) measured potentially mineralizable C and N during laboratory incubations of intact and crushed macroaggregates and intact microaggregates. Three pools of aggregate-associated soil organic matter were quantified: unprotected, protected and resistant C and N. Seech and Beauchamp (1988) studied the denitrification potential in different size aggregates and reported that the denitrification rate decreased as aggregate size increased and that the rate was related to substrate C availability. These findings contrast with those reported by Elliott (1986), and the discrepancies between the studies may be attributed to the method used to separate the soil into aggregate size fractions. Seech and Beauchamp (1988) argue that the dry sieving method they used would be less likely than wet sieving to remove soluble C compounds that contain the most readily metabolizable portion of soil organic C.

Bioassays of material fractionated by density may also give information about the mineralization/immobilization processes occurring concomitantly within the soil matrix. For example, Sollins et al. (1984) suggested that net N mineralization was greater from the heavy fraction than from the whole soil, because the light fraction immobilized N released from the heavy fraction when they were incubated together.

VI. Isotopic Methods

The redistribution of added substrate and residues among organic matter fractions, particle-size or density fractions, and aggregate-size fractions in soil is traceable with isotopes. Various radioisotopic techniques have been widely used since the 1960s. Recently, however, instrumentation has become more sophisticated and readily available, allowing widespread use of stable isotopes in soil organic matter studies. The type of isotopic measurement performed depends on the type of information required and the objectives of the experiment. The benefit of using isotopic methods is that they allow precise quantification of changes in organic matter during decomposition. Generally, ^{14}C is used to study short-term dynamics of organic matter, and the natural abundance of ^{13}C and ^{15}N is used to characterize the long-term status. Tables 3 and 4 provide an overview of some isotopic methods used to measure the turnover and storage of organic matter in fractions separated from agricultural soils.

A. Artificially Enriched Substrates

For finer levels of resolution, enrichment of a substrate with labelled C or N is needed to distinguish the added organic matter from soil organic matter, because the amount of C or N added is usually small relative to the native C or N present in soil. Substrates labelled with ^{14}C or ^{15}N that have been used in this way range from simple compounds, such as glucose and nitrate, to complex compounds, such as cellulose and proteins, to plant leaves and roots.

Table 3. Some selected studies using various fractionation and carbon isotopic techniques to quantify and partition organic matter (OM) in soil according to origin and turnover time

Site location and reference	Isotopic measurement	Fractionation	Purpose of the study	Some conclusions of the study in regard to organic matter storage
Canada (1)	$\Delta^{14}C$	Particle size, chemical	Evaluate significance of clay-humus complexes in C cycling.	Coarse clay OM contained oldest C; fine clay was much younger.
Denmark (2)	^{14}C	Particle size	Measure distribution of labelled and native C between particle size fractions.	Clay-bound OM more important in medium-term turnover; silt-bound OM participates in longer-term cycling.
France (3)	$\delta^{13}C$	Particle size	Estimate SOM turnover.	Coarse and fine particle size fractions contained youngest OM. Turnover of silt size fraction was slower.
Brazil and U.S. (4)	$\Delta^{14}C$	Density	Compare C dynamics in tropical and temperate soils.	Light fraction (LF) is a major contributor to annual flux of C into and out of soil.
U.S. (5)	^{14}C	Particle size, aggregate fractions	Measure C turnover in soil physical fractions.	Turnover of macroaggregates with partially processed $C \approx 1$ to 3 yr; for microaggregates with highly humified $C \approx 7$ yr.
Australia (6)	$\delta^{13}C$	Density	Assess structural and dynamic SOM properties.	Free LF had higher turnover than occluded materials.
Canada (7)	$\delta^{13}C$	Density, particle size	Estimate turnover of SOM and storage of corn residue C. Effect of tillage on distribution of corn-derived C.	LF OM had higher turnover than sand-size OM. Greater enrichment of corn-derived C in macro-organic matter (> 50 μm) than in either whole soil or microorganic matter (<50 μm).

References: (1) Anderson and Paul (1984); (2) Christensen and Sørenson (1985); (3) Balesdant et al. (1987); (4) Trumbore (1993); (5) Buyanovsky et al. (1994); (6) Golchin et al. (1995); (7) Gregorich et al. (1995), Angers et al. (1995).

Table 4. Some selected studies using various fractionation and nitrogen isotopic techniques to quantify and partition organic matter (OM) in soil according to origin and turnover time

Site location and reference	Isotopic measurement	Fractionation	Purpose of the study	Some conclusions of the study in regard to organic matter storage
Australia (1,2,3)	^{15}N	Particle size, density	Evaluate N immobilization and mineralization in soil fractions.	Silt fraction accumulates an increasing proportion of stable nitrogenous residues; rapid and extensive decrease of labelled material from fine clay-size OM.
Canada (4)	$\delta^{15}N$	Particle size	Evaluate changes in natural ^{15}N composition of particle size fractions in native and cultivated soils.	^{15}N abundance of fine-silt and coarse-clay associated OM unchanged after 60 yr. cultivation. Increases in ^{15}N abundance of sand and silt fractions due to reduced plant residue inputs.
Australia (5)	$\delta^{15}N$	Particle size	Determine natural enrichment of ^{15}N in different size fractions.	Humified clay-sized OM had higher $\delta^{15}N$ than less humified OM in sand and silt-sized fractions.
Denmark (6)	^{15}N	Particle size	Determine distribution of labelled and native N among paricle size fractions in soils with different textures.	Enrichment factor for labelled and native N decreased with increasing amounts of silt and clay. Clay enrichment higher for labelled than for native N; converse was true for silt.
Denmark (7)	^{15}N	Particle size	Determine distribution of labelled and native mineral-fixed NH_4 in soils with different textures.	In soil with finest texture about 25 to 33% of labelled N found as mineral-fixed NH_4 in clay and silt fractions.

References: (1) Ladd et al. (1977a,b); (2) Ladd and Amato (1980); (3) Amato and Ladd (1980); (4) Tiessen et al. (1984); (5) Ledgard et al. (1984); (6) Christensen and Sørenson (1986); (7) Jensen et al. (1989).

This approach has been used to study rapid mineralization processes (usually <5 yr) and the stabilization of decomposition products over the long term (Voroney et al., 1991). Cheshire and Mundi (1981) examined the distribution of labelled carbohydrates among size fractions from soils incubated with ^{14}C-labelled glucose (28 d), ryegrass (1 yr) and straw (4 yr). Christensen and Sørensen (1985) fractionated soils with different clay contents according to particle size after incubation for 5 to 6 yr with ^{14}C-labelled glucose, hemicellulose, or straw. Ladd et al. (1977a, b) incubated two soils of different texture with ^{15}N-nitrate and glucose or wheat straw and followed the changes in the distribution of the labelled soil organic matter in different particle-size and density fractions during the incubation.

Use of a ^{13}C- or ^{15}N-labelled substrate offers the potential to measure precise quantitative changes as well as to monitor the changes, as seen by ^{13}C or ^{15}N NMR, in chemical structure associated with the decomposition and stabilization of the added material. Baldock et al. (1990) reported a preferential accumulation of labelled soil organic matter in the clay fraction of a soil incubated with ^{13}C-labelled glucose and observed that the residual labelled C was predominantly in alkyl and O-alkyl structures.

B. ^{14}C Dating and Bomb ^{14}C

Accelerator mass spectrometry has greatly reduced the difficulties previously encountered in dating soil organic matter in samples with relatively small masses. This technique reduces the sample size necessary for ^{14}C measurements by a factor of 2000 over conventional counting methods (Goh, 1991). This refinement has allowed the determination of high-precision dates on as little as several hundred milligrams of sample, such as that of aggregate or size fractions separated from whole soils.

Radiocarbon dating of soils is based on the decay of ^{14}C in plant material, which had a $^{14}C/^{12}C$ ratio similar to that of atmospheric CO_2 at the time of its death and incorporation into the soil. The degree to which the $^{14}C/^{12}C$ ratio of soil organic matter differs from that of plant material from which it is derived reflects the "mean residence time" of C in soils (Paul et al., 1964). Dates for whole soils and soil fractions are expressed as mean residence times, because the annual input of new plant material into soils contaminates the soil with modern C. For the purposes of assessing soil organic matter dynamics, ^{14}C dating is done on samples that have not been enriched with bomb ^{14}C, as described below. These soils are usually archived samples that were taken prebomb (before 1960s) and stored in glass jars, or were obtained from under buildings. Methods used to calculate, report, and interpret ^{14}C age are given by Goh (1991).

Trumbore (1993) noted that several problems exist with using only ^{14}C-derived mean residence times of bulk soil organic matter to interpret soil C dynamics. These problems limit the usefulness of radiocarbon to study carbon pools less than several hundred years old. Separating soil into labile and stable constituents by aggregate, particle-size, or density fraction will help to elucidate more precisely the dynamics of organic matter within the soil matrix. Anderson and Paul (1984) radiocarbon dated particle-size fractions to investigate the long-term cycling of organic C in cultivated soils and observed that organic matter associated with coarse clays was oldest, with a much younger date for that associated with fine clays.

Detonation of thermonuclear devices in the 1950s and 1960s resulted in the enrichment of the atmosphere with ^{14}C. This spike input of bomb ^{14}C to the stratosphere subsequently entered the terrestrial ecosystem through plants and then recycled through animals, microorganisms, and soil organic matter (Goh, 1991). Measurement of the rate at which this pulse of ^{14}C moves through different fractions provides a unique opportunity to bridge the gap between the rapid turnover rates measured in labelling studies and the slow turnover rates measured in radiocarbon dating studies. Bomb ^{14}C is determined by comparing ^{14}C concentrations in soil sampled before 1960 (prebomb) with those in ^{14}C-enriched modern soils. The modern soil sample should be of the same soil type and should be collected near the location of the prebomb samples. Trumbore (1993) used bomb ^{14}C to show that the rapid turnover of organic matter with density <1.6 to 2.0 g cm^{-3} contributes a major component of the annual C flux into and out of soil.

C. Natural ^{13}C and ^{15}N Abundance

Plants with a C4 photosynthetic pathway have higher $^{13}C/^{12}C$ ratios ($\delta^{13}C$ of -12 to -14‰) than C3 plants ($\delta^{13}C$ -26 to -28 ‰). The use of natural ^{13}C abundance as a tracer is based on the fact that the ^{13}C content of soil organic matter corresponds closely to the ^{13}C content of the plant material from which it is derived and that there is negligible fractionation of ^{13}C during the decomposition of plant material in soil. Thus the introduction of C4 plants to a soil previously developed under C3 vegetation (or vice versa) results in that soil containing two isotopically different sources of C and provides a means of partitioning soil organic matter according to origin. This approach has been used to determine the storage of soil organic matter and turnover of plant residue C in particle-size fractions in temperate (Balesdent et al., 1987; 1988) and tropical (Martin et al., 1990; Desjardins et al., 1994) soils. Using this method, Golchin et al. (1995) concluded that free particulate organic matter formed a significant pool for soil organic matter turnover when a forest soil was converted to pasture.

Nitrogen enters soils either directly through nitrogen fixation and atmospheric deposition of nitrous oxides, or indirectly through the decomposition of plant and animal residues. Atmospheric N_2 has a ^{15}N abundance of 0.366% and is generally used as a standard The $\delta^{15}N$ values in living plants and residues, which vary from -9‰ to +8.0‰, increase quickly (to 10‰ or more) with increasing microbial transformation of the residues in litter (Andreux et al.,

1990), because of the microbial discrimination of ^{15}N during the catabolic processes occurring in soil. Thus successive microbial transformations of N-containing material causes a progressive increase in ^{15}N of the soil organic matter. Ledgard et al. (1984) observed that $\delta^{15}N$ values increased with decreasing particle size. Tiessen et al. (1984) reported similar results and concluded that by using ^{15}N abundance a pulse of low-enrichment litter can be followed through the soil systems into the silt fraction as well as into microbial products associated with the fine clays.

VII. Summary

Relatively simple and well-tested methodology is available for measuring mean organic storage in soil. In contrast, estimation of organic matter stored or sequestered in specific soil locations, such as within various sized aggregates, requires an array of methodologies and procedures. In most cases some form of soil disruption is used, such as vigorous shaking or sonication, followed by wet sieving, sedimentation, or densimetry to isolate soil aggregates or particles. Both chemical fractionation and biological assays of organic matter in isolated soil aggregates and particles provide useful information on the characteristic or type of stored organic matter and its potential bioavailability. Isotopic methods are powerful tools in the elucidation of organic matter redistribution, turnover, and residence time in soil.

Acknowledgements

The authors wish to thank Drs. H.H. Janzen and B.H. Ellert for helpful comments in the preparation of this chapter.

References

Amato, M. and J.N. Ladd. 1980. Studies of nitrogen immobilization and mineralization in calcareous soils. V. Formation and distribution of isotope-labelled biomass during decomposition of ^{14}C- and ^{15}N-labelled plant material. *Soil Biol. Biochem.* 12:405-411.

Anderson, D.W., E.A. Paul, and R.J. St. Arnaud. 1974. Extraction and characterization of humus with reference to clay-associated humus. *Can. J. Soil Sci.* 54:317-323.

Anderson, D.W. and E.A. Paul. 1984. Organo-mineral complexes and their study by radiocarbon dating. *Soil Sci. Soc. Am. J.* 48:298-301.

Andreux, F., C. Cerri, P.B. Vose, and V.A. Vitorello. 1990. Potential of stable isotope, ^{15}N and ^{13}C, methods for determining input and turnover in soils. p. 259-275. In: A.F. Harrison et al. (eds.), *Nutrient cycling in terrestrial ecosystems: field methods, application and interpretation*. Elsevier Applied Science, New York.

Angers, D.A. 1992. Changes in soil aggregation and organic carbon under corn and alfalfa. *Soil Sci. Soc. Am. J.* 56:1244-1249.

Angers, D.A., R.P. Voroney, and D. Côté. 1995. Dynamics of soil organic matter and corn residues as affected by tillage practices. *Soil Sci. Soc. Am. J.* (in press).

Atkinson, D. and L.A. Mackie-Dawson. 1991. Root growth: methods of measurement. p. 447-509. In: K.A. Smith and C.E. Mullins (eds.), *Soil analysis: physical methods*. Marcel Dekker, Inc., New York.

Baldock, J.A. and B.D. Kay. 1987. Influence of cropping history and chemical treatments on the water-stable aggregation of a silt loam soil. *Can. J. Soil Sci.* 67:501-511.

Baldock, J.A., J.M. Oades, A.M. Vassallo, and M.A. Wilson. 1990. Solid state CP/MAS ^{13}C N.M.R. analysis of particle size and density fractions of a soil incubated with uniformly labelled ^{13}C-glucose. *Aust. J. Soil Res.* 28:193-212.

Baldock, J.A., J.M. Oades, A.G. Waters, X. Peng, A.M. Vassallo, and M.A. Wilson. 1992. Aspects of the chemical structure of soil organic materials as revealed by solid-state ^{13}C NMR spectroscopy. *Biogeochemistry* 16:1-42.

Balesdent, J., A. Mariotti, and B. Guillet. 1987. Natural ^{13}C abundance as a tracer for soil organic matter dynamics studies. *Soil Biol. Biochem* 19:25-30.

Balesdent, J., G.H.Wagner, and A. Mariotti. 1988. Soil organic matter turnover in long-term field experiments as revealed by the carbon-13 natural abundance. *Soil Sc. Soc. Am. J.* 52:118-124.

Beare, M.H. and R.R. Bruce. 1993. A comparison of methods for measuring water-stable aggregates: implications for determining environmental effects on soil structure. *Geoderma* 56:87-104.

Beare, M.H., M.L. Cabrera, P.F. Hendrix, and D.C. Coleman. 1994. Aggregate-protected and unprotected organic matter pools in conventional- and no-tillage soils. *Soil Sci. Soc. Am. J.* 58:787-795.

Bremner, J.M. and D.A. Genrich. 1990. Characteristics of the sand, silt, and clay fractions of some Mollisols. p. 423-438. In: M.F. Boodt et al. (eds.), *Soil colloids and their associations in aggregates*. Plenum Press, New York.

Buyanovsky, G.A., M. Aslam, and G.H. Wagner. 1994. Carbon turnover in soil physical fractions. *Soil Sci. Soc. Am. J.* 58:1167-1173.

Cambardella, C.A. and E.T. Elliott. 1992. Particulate soil organic matter changes across a grassland cultivation sequence. *Soil Sci. Soc. Am. J.* 56:777-783.

Cameron, R.S. and A.M. Posner. 1979. Mineralisable organic nitrogen in soil fractionated according to particle size. *J. Soil Sci.* 30:565-577.

Carter, M.R. 1992. Influence of reduced tillage systems on organic matter, microbial biomass, macro-aggregate distribution and structural stability of the surface soil in a humid climate. *Soil Tillage Res.* 23:361-372.

Carter, M.R. and H.T. Kunelius. 1993. Effect of undersowing barley with annual ryegrasses or red clover on soil structure in a barley-soybean rotation. *Agric., Ecosyst. Environ.* 43:245-254.

Cheshire, M.V. and C.M. Mundie. 1981. The distribution of labelled sugars in soil particle size fractions as a means of distinguishing plant and microbial carbohydrate residues. *J. Soil Sci.* 32:605-618.

Christensen, B.T. 1987. Decomposability of organic matter in particle size fractions from field soils with straw incorporation. *Soil Biol. Biochem.* 19:429-435.

Christensen, B.T. 1992. Physical fractionation of soil and organic matter in primary particle size and density separates. *Adv. Soil Sci.* 20:1-90.

Christensen, B.T. and L.H. Sørensen. 1985. The distribution of native and labelled carbon between soil particle size fractions isolated form long-term incubation experiments. *J. Soil Sci.* 36:219-229.

Christensen, B.T. and L.H. Sørensen. 1986. Nitrogen in particle size fractions of soils incubated for five years with ^{15}N-ammonium and ^{14}C-hemicellulose. *J. Soil Sci.* 37:241-247.

Christensen, S. and B.T. Christensen. 1991. Organic matter available for denitrification in different soil size fractions: effect of freeze/thaw cycles and straw disposal. *J. Soil Sci.* 42:637-647.

Davidson, E.A. and I.L. Ackerman. 1993. Changes in carbon inventories following cultivation of previously untilled soils. *Biogeochemistry* 20:161-193.

Desjardins, T., F. Andreux, B. Volkoff, and C.C. Cerri. 1994. Organic carbon and ^{13}C contents in soils and soil size-fractions, and their changes due to deforestation and pasture installation in eastern Amazonia. *Geoderma* 61:103-118.

Ellert, B.H. and E.G. Gregorich. 1995. Management-induced changes in the actively cycling fractions of soil organic matter. p. 119-138. In: *Proceedings of the 8th North American Forest Soils Conference.* Gainsville, Florida.

Elliott, E.T. 1986. Aggregate structure and carbon, nitrogen, and phosphorus in native and cultivated soils. *Soil Sci. Soc. Am. J.* 50:627-633.

Elliott, E.T. and C.A. Cambardella. 1991. Physical separation of soil organic matter. *Agric. Ecosyst. Environ.* 34: 407-419.

Elliott, E.T., C.A. Palm, D.E. Reuss, and C.A. Monz. 1991. Organic matter contained in soil aggregates from a tropical chronosequence: correction for sand and light fraction. *Agric. Ecosyst. Environ.* 34:443-451.

Eswaran, H., E. Van Den Berg, and P. Reich. 1993. Organic carbon in soils of the world. *Soil Sci. Soc. Am. J.* 57:192-194.

Goh, K.M. 1991. Carbon dating. p. 125-145. In: D.C. Coleman and B. Fry (eds.), *Carbon isotope techniques.* Academic Press, Inc., New York.

Golchin, A., J.M. Oades, J.O. Skjemstad, and P. Clarke. 1995. Structural and dynamic properties of soil organic matter as reflected by ^{13}C natural abundance, pyrolysis mass spectrometry and solid-state ^{13}C NMR spectroscopy in density fractions of an Oxisol under forest and pasture. *Aust. J. Soil Res.* 33:59-76.

Gregorich, E.G. and B.H. Ellert. 1993. Light fraction and macroorganic matter in mineral soils. p. 397-407. In: M.R. Carter (ed.), *Soil sampling and methods of analysis*. Lewis Publishers, CRC Press, Boca Raton, Florida.

Gregorich, E.G., B.H. Ellert, and C.M. Monreal. 1995. Turnover of soil organic matter and storage of corn residue carbon estimated using natural ^{13}C abundance. *Can. J. Soil Sci.*75:161-167.

Gregorich, E.G., M.R. Carter, D.A. Angers, C.M. Monreal, and B.H. Ellert. 1994. Towards a minimum data set to assess soil organic matter quality in agricultural soils. *Can. J. Soil Sci.* 74:367-385.

Guggenberger, G., B.T. Christensen, and W. Zech. 1994. Land-use effects on the composition of organic matter in particle-size separates of soil: 1. Lignin and carbohydrate signature. *Eur. J. Soil Sci.* 45:449-458.

Hamblin, A.P. and D.J. Greenland. 1977. Effect of organic constituents and complexed metal ions on aggregate stability of some East Anglian soils. *J. Soil Sci.* 28:410-416.

Hsieh, Y.-P. 1992. Pool size and mean age of stable soil organic carbon in cropland. *Soil Sci. Soc. Am. J.* 56:460-464.

Imhof, M.P. 1988. A review of wet-sieving methodology. Technical Report No. 157, Victoria Dep. Agric., Melbourne, Victoria, Australia.

Jastrow, J.D. and R.M. Miller. 1991. Methods for assessing the effects of biota on soil structure. *Agric. Ecosyst. Environ.* 34:279-303.

Jensen, E.S., B.T. Christensen, and L.H. Sørensen. 1989. Mineral-fixed ammonium in clay- and silt-sized fractions of soils incubated with ^{15}N-ammonium sulphate for five years. *Biol. Fertil. Soils.* 8:298-302.

Juma, N.G. 1993. Interrelationships between soil structure/texture, soil biota/soil organic matter and crop production. *Geoderma* 57:3-30.

Juma, N.G. 1994. A conceptual framework to link carbon and nitrogen cycling to soil structure formation. *Agric. Ecosyst. Environ.* 51:257-267.

Kern, J.S. 1994. Spatial patterns of soil organic carbon in the contigous United States. *Soil Sci. Soc. Am. J.* 58:439-455.

Ladd, J.N., J.W. Parsons, and M. Amato. 1977a. Studies of nitrogen immobilization and mineralization in calcareous soils I. Distribution of immobilized nitrogen amongst soil fractions of different particle size and density. *Soil Biol. Biochem.* 9:309-318.

Ladd, J.N., J.W. Parsons, and M. Amato. 1977b. Studies of nitrogen immobilization and mineralization in calcareous soils II.Mineralization of immobilized nitrogen from soil fractions of different particle size and density. *Soil Biol. Biochem.* 9:319-325.

Ladd, J.N. and M. Amato. 1980. Studies of nitrogen immobilization and mineralization in calcareous soils. IV. Changes in the organic nitrogen of light and heavy subfractions of silt- and fine-clay size particles during nitrogen turnover. *Soil Biol. Biochem.* 12:185-189.

Ladd, J.N., R.C. Foster, and J.O. Skjemstad. 1993. Soil structure: carbon and nitrogen metabolism. *Geoderma* 56:401-434.

Ledgard, S.F., J.R. Freney, and J.R. Simpson. 1984. Variations in natural enrichment of ^{15}N in the profiles of some Australian pasture soils. *Aust. J. Soil Res.*22:155-164.

Lowe, L.E. 1993. Total and labile polysaccharide analysis of soils. p. 373-376. In: M.R. Carter (ed.), *Soil sampling and methods of analysis*. Lewis Publishers, CRC Press, Boca Raton, Florida.

Lowe, L.E. and A.A. Hinds. 1983. The mineralization of nitrogen and sulfur from particle size separates of gleysolic soil. *Can. J. Soil Sci*. 62:761-766.

Martin, A., A. Mariotti, J. Balesdent, P. Lavelle, and R. Vuattoux. 1990. Estimate of organic matter turnover rate in a savanna soil by ^{13}C natural abundance measurements. *Soil Biol. Biochem.* 22:517-523.

McGill, W.B., K.R. Cannon, J.A. Robertson, and F.D. Cook. 1986. Dynamics of soil microbial biomass and water-soluble organic C in Breton L after 50 years of cropping to two rotations. *Can. J. Soil Sci.* 66:1-19.

Monreal, C.M., E.G. Gregorich, M. Schnitzer, and D.W. Anderson. 1995. The quality of soil organic matter as characterized by solid CP/MAS ^{13}C NMR and Py-FIMS. p. 207-215. In: P.M. Huang et al. (eds.), *Environmental impacts of soil components interactions: Land quality, natural and anthropogenic organics*. Lewis Publishers, CRC Press, Boca Raton, Florida.

Moraes, J.L., C.C. Cerri, J.M. Melillo, D. Kicklighter, C. Neill, D.L. Skole, and P.A. Steudler. 1995. Soil carbon stocks of the Brazilian Amazon Basin. *Soil Sci. Soc. Am. J.* 59:244-247.

Oades, J.M., A.M. Vassallo, A.G. Waters, and M.A. Wilson. 1987. Characterization of organic matter in particle size and density fractions from a Red-brown Earth by solid-state ^{13}C N.M.R. *Aust. J. Soil Res.* 25:71-82.

Paul, E.A., C.A. Campbell, D.A. Rennie, and K.J. McCallum. 1964. Investigations of the dynamics of soil humus utilizing carbon dating techniques. *Trans. 8th Int. Congr. Soil Sci.* 3:201-208. Bucharest, Romania.

Schulten, H.-R. and P. Leinweber. 1991. Influence of long-term fertilization with farmyard manure on soil organic matter: Characteristics of particle-size fractions. *Biol. Fertil. Soils*. 12:81-88.

Seech, A.G. and E.G. Beauchamp. 1988. Denitrification in soil aggregates of different sizes. *Soil Sci. Soc. Am. J.* 52:1616-1621.

Skjemstad, J.O., L.J. Janik, M.J. Head, and S.G. McClure. 1993. High energy ultraviolet photo-oxidation: a novel technique for studying physically protected organic matter in clay- and silt-sized aggregates. *J. Soil Sci.* 44:485-499.

Sollins, P., G. Spycher, and C.A. Glassman. 1984. Net nitrogen mineralization from light- and heavy-fraction forest soil organic matter. *Soil Biol. Biochem.* 16:31-37.

Stevenson. F.J. and E.T. Elliott. 1989. Methodologies for assessing the quantity and quality of soil organic matter. p. 173-199. In: D.C. Coleman et al. (eds.), *Dynamics of soil organic matter in tropical ecosystems*. Univ. Hawaii Press, Honolulu.

Stone, E.L., W.G. Harris, R.B. Brown, and R.J. Kuehl. 1993. Carbon storage in Florida spodosols. *Soil Sci. Soc. Am. J.* 57:179-182.

Theng, B.K.G., K.R. Tate, and P. Sollins. 1989. Constituents of organic matter in temperate and tropical soils. p. 5-31. In: D.C. Coleman et al. (eds.), *Dynamics of soil organic matter in tropical ecosystems*. Univ. Hawaii Press, Honolulu.

Tiessen, H. and J.W.B. Stewart. 1983. Particle-size fractions and their use in studies of soil organic matter. II. Cultivation effects on organic matter composition in size fractions. *Soil Sci. Soc. Am. J.* 47:509-514.

Tiessen, H. and J.O. Moir. 1993. Total and organic carbon. p. 187-199. In: M.R. Carter (ed.), *Soil sampling and methods of analysis*. Lewis Publishers, CRC Press, Boca Raton, Florida.

Tiessen, H., R.E. Karamanos, J.W.B. Stewart, and F. Selles. 1984. Natural nitrogen-15 abundance as an indicator of soil organic matter transformations in native and cultivated soils. *Soil Sci. Soc. Am. J.* 48:312-315.

Tisdall, J.M. 1994. Possible role of soil microorganisms in aggregation in soils. *Plant Soil* 159:115-121.

Trumbore, S.E. 1993. Comparison of carbon dynamics in tropical and temperate soils using radiocarbon measurements. *Global Biogeochem. Cycles* 7:275-290.

Turchenek, L.W. and J.M. Oades. 1979. Fractionation of organo-mineral complexes by sedimentation and density techniques. *Geoderma* 21:311-343.

Van Veen, J.A. and P.J. Kuikman. 1990. Soil structural aspects of decomposition of organic matter by micro-organisms. *Biogeochemistry* 11:213-234.

Voroney, R.P., J.P. Winter, and E.G. Gregorich. 1991. Microbe/plant/soil interactions. p.77-99. In: D.C. Coleman and B. Fry (eds.), *Carbon isotope techniques*. Academic Press, Inc., New York.

Voroney, R.P., J.P. Winter, and R.P. Beyaert. 1993. Soil microbial biomass C and N. p. 277-286. In: M.R. Carter (ed.), *Soil sampling and methods of analysis*. Lewis Publishers, CRC Press, Boca Raton, Florida.

Index

Index